Tunneling Phenomena in Solids

Tunneling Phenomena in Solids

Lectures presented at the
1967 NATO Advanced Study Institute at Risö, Denmark

Edited by

ELIAS BURSTEIN

Department of Physics
University of Pennsylvania
Philadelphia, Pennsylvania

and

STIG LUNDQVIST

Institute of Theoretical Physics
Chalmers Tekniska Högskola
Göteborg, Sweden

PLENUM PRESS · NEW YORK · 1969

Library of Congress Catalog Card Number 69-12528

A Division of Plenum Publishing Corporation
227 West 17 Street, New York, N.Y. 10011

Printed in the United States of America

Foreword

The aim of this volume is to provide advanced predoctoral students and young postdoctoral physicists with an opportunity to study the concepts of tunneling phenomena in solids and the theoretical and experimental techniques for their investigation. The contributions are primarily tutorial in nature, covering theoretical and experimental aspects of electron tunneling in semiconductors, metals, and superconductors, and atomic tunneling in solids.

The work is based upon the lectures delivered at the Advanced Study Institute on "Tunneling Phenomena in Solids," held at the Danish A.E.C. Research Establishment, Risö, Denmark, June 19–30, 1967. Sponsored by the Danish Atomic Energy Commission, the Nordic Institute for Theoretical Physics (NORDITA), and the Science Affairs Division of NATO, with the cooperation of the University of Copenhagen, the Technical University of Denmark, Chalmers Institute of Technology, and the University of Pennsylvania, the lectures were presented by a distinguished panel of scientists who have made major contributions in the field. The relatively large number of lecturers was, in part, made possible by the close coordination of the Advanced Study Institute with the Second International Conference on Electron Tunneling in Solids, which was held at Risö on June 29, 30 and July 1, 1967, under the sponsorship of the U.S. Army Research Office–Durham. We are indebted to I. Giaever, E. O. Kane, J. Rowell, and J. R. Schrieffer for advice and assistance in planning the lecture program of the Institute.

The Institute was made possible through the active interest of an organizing group consisting of H. Højgard Jensen, A. Mackintosh, N. I. Meyer, M. Nielsen, and K. Saermark. The Danish Atomic Energy Commission supported the Institute financially and made available its facilities at the Risö Research Establishment. We take this opportunity to thank the Danish AEC officials for their gracious cooperation. In particular, we would like to express our gratitude to the Director of the Research Establishment, Professor T. Bjerge, for his kind interest and for being a generous and excellent host.

The heavy burden of making all the arrangements for the Institute was very ably undertaken by M. Nielsen and by the Information Officer of the Research Establishment, Mrs. J. Starcke. Dr. Nielsen and Mrs. Starcke kept the secretariat running smoothly at all times and helped the Directors of the Institute, as well as the participants, in solving the many problems that arose. We would like to express our warmest thanks to them for their patient and efficient work in following as closely as possible our intentions along a rather winding course toward a successful end.

E. Burstein, Director of the Institute
S. Lundqvist, Associate Director of the Institute

October 20, 1968

Contents

Chapter 1

Basic Concepts of Tunneling

E. O. Kane

Bell Telephone Laboratories
Murray Hill, New Jersey

A great many of the features of tunneling phenomena in solids are essentially of a one-dimensional nature. If the tunneling barrier extends in the x direction, the momentum in the y and z directions can usually be taken to be constants of the motion, and hence are merely fixed parameters.

In this introductory chapter we will describe some of the basic concepts of tunneling by studying purely one-dimensional problems, and we may anticipate that most of our results will have a wider range of applicability to real situations than one-dimensional problems usually have.

We will, in fact, study the simplest of all one-dimensional problems, namely, square barriers and square wells in one dimension. The Schrödinger equation will have the simple form

$$\left(\frac{p^2}{2m} + V\right)\psi = E\psi, \tag{1}$$

where V is a constant in a given region. The general solution of Eq. (1) has the well-known form

$$\psi(x) = ae^{ikx} + be^{-ikx}, \tag{2}$$

$$\hbar^2k^2/2m = E - V. \tag{3}$$

When $E - V > 0$ the wave functions are plane waves. When $E - V < 0$ we will write $k = i\varkappa$ and

$$\psi(x) = ae^{-\varkappa x} + be^{\varkappa x}. \tag{4}$$

The wave functions are now exponentially growing and decaying waves characteristic of barrier penetration problems. In the "square barrier" and

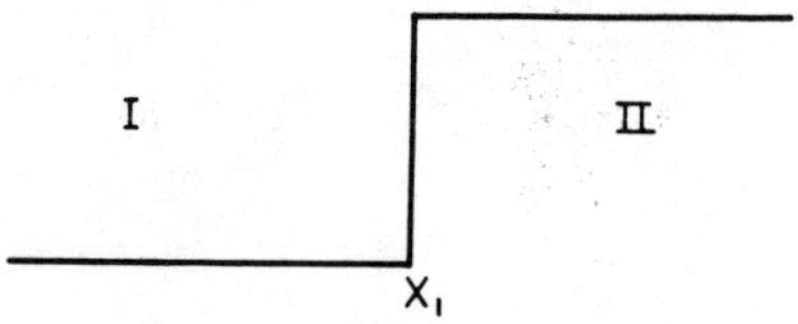

Fig. 1.

"square well" problem which we will discuss V changes abruptly from one constant value to another. The overall wave function is then constructed out of pieces having the form of Eq. (2) by matching ψ and $d\psi/dx$ at the discontinuities of V.

The basic matching problem can be done at a potential step as shown in Fig. 1. The matching conditions are then conveniently described as a 2×2 matrix R operating on the two-dimensional vectors

$$\begin{pmatrix} a_1 \\ b_1 \end{pmatrix} \quad \text{and} \quad \begin{pmatrix} a_2 \\ b_2 \end{pmatrix}$$

which describe the wave function in regions I and II, respectively. We may write

$$\begin{pmatrix} a_1 \\ b_1 \end{pmatrix} = R_1 \begin{pmatrix} a_2 \\ b_2 \end{pmatrix}, \tag{5}$$

$$R_1 = \frac{1}{2k_1} \begin{pmatrix} (k_1 + k_2) \exp[i(-k_1 + k_2)x_1] & (k_1 - k_2) \exp[i(-k_1 - k_2)x_1] \\ (k_1 - k_2) \exp[i(k_1 + k_2)x_1] & (k_1 + k_2) \exp[i(k_1 - k_2)x_1] \end{pmatrix}. \tag{6}$$

Of course, Eqs. (5) and (6) are valid for either real or imaginary values of k. The general problem of any number of square barriers or wells may then be succinctly described in terms of a chain of operators of the form R_j.

We turn first to the square barrier problem of Fig. 2. We may write the solution

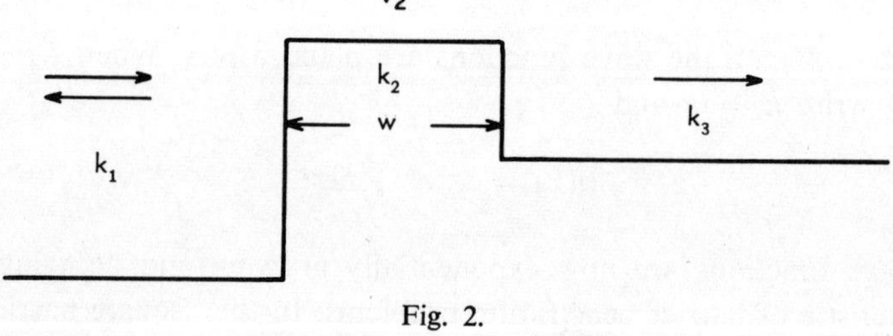

Fig. 2.

$$\begin{pmatrix} a_1 \\ b_1 \end{pmatrix} = R_1 R_2 \begin{pmatrix} a_3 \\ b_3 \end{pmatrix}. \tag{7}$$

Let us study the boundary condition of an electron incident from the left. There will be a reflected wave in region 1 but only a transmitted wave in region 3; hence $b_3 = 0$. We want a_3 in terms of a_1. Equation (7) then gives

$$a_1 = (R_1 R_2)_{11} a_3. \tag{8}$$

Explicit evaluation of (8) with the use of (6) results in

$$\begin{aligned} (R_1 R_2)_{11} = {} & [\exp(ik_3 x_2 - ik_1 x_1)](k_1^2 + \varkappa_2^2)^{1/2}(k_3^2 + \varkappa_2^2)^{1/2} \\ & \times (e^{i\alpha} e^{\varkappa_2 w} - e^{-i\alpha} e^{-\varkappa_2 w})/(4ik_1\varkappa_2) \end{aligned} \tag{9}$$

$$\begin{gathered} \alpha = \tan^{-1}(\varkappa_2/k_1) + \tan^{-1}(\varkappa_2/k_3) \\ \varkappa_2 = ik_2 \qquad w = x_2 - x_1 \end{gathered} \tag{10}$$

We are assuming real plane waves in regions 1 and 3 and tunneling in region 2. To simplify the formula, we will neglect $e^{-\varkappa_2 w}$ in comparison to $e^{\varkappa_2 w}$; in other words, we consider only a strongly attenuating barrier. Then we have

$$a_3 = \frac{4k_1\varkappa_2 \varphi e^{-\varkappa_2 w}}{(k_1^2 + \varkappa_2^2)^{1/2}(k_3^2 + \varkappa_2^2)^{1/2}} a_1, \tag{11}$$

$$\varphi = ie^{-i\alpha} \exp(ik_1 x_1 - ik_3 x_2); \tag{12}$$

φ is a phase factor of absolute value 1, which we ignore for the moment.

The quantity of physical interest that we wish to compute is the ratio of the transmitted current j_3 to the incident current j_1,

$$j_1 = (\hbar k_1/m) \mid a_1 \mid^2; \qquad j_3 = \hbar k_3/m \mid a_3 \mid^2. \tag{13}$$

We find

$$\frac{j_3}{j_1} = \frac{16 k_1 k_3 \varkappa_2^2}{(k_1^2 + \varkappa_2^2)(k_3^2 + \varkappa_2^2)} e^{-2\varkappa_2 w}, \tag{14}$$

$$\hbar^2 \varkappa_2^2/2m = V_2 - E. \tag{15}$$

The dominant feature of Eq. (14) is, of course, the barrier penetration factor $e^{-2\varkappa_2 w}$. In typical problems of interest this factor may be 10^{-5}–10^{-10}, so that it tends to dominate the prefactor. In more realistic problems this factor can seldom be calculated accurately, so that usually any experimental determination of the prefactor is doubtful or impossible.

Tunneling is one of the simplest examples of a truly quantum mechanical phenomenon. In classical physics a particle whose energy is less than the height of a potential barrier can never pass through it.

We note that Eq. (14) shows that the barrier transmission is symmetric in the indices 1 and 3. This means that the barrier transmission is the same for currents incident from the right or left, as we would expect. We may also compute the incident current per unit energy if we assume that the states on the left and right are quantized in a box and occupied with an occupancy factor $f(E)$. The incident current is given by

$$\Delta j_1(E) = |a|^2 \frac{\hbar k_1}{m} \frac{dn}{dE} \Delta E f_1(E). \tag{16}$$

Box normalization gives $|a|^2 = 1/L_1$, where L_1 is the length of the box. The density of states in one dimension without spin is

$$dn/dE = mL_1/2\pi\hbar^2 k_1; \tag{17}$$

hence (16) becomes

$$\Delta j_1(E) = [f_1(E)/2\pi\hbar]\,\Delta E. \tag{18}$$

In equilibrium $f_1(E) = f_3(E)$; hence the current per unit energy incident on the barrier from either side of the barrier is the same. By the symmetry of Eq. (14), the transmitted currents per unit energy are also the same, and equilibrium is maintained, as we should expect.

We may also note that Eq. (14) shows that the transmitted current goes continuously to zero as either k_1 or k_3 approaches zero. This is in contrast to simple calculations of tunneling in *p–n* junctions, which showed a discontinuous change in the transmitted current at the band edge ([1]). A more careful treatment by Shuey ([2]) showed that in the junction problem also the current goes continuously to zero at the band edge.

We now investigate a little more closely the character of the attenuating states $ae^{-\varkappa x}$ and $be^{+\varkappa x}$ in the barrier. We use the expression for the particle current operator

$$j(x) = \frac{-i\hbar}{2m}\left(\psi^*(x)\frac{d\psi}{dx} - \psi(x)\frac{d\psi^*}{dx}\right). \tag{19}$$

We see immediately that $ae^{-\varkappa x}$ or $be^{\varkappa x}$ by themselves carry zero current. In combination

$$j(x) = (i\hbar\varkappa/m)(ab^* - ba^*). \tag{20}$$

We can use this property to develop an effective Hamiltonian theory of tunneling in a manner first described by Bardeen ([3]).

The smallness of the right-hand sides of Eqs. (11) or (14) suggests that a perturbation treatment of tunneling would be very appropriate. However, there is no obvious way of introducing a term in the Hamiltonian which can be treated as small. Bardeen took a different approach. Instead of introducing states which are the exact solutions of an approximate Hamiltonian, he introduced approximate solutions of the exact Hamiltonian. Following Bardeen, we choose the states

$$\psi_r(x) = b_{2r} e^{\varkappa_2 x}; \qquad x \le x_2, \tag{21}$$

$$\psi_l(x) = a_{2l} e^{-\varkappa_2 x}; \qquad x \ge x_1; \tag{22}$$

ψ_r is to be matched to the correct solution for $x \ge x_2$ but is to decay in the region $x \le x_1$ instead of satisfying the Schrödinger equation. Similarly, ψ_l continues to decay for $x \ge x_2$. Then ψ_r is a correct solution of H for $x \ge x_1$ and ψ_l is correct for $x \le x_2$. We assume an electron is initially in ψ_l and compute its transition rate into state ψ_r. We write

$$\psi = c(t)\psi_l e^{-iE_l t} + d(t)\psi_r e^{-iE_r t} \tag{23}$$

and substitute in the Schrödinger equation

$$\begin{aligned} i\dot{c}\psi_l e^{-iE_l t} + c\psi_l E_l e^{-iE_l t} + i\dot{d}\psi_r e^{-iE_r t} + d\psi_r E_r e^{-iE_r t} \\ = ce^{-iE_l t} H\psi_l + de^{-iE_r t} H\psi_r. \end{aligned} \tag{24}$$

Since $c \approx 1$, $d \approx 0$, we will set $d = 0$, $\dot{c} = 0$, $c = 1$ in Eq. (24) [$\dot{c} \approx 0$ follows from normalization, $(d/dt)(cc^* + dd^*) = 0$]. Equation (24) then becomes

$$i\dot{d}\psi_r e^{-iE_r t} = (H - E_l)\psi_l e^{-iE_l t}, \tag{25}$$

or

$$i\dot{d} = \int \psi_r^*(H - E_l)\psi_l \, dx \exp[i(E_r - E_l)t]. \tag{26}$$

We assume ψ_l and ψ_r are normalized.

If $H = H_0 + H_1$, as in ordinary perturbation theory, and $H_0\psi_l = E_l\psi_l$, we have

$$\int \psi_r^*(H - E_l)\psi_l \, dx = \int \psi_r^* H_1 \psi_l \, dx. \tag{27}$$

Accordingly, we identify

$$T_{rl} = \int \psi_r^*(H - E_l)\psi_l \, dx \tag{28}$$

as the effective matrix element for tunneling.

The integral in Eq. (28) gives a nonzero contribution only for $x > x_2$. We can symmetrize Eq. (28) by writing

$$T_{rl} = \int_{x_B}^{\infty} \psi_r^*(H - E_l)\psi_l - \psi_l(H - E_r)\psi_r^* \, dx, \qquad x_1 \leq x_B \leq x_2, \tag{29}$$

since the added term is zero for the range of integration $x \geq x_1$. The lower limit of the integral can take any value in the barrier since the integrand is zero there. Integrating Eq. (29) by parts gives

$$T_{rl} = -\frac{\hbar^2}{2m}\left(\psi_r^* \frac{d\psi_l}{dx} - \psi_l \frac{d\psi_r^*}{dx}\right)_{x_B}, \tag{30}$$

or

$$T_{rl} = -i\hbar j_{rl}, \tag{31}$$

where j_{rl} is the current operator as given in Eq. (19) evaluated at any point in the barrier. Using Eqs. (21) and (22), we have

$$T_{rl} = (\hbar^2 \varkappa_2/m) b_{2r}^* a_{2l}. \tag{32}$$

We will now use this tunneling matrix element to calculate the tunnel current from the effective Hamiltonian viewpoint. Noting that $a_{2r} = b_{2l} = 0$, we use Eq. (5) to relate a_{1l} to a_{2l} and b_{2r} to a_{3r}. This gives

$$|T_{rl}|^2 = \frac{\hbar^4 \varkappa_2^2}{m^2} \frac{16 k_1^2 k_3^2 |a_{1l}|^2 |a_{3r}|^2 e^{-2\varkappa_2 w}}{(k_1^2 + \varkappa_2^2)(k_3^2 + \varkappa_2^2)}. \tag{33}$$

The current incident on the barrier, j_l, is

$$j_l = (\hbar k_1/m) |a_{1l}|^2. \tag{34}$$

The transmitted current, j_t, is given by

$$j_t = w = \frac{2\pi}{\hbar} |T_{rl}|^2 \frac{dn}{dE_r}, \tag{35}$$

using the "golden rule" of first-order time-dependent perturbation theory. Using Eq. (17) we have

$$|a_{3r}|^2 \frac{dn}{dE_r} = \frac{m}{2\pi\hbar^2 k_3}. \tag{36}$$

Substituting (33), (34), and (36) in (35), we verify that we obtain Eq. (14). Hence the effective Hamiltonian approach gives the same result for the tunnel current as the simple matching procedure.

The "effective tunneling Hamiltonian" provides a justification for the often-quoted assumption that the tunnel current is proportional to the density of states. However, we see from the matrix element of Eq. (33) that other terms in the exponential prefactor are equally important, so that the final result in Eq. (14) does not bear any simple relation to the density of states in Eq. (36). Hence the use of tunneling measurements as a quantitative probe of the density of states must be regarded with caution. However, it is a well-known fact that superconducting tunneling does quite accurately reflect the density of states. This has been justified theoretically, using the effective tunneling Hamiltonian, by Bardeen and others. This is discussed a great deal more elsewhere in this book.

We turn now to the much discussed question of how long it takes an electron to tunnel through a barrier. We refer to Fig. 2 and represent the electron in region 1 by a wave packet with average momentum k_1 and energy $E_1 = E(k_1)$:

$$\psi_i(x, t) = \exp[i(k_1 x - E_1 t)/\hbar] \times \int_{-\infty}^{\infty} \exp i(k - k_1)(x - v_1 t)] \exp -(k - k_1)^2]\, dk, \quad (37)$$

where we have used

$$E(k) \approx E(k_1) + (k - k_1)(dE/dk)_{k_1}, \quad (38)$$

$$v \equiv \frac{1}{\hbar}\frac{dE}{dk} = \frac{\hbar k}{m}. \quad (39)$$

Equation (39) is the usual definition of the group velocity. Carrying out the integration in Eq. (37) gives

$$\psi_i(x, t) = \exp[i(k_1 x - E_1 t)/\hbar] \exp[-\tfrac{1}{4}(x - vt)^2]. \quad (40)$$

Equation (40) shows that the packet moves with group velocity v, in agreement with what we expect from the classical limit. Since the kinetic energy of the wave packet is clearly positive, we cannot use it to localize the electron in the barrier region because the electron would then automatically have an energy greater than the barrier height. What we can do is look at the packet in region 3 beyond the barrier and compute its time dependence there. We use Eq. (11) to get the transmitted packet in terms of the incident packet. To simplify matters, we set $k_3 = k_1$. Then Eq. (12) shows that the plane wave $\exp\{i[kx - E(k)t/\hbar]\}$ on the left of the junction has the form $Aie^{-i\alpha}\exp\{i[k(x - w) - E(k)t/\hbar]\}$ on the right. Ignoring the phase term $ie^{-i\alpha}$, which does not depend on the junction thickness, we see that the electron has "lost," or rather failed to accumulate, the phase ikw in passing through the barrier. This is, of course, due to the real character of the x dependence in the barrier region as given in Eq. (4). If we ignore the k dependence of $Ae^{-i\alpha}$, the transmitted wave packet ψ_t can be written

$$\psi_t(x_1t) = A\exp\{i[k_1(x - w) - Et/\hbar]\}\exp[-\tfrac{1}{4}(x - w - vt)^2]. \quad (41)$$

The lack of phase change in traversing the barrier has advanced the transmitted packet by the junction width w relative to the incident packet. In other words, it appears that the packet has traversed the barrier in zero time or with infinite velocity.

There is no contradiction between this relation and relativistic limitations, because it was derived from the Schrödinger equation, which is only exact if the speed of light is made to approach infinity. This curious result is another way in which tunneling defies classical analogy.

It should not be concluded from this result that tunneling is a very rapid process. It is, in fact, very slow, as indicated by the very weak tunneling matrix element in Eq. (33). We can indicate the slowness of the tunneling process by solving the two-well problem shown in Fig. 3.

We use Eqs. (5) and (6) to write

$$\begin{pmatrix} a_1 \\ b_1 \end{pmatrix} = R_1R_2R_3R_4\begin{pmatrix} a_5 \\ b_5 \end{pmatrix}. \quad (42)$$

The eigenvalue condition results from the requirement that only attenuating states exist in regions 1 and 5. That is, $a_1 = b_5 = 0$. This gives the eigenvalue equation

$$(R_1R_2R_3R_4)_{11} = 0. \quad (43)$$

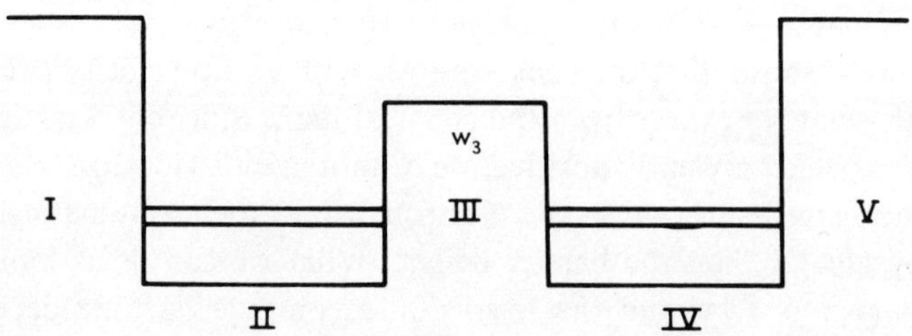

Fig. 3.

To simplify the result, we will assume that the barriers in regions 1 and 5 are infinitely high. We also assume $w_4 = w_2$, $k_2 = k_4 \equiv k$, $k_3 \equiv i\varkappa$. The solution of the eigenvalue equation (43) can then be found to be

$$e^{i2kw_2} = -\frac{(e^{\varkappa w_3} - e^{-\varkappa w_3}) \pm 2i \sin 2\alpha}{e^{\varkappa w_3}e^{-2i\alpha} - e^{-\varkappa w_3}e^{2i\alpha}}, \tag{44}$$

$$\alpha = \tan^{-1}(\varkappa/k). \tag{45}$$

If we neglect $e^{-\varkappa w_3}$ compared to $e^{\varkappa w_3}$, the eigenvalue equation becomes

$$k^0 w_2 = \alpha + \tfrac{1}{2}\pi + n\pi, \tag{46}$$

which is the eigenvalue condition for a single well. Solving (44) to next lowest order in $e^{-\varkappa w_3}$ gives

$$\delta k\, w_2 = \pm \sin 2\alpha\, e^{-\varkappa w_3}, \tag{47}$$

$$\delta k \equiv k - k^0. \tag{48}$$

The effect of the tunneling transitions is to split the degenerate levels of the unconnected wells by the amount

$$2\,\delta E = 2\hbar^2 k\, \delta k/m. \tag{49}$$

Let ψ_{II} and ψ_{IV} refer to the wave function in wells II and IV, respectively. Then $\psi_{\text{II}} \pm \psi_{\text{IV}}$ has energy $E_0 \mp \delta E$. If we substitute in the time-dependent Schrödinger equation,

$$\psi = \exp(-\,iE_0t/\hbar)\,[(\psi_{\text{II}} + \psi_{\text{IV}})\exp(+\,i\,\delta E\,t/\hbar) + (\psi_{\text{II}} - \psi_{\text{IV}})\exp(-\,i\,\delta E\,t/\hbar)], \tag{50}$$

$$\psi = 2\exp(-iE_0t/\hbar)\left(\cos\frac{\delta E\,t}{\hbar}\,\psi_{\text{II}} + i\sin\frac{\delta E\,t}{\hbar}\,\psi_{\text{IV}}\right). \tag{51}$$

We have chosen phases such that the electron is in well II at time $t = 0$. After a time $t = \frac{1}{2}\hbar\pi/\delta E$ the electron will be entirely in well IV. Since $\delta E \propto e^{-\varkappa w_3}$, the rate of exchange of energy between the wells is very slow. Hence by this criterion, tunneling is a slow process. We believe that this is the applicable criterion in most physical problems.

As one final application of the square barrier model we discuss the case of two tunnel barriers separated by a well. Generally speaking, as one might expect, the tunneling attenuation factor of two barriers is the product of the attenuation factors for each. For a resonating condition in the inter-

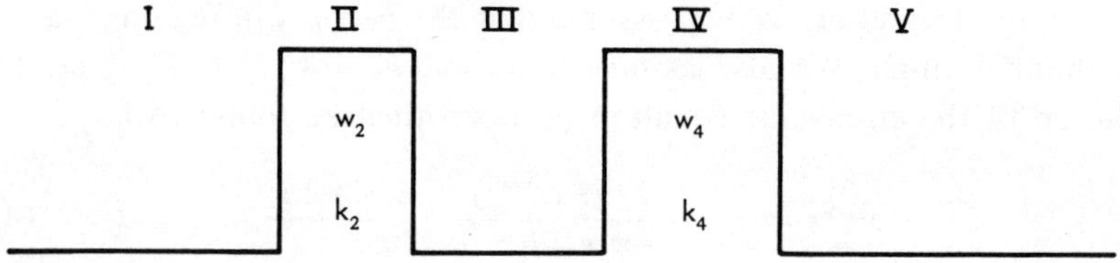

Fig. 4.

mediate well the electron may pass through both barriers without any attenuation.

The arrangement of barriers is shown in Fig. 4. The profile is just that of Fig. 3 upside down. Equation (42) remains applicable. If we assume a wave incident from the left and transmitted on the right, the ratio of transmitted to incident current is given by

$$j_t/j_i = |\, a_5\, |^2 k_5 / |\, a_1\, |^2 k_1\,, \tag{52}$$

$$a_1 = (R_1R_2R_3R_4)_{11}a_5\,, \tag{53}$$

$$\frac{j_t}{j_i} = \frac{2^8 k_1 \varkappa_2{}^2 k_3{}^2 \varkappa_4{}^2 k_5/K}{(k_1{}^2 + \varkappa_2{}^2)(\varkappa_2{}^2 + k_3{}^2)(k_3{}^2 + \varkappa_4{}^2)(\varkappa_4{}^2 + k_5{}^2)}\,, \tag{54}$$

$$\begin{aligned} K = {} & \exp(\varkappa_2 w_2 + \varkappa_4 w_4)\,[\exp i(-\varphi_1 + \varphi_2 + \varphi_3 + \varphi_4 + \varphi_5) \\ & - \exp i(\varphi_1 + \varphi_2 - \varphi_3 - \varphi_4 + \varphi_5)] \\ & + \exp(\varkappa_2 w_2 - \varkappa_4 w_4)\,[-\exp i(-\varphi_1 + \varphi_2 + \varphi_3 - \varphi_4 - \varphi_5) \\ & + \exp i(\varphi_1 + \varphi_2 - \varphi_3 + \varphi_4 - \varphi_5)] \\ & + \exp(-\varkappa_2 w_2 + \varkappa_2 w_4)\,[-\exp i(-\varphi_1 - \varphi_2 - \varphi_3 + \varphi_4 + \varphi_5) \\ & + \exp i(\varphi_1 - \varphi_2 + \varphi_3 - \varphi_4 + \varphi_5)] \\ & + \exp(-\varkappa_2 w_2 - \varkappa_4 w_4)\,[\exp i(\varphi_1 - \varphi_2 - \varphi_3 - \varphi_4 - \varphi_5) \\ & - \exp i(-\varphi_1 - \varphi_2 + \varphi_3 + \varphi_4 - \varphi_5)]\,, \end{aligned} \tag{55}$$

$$\varphi_1 = k_3 w_3\,, \tag{56}$$

$$\varphi_2 = \tan^{-1}(\varkappa_2/k_1)\,, \tag{57}$$

$$\varphi_3 = \tan^{-1}(\varkappa_2/k_3)\,, \tag{58}$$

$$\varphi_4 = \tan^{-1}(\varkappa_4/k_3)\,, \tag{59}$$

$$\varphi_5 = \tan^{-1}(\varkappa_4/k_5)\,. \tag{60}$$

The dominant factor in Eq. (55) is, of course, $\exp(\varkappa_2 w_2 + \varkappa_4 w_4)$, which leads to an attenuation equal to the product of the attenuation of the individual barriers, as one might expect. This term can clearly be made to cancel by adjusting k_3 for resonance, namely

$$\varphi_1 = \varphi_3 + \varphi_4 + n(2\pi). \tag{61}$$

When Eq. (61) is satisfied the value of K will be dominated by the next largest term. Let us assume $\varkappa_2 w_2 > \varkappa_4 w_4$. Then $K \propto \exp(\varkappa_2 w_2 - \varkappa_4 w_4)$ and the barrier attenuation is the ratio of the individual barrier factors. We do not expect to be able to cancel the coefficient of $\exp(\varkappa_2 w_2 - \varkappa_4 w_4)$ at the same time that the coefficient of $\exp(\varkappa_2 w_2 + \varkappa_4 w_4)$ is zero, because this would result in a transmitted current larger than the incident current. It is easily shown that simultaneous cancellation implies

$$\varphi_4 = n\pi/2, \tag{62}$$

which in turn implies $k_3 = 0$ or $\varkappa_4 = \infty$, and the limit $k_3 \to 0$ or $\varkappa_4 \to \infty$ would have to be studied more carefully.

However, if the two penetration constants are equal, i.e., $\varkappa_2 w_2 = \varkappa_4 w_4$, then $K \approx 1$, and the transmission of the barriers is independent of their width. There is no barrier attenuation in this case, there is only reflection due to mismatch at the various edges. An effect of this sort might conceivably be important in tunneling. Resonant states at a surface or heterojunction would correspond most closely to the one-dimensional model. Impurity levels could also give effects like this, but the geometry is not one-dimensional, so the calculations would be much more difficult.

The resonant transmission through barriers is a familiar phenomenon. The Kronig–Penney model of a one-dimensional crystal is a series of identical barriers and square wells. Here we know we have allowed bands of perfect transmission separated by forbidden bands of attenuation. The attenuating case can be thought of as a product of the attenuation factors for each barrier the electron passes through.

REFERENCES

1. E. O. Kane, J. Appl. Phys. 32:83 (1961).
2. R. T. Shuey, Phys. Rev. 137:A1268 (1965).
3. J. Bardeen, Phys. Rev. Letters 6:57 (1961).

Chapter 2

WKB Methods

W. Franz

Institute of Theoretical Physics
University of Münster

As is well known, the basic principle of classical mechanics may be stated as follows: The mechanical path between two points 1 and 2 in space-time is a geodetic line in a suitable Finsler geometry, or, mathematically speaking,

$$\delta \int_1^2 dS = 0, \tag{1}$$

where the line element dS is a given function of coordinates q_k and their differentials dq_k, dS being homogeneous of first order in the dq_k:

$$dS = \sum_k \frac{\partial dS}{\partial dq_k} dq_k \equiv \sum_k p_k \, dq_k. \tag{2}$$

Here conjugate momenta p_k have been defined as usual by $p_k = \partial(dS)/\partial(dq_k)$ The q_k being chosen to be **x**, t, Eq. (2) gives

$$dS = \mathbf{p} \cdot d\mathbf{x} - E \, dt, \tag{3}$$

where **p** is the momentum and E the total energy. The connection to wave mechanics is now made by stating that the mechanical path is to be equal in some sense to the optical path, i.e., $S \approx \varphi$, where φ is the optical phase. At this point a universal constant is to be introduced (which is Planck's constant)

$$S = \hbar \cdot \varphi. \tag{4}$$

From this one gets the wave functions

$$\psi = e^{i\varphi} = \exp\left(\frac{i}{\hbar} \int dS\right) = \exp\left[\frac{i}{\hbar} \int (\mathbf{p} \cdot d\mathbf{x} - E \, dt)\right]. \tag{5}$$

For constant momentum (5) is a solution of

$$\left(\frac{d^2}{dx^2} + \frac{p^2}{\hbar^2}\right)\psi = 0, \tag{6}$$

where the one-dimensional case has been taken for simplicity. In case p depends on x, (5) is no longer an exact solution of (6), but it is still an approximate solution if p varies slowly with x. This now is the point where the WKB method starts trying a solution of (6) for x dependent p in the form

$$\psi(x) = T(x)\exp\left[\frac{i}{\hbar}\int (p\,dx - E\,dt)\right] = T(x)\exp\left[\frac{i}{\hbar}S(k)\right], \tag{7}$$

which differs from (5) by an amplitude factor $T(x)$. As we are concerned with the case of constant E only, the factor $\exp(-iEt/\hbar)$ will be dropped in the following.

With expression (7) Eq. (6) gives

$$\hbar^2 T + i\hbar(2T'\cdot S' + T'\cdot S'') - TS'^2 + Tp^2 = 0, \tag{8}$$

where the prime denotes differentiation with respect to x. Treating $\hbar^2$ as a small number, one may drop the first term in (8). One then gets

$$\begin{gathered} S = \int p(x)\,dx, \qquad T(x) = [S'(x)]^{-1/2} = p(x)^{-1/2}, \\ \psi(x) = p^{-1/2}\exp\left(\frac{i}{\hbar}\int p\,dx\right). \end{gathered} \tag{9}$$

The question now is how to tackle the singularity at $p = 0$.

LANGERS' METHOD

If $p^2(x_0) = 0$, then p^2 in the neighborhood of x_0 is given as

$$p^2(x) = a(x - x_0), \qquad a \equiv [dp^2(x)/dx]_{x=x_0}. \tag{10}$$

The solution of (6) may then in the vicinity of x_0 be expressed by an Airy integral as

$$\psi(x) = Ai[(3a/\hbar^2)^{1/3}(x - x_0)]. \tag{11}$$

The Airy integral is defined as

$$Ai(u) = \int_C dt \exp i(t^3 - ut), \tag{12}$$

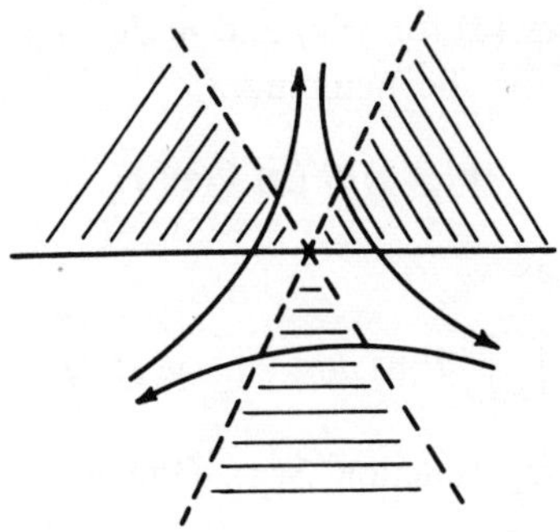

Fig. 1. Possible contours for Airy integrals.

where as contour C one of the three shown in Fig. 1 may be chosen. One thus has three solutions of (11) according to the three possible contours C of the integral (12). A linear combination of these solutions can be linked asymptotically to the solution (9).

THE METHOD OF MILLER AND GOOD

In the method of Miller and Good one starts from a model problem which can be solved exactly:

$$\frac{d^2\Phi}{dS^2} + \frac{P^2(S)}{\hbar^2}\,\Phi(S) = 0, \tag{13}$$

where $P(S)$ is a function which has as many zeros of the same order as $p(x)$ has but is of such a form as to allow Eq. (13) to be solved exactly. The model problem (13) is now connected to the real problem (6) by mapping the independent variable x of the real problem into the variable S of the model. One sets

$$\psi(x) = T(x)\Phi[S(x)]. \tag{14}$$

From (14) one gets

$$\psi''(x) = T''\Phi[S(x)] + (2T'S' + TS'')\frac{d\Phi}{dS} + TS'^2\frac{d^2\Phi}{dS^2}. \tag{15}$$

With $T(x) = S'(x)^{-1/2}$ and (13) this reduces to

$$\psi''(x) = \left\{T''(x) - TS'^2\frac{P^2[S(x)]}{\hbar^2}\right\}\Phi[S(x)]. \tag{16}$$

Putting this into the Eq. (6) for $\psi(x)$ and neglecting the term $\hbar^2 T''(x)$, one obtains the condition for the mapping,

$$P[S(x)]S'(x) = p(x),$$

which can be written as

$$\int_{S(x_0)}^{S(x)} P(\sigma)\, d\sigma = \int_{x_0}^{x} p(\xi)\, d\xi, \tag{17}$$

which gives the mapping function $S(x)$. One now has a solution of (6) in the form

$$\psi(x) = [S'(x)]^{-1/2}\Phi[S(x)]. \tag{18}$$

This is continuous if $S'(x) \neq 0$ even at the zeros of $p(x)$. According to Eq. (17), this means: if $p(x)$ has a zero of order n at x_0 then $P(S)$ must have a zero of order n at $S(x_0)$.

EXAMPLES

1. If $p^2(x)$ has no zero, one simply takes $P(S) = 1$ and gets the original WKB result (9).

2. $p^2(x)$ has a zero at x_0 of first order, $p^2(x) = a(x - x_0)$. Now one can choose $P^2(S)$ as $(\hbar^2/3)S$. With this choice of $P(S)$ one gets from (13)

$$\Phi(S) = Ai(S); \tag{19}$$

Eq. (17) now gives

$$\begin{aligned} S(x) &= (3a/\hbar^2)^{1/3}(x - x_0) \\ S'(x_0) &= (3a/\hbar^2)^{1/3} \neq 0. \end{aligned} \tag{20}$$

With (19) and (20) one gets from (18) the solution

$$\psi(x) = (\hbar^2/3a)^{1/6}\, Ai[(3a/\hbar^2)^{1/3}(x - x_0)],$$

which is the Langer result.

3. The case of barrier tunneling where $p(x)$ has two zeros at x_1 and x_2: In this case one uses the model of a harmonic oscillator

$$P^2(S) = S^2 - C^2,$$

C being a constant. Equation (17) then reads

$$i\int_{-C}^{S} (C^2 - \sigma^2)^{1/2} = \int_{x_1}^{x} p(\xi)\, d\xi.$$

Here we mapped $x = x_1$ into $S = -C$. In order to also map $x = x_2$ into $S = +C$, we have to choose

$$C^2 = (2/\pi i) \int_{x_1}^{x_2} p(\xi)\, d\xi\,.$$

For the transmission coefficient through the barrier Miller and Good derive

$$D = \left\{1 + \exp\left[-2i \int_{x_1}^{x_2} p(\xi)\, d\xi\right]\right\}^{-1}.$$

If the energy of the impinging particle is less than the barrier height, then $p(x)$ is imaginary between x_1 and x_2 and D is real. If E is above the barrier, then there are no more real zeros x_1, x_2 of $p(x)$ but there are imaginary ones. So D is real again. It is seen that the transmission coefficient calculated by the Miller–Good method is still valid even in the scattering case, where the energy of the particle is above the barrier height and there are no classical turning points.

REFERENCES

1. S. Flügge and H. Marschall, Rechenmethoden der Quantenmechanik, Springer-Verlag, Berlin, 1965, p. 152ff.
2. L. I. Schiff, Quantum Mechanics, McGraw-Hill Book Co., New York, 1955, p. 184ff.
3. S. C. Miller, Jr., and R. M. Good, Jr., Phys. Rev. 91:174–179 (1953).

Chapter 3

Metal–Insulator–Metal Tunneling

I. Giaever

General Electric Research and Development Center
Schenectady, New York

INTRODUCTION

In principle, it is easy to perform a tunneling experiment. Imagine the simple DC circuit shown in Fig. 1, consisting of two parallel capacitor plates connected to a battery. There are also an ammeter, a voltmeter, and some resistors in the circuit so that we can measure the current through the capacitor and both measure and vary the voltage applied to the capacitor plates. Normally, we do not expect a DC current to flow between the capacitor plates; however, if the capacitor plates are spaced very close together (on the order of 50 Å), we find that a DC current is easily detect-

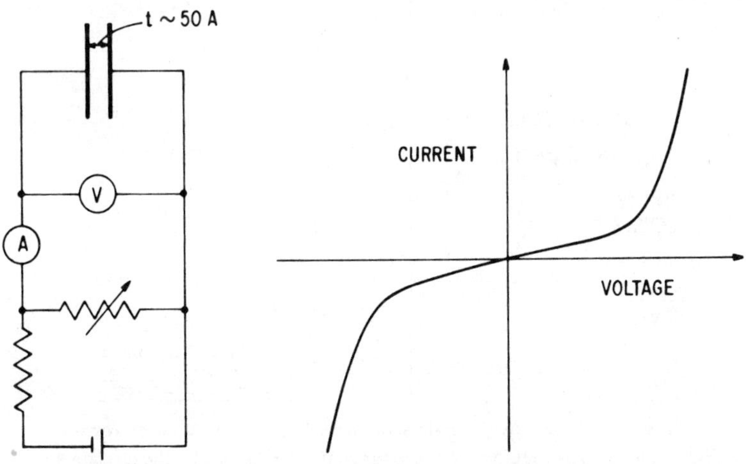

Fig. 1. Schematic drawing of a tunneling experiment. If the capacitor plates are spaced about 50 Å apart or less, a tunnel current will be easily observable. The current–voltage characteristic will be nearly symmetric about zero, linear at low voltages (below ~ 0.1 V), and nonlinear at higher voltages.

able with an ordinary ammeter, and we may get a current–voltage characteristic somewhat as shown in Fig. 1. The DC current flowing through the thin insulator between the capacitor plates is a tunneling current.

MODEL

From experience we know that a DC current will not flow between the plates of an ordinary capacitor, and the reason we get a current when the plates are spaced very close together is that quantum mechanical phenomena become important. Let us briefly review the simple free-electron model of a metal. The metal is viewed simply as a potential box filled with electrons up to some energy, E_F, called the Fermi energy. This is a consequence of Pauli's principle, which states that only two electrons (because of the electron spin) can occupy each level. Since the electrons do not normally escape from the metal, there must be some energy, φ, required to remove an electron. This energy is called the work function. In our experi-

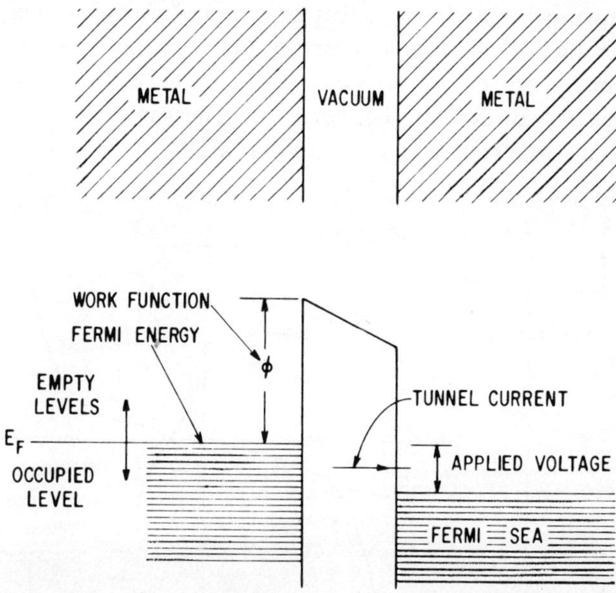

Fig. 2. Above: two metals separated by a thin vacuum region. Below: an energy diagram of the arrangement. Classically, the electrons have insufficient energy to enter the vacuum region; however, because electrons obey quantum mechanics they may tunnel through the vacuum region. Only the electrons on the left, which face empty states on the right, can tunnel and contribute to the current.

ed onto the glass with a soldering iron. If the pure In has been previously heated above its melting point in air for about 24 hr, it sticks nicely to the glass slide.

The glass slide is placed over a metal mask such that only the desired areas are exposed and coated with metal. The In solder helps to separate the glass slide from the mask so that the films will not be scratched or damaged when the slide is removed.

After a metal film is deposited the slide is removed from the vacuum system and the metal film is allowed to oxidize. After 50 Å or so of oxide has formed the slide is returned to the evaporator for a second evaporation. The completed sample is shown in Fig. 4 (*d*). Wires are connected to the In dots and an approximately four-terminal measurement is made. It is important to note that only an approximate four-terminal measurement is possible; we shall return to this point later.

The evaporated metal films are in general quite dirty. In a vacuum of 10^{-5} mm Hg the glass slide is being struck as often by foreign gas atoms and molecules as by the metal atoms; thus, unless the vacuum is appreciably better than that, it does not pay to expend too much effort cleaning the glass slide. The microscope slide can generally be taken directly from the box in which it is purchased.

I have found that several metals will form a good, tough, insulating oxide. I would rate Al, Cr, Ni, Mg, Nb, Ta, Sn, and Pb, in that order, as relatively easy to use; Cu, La, Co, V, and Bi as difficult; and Ag, Au, and In as impossible. The forming of thin oxide films tends to be more an art than a science; my films have all been oxidized in ordinary air at room temperatures. It is likely that different procedures will produce different results.

MEASUREMENTS

Let us discuss more carefully the current which flows through the barrier. The current for small voltages is given by

$$I \approx V \exp[-(2t/\hbar)(\varphi m)^{1/2}].$$

For very high voltages the shape of the barrier will change, and the current will be given by the Fowler–Nordheim formula developed for emission into vacuum. For intermediate voltages various formulas have been developed to connect the two limits. A typical current–voltage characteristic is shown in Fig. 5 for a sandwich made of Ni–NiO–Pb. It is seen that the curves are not quite symmetric when the voltage is reversed.

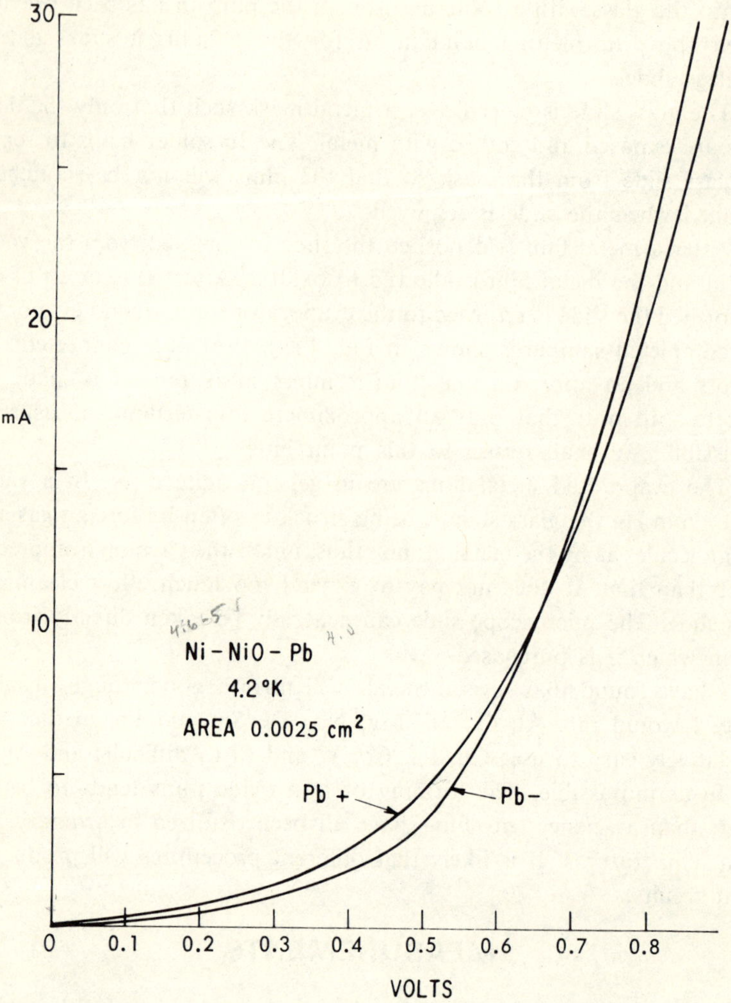

Fig. 5. The current–voltage characteristic of a Ni–NiO–Pb sandwich. Note that the characteristic is linear at low voltages, and that the characteristic is slightly polarity dependent.

In order to calculate the current to compare with experiments one must know the junction thickness, t, the dielectric constant, ε, the junction area, A, the barrier height, φ, and shape, and the effective mass, m^*, of the electron in the oxide. The dielectric constant of a 50-Å film has not been measured, but we use the bulk value, which is probably not too far off. The apparent area can be determined easily, and therefore the thickness can be calculated from the measured junction capacitance, C,

$$t = 4\pi\varepsilon A/C.$$

We should remember that this gives us only some average thickness, and that tunneling is very sensitive to thin spots. Nevertheless, this should give a reasonable value for the thickness. Information obtainable from the Josephson effect shows that the junction thickness is surprisingly uniform. The barrier height can be measured by photoelectric emission, but the barrier shape cannot be determined. We must remember that an error of a few per cent in determining the barrier thickness or the barrier height may give errors of two orders of magnitude in the current. For this reason the effective mass is often used as a "fudge factor" (the effective mass appears in the exponent, and it is more attractive to change the mass than Planck's contant!). I think at this stage of the development it is fair to say that there exists only a qualitative agreement between tunneling theory and experiments.

ARTIFICIAL BARRIERS

The fabrication method described above relies upon a natural grown oxide layer as the insulating barrier separating the two metal films. It would be very convenient if we somehow could make an artificial insulating barrier. Let me first describe how I think such an experiment should be performed. One should start with a Au film to insure that no natural oxide is present. Then somehow the artificial barrier is deposited, and finally a superconductor is used for the top electrode. The reason for the superconducting electrode is that the current–voltage characteristic undergoes a characteristic change when the superconducting film goes from its normal state to the superconducting state, thereby providing a critical test for the existence of tunneling current. (Tunneling between superconductors is discussed in detail in Chapter 19 in this volume.)

Unfortunately, to my knowledge, none of the existing experiments with artificial barriers have been done in this way. Several techniques for producing artificial barriers have been proposed and tried, such as (1) evaporating aluminum and oxidizing to completion; (2) evaporating a thin insulating film; (3) applying Langmuir films such as barium stearite; and (4) polymerizing various oil films. I have also tried all these methods in attempts to perform the proper tunneling experiment outlined above, and have invariably met with failure. In my opinion no convincing evidence exists which shows that it is possible to fabricate an artificial barrier thin enough to tunnel through at low voltages.

HEATING EFFECTS

Let me describe one curious result I obtained trying to use Ge as an artificial barrier. Every time I used Au–Ge–Pb in that sequence of evaporation the experiment failed, i.e., I got short circuits. When I used Pb–Ge–Pb, however, there was definitely tunneling at low voltages, and the whole curve is shown in Fig. 6. The characteristic looks very interesting; unfor-

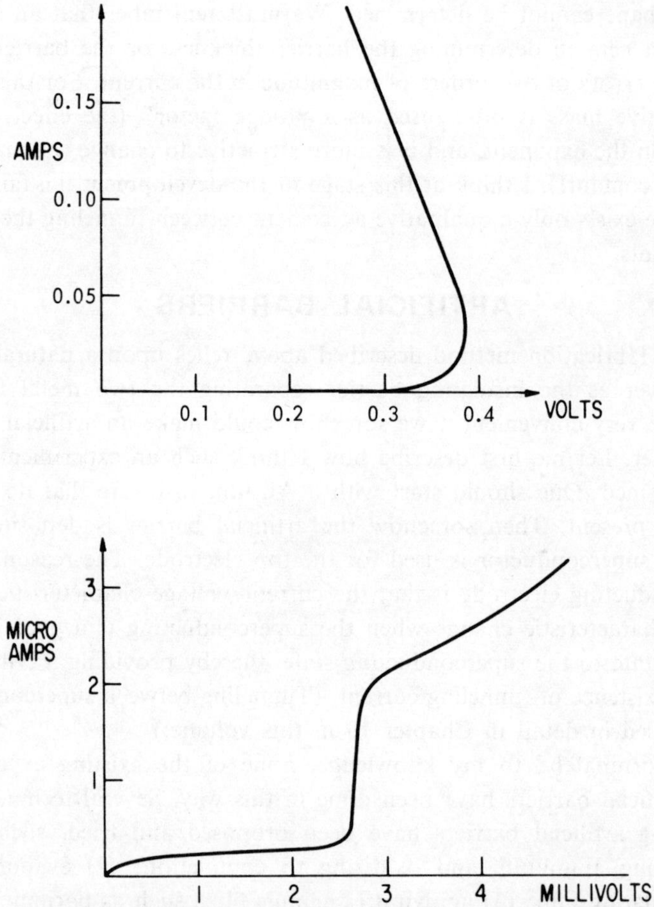

Fig. 6. The current–voltage characteristic of a Pb–Ge–PbO–Pb sandwich. The lower curve shows the characteristic energy gap in Pb, i.e., we deal with tunneling current. The upper curve shows a negative resistance, which is due to heating of thin Pb bridges sticking through the Ge (see Fig. 7). Thus it is possible to have tunneling at low voltages and thermionic emission at high voltages.

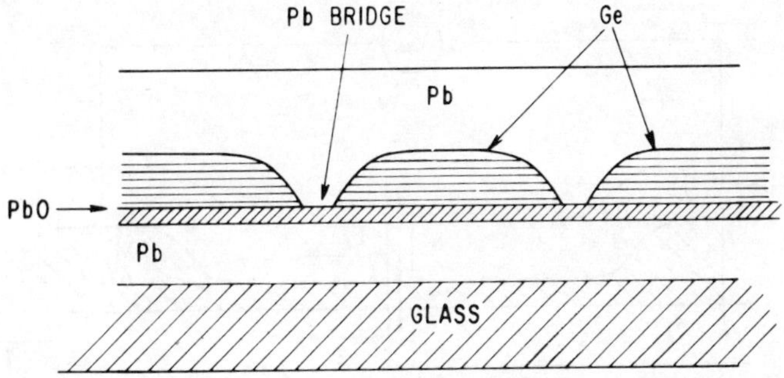

Fig. 7. Schematic of a Pb–Ge–PbO–Pb sandwich. First the Pb is evaporated, and a thin oxide layer inadvertently forms. Next the Ge is evaporated, and it does not form a continuous layer, so that the second layer of Pb reaches through the Ge in certain spots. At high voltages the Pb bridges get hot, and it is known from other experiments that the nonlinearity observed at high voltages is simply due to a heating effect.

tunately, it turns out to be a rather trivial heating effect. By postulating that the Ge does not form a continuous layer, we get a structure shown in Fig. 7, where thin bridges of Pb are sticking through the Ge, but separated from the first Pb film by PbO. At low voltages the current flows through the bridges and tunnels through the oxide. As the current is increased, the bridges get hot and a large amount of the current will now get across by thermionic emission. It is easy to adjust the parameters in the problem to get curves akin to the experimentally observed result. While this may be an extreme example, it is well worth remembering that local heating, even at helium temperatures, can produce large effects of rather trivial origin in tunnel junctions.

THE FOUR-PROBE PROBLEM

It is not generally recognized that, barring the use of superconductors, it is impossible to measure the exact current–voltage characteristic of a tunneling sample. Let us consider a simple calculation which illustrates the point, assuming that we have a sample as shown in Fig. 8. We have picked this simple geometry to be able to easily calculate the results; the conclusions will be similar no matter which geometry we use. The currents and voltages along the films will be a function of the position along the film. The total resistance along the film is R_F, while the tunneling resistance is R_T. We assume that the tunneling resistance is constant, i.e., independent

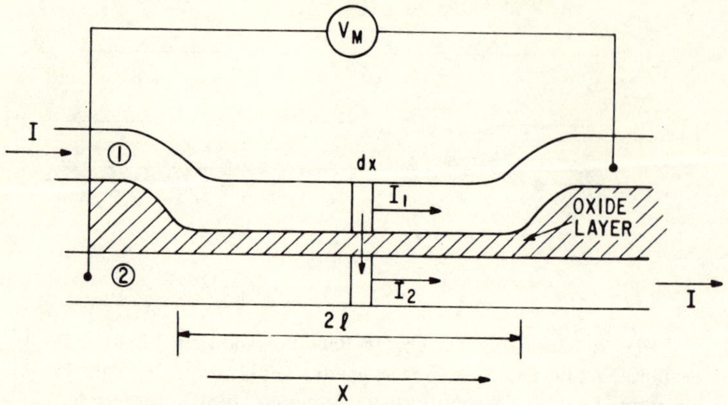

Fig. 8. A schematic arrangement of the two films forming the capacitor. The currents, as well as the potentials, will be a function of positions along the junction. For this simple geometry the measured resistance R_M is equal to the true tunneling resistance R_T only as long as the resistance along the films $R_F \ll R_T$.

of voltage, which, as seen from Fig. 5, is not true. We then obtain the following set of equations:

$$(R_F/2l)I_1 = -\,dV_1/dx\,,$$

$$(R_F/2l)I_2 = -\,dV_2/dx\,,$$

$$I_1 + I_2 = I\,,$$

and

$$V_1 - V_2 = R_T 2l\,dI_2/dx\,,$$

where $2l$ is the total length of the junction, I_1 and I_2 the currents along the two films, and V_1 and V_2 the voltages along the films. Since the measured values are $I_M = I$ and $V_M = V_1(2l) - V_2(0)$, we obtain

$$\frac{V_M}{I_M} = R_T\left\{\left(\frac{R_F}{2R_T}\right)^{1/2}\frac{\cosh(R_F/2R_T)^{1/2}}{\sinh(R_F/2R_T)^{1/2}} - \frac{R_F}{2R_T}\right\}.$$

If $R_F > R_T$, this result may be negative, and this happens very often experimentally. Thus we should be very critical of all experiments where $R_F \approx R_T$. At high voltages R_T becomes very small, so that an experiment may give correct results at low voltages and nonsense at high voltages. By using superconductors, $R_F = 0$, and these problems are entirely avoided.

APPLICATIONS

To date the actual applications of metal–insulator–metal films have been rather disappointing. The two most often recurring suggestions are the simple three-film sandwich and the cold cathode depicted in Fig. 9. The hope is that the three-film sandwich will work as a triode. Electrons first tunnel through one oxide layer, exist as hot electrons in the middle metal layer, then tunnel through the second oxide layer into the third film. To my knowledge, no convincing experimental evidence exists for the successful operation of such a three-film sandwich this way, and as far as I know, nobody even expects theoretically to get more than a very small fraction of the current through the second oxide film, so that a practical triode seems remote. The cold cathode is somewhat more promising in

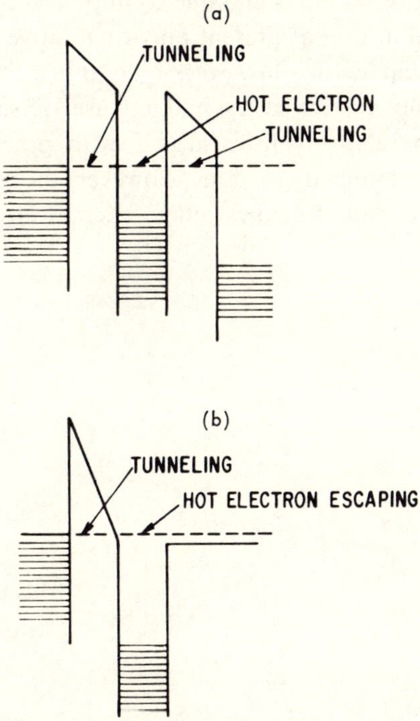

Fig. 9. Practical applications suggested have been the three-layer sandwich shown in (*a*). The electron has to tunnel through the second oxide layer without being thermalized in the middle metal layer. A variation of this is the cold cathode shown in (*b*), where the hot electrons escape into vacuum.

that the electrons do indeed come out into the vacuum; however, undoubtedly in some of the reported cases at least, the current has been due to thermionic emission. The difficulties are connected with the fact that when an oxide film becomes thick enough to support the necessary voltage, which has to be larger than the work function, the tunneling current becomes small.

SUMMARY

It is very easy both to fabricate a tunnel junction and to perform an experiment. However, since we rely upon insulating films rather than vacuum, we should be aware of the possibilities of errors due to other conduction mechanisms. Large errors in tunneling characteristics can also come about due to Joule heating of the films and due to imperfect four-probe measurements. It is my opinion that at present only qualitative agreement between theory and experiment exists; however, I do think that the experiments could be considerably improved by making use of superconductors. To now there has been relatively little success with practical active devices made with metal–insulating–metal films; however, they may eventually find uses as passive *RC* circuits for low-voltage microelectronics.

Chapter 4

Theory of Metal–Barrier–Metal Tunneling

C. B. Duke

Research and Development Center
General Electric Co.,
Schenectady, New York

Our objective in this chapter is to illustrate the calculation of the general features of the current–voltage (I–V) characteristics of a metal–insulator–metal (M–I–M) tunnel junction. In particular, no attempt is made to give either a review or a comprehensive survey of the extensive literature on this topic,* and our attention is confined to the use of the simplest models on the basis of which the I–V characteristic may be evaluated.

The first step in constructing a suitable model to describe a M–I–M junction is the specification of the energy (bias) scale on which we wish to describe the I–V characteristic. For example, many-body effects, such as phonon excitation in the insulator or superconductivity in either of the metals, can cause rapid variations in the I–V characteristics on a bias scale of the order of meV. In this chapter we do not consider such fine structure in the I–V characteristic but rather attempt to give a description of variations on an energy scale of the order of hundreds of meV. This restriction permits us to use a noninteracting-electron model of the tunneling process. In such a model the interactions of the electrons both among themselves and with internal degrees of freedom (e.g., phonons) in both the barrier and the bulk metals are neglected.

A noninteracting-electron model of the tunneling process is not yet the historical "free" electron model because the electrons move in a periodic potential in the metals and in a complicated, disordered-ion potential in the insulating barrier. In the free-electron model the effect of the periodic potential in the metals is not analyzed in detail but rather is approximated

* Such a review has given by C. B. Duke, *Tunneling in Solids*, Academic Press, to be published.

by considering that the potential only changes an electron's energy–momentum dispersion relation from that of a free electron, $E(k) = \hbar^2k^2/2m$, to some more general form. More precisely, the free-electron model is based on the use of the effective-mass approximation for the electronic wave functions and the imposition of boundary conditions on the envelope function rather than the exact one-electron eigenfunctions [see, e.g., W. A. Harrison ([14]); D. J. BenDaniel and C. B. Duke ([9])].

These considerations form the outline of a justification for the application by L. Nordheim ([1]) and R. H. Fowler and L. Nordheim ([2]) of the free-electron model to describe the field emission of electrons from a metal into a vacuum ([3,4]). However, in a consideration of M–I–M junctions, a discussion of the mechanism of current flow through the insulator still must be given. The usual potential-barrier model of M–I–M tunneling results from replacing the exact potential in the insulator by a local average value which varies (slowly) in the junction due to macroscopic electric fields. A minor variant of this model is given by treating the insulator as a perfectly periodic structure and applying the effective-mass approximation to describe an electron's motion in the resulting periodic potential. Both models ignore the fact that in thin junctions or junctions containing charged impurities the effects of fluctuations from the average potential are likely as important as the average potential itself in determining the current flow through the junction. A limiting case illustrating this defect is that of the thin "Josephson junctions," discussed later in this volume, which often are only a monolayer thick.

For the remainder of this chapter we consider only the "one-electron potential-barrier" model of a M–I–M junction. The limitations of the model must be considered when comparisons of its predictions with experimental results are made. The model contains at least three parameters (average barrier height, barrier thickness, and electronic effective mass inside the barrier) and in most applications considerably more are used. The demonstration that these three (or more) parameters can be chosen to fit experimental *I–V* characteristics does not suffice to validate the application of the model to describe the experiment. Such validation is provided if and only if all the parameters needed to describe both the shape and magnitude of the *I–V* characteristic are determined independently of the measurement of this characteristic, and predictions of the model are subsequently verified by experiment. No such test of the model has been performed. The success of its application to parametrize the gross features of *I–V* characteristics is reported in many of the references given in this volume. This success indicates the qualitative veracity of the idea that the insulating barrier behaves similarly to a vacuum, but should not be misinterpreted as indicating the quantitative applicability of the model.

The literature on the free-electron potential-barrier model contains discussions of three topics: (a) derivation of general formulae for the current, (b) evaluation of the barrier penetration probability, and (c) discussion of band structure effects on the current. We proceed by first reviewing the derivation of the formula for the current. Next a brief outline is given of the extent to which the model predicts that the bias-dependence of the tunneling current reflects the details of either the metal or insulator band structure. The chapter concludes with a discussion of the qualitative features of the I–V characteristic as a function of bias and temperature for a case in which the Fermi energies of both the left- and right-hand metals exceed the height of the barrier at zero bias.

Consider the M–I–M barrier shown in Fig. 1. The probability that an electron state of energy E is occupied is given by

$$f_L(E_L) = \{1 + \exp[(E_L - E_{FR})/kT]\}^{-1}, \tag{1a}$$

$$f_R(E_R) = \{1 + \exp[(E_R - E_{FR})/kT]\}^{-1}, \tag{1b}$$

with $E = 0$ taken, by definition, to lie at the bottom of the conduction band of the right-hand metal. Taking the plane of the junction to lie normal to the x direction, the current per unit area j from right to left is given by

$$j_{RL} = -e \int \frac{d^3k_R}{(2\pi)^3} 2v_{xR} f_R(E_R)[1 - f_L(E_R + eV)] D_{RL}(E_R). \tag{2}$$

The integral runs over all k_{yR} and k_{zR} but only over $v_{xR} < 0$, i.e., over

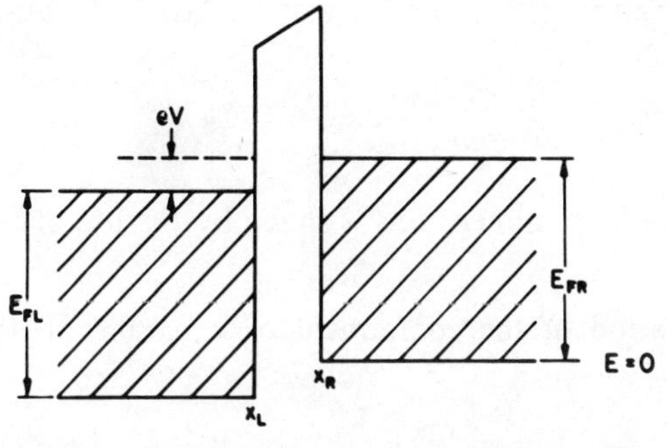

Fig. 1. M–I–M potential barrier model with a bias $+ V$ on the left-hand metal. E_{FL} and E_{FR} denote the Fermi energies on the left- and right-hand sides, respectively. Electrons are shown at zero temperature.

electrons headed toward the junction. The extra 2 is for spin because

$$\langle j \rangle = \sum_{\text{momenta, spin}} \langle \mathbf{k} m_s \mid e v_x \mid \mathbf{k} m_s \rangle, \tag{3}$$

the $f_R(E)$ factor is the probability that the initial state is occupied and the $(1 - f_L)$ factor is the probability that the final state is unoccupied. The $D(E)$ denotes the probability that the electron penetrates the barrier. A similar expression gives the current per unit area from left to right:

$$j_{LR} = -e \int \frac{d^3k_L}{(2\pi)^3} 2v_{xL} f_L(E_R + eV)[1 - f_R(E_R)] D_{LR}(E_R). \tag{4}$$

The total current is the sum of these two expressions. To add them we must (*a*) relate D_{RL} to D_{LR}, and (*b*) relate $v_{xL}\, d^3k_L$ to $v_{xR}\, d^3k_R$. Kane discussed (*a*) in Chapter 1. We recall that the two linearly independent solutions to Schrödinger's equation for $E_x > 0$ are

$$\begin{aligned} \psi^{(1)}_{E_x}(x) &= S_{RL} \exp(-ik_{xL}x), & x < x_L, \\ &= \exp(-ik_{xR}x) + R_{RR} \exp(ik_{xR}x), & x > x_R, \end{aligned} \tag{5a}$$

$$\begin{aligned} \psi^{(2)}_{E_x}(x) &= \exp(ik_{xL}x) + R_{LL} \exp(-ik_{xL}x), & x < x_L, \\ &= S_{LR} \exp(ik_{xR}x), & x > x_R, \end{aligned} \tag{5b}$$

$$\begin{aligned} E_x &\equiv \hbar^2 k_{xR}^2 / 2m_R, \qquad \text{(parabolic band model)}, \\ &= (\hbar^2 k_{xL}^2 / 2m_L) + E_{FR} - E_{FL}, \end{aligned} \tag{5c}$$

and that in terms of these solutions

$$D_{LR} = \frac{v_{xR}}{v_{xL}} \mid S_{LR} \mid^2 = \frac{v_{xL}}{v_{xR}} \mid S_{RL} \mid^2 = D_{RL}, \tag{6a}$$

$$v_{xL} = \frac{1}{\hbar} \frac{\partial E(\mathbf{k}_L)}{\partial k_{xL}}. \tag{6b}$$

The transition to three dimensions is made by the use of two conservation laws:

1. Conservation of the component of **k** parallel to the plane of the junction:

$$\mathbf{k}_{\parallel,L} = \mathbf{k}_{\parallel,R} = \mathbf{k}_{\parallel,b}. \tag{7}$$

2. Conservation of the *total* energy of the tunneling electron

$$E_L = E_R \equiv E. \tag{8}$$

From Eq. (6) we find that D_{LR} equals D_{RL}, and, further, that it depends only on the x component of the energy in the case that we use a separable model of the energy in the metals:

$$D_{LR}(E) = D_{LR}(E_x) = D_{RL}(E_x)\,, \tag{9a}$$

$$E_i = E_{xi} + E_{\parallel,i}. \tag{9b}$$

For example, in a simple effective-mass model, E_{xi} is related to E and $\mathbf{k}_{\parallel,L}$ by the equations:

$$\begin{aligned} E_{xR} &= \frac{\hbar^2 k_{xR}^2}{2m_R}, \qquad E_{\parallel,R} = \frac{\hbar^2}{2m_R}(k_{yR}^2 + k_{zR}^2), \\ E_{xL} &= \frac{\hbar^2 k_{xL}^2}{2m_L} + E_{FR} - E_{FL}, \qquad E_{\parallel,L} = \frac{\hbar^2}{2m_L}(k_{yL}^2 + k_{zL}^2). \end{aligned} \tag{10}$$

To consider the Jacobian relations (*b*) given below Eq. (4) we note that from Eqs. (6*b*), (7), and (8), respectively,

$$v_{xi}\, dk_{xi} = (1/\hbar)\, dE_i\,, \tag{11a}$$

$$d^2k_{\parallel,L} = d^2k_{\parallel,R} \equiv d^2k_{\parallel}\,, \tag{11b}$$

$$dE_L = dE_R. \tag{11c}$$

Thus we can combine Eqs. (2) and (4), noting that

$$f_R(E)[1 - f_L(E + eV)] - f_L(E + eV)[1 - f_R(E)] = f_R(E) - f_L(E + eV)\,, \tag{12}$$

to give

$$j = j_{LR} - j_{RL} = \frac{2e}{h}\int dE\,[f_R(E) - f_L(E + eV)] \int \frac{d^2k_{\parallel}}{(2\pi)^2}\, D(E, \mathbf{k}_{\parallel}). \tag{13}$$

For our simple model of a potential barrier separated by two effective-mass metals we get

$$D(E, \mathbf{k}_{\parallel}) = D(E_{xR}) = D[E - (\hbar^2 k_{\parallel}^2/2m_R)]. \tag{14}$$

Because of the dependence of D only on E_x the independent integration variables in Eq. (13) for field emission have typically been taken as E_x and $\mathbf{k}_{\parallel}$ rather than E and $\mathbf{k}_{\parallel}$. Then the $\mathbf{k}_{\parallel}$ integral can be performed independent of the form for D. For thermionic emission this sequence of steps is particularly convenient and is that taken in most elementary texts. The presence of the exclusion principle restrictions on both sides of the junction render the form (13) most convenient in M–I–M tunneling considerations. The use of Eqs. (7) and (8) can lead to the use of differing values of $E_{x,i}$ in the various regions of the junction if the values of m_i for motion parallel to the plane of the junction are not equal. This fact is accounted for in the evaluation of $D(E, \mathbf{k}_{\parallel})$ by use of the appropriate values of k_{xL} in Eq. (10) and in the barrier.

The calculation of the barrier penetration probability has been discussed extensively in the literature. The analyses vary both with regard to the approximations used (e.g., use of WKBJ approximation or an exact one-dimensional calculation) and with regard to the sophistication of the potential model used for the barrier (e.g., the simplest barrier is the square well, incorporating the potential gradient in the barrier gives a trapezoidal barrier, image force corrections and impurity potentials in the barrier lead to still more complicated potentials). For M–I–M barriers Simmons discusses the image-force corrections in Chapter 10, and ([5]) contain reviews of the earlier literature. The effects of impurities in the junction have been discussed by Schmidlin ([6]) and by Scalapino and Marcus ([7]), whose work also contains earlier references. Multiple-barrier interference effects are treated by Iogansen ([8]). Dependence of the barrier on the bias is discussed in Simmon's papers and by BenDaniel and Duke ([9]), with references to earlier work given in these papers. The influence of the dispersion relation in the barrier oxide has been analyzed recently by Lewicki and Mead ([10]). The analogous results in metal–vacuum field emission have been briefly reviewed by Duke and Alferieff ([3]), except for the work of Itskovich ([11,12]) on the effect of $\mathbf{k}_{\parallel}$ conservation on the "observed" work functions by different techniques and from different crystalline faces. These references are by no means exhaustive, but their number and variety give some indication of the state of development of the consequences of the potential-barrier model. For the remainder of this chapter I will consider only the simplest barrier, the square barrier, and use it to illustrate the consequences of Eq. (13).

Using a square-barrier model, we can learn a great deal about the bias dependence of the current in Eq. (13) by just considering the form of D without evaluating the integrals. First, we note that the shape of the barrier depends explicitly on the bias, an shown in Fig. 2. Thus if we take an "effective" square barrier we must select the bias-dependent height

$$V_b(x) = V_0\theta(x - x_L)\theta(x_R - x); \quad V_0 = V_L + E_{FR} + \tfrac{1}{2}(V_b - |eV|), \quad (15)$$

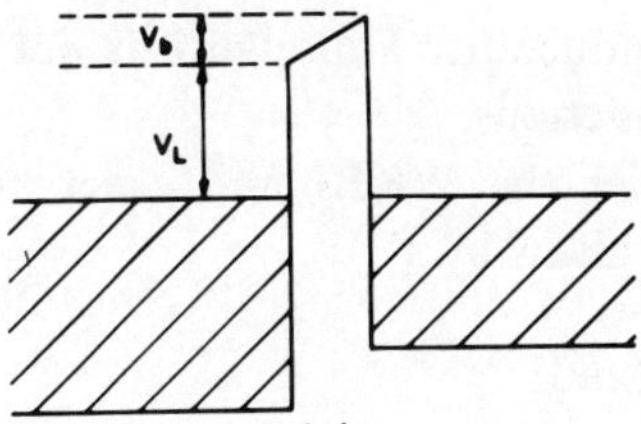

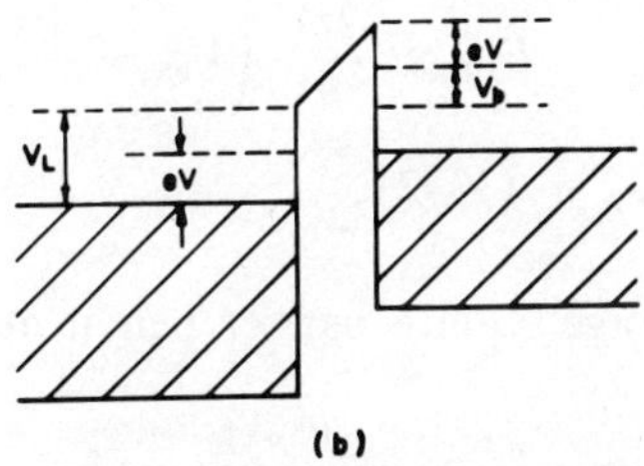

Fig. 2(*a*). Barrier at zero bias.
2(*b*). Barrier at bias eV.

where V_b is the "built-in" voltage drop across the barrier. This result tells us that D depends explicitly upon the bias, so that at zero temperature the current per unit area is given by

$$j = \frac{2e}{h} \int_{E_{FR}-eV}^{E_{FR}} dE \int \frac{d^2k_\parallel}{(2\pi)^2} D[eV; (E - \hbar^2 k_\parallel{}^2/2m_R)], \tag{16}$$

$$j(eV < 0) = -j(eV > 0).$$

The differential tunnel conductance G is defined by

$$\begin{aligned} G &\equiv dj/d(eV) \\ &= \frac{2e}{h} \left\{ \int_{E_{\parallel,R} \le E_{FR}-eV} \frac{d^2k_\parallel}{(2\pi)^2} D[eV; (E_{FR} - eV - \hbar^2 k_\parallel{}^2/2m_R)] \right. \\ &\quad \left. + \int_{E_{FR}-eV}^{E_{FR}} dE \int \frac{d^2k_\parallel}{(2\pi)^2} \frac{\partial}{\partial(eV)} D[eV; (E - \hbar^2 k_\parallel{}^2/2m_R)] \right\}. \end{aligned} \tag{17}$$

The first term in (17) is due to the increased number of electron states made available for tunneling due to an incremental increase in bias. The second term is due to the change in barrier penetration probability due to the change in the barrier with bias. For some tunnel junctions [e.g., metal–semiconductor contacts—see, e.g., ([13])]. The second term can dominate the observed conductance although it is not always important near zero bias in M–I–M junctions.

For a square barrier the WKBJ and exact "thick-junction" barrier penetration factors are given by ([14]):

$$D_{WKBJ} = \exp[-2\varkappa w]; \tag{18a}$$

$$w = x_R - x_L; \tag{18b}$$

$$\varkappa^2 = \frac{2m_b}{\hbar^2}[V_0 - E_x] = \frac{2m_b}{\hbar^2}\left(V_{00} - \frac{|eV|}{2} - E_{xb}\right), \tag{18c}$$

$$V_{00} = V_L + E_{FR} + (V_b/2), \tag{18d}$$

= the average square barrier height at zero bias;

$$D = \left[\frac{16v_{xL}v_{xR}v_{xb}^2}{v_{xb}^4 + v_{xb}^2(v_{xL}^2 + v_{xR}^2) + v_{xL}^2v_{xR}^2}\right]D_{WKBJ}; \tag{19a}$$

$$v_{xb} \equiv \hbar\varkappa/m_b; \tag{19b}$$

where we have used parabolic one-electron energy bands centered about an arbitrary value of k_x.

Let us next investigate the consequences on D of having the energy bands in the right-hand metal centered about a nonzero $\mathbf{k}_{\parallel,0}$ value so that, e.g.,

$$E_R = \frac{\hbar^2 k_{xR}^2}{2m_R} + \frac{\hbar^2}{2m_R}(\mathbf{k}_\parallel - \mathbf{k}_{\parallel,0})^2. \tag{20}$$

For a small E_{FR} we get a Fermi surface for which a cross section in the Brillouin zone looks like that shown in Fig. 3. We see that $\mathbf{k}_\parallel$ has a minimum value $\mathbf{k}_{\parallel,\min}$. Thus, as for an oxide with energy bands centered at $\mathbf{k} = 0$,

$$\varkappa^2 = (2m_b/\hbar^2)[V_0 - E + (\hbar^2 k_\parallel{}^2/2m_b)]. \tag{21}$$

Such a nonzero minimum value of $\mathbf{k}_\parallel$ makes the barrier penetration factor small relative to its value if the normal to the junction lies in the $\mathbf{k}_{0,\parallel}$ direction. This case illustrates one type of influence of the band structure on the tunneling current which is unrelated to any kind of "density-of-states" effect, but depends on the assumptions of (1) $\mathbf{k}_\parallel$ conservation (i.e., specular reflection) and (2) a "simple" parabolic energy band in the oxide. An experimental consequence of this effect would be that for a M–I–Bi (or Sb) junction normal to the trigonal axis of the semimetal the current

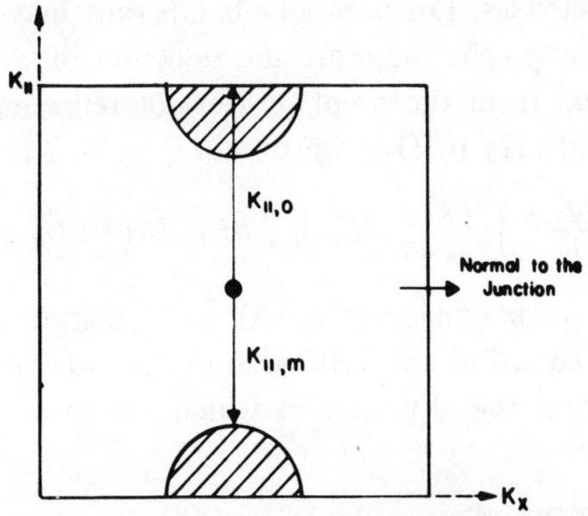

Fig. 3. Cross section of a hypothetical Brillouin zone for which the one-electron energies are given by Eq. (20) and for which the filled portions are shaded.

at forward bias ($eV > 0$) due to the "hole" band is much greater than the current at reverse bias ($eV < 0$) due to the "electron" bands. Such an effect has not been unambiguously observed. The consequences of this idea for work functions are discussed by Itskovich ([11]). The Bi case is discussed by BenDaniel and Duke ([15]).

We next discuss the question, analyzed by Harrison ([14]), of whether or not structure in the density of states of either the left- or right-hand metal is reflected by structure in the tunnel conductance G. First note that Eq. (13) above is identical with Eq. (11) in Harrison's paper provided that we use the WKBJ approximation for D and ignore the difference between v_{xL} and v_{xR}. Thus our conclusions may be regarded as either explicitly or implicitly contained in that paper. For an elliptical energy band it is often convenient to use

$$E_{\parallel,i} = \hbar^2 k_\parallel{}^2/2m_i\,, \tag{22a}$$

$$\int \frac{d^2k_\parallel}{(2\pi)^2} = \int \varrho_{\parallel,i}(E_{\parallel,i})\,dE_{\parallel,i}\,, \tag{22b}$$

$$\varrho_{\parallel,i}(E_{\parallel,i}) = m_i/2\pi\hbar^2. \tag{22c}$$

The $\varrho_\parallel$ which is most natural to use is that characteristic of the lowest density metal because the factor

$$E_{xb} = \hbar^2 k_\parallel{}^2/2m_b\,, \tag{23}$$

which appears in Eqs. (18c) and (21) does not distinguish between the left and right-hand electrodes. For parabolic bands only mass ratios distinguish between the various $\varrho_{\parallel}(E)$; therefore the selection of which one to use is basically immaterial from the point of view of reflecting structure in the conductance. Using (22) in (16) we obtain

$$j = \frac{2e\varrho_{\parallel,R}}{h} \int_{E_{FR}-eV}^{E_{FR}} dE_R \int_0^{E_R} dE_{\parallel,R} D(eV; E_R - E_{\parallel,R}), \tag{24}$$

where we used $\varrho_{\parallel,R}$ for convenience with the integration limits. Equation (24) is the basic equation for estimates of the various density-of-states effects. Consider first the situation in which the bias eV is near E_{FR} as shown in Fig. 4.

Case I: WKBJ Approximation for D. Approximate D by a constant D_0 to get

$$j \approx \frac{2eD_0\varrho_{\parallel,R}}{h} \int_{E_{FR}-eV}^{E_{FR}} E_R \, dE_R = \frac{eD_0\varrho_{\parallel,R}}{h} [E_{FR}^2 - (E_{FR} - eV)^2]. \tag{25a}$$

Thus for bias near a band edge, $eV \approx E_{FR}$,

$$G \approx \frac{2eD_0\varrho_{\parallel,R}}{h} (E_{FR} - eV)\theta(E_{FR} - eV). \tag{25b}$$

We see that for $eV > E_{FR}$ only the (neglected) reduction in D with increasing eV gives rise to a nonzero conductance.

Case II: Sharp Junction Approximation. In this case for bias near a band edge $v_{xR} \approx 0$ for the tunneling electrons, so we write Eq. (19a) as

$$D \approx (E_{xR}/V_{00})^{1/2} D_0 = (E_R - E_{\parallel,R})^{1/2} D_0 / V_{00}^{1/2}, \tag{26a}$$

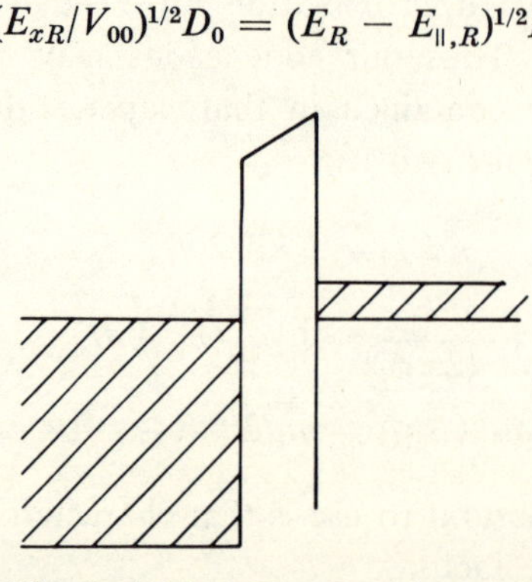

Fig. 4. A bias $eV \approx E_{FR}$ gives the junction potential shown in which the bottom of the band of the right-hand metal is at the Fermi energy of the left-hand metal.

giving

$$j \approx \frac{2eD_0\varrho_{\parallel,R}}{hV_{00}^{1/2}} \frac{2}{3} \int_{E_{FR}-eV}^{E_{FR}} E_R^{3/2}\, dE_R$$

$$= \frac{8}{15} \frac{eD_0\varrho_{\parallel,R}}{hV_{00}^{1/2}} \{E_{FR}^{5/2} - (E_{FR} - eV)^{5/2}\}, \tag{26b}$$

$$G \approx \frac{4}{3} \frac{eD_0\varrho_{\parallel,R}}{hV_{00}^{1/2}} (E_{FR} - eV)^{3/2}. \tag{26c}$$

Schematically we get the conductance shown in Fig. 5 when we neglect to consider the effect of the explicit dependence of D on eV.

The results obtained with these approximations turn out to be quite insensitive to refinements in the analysis, as discussed in detail by BenDaniel and Duke ([9]). These results form the basis for the oft-quoted remark that density-of-states effects do not give rise to structure in the conductance for values of the bias near a band edge in either metal. We note that near $eV = 0$ both the WKBJ and sharp-junction models predict at least a linear dependence of the conductance on the bias so that the junction is only approximately ohmic.

Our next calculation consists of an estimation of the shape of the conductance curve near zero bias including the effect of bias on the barrier penetration factor. We expand $\varkappa$ about $E = E_{FR}$, $E_\parallel = 0$ and consider that $m_b = m_R$ so that $E_{\parallel,b} = E_{\parallel,R}$ to get:

$$\varkappa = \left[\frac{2m_b}{\hbar^2} [V_0 - E_{FR} - (E - E_{FR}) + E_{\parallel,R}]\right]^{1/2}$$

$$\approx \varkappa_1 \left[1 - \frac{E - E_{FR}}{2(V_0 - E_{FR})} + \frac{E_{\parallel,R}}{2(V_0 - E_{FR})}\right]$$

$$\approx \varkappa_1 \left[1 - \frac{E - E_{FR}}{2E_0\varkappa_1 w} + \frac{E_{\parallel,R}}{2E_0\varkappa_1 w}\right], \tag{27a}$$

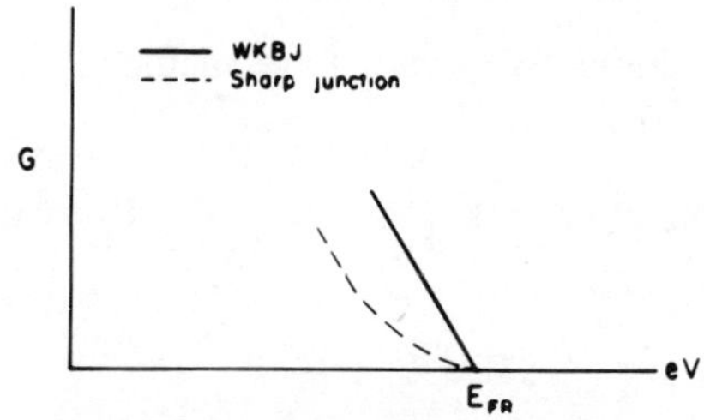

Fig. 5. Conductance for a single electron band for biases such that its bottom passes the Fermi energy of the other metal. The dependence of the barrier penetration probability on eV is neglected.

$$\varkappa_1^2 = 2m_b(V_0 - E_{FR})/\hbar^2 \,, \tag{27b}$$

$$E_0 = \frac{V_0 - E_{FR}}{\varkappa_1 w} = \frac{V_L + \frac{1}{2}(V_b - |eV|)}{\varkappa_1 w} \,. \tag{27c}$$

Inserting this approximation into Eq. (24) gives

$$\begin{aligned} j &= \frac{2e\varrho_{\|,R}}{h} \int_{-eV}^{0} dx \exp(-2\varkappa_1 w) e^{+x/E_0} \int_0^{x+E_{FR}} \exp(-E_{\|,R}/E_0)\, dE_{\|,R} \\ &= \frac{-2e\varrho_{\|,R}E_0}{h} \exp(-2\varkappa_1 w) \int_{-eV}^{0} dx\, [\exp(-E_{FR}/E_0) - e^{x/E_0}]. \end{aligned}$$

For $E_{FR} \gtrsim E_0$ the predominance of the e^{x/E_0} term indicates that only $E_x \approx E_F \Rightarrow E \approx E_F$, $E_{\|} \approx 0$ are important regions in the integral. Integrating again, we obtain

$$\begin{aligned} j = {} & \frac{-2eE_0^2\varrho_{\|,R} \exp(-2\varkappa_1 w)}{h} \\ & \left\{ \frac{eV}{E_0} \exp(-E_{FR}/E_0) - \left[1 - \exp\left(-\frac{eV}{E_0} \right) \right] \right\}. \end{aligned} \tag{28}$$

Let us examine some "typical" numbers. For a 20-Å junction and 4-eV barrier height $V_0 - E_F$:

$$\varkappa_1 \approx 10^8 \text{ cm}^{-1} \,,$$

$$\varkappa_1 w \approx 20 \,,$$

$$E_0 = (4/20)\text{eV} = 200 \quad \text{meV} \,.$$

Thus for $eV \lesssim 100$ meV,

$$j \approx \frac{2eE_0\varrho_{\|,R}}{h} \exp(-2\varkappa_1 w) eV [1 - \exp(-E_{FR}/E_0)] \,,$$

and the junction is ohmic. We could have included the eV term in (18c) in the expansion (27a) to get

$$\varkappa_0^2 = \frac{2m_b}{\hbar^2} \left(V_L + \frac{1}{2} V_b \right); \qquad E_0 = \frac{V_L + (V_b/2)}{\varkappa_0 w} \,, \tag{29a}$$

$$\begin{aligned} j &= \frac{2e}{h} E_0^2 \varrho_{\|,R} \exp(-2\varkappa_0 w) \exp(|eV|/2E_0) \\ &\quad \times \left[1 - \exp(-eV/E_0) - \frac{eV}{E_0} \exp(-E_{FR}/E_0) \right] \\ &\xrightarrow{E_0 \gg E_{FR}, eV} \frac{e}{h} \varrho_{\|,R} \exp(-2\varkappa_0 w) \exp(|eV|/2E_0) [E_{FR}^2 - (E_{FR} - eV)^2] \,. \end{aligned} \tag{29b}$$

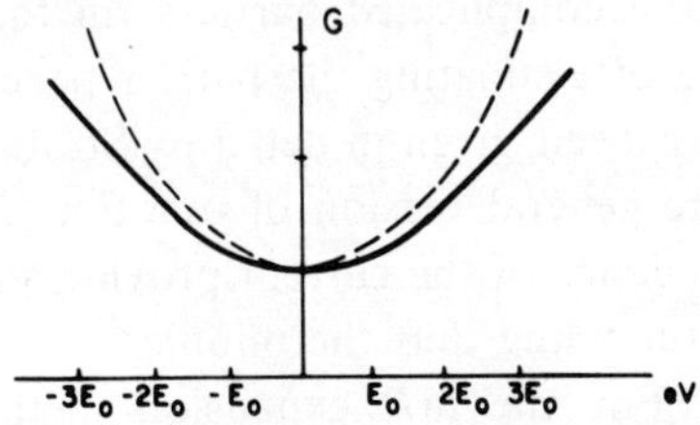

Fig. 6. Conductance versus bias in the quadratic approximation. The exact conductance is shown in dashed lines. $E_0 \approx 100$–500 meV for metals.

We thus establish the connection between our previous constant-D_0 approximations near $eV \approx E_{FR}$ and our zero-bias approximations. If $E_{FR} \ll E_0$, the conductance looks like Fig. 4. But since $E_0 \approx 200$ meV this is not the case in metals. Thus we consider $E_0 \ll E_{FR}$. In this case the conductance is

$$G \approx \frac{2eE_0\varrho_{\parallel,R}\exp(-2\varkappa_0 w)}{h}\cosh\frac{eV}{2E_0}, \tag{30}$$

as shown schematically in Fig. 6. Evidently, combining Figs. 5 and 6 would give a conductance like that shown in Fig. 7, which is not observed in practice due to the presence of higher- and lower-lying bands in both metals at $eV > E_{FR}$. The symmetry of the conductance about $V = 0$ is a consequence of the square-barrier approximation given by Eq. (15). For a trapezoidal barrier like that shown in Fig. 2 the minimum in the conductance is offset from zero bias by an amount small compared to V_b, i.e. $V_{\text{offset}} \sim V_b/\varkappa_0 w$.

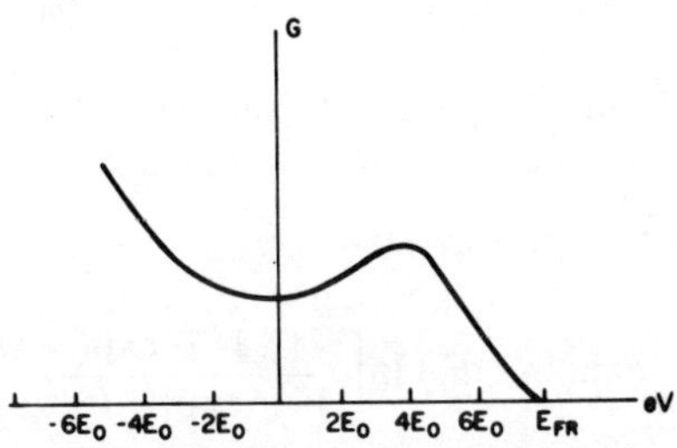

Fig. 7. Schematic of the conductance in a M–I–M junction for $E_{FR} \ll E_{FL}$ and using a one-band model for both metals.

Extensions to more complicated barriers (in the WKBJ approximation) of the technique of evaluating the barrier penetration probability D by expanding $\ln D$ have been given in detail by Stratton ([16]). Equation (23) in his paper is a more general version of our Eq. (30). Measurements of the temperature dependence of the current provide a technique for distinguishing between the tunneling and thermionic emission for both M–I–M and other junctions. For the $\ln D$ expansion methods the temperature dependence can be obtained in closed form for the barrier-penetration models $D(E, \mathbf{k}_\parallel) = D(E_x)$. The derivation of the temperature dependence of j for parabolic energy bands centered about $\mathbf{k}_\parallel = 0$ is simple in the case that only linear terms in E and $E_\parallel$ are retained in Eq. (27a). It is convenient to write Eqs. (24) and (27) as:

$$
\begin{aligned}
j &= \frac{2e\varrho_{\parallel,R}}{h} \int_0^\infty D(eV, E_{xR})\, dE_{xR} \int_0^\infty dE_{\parallel,R} \\
&\quad \times \{f(E_{\parallel,R} + E_{xR}) - f(E_{\parallel,R} + E_{x,R} + eV)\} \\
&= \frac{2e\varrho_{\parallel,R}kT}{h} \int_0^\infty D(eV, E_{xR})\, dE_{xR} \\
&\quad \times \left\{ \ln\left[\frac{e^{E_{\parallel,R}/kT}}{1 + e^{E_{\parallel,R}/kT} \exp[(E_{xR} - E_{FR})/kT]}\right] \right. \\
&\qquad \left. - \ln\left[\frac{e^{E_{\parallel,R}/kT}}{1 + e^{E_{\parallel,R}/kT} \exp[(E_{xR} - E_{FR} + eV)/kT]}\right] \right\}_0^\infty \\
&= \frac{2e\varrho_{\parallel,R}kT}{h} \int_0^\infty dE_{x,R} D(eV, E_{xR}) \\
&\quad \times \ln\left[\frac{1 + \exp[-(E_{xR} - E_{FR})/kT]}{1 + \exp[-(E_{xR} - E_{FR} + eV)/kT]}\right]
\end{aligned}
\tag{31}
$$

$$
\ln[D(eV, E_{xR})] \cong -2\varkappa_1 w + \left(\frac{E_{xR} - E_{FR}}{E_0}\right). \tag{32}
$$

Thus substituting (32) in (31), we get [see also ([17])]:

$$
\begin{aligned}
j &= \frac{2ekT\varrho_{\parallel,R}}{h} \exp(-2\varkappa_1 w) \\
&\quad \times \int_{-E_{FR}}^\infty \exp(+x/E_0) \ln\left[\frac{1 + \exp(-x/kT)}{1 + \exp[-(x + eV)/kT]}\right] dx.
\end{aligned}
$$

We first integrate by parts and then make the variable substitution $y = e^{+x/kT}$ so that, taking $E_{FR}/kT \to \infty$, we find

$$
\begin{aligned}
j(eV) &= \frac{2e\varrho_{\|,R}E_0}{h}\exp(-2\varkappa_1 w) \\
&\quad\times \int_{-\infty}^{\infty}\exp(+x/E_0)\,dx\left[\frac{1}{1+\exp(+x/kT)}-\frac{1}{1+\exp[+(x+eV)/kT]}\right] \\
&= \frac{2e\varrho_{\|,R}E_0 kT}{h}\exp(-2\varkappa_1 w)\,[1-\exp(-eV/E_0)]\int_0^{\infty}\frac{y^{[(kT/E_0)-1]}\,dy}{1+y} \\
&= \frac{2e\varrho_{\|,R}E_0 kT}{h}\exp(-2\varkappa_1 w)\,(1-\exp(-eV/E_0))\frac{\pi}{\sin(\pi kT/E_0)} \\
&= \frac{2e\varrho_{\|,R}E_0^2}{h}\left(\frac{\pi kT/E_0}{\sin(\pi kT/E_0)}\right)\exp(-2\varkappa_1 w)\,(1-\exp(-eV/E_0)) \\
&= \frac{4e\varrho_{\|,R}E_0^2}{h}\left(\frac{\pi kT/E_0}{\sin(\pi kT/E_0)}\right)\exp(-2\varkappa_0 w)\sinh(eV/2E_0), \qquad (33)
\end{aligned}
$$

which is essentially Eq. (14) in Stratton ([16]).

We have confined our attention to the "small" bias region by keeping only linear terms in V in expansions of barrier penetration exponents. Evidently, for bias value $V \geq V_L$ the barrier becomes triangular and the nature of the conductance curve changes to that analogous to field emission into a vacuum. At such large bias values the details of the potential play a more important role in determining the shape of the theoretical I–V curves. This sensitivity suggests caution in interpreting high-bias data using the model because tunneling through potential fluctuations due to impurities and irregularities in the surface, which are not incorporated in the simple potential-barrier model, becomes relatively more important due to the effects of local heating and locally inhomogeneous fields. For small values of the bias, calculations of the current through impurities, following the methods developed in the chapter on "tunnel-conductance due to energy loss mechanisms in the barrier," exhibits an exponential bias dependence like that shown in Fig. 6. Therefore, in this small bias limit such impurity processes have the primary effect of renormalizing the prefactor of Eq. (30). In many data analyses this prefactor does not enter the analysis because only ratios of currents at different temperatures or barrier thicknesses are considered [see, e.g., R. Stratton, G. Lewick, and C. A. Mead ([18])].

REFERENCES

1. L. Nordheim, Z. Phys. 46:833 (1928).
2. R. H. Fowler and L. Nordheim, Proc. Roy. Soc. (London) A119:173 (1928).
3. C. B. Duke and M. E. Alferieff, J. Chem. Phys. 46:923 (1967).

4. R. H. Good, Jr. and E. W. Müller, Handbuch der Physik, Vol. 21 (S. Flugge, ed.), Springer-Verlag, Berlin, 1956, p. 176.
5. J. G. Simmons, J. Appl. Phys. 34:1793 (1963); 34:2581 (1963); 35:2472 (1964).
6. F. W. Schmidlin, J. Appl. Phys. 37:2823 (1966).
7. D. J. Scalapino and S. M. Marcus, Phys. Rev. Letters 18:459 (1967).
8. L. V. Iogansen, Soviet Phys.—Uspekhi 8:413 (1965).
9. D. J. BenDaniel and C. B. Duke, Phys. Rev. 152:683 (1966).
10. L. Lewicki and C. A. Mead, Phys. Rev. Letters 16:939 (1966).
11. F. I. Itskovich, Soviet Phys.—JETP 23:945 (1966).
12. F. I. Itskovich, Soviet Phys.—JETP 24:202 (1967).
13. J. W. Conley, C. B. Duke, G. D. Mahan, and J. J. Tiemann, Phys. Rev. 150:466 (1966).
14. W. A. Harrison, Phys. Rev. 123:85 (1961).
15. D. J. BenDaniel and C. B. Duke, Phys. Rev. 160:679 (1967).
16. R. Stratton, J. Phys. Chem. Solids 23:1177 (1962).
17. E. L. Murphy and R. H. Good, Jr., Phys. Rev. 102:1464 (1956).
18. R. Stratton, G. Lewick, and C. A. Mead, J. Phys. Chem. Solids 27:1599 (1967).

Chapter 5

Tunneling

L. Esaki*

IBM Watson Research Center
Yorktown Heights, New York

INTRODUCTION

The concept of particle tunneling is nearly as old as quantum mechanics itself. Once an electron is represented by the wave function, it penetrates into a classically forbidden region, and can tunnel through a reasonably thin potential barrier without any real "tunnel." In other words, according to the quantum mechanics, an electron passes through the barrier without acquiring enough energy to pass over the top.

In solids, even electrons strongly bound to an atom or in any trap state may not remain permanently at the same site but may hop from one site to another, mainly due to tunneling. This process has been invoked to explain certain transport phenomena in the bulk of semiconductors and also in molecular crystals. In the latter case, electrons are considered to pass from one molecule to a neighboring molecule under an externally applied electric field.

In this talk, however, we are primarily interested in the macroscopic potential barriers for electron tunneling between two conductors rather than elementary tunneling process in atomic field. These barriers usually are made of the forbidden energy gap in the semiconductor or in the insulator, and the top or the bottom of the barrier is to be located at the bottom of the conduction band or at the top of the valence band. Before going into the main subjects, we will briefly sketch the historical development of tunneling theories and experiments in the last forty years.

In 1928 theoretical physicists believed that tunneling could occur by the distortion, lowering, or thinning of a potential wall under an externally

* Sponsored in part by the Army Research Office, Durham, N.C., under Contract DA-31-124-ARO-D-478.

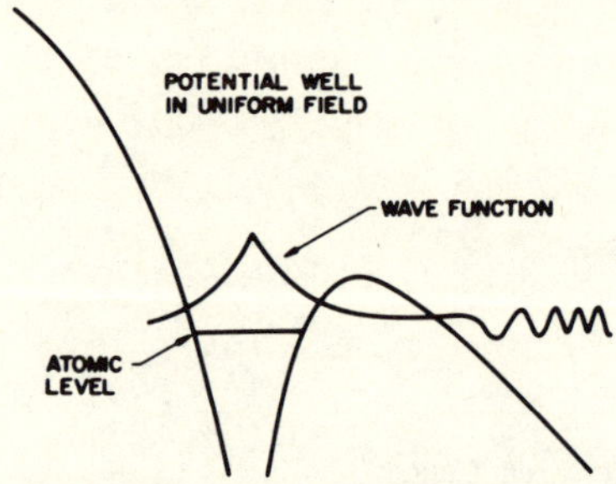

Fig. 1. Distortion of a potential well by a large external field.

applied high electric field. Oppenheimer (1) pointed out that the Coulombic potential wall which binds an atomic electron could be distorted by a large electric field, so that an electron could see a finite potential barrier through which it could tunnel, as shown in Fig. 1. Fowler and Nordheim (2) explained, on the basis of electron tunneling, the main features of the phenomenon of electron emission from cold metals by high external electric field, which had been unexplained since its observation by Lilienfeld in 1922 (3). They proposed a one-dimensional model. Metal electrons are confined by a potential wall whose height is determined by the work function φ plus the Fermi energy E_f, and the wall thickness is substantially decreased with an externally applied high electric field, allowing electrons to tunnel across the potential wall, as shown in Fig. 2. They successfully derived the well-known Fowler–Nordheim formula:

$$J = AF^2 \exp[-\ 4(2m)^{1/2}\varphi^{3/2}/3\hbar F].$$

The next and almost immediate application of these ideas came in the model for α decay as a tunneling process put forth by Gamow (4) and Gurney and Condon (5).

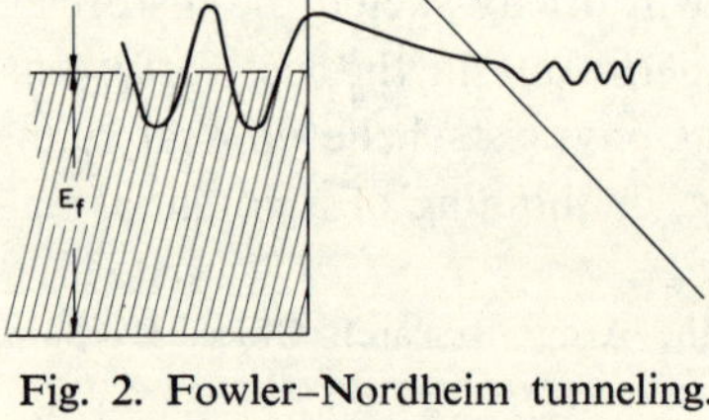

Fig. 2. Fowler–Nordheim tunneling.

Beyond this point tunneling phenomena in solids may be thought of as branching into two categories: (1) tunneling in conductor–insulator–conductor systems (M–I–M, M–I–S, Schottky junctions, etc.), and (2) interband tunneling in one and the same semiconductor.

In the former category Frenkel ([6]) in 1930 proposed that the anomalous temperature independence of contact resistance between metals could be explained in terms of tunneling across a narrow vacuum separation. Holm and Meissner ([7]) then did careful measurements of contact resistances and showed that the magnitude and temperature independence of the resistance of insulating surface layers were in agreement with an explanation of tunneling through a vacuum-like space. These measurements probably constitute the first correctly interpreted observations of tunneling currents in solids ([8]).

It is interesting to note a general tendency existing in those early days of the quantum mechanics: attempts were made to explain every junction effect in terms of tunneling. These efforts were not always successful. In 1932 quantum mechanical tunneling was applied to rectifiers of metal–semiconductor contact such as selenium or cuprous oxide ([9]). Figure 3 shows the simplified energy diagram from which the following well-known current–voltage relationship was derived:

$$J = J_s[\exp(eV/kT) - 1].$$

Apparently, this theory was accepted for a number of years until it was finally realized that it predicted rectification in the wrong direction for the ordinary diodes. Another example of the wrong application may be the explanation given in the early 1950's of reverse breakdown in the *p–n* junction in terms of tunneling, which will be mentioned later.

In 1934 Zener ([10]) proposed interband tunneling, or internal field emission, as an explanation for dielectric breakdown. He calculated the rate of transitions from a filled band to a next-higher unfilled band by the

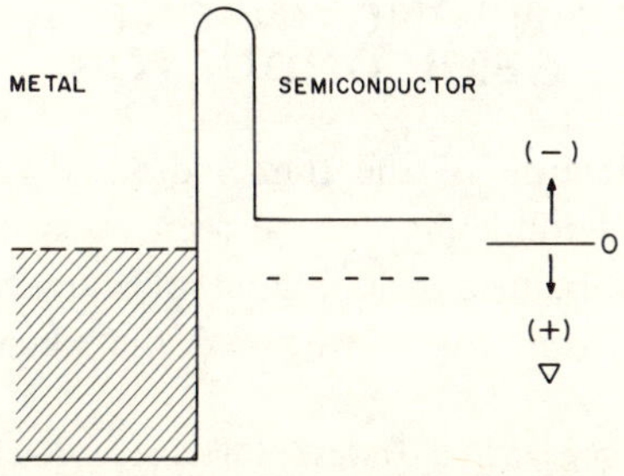

Fig. 3. A tunnel model junction.

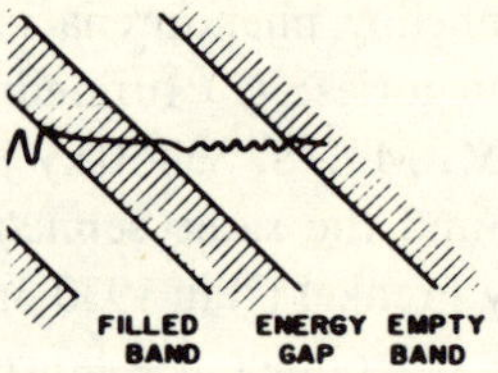

Fig. 4. Zener tunneling: the rate of transition proportionate to exp $(-maE_g^2/4\hbar^2F)$ or alternative form $\exp(-m^{1/2}Eg^{3/2}/2^{3/2}hF)$.

application of an electric field as shown in Fig. 4. In effect, he showed that an energy gap could be treated in the manner of a potential barrier. This approach was refined by Houston ([11]) in 1940. The Zener mechanism in dielectric breakdown has never been proved to be important in reality. If a high electric field is applied to the bulk crystal of a dielectric or a semiconductor, hot-electron effects such as impact ionization, avalanche,* etc., precede tunneling, and thus the field never reaches a critical value for tunneling. In other words, observation of tunneling phenomena, as discussed here (band-to-band tunneling), has been limited to particular types of junctions: narrow *p–n* junctions, narrow Schottky junctions, conductor–thin-insulator–conductor junctions, etc. This fact always gives rise to the following problems:

1. Spatially varying electric field across the junctions: $F \neq$ const.
2. Boundary problems at the interface: continuity of ψ and $d\psi/dx$, extra scattering due to contamination or defects, etc.
3. Experiments suffer from technological difficulties of junction formation.

All these mean that idealistic theories could be far from actual experimental situation.

INTERBAND TUNNELING IN ONE AND THE SAME SEMICONDUCTOR

Since the introduction of the tunnel diode ([12]) in 1957 it has been known that interband tunneling can be seen most easily and almost exclusively in the high built-in field region of narrow semiconductor *p–n* junctions. Semiconducting engineering and the phenomenological theory

* An avalanche process, similar to a Townsend discharge, results from many successive electron–hole pair productions by field-accelerated hot carriers.

of semiconductors involved in the diode can be traced back to that developed in conjunction with the invention of the transistor. In the late 1940's and the early 1950's high-quality single crystals of Ge and Si were grown for the first time, and technologies to form the *p–n* junctions were established (see Fig. 5). The alloy junction method gives the sharpest impurity gradient, which is favorable for tunneling, among the three methods: grown junction method, alloy junction method, and diffused junction method. Poisson's equation states that

$$d^2\varphi/dx^2 = -\varrho/K\varepsilon_0,$$

where φ is the electrostatic potential, ϱ is the charge density, K is the dielectric constant, and ε_0 is the permittivity of free space, 8.86×10^{-14} F/cm. If this equation is solved at proper boundary conditions, one will get ϱ,

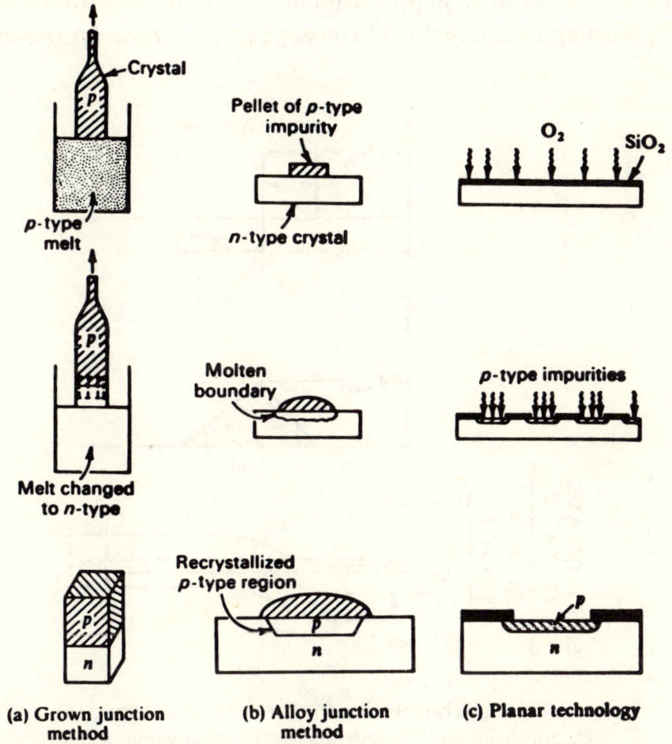

Fig. 5. Semiconductor technologies. From A. S. Grove, Physics and Technology of Semiconductor Devices, John Wiley and Sons, New York, 1967, p. 2.

F (electric field), and φ as a function of x as shown in Fig. 6 for no applied voltage and in Fig. 7 for a reverse bias voltage V_R. If one side is much more heavily doped than the other side, the potential in the space charge region is of the parabolic type. In this reverse-bias situation McAfee *et al.* ([13]) applied a modified Zener theory and asserted that low-voltage breakdown in Ge diodes (specifically, they showed a 10-V breakdown) resulted from interband tunneling from the valence band in the *p*-type to the empty conduction band in the *n*-type. The work of McAfee *et al.* inspired a number of other investigations of breakdown in *p–n* junctions. Results of those later studies ([14]) indicated that most Ge junctions broke down by avalanche (electron–hole pair generation), the original claim of tunneling became controversial and finally was judged incorrect, but by that time the name "Zener diodes" had already been given to the low-breakdown Si diodes. Actually, these diodes are almost always, avalanche diodes.

At this point the emphasis seemed to be shifted from determining the mechanism of breakdown in practical junctions to demonstrating the operation of tunneling or internal field emission in junctions. To this end, Chy-

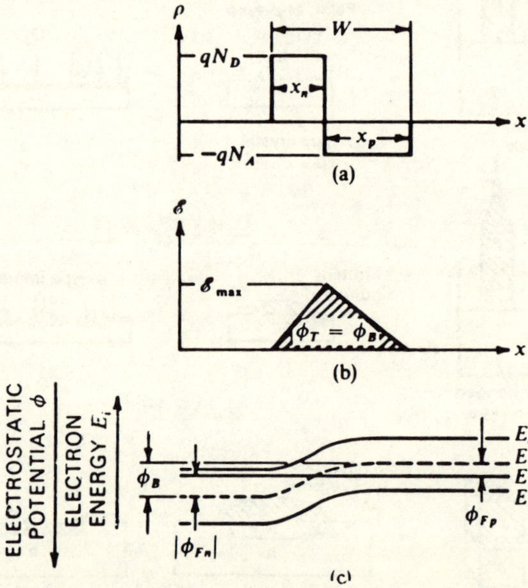

Fig. 6. Schematic illustration of the (*a*) charge, (*b*) electric field, and (*c*) potential distribution within a step junction in equilibrium. From A. S. Grove, Physics and Technology of Semiconductor Devices, John Wiley and Sons, New York, 1967, p. 156.

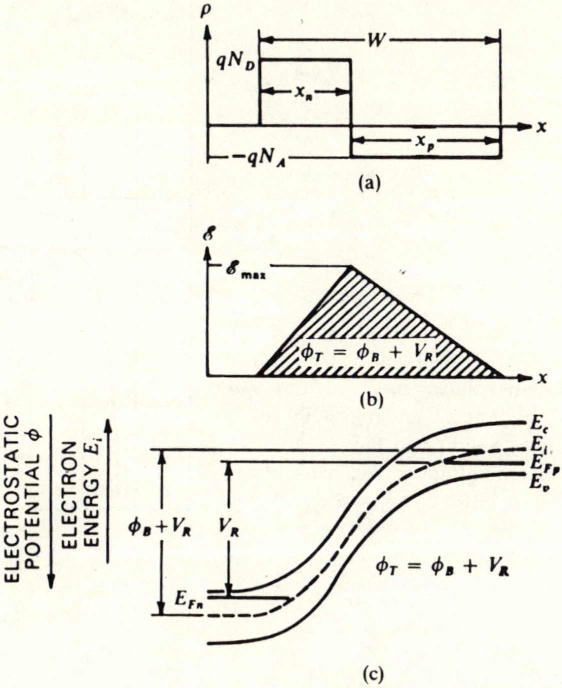

Fig. 7. Schematic illustration of the (*a*) charge distribution, (*b*) electric field, and (*c*) potential distribution within a reverse-biased step junction. From A. S. Grove, Physics and Technology of Semiconductor Devices, John Wiley and Sons, New York, 1967, p. 162.

noweth and McKay ([15]) in 1957 examined some narrow (~400 Å) Si-diffused junctions and concluded that they had finally observed tunneling.

Around the same time we studied interband tunneling with Ge alloy (step) junctions ([12]) where the impurity concentrations, N_A of the acceptor and N_D of the donor, were made sufficiently high (10^{18}–10^{19} cm^{-3}) to get narrow junction width, less than 200 Å. Figure 8 shows the current–voltage curve for $N_A \approx 10^{18}$ cm^{-3} and $N_D \approx 10^{19}$ cm^{-3}; the curve is almost symmetric with respect to the voltage. The breakdown voltage of the junction is far less than the threshold for electron–hole pair production: ~1.5 eV for Ge and ~2.5 eV for Si ([14]). This voltage keeps decreasing with increase in the impurity concentration. Figure 9 shows a backward diode in which the breakdown voltage in the usual sense is virtually zero and the reverse direction is far more conductive than the forward direction. When this unit was cooled we saw the negative resistance for the first time. With

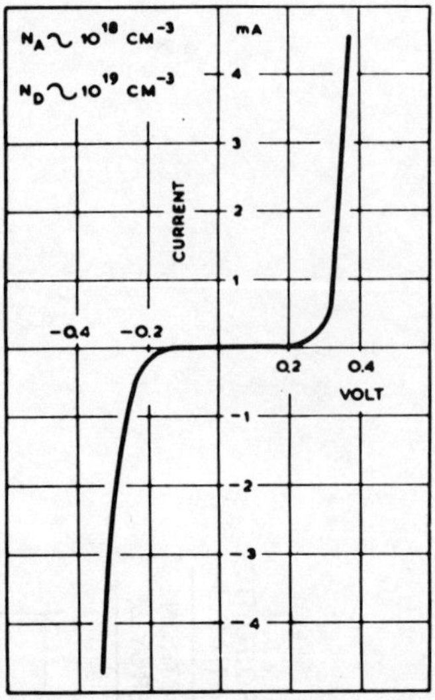

Fig. 8. Current–voltage curve of the *p–n* junction with $N_A \approx 10^{18}$ cm^{-3} and $N_D \approx 10^{19}$ cm^{-3}. From L. Esaki, in: Solid State Physics in Electronics and Telecommunications, Academic Press, New York, 1960, p. 518.

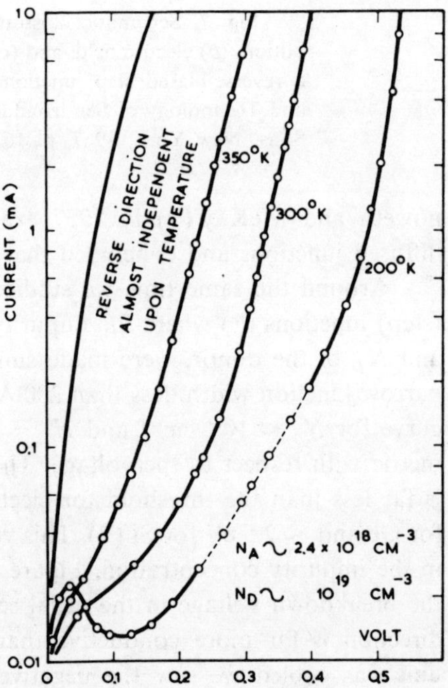

Fig. 9. Current–voltage curve of the *p–n* junction with $N_A \approx 2.4 \times 10^{18}$ cm^{-3} and $N_D \approx 10^{19}$ cm^{-3}. From L. Esaki, in: Solid State Physics in Electronics and Telecommunications, Academic Press, New York, 1960, p. 518.

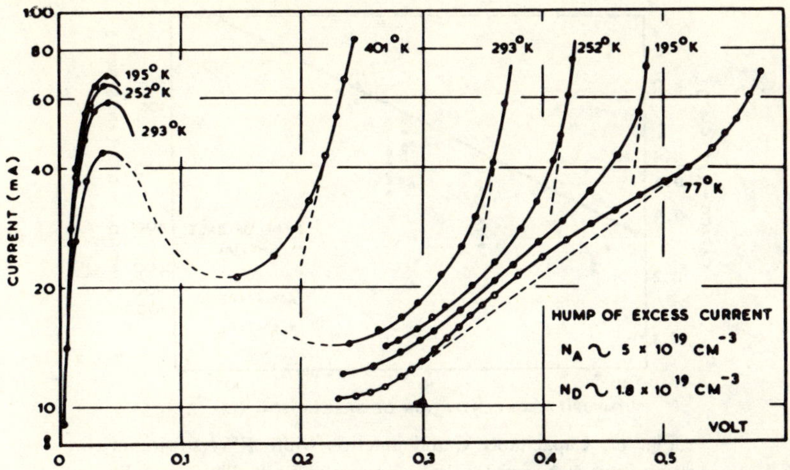

Fig. 10. Current–voltage curve of the *p–n* junction with $N_A \approx 5 \times 10^{19}$ cm^{-3} and $N_D \approx 1.8 \times 10^{19}$ cm^{-3}. From L. Esaki, in: Solid State Physics in Electronics and Telecommunications, Academic Press, New York, 1960, p. 518.

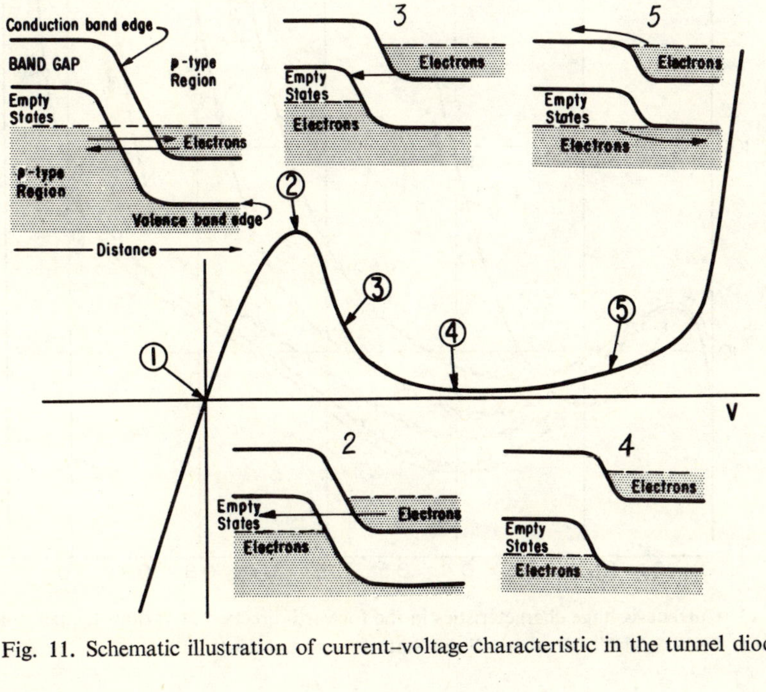

Fig. 11. Schematic illustration of current–voltage characteristic in the tunnel diode.

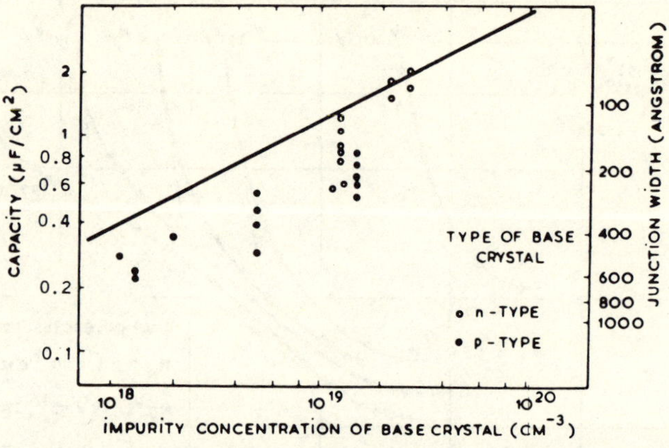

Fig. 12. Capacitance C and junction width W versus impurity concentration N. From L. Esaki, in: Solid State Physics in Electronics and Telecommunications, Academic Press, New York, 1960, p. 520.

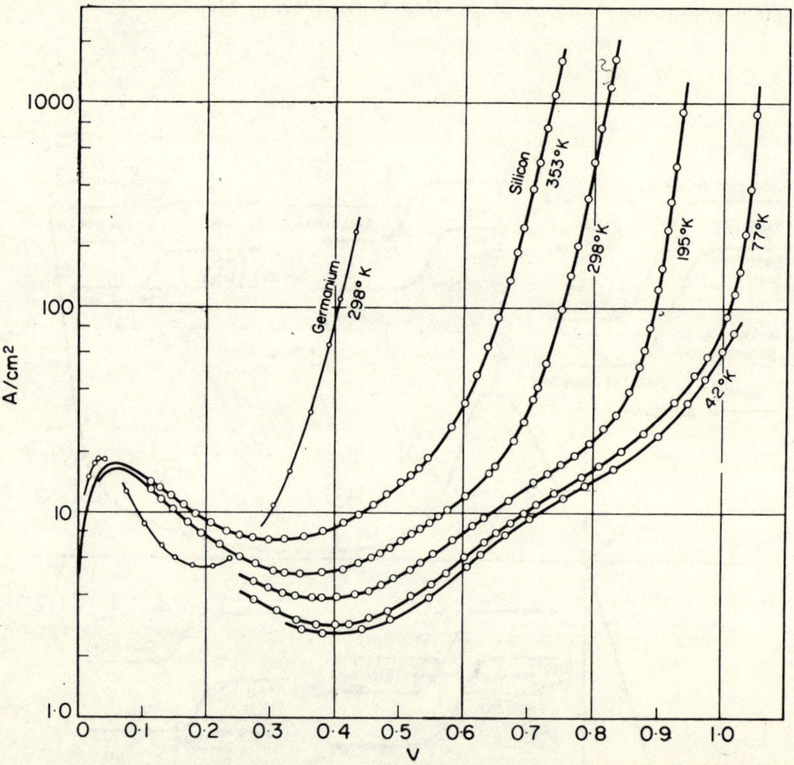

Fig. 13. Current–voltage characteristics in the forward direction at various temperatures. From L. Esaki and Y. Miyahara, Solid-State Electron. 1:13 (1960).

further increase of impurities up to more than 10^{19} cm^{-3} the negative resistance appeared clearly at all temperatures, as shown in Fig. 10.

The characteristic was analyzed in terms of interband tunnel current, excess (tunnel) current, and conventional thermal current, as shown in Fig. 11. The Fermi levels in those heavily doped semiconductors penetrate deeply inside the bands. One can see that as the bias is increased, the bands, which overlap each other at zero bias, become uncrossed. Since the energy should be conserved in tunneling, the current obviously should decrease as the bands become uncrossed. The excess current could be explained in terms of tunneling through mid-gap levels ([16]).

The junction width W of these junctions is given by

$$W = \left(\frac{2K\varepsilon_0}{q} \frac{N_A + N_D}{N_A N_D} \phi\right)^{1/2}.$$

Figure 12 shows values calculated from the above equation as well as those determined by the capacitance measurements. The results indicate built-in field strengths in the junctions as high as 5×10^5 to 10^6 V/cm.

TABLE I

	Energy gap (eV)		Effective mass ratios*		Static dielectric constant
	0 °K	300 °K	Electrons	Holes	
PbTe	0.185	0.30	0.024 (m_t) 0.27 (m_l)	0.024 (m_t) 0.27 (m_l)	412
InSb	0.235	0.17	0.0155	0.02 (l) 0.4 (h)	17.9
InAs	0.43	0.35	0.025	0.025 (l) 0.41 (h)	14.6
Ge	0.785	0.66	0.082 (m_t) 1.64 (m_l)	0.04 (l) 0.3 (h)	16.0
GaSb	0.81	0.72	0.047	0.052 (l) 0.35 (h)	15.7
Si	1.21	1.08	0.19 (m_t) 0.98 (m_l)	0.16 (l) 0.5 (h)	12.0
InP	1.41	1.27	0.073	—	12.4
GaAs	1.53	1.40	0.07	0.12 (l) 0.68 (h)	13.1
GaP	2.4	2.25	0.35	—	10.2

* Notation: l, light holes; h, heavy holes; m_t, transverse effective mass; m_l, longitudinal effective mass.

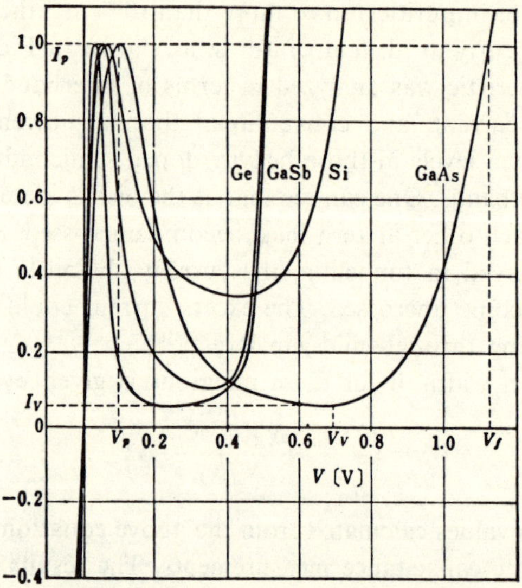

Fig. 14. Comparison of current-voltage characteristics in the Ge, GaSb, Si, and GaAs tunnel diodes. The peak current I_p is normalized.

We soon succeeded in fabricating the silicon tunnel diode ([17]) with similar techniques. The characteristics are shown in Fig. 13. Other semiconductor materials were quickly explored to obtain the tunnel diodes; InSb ([18]), GaAs ([19]), InAs ([20]), PbTe ([21]), GaSb ([22]), SiC ([23]), etc. Table I shows the material constants and Fig. 14 compares these at the normalized peak current for Ge, GaSb, Si, and GaAs. The tunneling characteristics (including the current density) depend on the band structure and other material constants, and thus detailed analyses of refined data have yielded a great deal of information about the properties of heavily doped semiconductors as well as a detailed understanding of the tunneling process itself.

When the silicon tunnel diode was cooled down to 4.2°K a significant fine structure was found on the current–voltage curve, as shown in Fig. 15 ([24]). We were surprised that four characteristic values agreed almost exactly with the four phonon energies obtained from the optical absorption spectra of pure silicon or analysis of intrinsic recombination radiation in 1958 ([25]) (their assignment was not correct at that time), indicating the existence of the indirect (phonon-assisted) tunneling ([26]) due to the band structure of silicon.

The announcement of the tunnel diode has stimulated not only experimental investigation but also theoretical studies of interband tunneling

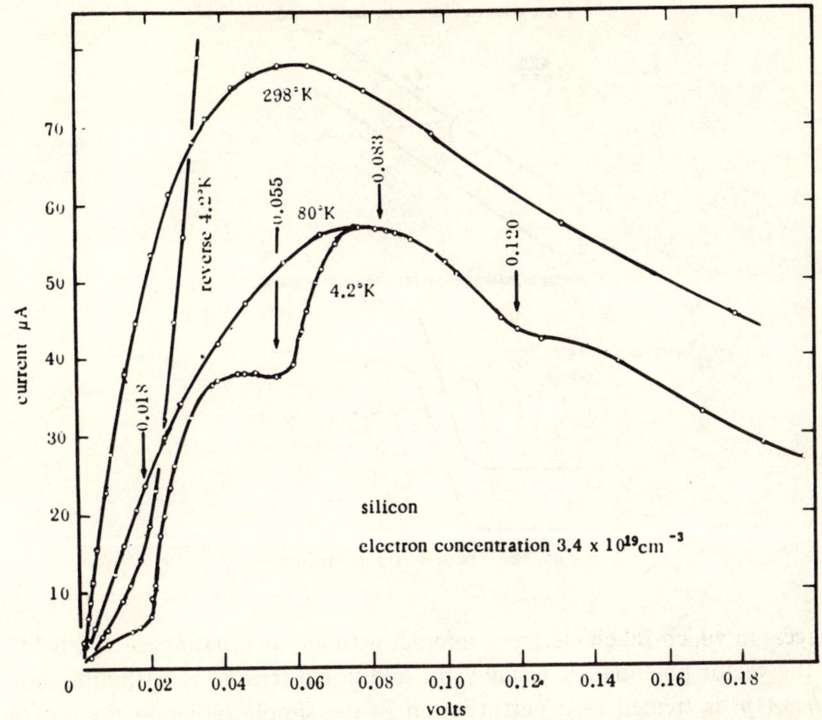

Fig. 15. Current–voltage characteristics in the Si tunnel diode at 4.2, 80, and 298 °K.

[see, e.g., ([2])]. The theories may be classified into two schools: one traditional school ([10,11,28–30]), and the other a relatively new school ([31–33]). The former assumes a constant and modestly high electric field. Under this field electrons cycle repeatedly through the Brillouin zone according to the equation of motion: $d\mathbf{k}/dt = eF/\hbar$, with the cycle time $\tau = 2\pi\hbar/eFa$, where a is the lattice constant. The rate of leaking into the adjacent band is greatest when the $\mathbf{k}$ vector is at the band edge. In the latter theory the tunneling probability is essentially defined in terms of the formalism of the scattering theory. Incident and out-going electrons are represented by the Bloch waves in respective bands, and the energy conservation in "elastic" tunneling is to be applied. Therefore one may say that the former treats tunneling inside the junction region or the high-field region, whereas the latter looks at tunneling from the outside regions (Fig. 16).

In the simple theory of interband tunneling Bloch electrons interact with a "steady electric field" E, which contributes to a term eEr to the Hamiltonian. This can be compared with the interband optical absorption

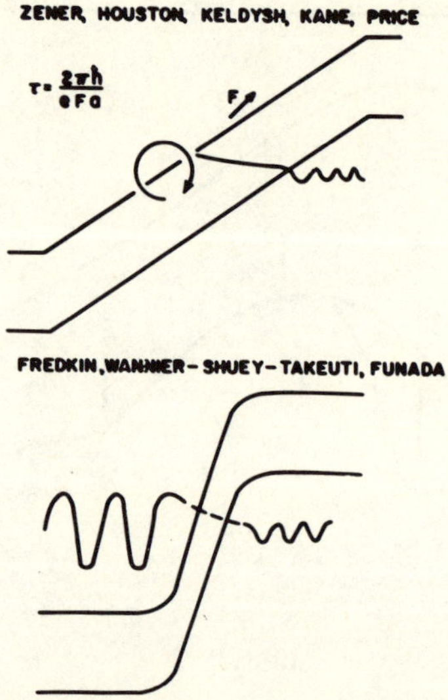

Fig. 16. Theories on tunneling.

process in which Bloch electrons interact with an "alternating electric field" of the vector potential A. Usually the additional term to the Hamiltonian, $(e/m)A \cdot p$, is treated as a perturbation in the simple radiation theory. In both cases one ought to derive the matrix element governing the transition probability. The phonon bumps observed in the tunneling current will correspond to phonon effects in the absorption spectra or recombination radiation spectra. The magnetotunneling and the pressure effect on tunneling will also be considered to be the counterparts of the magneto-optical effect and the pressure dependence of the optical effects, respectively. The Franz–Keldysh effect ([34]) and photon-assisted tunneling ([35]) are worth mentioning here, since they are really a combination of tunneling and photon absorption or photon emission under a steady electric field.

TUNNELING THROUGH A THIN INSULATOR SANDWICHED BETWEEN TWO CONDUCTORS (TUNNELING SPECTROSCOPY)

Figure 17 illustrates two rather extreme cases in the thin-insulating-barrier tunneling: the sharp boundary and the graded boundary cases. In the former the potential changes sharply as a function of position, whereas in the latter the potential changes slowly, and hence the fractional change

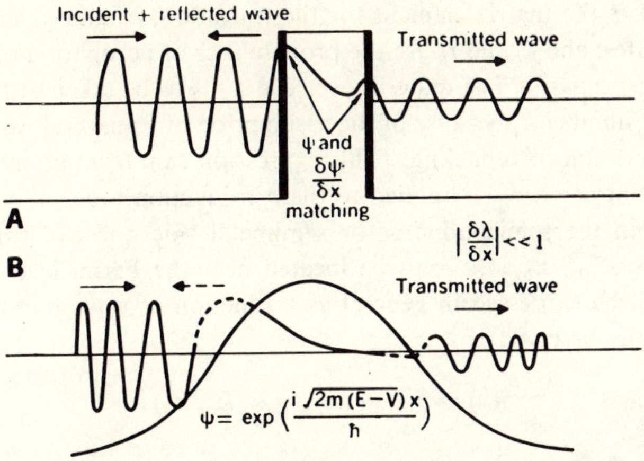

Fig. 17. A, sharp boundary; B, graded boundary in the thin–insulating–barrier tunneling.

in the wavelength is small over a distance of a wavelength. In the case of the sharp boundary electron waves are reflected by such a discontinuity, just as light waves are, where the wave function as well as its derivative should be matched. With the graded boundary, however, the slowly changing potential, where the so-called WKB approximation can be applied, is analogous to a slowly changing index of refraction, and we do not have any clear reflected electron wave, although its path may be curved as a result of refraction.

Figure 18 shows a schematic energy diagram of a tunnel junction of a thin insulating layer c sandwiched between a degenerate semiconductor or a semimetal (left) and a metal (right). Now, tunneling current between the band a and the band b is given by [see, e.g., ([36])]:

$$j_{ab} = \frac{4\pi e}{\hbar} \sum_{k_t} \int_{-\infty}^{\infty} | M_{ab} |^2 \varrho_a \varrho_b (f_a - f_b) \, dE, \tag{1}$$

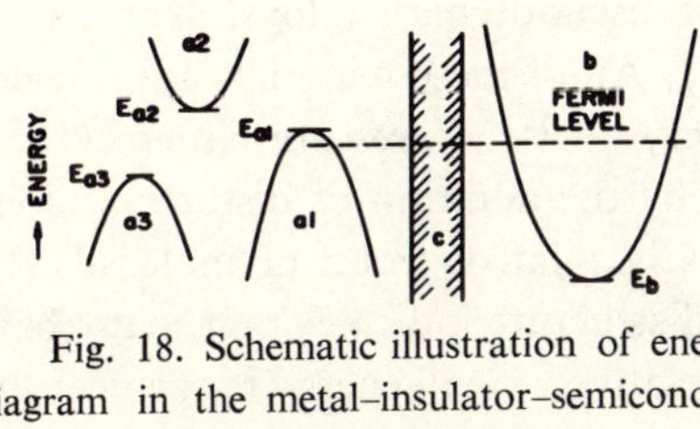

Fig. 18. Schematic illustration of energy diagram in the metal–insulator–semiconductor or –semimetal tunnel junction.

where M_{ab} is the matrix element for the transition, ϱ_a and ϱ_b are the densities of states, and f_a and f_b are the probabilities of occupation of the states a and b, respectively. The integral over energy is to be taken at fixed transverse wave number k_t because of the assumption of transveral wave number (k_t) conservation in tunneling. (This expression can, from an independent-particle point of view, be applied to the p–n junction too).

Since in the semiconductor or semimetal side one can expect many band edges, E_{a1}, E_{a2}, ..., to be located near the Fermi level, tunneling current will be expressed in general as a function of wave numbers of the three regions involved:

$$j(v) = j(k_{a1}, k_{a2}, \ldots, k_c, k_b), \tag{2}$$

where V is the applied voltage.

Now, if we make a WKB approximation,

$$| M_{ab} |^2 \varrho_a \varrho_b \rightarrow \exp(-2 \int | k_{cx} | dx),$$

and thus the current is given simply by

$$j(v) = j(E_{a1}, E_{a2}, \ldots, k_c), \tag{3}$$

assuming $E_{ai} \ll E_b$. If both sides are metals, we can simplify further, as

$$j(v) = j(k_c). \tag{4}$$

The equations (2)–(4) indicate that structure in the current–voltage characteristics or in the conductance versus voltage curves of the tunnel junctions will provide information pertinent to the band structure of materials involved. In some cases band bending near the interface ([37]), surface states ([38]), and impurity levels ([39]) and other scattering centers ([40]) will also give rise to anomalies in the curves; these will not be discussed here.

We have initiated a study of band structure of the group V elements and their alloys ([41]), such as Bi ([42]) and BiSb alloys ([43]), and of IV–VI compounds and their pseudobinary alloys, such as SnTe ([44]), GeTe ([45]), and $(PbTe)_x(SnTe)_{1-x}$. All of these materials are isovalent (average valence five), for which the crystal structures are either cubic or deviate in some small way from that by the addition of distortion along the trigonal axis, and thus form a closely related group of materials ([46]). Theoretical consideration of the band structure indicates that many band edges are expected to be located in a relatively small energy range near the Fermi energy ([47]).

Bismuth and antimony are the most common semimetals and they make a continuous solid solution. Galvanomagnetic measurements for some BiSb alloys indicate that they are semiconductors with energy gaps from zero to ~ 0.03 eV, depending on composition, from about 5 to about 40 at. % Sb in Bi. Both SnTe and GeTe are usually nonstoichiometric and highly *p*-type owing to high concentrations of Sn or Ge vacancies. In our studies a clear answer has been given about the uncertainty over whether or not these alloys are semiconductors or semimentals.

Equation (4) indicates the present technique will give directly information on the imaginary wave number k_c in the insulator. We have made an attempt to use magnetically ordered insulators such as EuSe ([48]), EuS, etc. (the Eu-chalcogenides) as a tunneling barrier. In these M–I–M magnetic tunnel junctions it has been found that the spin ordering in the insulators significantly changes the k_c involved, and the results have been explained in terms of a shift of the energy level of the *d* band and thus the tunneling barrier height.

The objective of our present study is generally to determine the position of band edges with respect to the Fermi level by analyzing the tunnel current. We call this type of study "tunneling spectroscopy."

Thin-insulating-barrier tunnel junctions are usually fabricated by successive evaporations. In Bi or BiSb-alloy tunnel junctions an insulating layer is formed on a cleaved surface of single crystals with a counter metal electrode evaporated over the insulator. We used vapor-deposited aluminum oxide or an oxidized surface layer of crystals as the tunneling barrier. In Al–Al_2O_3–SnTe or –GeTe junctions semiconductors are evaporated onto an oxidized evaporated stripe of Al. Substrates of either quartz or sapphire are used. The evaporated Al stripes are oxidized at room temperature or higher (up to 400°C). The substrates are kept at room temperature or slightly above for SnTe and at temperatures 150–350°C for GeTe. The higher the substrate temperature, the lower the carrier concentration. Magnetic tunnel junctions were constructed simply by successive evaporation: metal, magnetic insulator, metal, on a heated sapphire substrate without any oxidation process. The area of the junctions is of the order of 10^{-4} cm^2.

The tunneling current I, dI/dV, and d^2I/dV^2 were plotted as functions of applied voltage V at low temperatures as well as at room temperature.

Figure 19 shows a plot of conductance versus voltage at 2°K for a BiSb alloy junction of 12 at. % Sb in Bi. We have also made measurements on BiSb alloys of 3.6 and 13.5 at. % alloys. A large dip around 10–15 mV, seen here, is characteristic of 12 and 13.5% alloys, clearly indicating the existence of the energy gap. These BiSb-alloy single crystals are *n*-type, with a carrier concentration of 7–9 $\times$ 10^{15} cm^{-3}. The voltage at the bottom

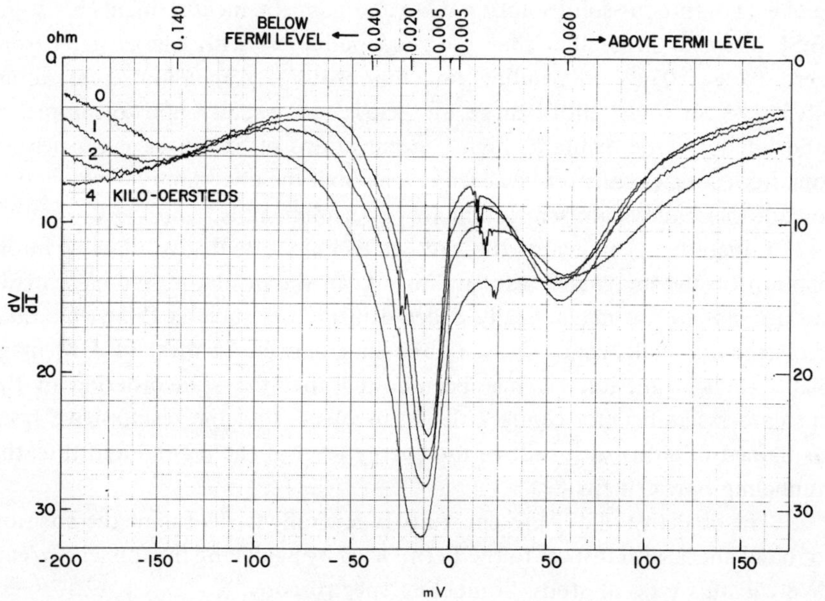

Fig. 19. Conductance versus voltage plot at 2 °K for BiSb alloy tunnel junction. From Esaki and Stiles ([43]).

of the dip may be equated to the half-energy gap plus the Fermi energy. Since the Fermi energy is 1 meV or less, the energy gap is derived to be 20–30 meV, which is in good agreement with the previous measurements.

Figures 20 and 21 show plots of conductance for a pure Bi junction

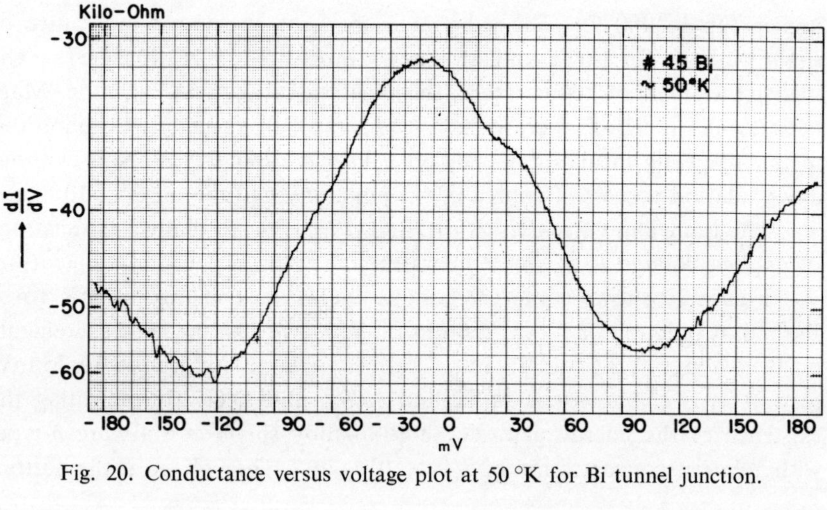

Fig. 20. Conductance versus voltage plot at 50 °K for Bi tunnel junction.

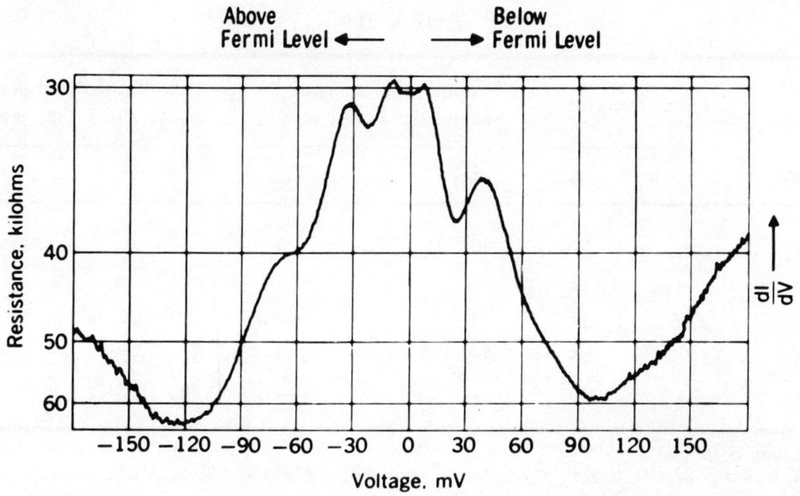

Fig. 21. Conductance versus voltage plot at 2 °K for Bi tunnel junction. From Esaki and Stiles ([42]).

at 50°K and 2°K, respectively. The latter curve, showing a fine structure, is considered as a sum of the conductances from many hole- and electron-bands, and is to be analyzed by decomposing into components. The shape

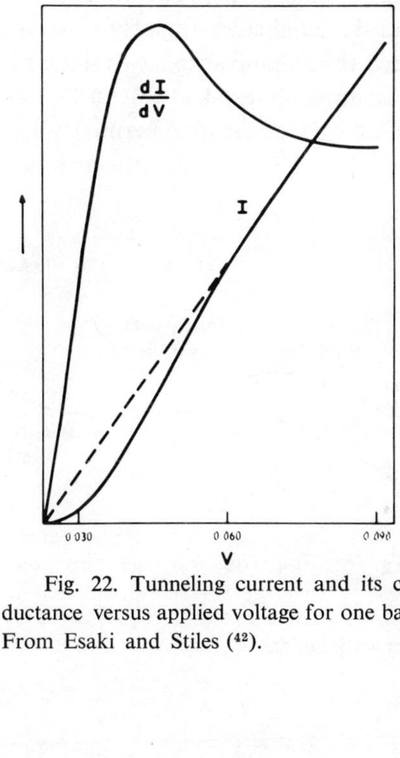

Fig. 22. Tunneling current and its conductance versus applied voltage for one band. From Esaki and Stiles ([42]).

TABLE II*

Ref.†	Fermi energies		Band edge energies below the Fermi level			Band edge energies above the Fermi level		
	E_{n0}	E_{p0}	E_{n1}	E_{n2}	E_{n3}	E_{p1}	E_{p2}	E_{p3}
a	25	—	40	—	—	—	—	—
b	—	12	—	—	—	—	—	—
c	27.6	11	42.9	—	—	—	—	—
d	15	15	35	105	150	30	60	135 $\pm$10
e	—	—	—	—	—	30	45	—

* Energies in meV.

† a. Brown *et al.* (54). b. Kao (55). c. Smith *et al.* (56). d. Esaki and Stiles (42). e. Antcliffe and Bate (49).

of the conductance curve due to one band is arrived at by subtracting the actual conductance from a continuation of the background conductance. This method was undertaken for the peak at ~ 37 mV in Fig. 21. The result is shown in Fig. 22, where a reconstructed plot of the tunneling current versus voltage is also given. The tunneling current apparently shows an anomaly near the band edge—namely, a deviation from the linear relation. We constructed the conductance curve similar to the observed one using a series of conductance curves like those in Fig. 22 and thus detected many band edges below and above the Fermi level; these are listed in Table II. Antcliffe and Bate (49) recently detected two bands above the

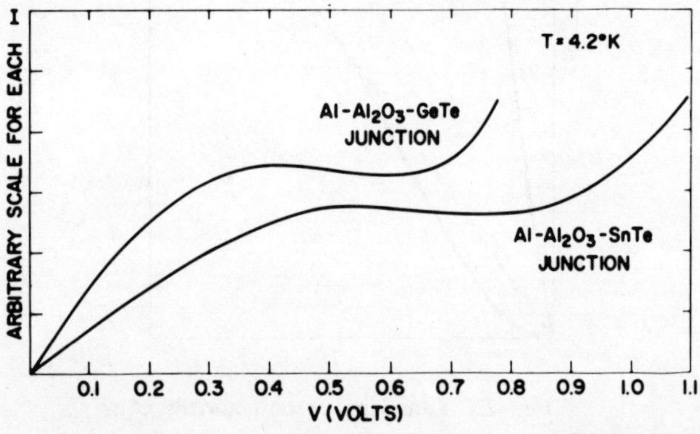

Fig. 23. Current–voltage characteristics at 4.2 °K for SnTe and GeTe tunnel junctions.

Fermi level by a study of the Shubnikov–deHaas effect; these seem to be in agreement with our studies.

It should be mentioned that many units we have made showed either no structure or weak stucture. The fine structure mentioned here seems to suffer considerably from surface preparation as well as thin oxide formation.

The SnTe and GeTe junctions, formed on a oxidized Al stripe, are relatively easy to produce, although all materials involved are polycrystalline. The evaporated SnTe and GeTe layers of $\sim$ 4000 Å thick are highly *p*-type, with typical carrier concentrations about 8×10^{20} and 2×10^{20} cm^{-3}, respectively. Figure 23 shows the current–voltage curves at 4.2°K of typical SnTe and GeTe tunnel junctions. It is interesting that the shape of the curves is reminiscent of that of the semiconductor tunnel junction. As mentioned before, the negative resistance of the tunnel diode is explained on the basis that the bands, which overlap each other at zero bias, become uncrossed as the bias is increased. On the other hand, this case can be explained by a voltage dependence of the tunneling probability in addition to an effect due to the energy gap of the semiconductor. Figure 24 illustrates the energy diagrams of the SnTe junction at zero bias and at an applied voltage V. As seen in the figure, until the bias voltage to the metal is increased up to the level corresponding to the top of the valence band in the semiconductor side the tunneling current continues to increase with increase in the applied voltage. When the bias voltage is further increased the total number of available electrons for tunneling is more or less unchanged, whereas the barrier height for tunneling is raised, resulting in decrease in the tunneling probability. Thus a negative resistance appears

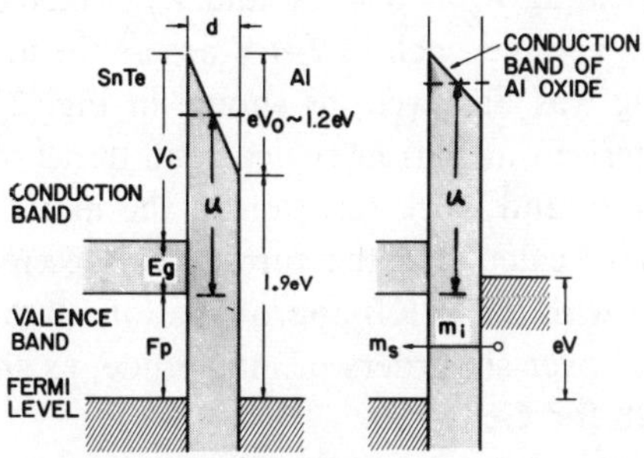

Fig. 24. Energy diagrams of the SnTe junction at zero voltage and at an applied voltage V (right). The GeTe junction is also very similar to this.

in the current–voltage curve. When the bias voltage to the metal becomes higher than the level corresponding to the bottom of the conduction band in the semiconductor a new tunneling current to the conduction band starts, and thus one sees the current again increasing with the voltage.

In this case it should be emphasized that the imaginary k in the insulator is assumed to be determined only by the bottom of the conduction band in the insulator instead of by both the conduction and valence bands or by the valence band. This assumption anticipates that the conduction band edge is closer to the metal Fermi energy than the valence band edge. It is also interesting to note that low-energy electrons of small E have a larger tunneling probability than high-energy electrons, and effective mass must flip its sign when the electrons cross the boundary between the highly p-type semiconductor and the insulator because of the opposite curvature in the E–k relationship.* This can be compared with the p–n junction tunneling case, for which the effective mass gradually changes its sign across the junction.

The tunneling current due to a particular band as a function of the applied voltage was calculated from Eq. (1) with either the WKB approximation or the sharp boundary condition. Figure 25 shows experimental as well as theoretical plots of conductance versus voltage for a typical SnTe tunnel junction. An attempt was made to obtain the closest fit of the calculated conductance to the observed one by adjusting the values of the Fermi energy F_p and the energy gap E_g. Here we have selected $F_p = 0.6$ eV and $E_g = 0.3$ eV at 4.2°K. A broken line (C) was obtained with the WKB approximation, and two broken lines (A) and (B) were obtained from a calculation using exact matching of the wave functions at the sharp boundary. The latter curves show no cusp at $eV = F_p$. The same technique was applied to GeTe, yielding $F_p = 0.45$ eV and $E_g = 0.20$ eV at 4.2°K.

At high applied voltage, such as 2–4 V across the junction, the Fowler–Nordheim tunneling was observed, as shown in Fig. 26. This is primarily intended to characterize our barrier structure in terms of energy differences in the electron affinity and work function of the materials involved ([50]).

Recently we have calculated the current–voltage relationship over the entire voltage range with the WKB approximation, finding good agreement with observations of over six orders of magnitude, as shown in Fig. 27 for SnTe and in Fig. 28 for GeTe ([51]).

* Definition of the reciprocal effective mass tensor:

$$\left(\frac{m_0}{m^*}\right)_{\alpha\beta} = \frac{m_0}{\hbar^2}\,\frac{\partial^2 E}{\partial k_\alpha \partial k_\beta}.$$

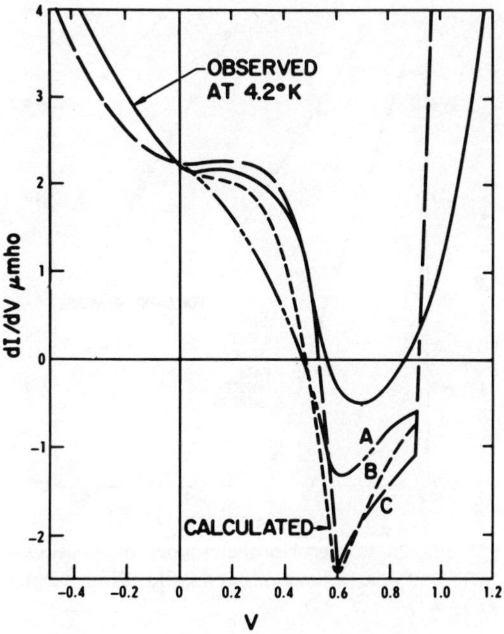

Fig. 25. Conductance versus voltage in the SnTe junction. Solid line, observed at 4.2 °K; C line, calculated using WKB approximation; A and B lines, calculated using exact matching of wave function.

In both SnTe and GeTe the energy gap tends to be narrowed with increase in temperature. In order to measure the temperature dependence of the semiconductor energy gap below 100°K, the experiments were performed by passing a constant current through the tunnel junction and measuring the voltage drop across the junction. The current and voltage were measured to 1 part in 10^5. Figure 29 is a plot of the results for two types of SnTe and one type of GeTe unit. Experiments were also carried out in which the temperature dependence of the conductance zeros was measured. These indicated that the Fermi energy is relatively independent of temperature and that the results were consistent with the constant-current measurements. All results obtained are summarized in Table III.

By cooling the GeTe tunnel junction below 0.5°K, the superconducting energy gap of GeTe with the carrier concentration 1.2×10^{21} cm^{-3} has been measured to be 0.15 meV below 0.1°K ([52]), as shown in Fig. 30. The transition temperature T_c of this material is about 0.4°K, and therefore the ratio of the energy gap to kT_c becomes 4.3, which seems reasonable.

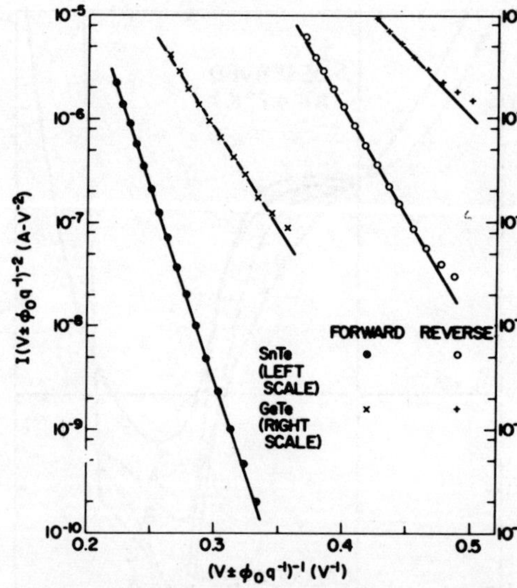

Fig. 26. Fowler–Nordheim plots for Al–Al_2O_3–SnTe and –GeTe junctions at 4.2 °K. From Chang *et al.* ([51]).

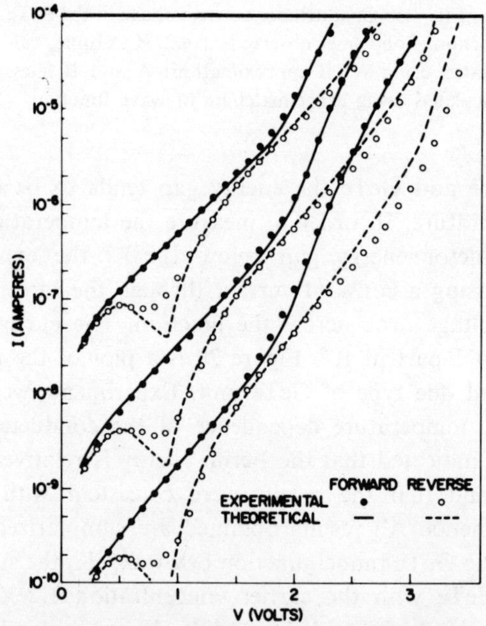

Fig. 27. Current–voltage characteristics for Al–Al_2O_3–SnTe junctions with various oxide thickness at 4.2 °K. From Chang *et al.* ([5]).

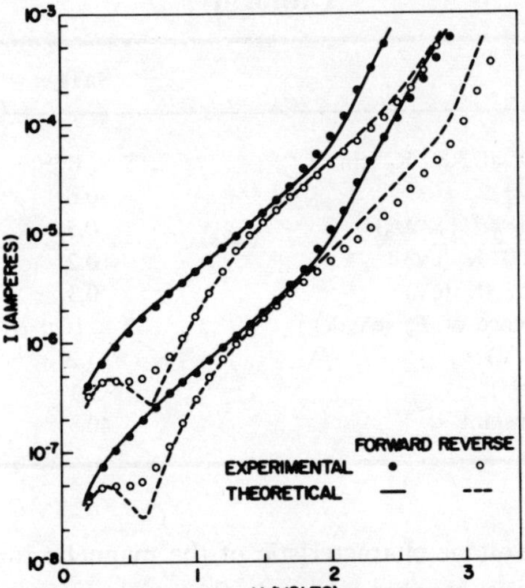

Fig. 28. Current–voltage characteristics for Al–Al_2O_3–GeTe junctions with various oxide thickness at 4.2 °K. From Chang *et al.* ([51]).

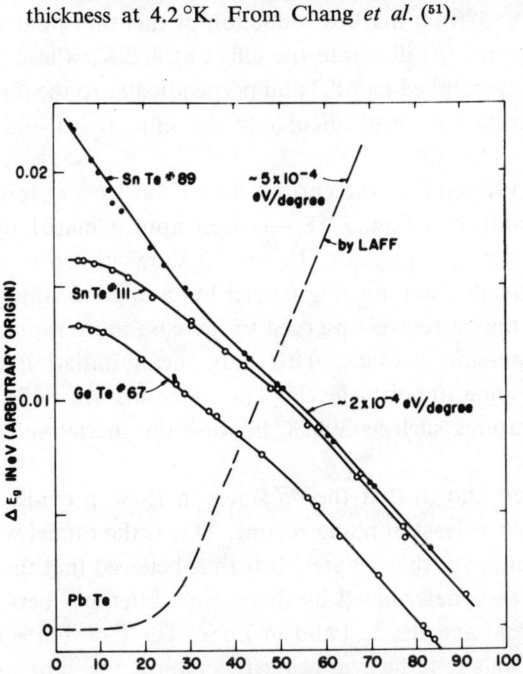

Fig. 29. Temperature variation of energy gap for SnTe and GeTe.

TABLE III

	SnTe	GeTe
Hole concentration n at 300 °K (cm^{-3})	8×10^{20}	2×10^{20}
Fermi energy F_p (eV)	0.6	0.4
Density-of-state mass m^* ($\times M_0$)	0.5	0.3
Energy gap E_g at 300 °K (eV)	0.2	0.1
at 4.2 °K (eV)	0.3	0.2
Temperature dependence of E_g (eV/deg)	2×10^{-4}	2×10^{-4}
Electron affinity χ (eV)	4.2	4.2
Work function W (eV)	5.1	4.8
Optical dielectric constant ε	40	35

The current–voltage characteristic of the magnetic tunnel junctions at low temperatures is somewhat similar to that of the so-called Zener diode—namely, a sharp increase in current near a certain threshold voltage. When magnetic fields are applied at temperatures below the magnetic transition temperature ($\sim$ 5°K) a marked reduction of this threshold voltage occurs. Figures 31 (*a*) and (*b*) illustrate the effect at 4.2°K, where magnetic fields up to 20 KOe are applied parallel and perpendicular to the film, respectively. [Since the current I is perpendicular to the film, $H \perp I$ and $H \parallel I$, respectively([57]).]

We have verified that the current flow I—at least at low current levels and at temperatures below 77°K—is well approximated by the Fowler–Nordheim tunneling expression. Figure 32 shows this plot and also illustrates schematically the tunneling barrier at an applied voltage V. At higher current levels the current is observed to increase more rapidly with voltage than this expression predicts. This may mean initiation of the impact ionization of some impurity levels. The junctions are more conductive at higher temperatures such as 300°K because the thermionic tunnel current dominates.

It has been known that the 5*d* states in EuSe provide a narrow conduction band or at least hopping regime. This is the tunnel which I described at the beginning of this chapter. It is thus believed that the barrier height φ at the cathode is determined by the energy difference between the Fermi level in the metal and the 5*d* band in EuSe. The fact that φ is found to be a function of magnetic field appears to confirm the shift of the 5*d* band due to the spin ordering in the 4*f* electrons with respect to the metal Fermi level, and thus to the vacuum level. On the assumption that the effective

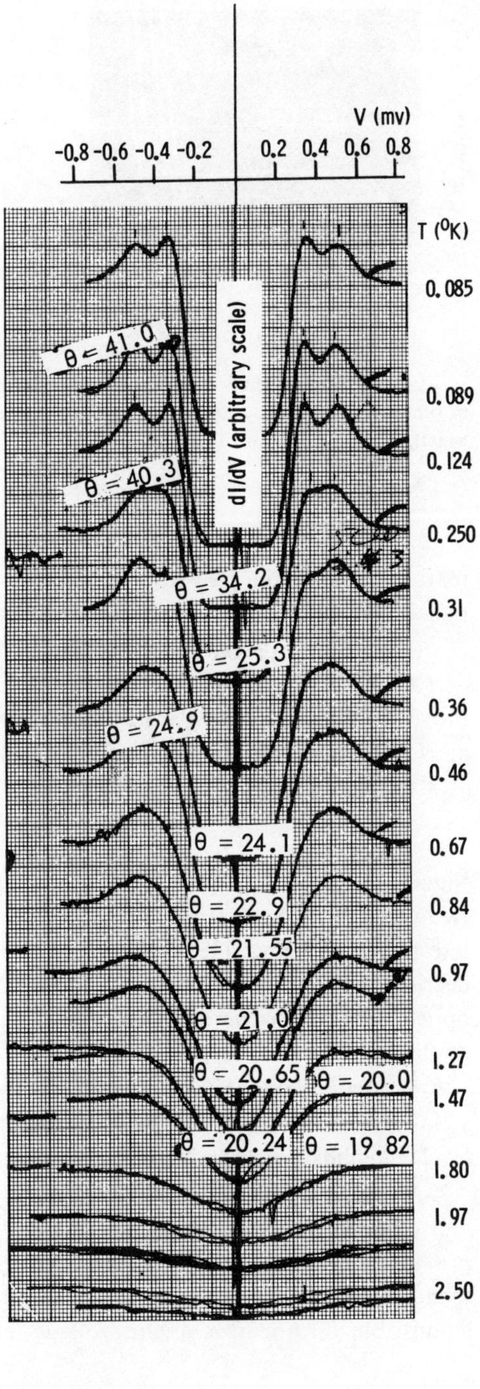

Fig. 30. Conductance versus voltage plots at temperatures between 2.5 and 0.085 °K for the $Al–Al_2O_3–GeTe$ tunnel junction. At temperatures above 0.5 °K the structure is entirely due to superconducting Al, whereas at temperatures below 0.5 °K additional structure is observed which is due to superconducting GeTe. From Stiles *et al.* ([52]).

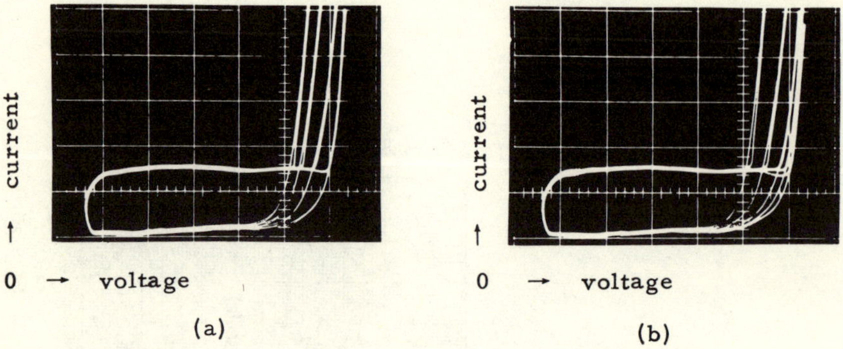

Fig. 31. The current–voltage traces at 4.2°K; 1 V/div in the abscissa and 10 uA/div in the ordinate (large loops in the figures are due to circuit reactance). (*a*) $H \perp I$, right to left, $H = 0$, 2.5, t, and 10 KOe; (*b*) $H \parallel I$, right to left, $H = 0$, 2.5, 5, 10, and 20 KOe.

mass is equal to the free-electron mass, the slope of the Fowler–Nordheim plot in Fig. 32 gives $\varphi \approx 0.5$ eV in the case of Al electrodes.

The relative decreases in the barrier height with magnetic field H are derived with reasonable accuracy from the above Fowler–Nordheim expression. With the voltages V_0 and V_H, which give the same current level at $H = 0$ and $H = H$, $\varphi(H)/\varphi(0)$ is given by

$$\varphi(H)/\varphi(0) = (V_H/V_0)^{2/3}.$$

Figure 33 shows $1 - [\varphi(H)/\varphi(0)]$ as a function of magnetic field for the two cases $H \perp I$ and $H \parallel I$. The difference between the two curves can be interpreted as being due to the demagnetizing effect in the film. The large decrease in the barrier height is indeed reminiscent of a shift of all 4.2°K optical absorption curves as much as 10% toward longer wavelengths with applied magnetic fields.

We have demonstrated that the technique of tunneling spectroscopy yields not only corroborative but also new information on the energy levels in solids: semiconductors, semimetals, and insulators.

We believe that this technique has a great potential for band-structure and other electronic state studies in solids; at least tunneling studies will provide substantial complementary information to the optical measurements, which often give precise values for interband and intraband transitions, but do not have a zero voltage (energy) reference.

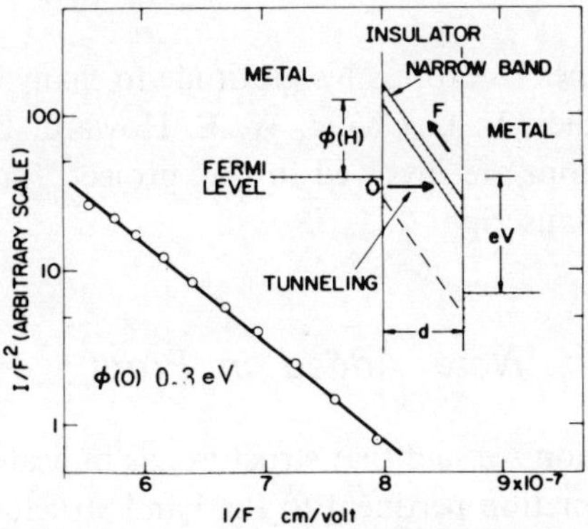

Fig. 32. The Fowler–Nordheim plot, I/F^2 versus $1/F$, and the energy scheme of the tunnel junction at an applied voltage V.

Here we have used two types of tunnel junction; one with monocrystalline materials and the other with polycrystalline materials. The former junctions seem to provide more detailed information, but stable junctions on monocrystalline solids are not always easy to find with our present fabrication method; the fine structure in the conductance curve seems ot suffer considerably from surface preparation as well as thin oxide formation. The latter junctions are relatively easy to produce, although their results are limited by the polycrystalline nature of materials used.

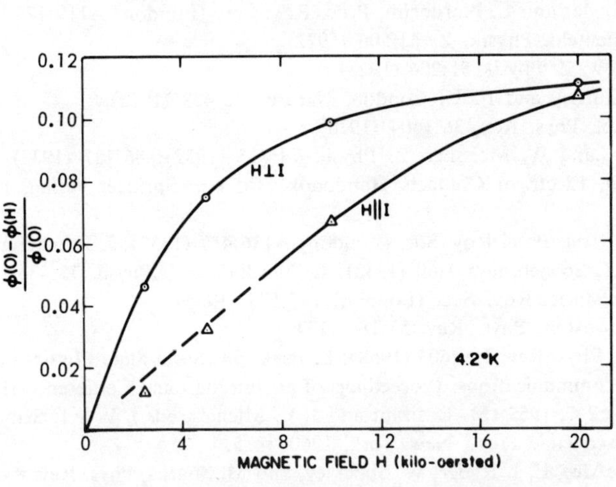

Fig. 33. The relative change of the barrier height as a function of magnetic field for the $H \perp I$ and $H \parallel I$ cases.

ACKNOWLEDGMENTS

The author wishes to express his gratitude to many colleagues at IBM, particularly P. J. Stiles, L. L. Chang, W. E. Howard, D. F. O'Kane, and N. Wiser, all of whom are involved in this project, for their cooperation in preparing this manuscript.

Note Added in Proof

In the final section we said that structure, as indicated by Eqs. (1)–(3), could provide information pertinent to the band structure of the materials involved. Recently, however, we found evidence of phonon-assisted tunneling in Bi tunnel junctions, represented by remarkably fine structure in the conductance–voltage curves. This process is particularly important near the onset of a new band, where it gives rise to a conductance peak at a voltage corresponding to the band edge plus the relevant phonon energy. Thus it has turned out that phonon-assisted tunneling plays a key role in determining band edge energies in Bi by tunneling spectroscopy ([53]).

REFERENCES

1. J. R. Oppenheimer, Phys. Rev. 31:66 (1928); Proc. Natl. Acad. Sci. U.S. 14:363 (1928).
2. R. H. Fowler and L. Nordheim, Proc. Roy. Soc. (London) A119:173 (1928).
3. J. E. Lilienfeld, Physik. Z. 23:506 (1922).
4. G. Gamow, Z. Physik 51:204 (1928).
5. R. W. Gurney and E. U. Condon, Nature 122:439 (1928).
6. J. Frenkel, Phys. Rev. 36:1604 (1930).
7. R. Holm and W. Meissner, Z. Physik 74:715 (1932); 86:787 (1933).
8. R. Holm, Electrical Contacts Handbook, 3rd ed., Springer-Verlag, Berlin, 1958, p. 122.
9. A. H. Wilson, Proc. Roy. Soc. (London) A136:487 (1932); J. Frenkel and A. Joffe, Physik. Z. Sowjetunion 1:60 (1932); L. Nordheim, Z. Physik 75:434 (1932).
10. C. Zener, Proc. Roy. Soc. (London) 145:523 (1934).
11. W. V. Houston, Phys. Rev. 57:184 (1940).
12. L. Esaki, Phys. Rev. 109:603 (1958); L. Esaki, in: Solid State Physics in Electronics and Telecommunications, Proceedings of an International Conference Held in Brussels, June 2–7, 1958 (M. Desirant and J. L. Michels, eds.), Vol. 1, Semiconductors, Part I, Academic Press, New York, 1960, p. 514.
13. K. B. McAfee, E. J. Ryder, W. Shockley, and M. Sparks, Phys. Rev. 83:650 (1951).
14. K. G. McKay and K. B. McAfee, Phys. Rev. 91:1079 (1953); K. G. McKay, Phys. Rev. 94:877 (1954); S. L. Miller, Phys. Rev. 99:1234 (1955).

15. A. G. Chynoweth and K. G. McKay, Phys. Rev. 106:418 (1957); A. G. Chynoweth. W. L. Feldmann, C. A. Lee, R. A. Logan, G. L. Pearson, and P. Aigrain, Phys, Rev. 118:425 (1960).
16. T. Yajima and L. Esaki, J. Phys. Soc. Japan 13:1281 (1958).
17. L. Esaki and Y. Miyahara, Solid-State Electron 1:13 (1960).
18. R. L. Batdorf, G. C. Dacey, R. L. Wallace, and D. J. Walsh, J. Appl. Phys. 31:613 (1960).
19. N. Holonyak and I. A. Lesk, Proc. IRE 48:1405 (1960).
20. H. P. Kleinknecht, Solid-State Electron. 2:133 (1961).
21. R. H. Rediker and A. R. Calawa, J. Appl. Phys. 32:2189 (1961).
22. L. Armstrong, IRE Trans. ED-9:114 (1962).
23. R. F. Rutz, IBM J. Res. Develop. 8:539 (1964).
24. L. Esaki and Y. Miyahara, unpublished; L. Esaki and Y. Miyahara, Solid-State Electron. 1:13 (1960); N. Holonyak, I. A. Lesk, R. N. Hall, J. J. Tiemann, and H. Ehrenreich, Phys. Rev. Letters 3:167 (1959).
25. G. G. Macfarlane, T. P. McLean, J. E. Quarrington and V. Roberts, J. Phys. Chem. Solids 8:388 (1959); J. R. Haynes, M. Lax, and W. F. Flood, J. Phys. Chem. Solids 8:392 (1959).
26. L. V. Keldysh, Zh. Eksperim. i Teor. Fiz. 34:962 (1968) [English translation: Soviet Phys.—JETP 7:665 (1958)]; E. O. Kane, J. Appl. Phys. 32:83 (1961).
27. E. O. Kane, "Report on Symposium on Electron Tunneling in Solids (Philadelphia, 1961)," J. Phys. Chem. Solids 23:173 (1962); also Proc. International Conference on Semiconductor Physics, Prague, 1960, Academic Press, New York, 1961.
28. L. V. Keldysh, Zh. Eksperim. i Teor. Fiz. 33:994 (1957) [English translation: Soviet Phys.—JETP 6:763 (1958)].
29. E. O. Kane, J. Phys. Chem. Solids 12:181 (1959); J. Appl. Phys. 32:83 (1961).
30. P. J. Price and J. M. Radcliffe, IBM J. Res. Develop. 3:364 (1959).
31. D. R. Fredkin and G. H. Wannier, Phys. Rev. 128:2054 (1962).
32. R. T. Shuey, Phys. Rev. 137:A1268 (1965).
33. Y. Takeuti and H. Funada, J. Phys. Soc. Japan 20:1854 (1965).
34. W. Franz, Z. Naturforsch. 13:484 (1958); L. V. Keldysh, Zh. Eksperim. i Teor. Fiz. 34:1138 (1958) [English translation: Soviet Phys.—JETP 7:788 (1958)].
35. J. I. Pankove, Phys. Rev. Letters 9:283 (1962).
36. W. A. Harrison, Phys. Rev. 123:85 (1961).
37. D. J. BenDaniel and C. B. Duke, private communication.
38. P. V. Gray, Phys. Rev. Letters 9:302 (1962); Phys. Rev. 140:A179 (1965).
39. J. Appelbaum, Phys. Rev. Letters 17:91 (1966); P. W. Anderson, Phys. Rev. Letters 17:95 (1966).
40. R. C. Jaklevic and J. Lambe, Phys. Rev. Letters 17:1139 (1966).
41. L. Esaki, "Proc. International Conference on Semiconductor Physics, Kyoto, 1966", J. Phys. Soc. Japan 21:589 (1966).
42. L. Esaki and P. J. Stiles, Phys. Rev. Letters 14:902 (1965).
43. L. Esaki and P. J. Stiles, Phys. Rev. Letters 16:574 (1966).
44. L. Esaki and P. J. Stiles, Phys. Rev. Letters 16:1108 (1966).
45. P. J. Stiles, L. Esaki, and W. E. Howard, presented at 10th International Conference of Low Temperature Physics, Moscow, 1966.
46. M. Cardona and D. L. Greenaway, Phys. Rev. 133:A1685 (1964).
47. M. H. Cohen, L. M. Falicov, and S. Golin, IBM J. Res. Develop. 8:215 (1964).

48. L. Esaki, P. J. Stiles, and S. von Molnar, unpublished.
49. G. A. Antcliffe and R. T. Bate, Phys. Letters 23:622 (1966).
50. L. L. Chang, P. J. Stiles, and L. Esaki, IBM J. Res. Develop. 10:484 (1966).
51. L. L. Chang, P. J. Stiles, and L. Esaki, J. Appl. Phys. 38:4440 (1957).
52. P. J. Stiles, L. Esaki, and J. F. Schooley, Phys. Letters 23:206 (1966).
53. L. Esaki, L. L. Chang, P. J. Stiles, D. F. O'Kane, and N. Wiser, Phys. Rev. 167:637 (1968); Erratum: Phys. Rev. 171:1114 (1968).
54. R. N. Brown, J. C. Mavroides, and B. Lax, Phys. Rev. 129:2055 (1963).
55. Y. H. Kao, Phys. Rev. 129:1122 (1963).
56. G. E. Smith, G. A. Baraff, and J. M. Rowell, IBM J. Res. Develop. 8:233 (1964).
57. L. Esaki, P. J. Stiles, and S. von Molnar, Phys. Rev. Letters 19:852 (1967).

Chapter 6

Interband Tunneling

E. O. Kane and **E. I. Blount**

Bell Telephone Laboratories,
Murray Hill, New Jersey

We will begin our discussion of interband tunneling by considering the square barrier problem for a solid as shown in Fig. 1. Here c and v represent the edges of the conduction band and valence band, respectively, with the forbidden band between them. The straight line represents the constant energy of the electron. In region 1 the electron has band energy

$$\mathscr{E}(k) = E_c + E_1, \tag{1}$$

where E_c represents the conduction band edge and E_1 is positive. This equation determines a real value of k and corresponds to an eigenfunction of the Bloch form

$$\psi(x) = e^{ikx} u_{c,k}(x), \tag{2}$$

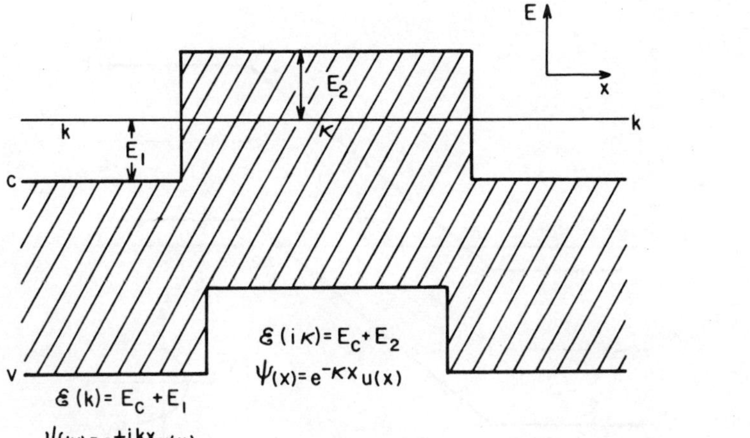

Fig. 1.

where $u_c(x)$ is the cell periodic part of the conduction-band wave function. The solution is analogous to a plane wave which propagates without attenuation. In the barrier region the band energy of the electron is given by

$$\mathscr{E}(i\varkappa) = E_c + E_2, \tag{3}$$

where E_2 is negative. In this forbidden region k is pure imaginary (in the simplest cases) and the eigenfunctions have the form

$$\psi(x) = e^{-\varkappa x} u_{c,\varkappa}(x). \tag{4}$$

These states do not propagate, but instead attenuate rapidly, so that such states play no role in the bulk properties of a perfect crystal. However, near a surface or in a tunneling barrier these states are obviously important. The magnitude of $\varkappa$ is the dominant consideration. To find this we need to know $\mathscr{E}(k)$ for complex k as well as real k. Because of the properties of analytic functions, if we have an analytic expression for $\mathscr{E}(k)$ which is valid for real k, then by merely inserting a complex k we get the correct expression for complex k as well. Thus in a sense, tunneling band structure requires no extra work on the part of the band theorist.

If we now go to a constant-field type of barrier, as shown in Fig. 2, we would expect that if the slope were gentle enough, we could describe the solution locally in terms of functions of the form

$$\begin{aligned} \psi(x) &= \exp[- \textstyle\int \varkappa(x)\, dx]\, u_{c\varkappa(x)} \\ \mathscr{E}[i\varkappa(x)] &= E_c + E(x) \end{aligned} \tag{5}$$

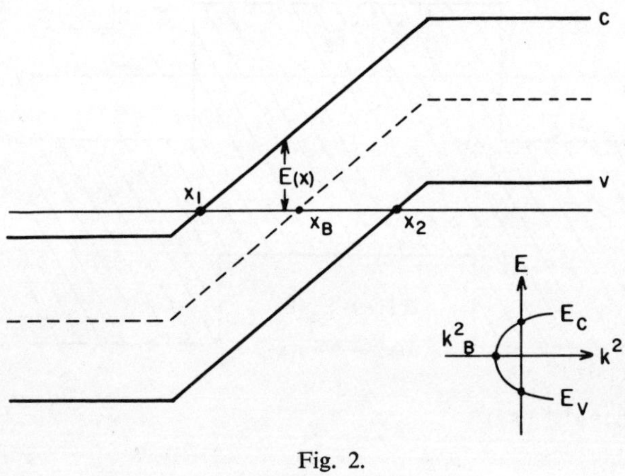

Fig. 2.

This is the semiclassical or WKB approach, and it turns out to be quite a good description. We will not discuss the mathematics of the WKB approach any further here, but we will use it extensively to interpret the results we obtain by methods which are mathematically easier but physically less intuitive than the WKB method.

For the case shown in Fig. 2 we see that the electron can tunnel through the forbidden band and appear in an allowed state in the valence band. It turns out that in many important cases the conduction and valence band are connected at a branch point in complex k space. The electron then passes smoothly from one band to the other through this branch point. When this can occur we say that we have direct tunneling or "branch point" tunneling. This is the main subject of the present chapter.

We now turn to the simplest model in which we can give an accurate discussion of "branch point" tunneling, namely, the "two-band" model.

The simplest derivation of this model is a purely one-dimensional approach based on free electrons, where we introduce the simple periodic potential

$$V(x) = V_0(e^{iKx} + e^{-iKx}), \tag{6}$$

with V_0 real and small. If V_0 is small enough, we need only consider the free-electron basis states

$$\begin{aligned} \psi_a(x) &= \exp[i(k + \tfrac{1}{2}K)x] \\ \psi_b(x) &= \exp[i(k - \tfrac{1}{2}K)x]. \end{aligned} \tag{7}$$

The Hamiltonian matrix then is 2×2 with the form

$$\mathscr{H} = \begin{pmatrix} \hbar^2(k + \frac{1}{2}K)^2/2m & V_0 \\ V_0 & \hbar^2(k - \frac{1}{2}K)^2/2m \end{pmatrix}. \tag{8}$$

We further simplify the Hamiltonian by assuming $k \ll K$ and ignoring K^2 on the diagonal, which only determines the energy zero. Equation (8) then becomes

$$\mathscr{H} = \begin{pmatrix} sk & V_0 \\ V_0 & -sk \end{pmatrix}, \tag{9}$$

$$s = \hbar^2 K/2m. \tag{10}$$

The eigenvalues of Eq. (9) are

$$\mathscr{E}_{\pm}(k) = \pm (V_0^2 + s^2k^2)^{1/2}. \tag{11}$$

This is the well-known "two-band" formula for $\mathscr{E}(k)$, and was first used in tunneling problems by Franz. It is actually more general than the one-dimensional derivation suggests. It provides a very good approximation to the band structure of narrow-gap direct semiconductors with the understanding that k is the absolute value of k, not just the component of k in the x direction.

Equations (9) and (11) are, mathematically speaking, identical to the Dirac equation, so that the considerable literature on the Dirac problem is usable here. This fact was used recently by Aronov and Pikus, who treated the magnetic field problem in tunneling by making a Lorenz transformation to eliminate the magnetic field.

The correspondence with the Dirac equation requires the identification of the fundamental parameters

$$\begin{aligned} s &= \hbar c\,, \\ V_0 &= mc^2. \end{aligned} \tag{12}$$

We want to add the energy,

$$F(x - x_0) = Fi\frac{d}{dk} - Fx_0\,,$$

to the Hamiltonian to represent a constant electric force in the x direction. With this, $\mathscr{H}$ then becomes $\mathscr{H}_F$, where

$$\mathscr{H}_F = \begin{pmatrix} sk + iF(d/dk) - Fx_0 & V_0 \\ V_0 & -sk + iF(d/dk) - Fx_0 \end{pmatrix}. \tag{13}$$

Equation (13) is written in momentum space and x_0 is arbitrary. We want to solve the eigenvalue equation

$$\mathscr{H}_F\psi = E\psi. \tag{14}$$

We choose the origin x_0 to be

$$Fx_0 = -E \tag{15}$$

to simplify the equations. Equation (14) is then

$$\left(sk + iF\frac{d}{dk}\right)\psi_1(k) + V_0\psi_2(k) = 0\,, \tag{16}$$

$$V_0\psi_1(k) + \left(-sk + iF\frac{d}{dk}\right)\psi_2(k) = 0; \tag{17}$$

ψ_2 can be eliminated to give

$$\frac{d^2}{dk^2}\psi_1(k)+\left(\frac{s^2}{F^2}k^2+\frac{V_0{}^2-isF}{F^2}\right)\psi_1(k)=0. \tag{18}$$

If we choose dimensionless variables

$$k=\sqrt{i}\,\beta\sigma,\qquad \beta=\sqrt{F/2s},\qquad \sqrt{i}\equiv(1+i)/\sqrt{2}, \tag{19}$$

Eq. (18) becomes

$$\frac{d^2}{d\sigma^2}\psi_1+\left(\frac{1}{2}+\frac{iV_0{}^2}{2sF}-\frac{\sigma^2}{4}\right)\psi_1=0. \tag{20}$$

The solutions to Eq. (20) may be written in terms of Weber functions:

$$\psi_1(\sigma)=aD_{i\mu}(\sigma)+bD_{i\mu}(-\sigma), \tag{21}$$

$$\mu=V_0^2/2sF. \tag{22}$$

To complete the solution we need to express $\psi_2(\sigma)$ in terms of $\psi_1(\sigma)$. Using Eqs. (16) and (19) we find

$$\psi_2(\sigma)=-\frac{\sqrt{i}\,F}{V_0\beta}\left(\frac{d}{d\sigma}+\frac{1}{2}\sigma\right)\psi_1(\sigma). \tag{23}$$

Using the identity

$$\frac{d}{d\sigma}D_m(\sigma)+\frac{1}{2}\sigma D_m(\sigma)=mD_{m-1}(\sigma) \tag{24}$$

as well as Eqs. (21) and (22), Eq. (23) may be written

$$\psi_2(\sigma)=(\sqrt{\mu}/\sqrt{i})[aD_{i\mu-1}(\sigma)-bD_{i\mu-1}(-\sigma)]. \tag{25}$$

The pair of Eqs. (21) and (25) are then the general solutions of Eq. (14).

We need to know the asymptotic forms of Eqs. (16) and (20) for large values of the argument. We write

$$\sigma\equiv\pm q/\sqrt{i}, \tag{26}$$

where q is real and positive. The asymptotic forms of the Weber functions, D, can then be written

$$\begin{aligned} D_{i\mu}(\sigma)&\approx\exp[i(\tfrac{1}{4}q^2+\mu\ln q)]\,e^{\pi\mu/4},\\ D_{i\mu-1}(\sigma)&\approx 0, \end{aligned} \tag{27}$$

$$\sigma = q/\sqrt{i}; \qquad \arg\sigma = -\pi/4;$$

$$D_{i\mu}(\sigma) \approx \exp[i(\tfrac{1}{4}q^2 + \mu\ln q)]\, e^{-3\pi\mu/4}, \tag{28}$$

$$D_{i\mu-1}(\sigma) \approx \frac{\sqrt{2\pi}\exp\ -i(\tfrac{1}{4}q^2 + \mu\ln q)e^{-\pi\mu/4}}{\Gamma(-i\mu+1)},$$

$$\sigma = -q/\sqrt{i}; \qquad \arg\sigma = 3\pi/4.$$

For small fields μ in Eq. (22) is very large and we can use the asymptotic form of the Γ function

$$\Gamma(-i\mu+1) \approx \sqrt{2\pi\mu}\, e^{-\pi\mu/2}\exp[i(\mu - \mu\ln\mu - \tfrac{1}{4}\pi)]. \tag{29}$$

With the use of Eqs. (27)–(29), Eqs. (21) and (25) can be seen to have the asymptotic forms

$$\begin{aligned} \psi_1(+k) &\approx \varphi(q)(ae^{\pi\mu/4} + be^{-3\pi\mu/4}), \\ \psi_1(-k) &\approx \varphi(q)(ae^{-3\pi\mu/4} + be^{\pi\mu/4}), \\ \psi_2(+k) &\approx \varphi^*(q)f(\mu)(-b)e^{\pi\mu/4}, \\ \psi_2(-k) &\approx \varphi^*(q)f(\mu)ae^{\pi\mu/4}; \end{aligned} \tag{30}$$

$$\begin{aligned} \varphi(q) &= \exp[(iq^2/4) + i\mu\ln q], \\ f(\mu) &= \exp(-i\mu + i\mu\ln\mu). \end{aligned} \tag{31}$$

To determine the tunneling transmission we must relate these asymptotic forms to the physical boundary conditions. First we note that the solution $\psi_1(k)$ corresponds to an electron in the upper energy band for k and in the lower energy band for $-k$. Denoting the upper band by $\psi_+(k)$ and the lower by $\psi_-(k)$, we have the relations

$$\begin{aligned} \psi_+(+k) &= \psi_1(+k), \\ \psi_-(+k) &= \psi_2(+k), \\ \psi_+(-k) &= \psi_2(-k), \\ \psi_-(-k) &= \psi_1(-k). \end{aligned} \tag{32}$$

We wish to consider a wave packet in the upper band incident from the left. This state will correspond to $\psi_+(+k)$. There will then be a reflected wave in the upper band corresponding to $\psi_+(-k)$. There will also be a transmitted wave in the lower band. Since $d\mathscr{E}_-/dk = -s\operatorname{sgn}k$ for large k in the lower band, the transmitted wave will have negative k vector, and hence it will correspond to $\psi_-(-k)$. Our boundary condition is then $\psi_-(+k) = 0$, i.e., no wave packet moving to the left in the lower band. Using Eqs. (32) and (30) this boundary condition requires $b = 0$. The incident current j_i is $|\psi_+(+k)|^2 \, d\mathscr{E}_+/dk$, and the transmitted current j_t is $|\psi_-(-k)|^2 \, d\mathscr{E}_-/dk$. Equations (30) with $b = 0$ show

$$j_i/j_t = e^{-2\pi\mu}. \tag{33}$$

To express this result in terms of the band gap E_G and the effective mass m^* we use Eq. (11) to write

$$E_G = 2V_0, \tag{34}$$

$$m^* = \hbar^2 V_0/s^2. \tag{35}$$

Equation (22) then gives

$$2\pi\mu = \pi E_G^{3/2} m^{*1/2}/(2\sqrt{2}\hbar F). \tag{36}$$

The exponential factor is the same as the perturbation theory result, but the prefactor is unity rather than $(\pi/3)^2$. Since the present treatment is more accurate, the factor $(\pi/3)^2$ is seen to be an artifact of the perturbation theory.

Our treatment of the Dirac equation follows quite closely unpublished work of E. I. Blount. The result is valid for the three-dimensional Dirac equation as well as for one dimension, since the mathematics are identical. The origin of the Dirac equation in three dimensions is more complicated than the simple one-dimensional derivation in Eqs. (6)–through (11), however.

We now turn to the more conventional description of band electrons in electric fields. Using the crystal momentum representation, the eigenvalue equation is written

$$\left(\mathscr{E}_n(k) - E + iF\frac{d}{dk}\right)q_n(k) + \sum_{n'} FX_{nn'}(k)q_{n'}(k) = 0. \tag{37}$$

In the absence of a perturbing electric field the eigenstates of Eq. (37) are Bloch functions $\psi_{nk0}(\mathbf{r})$ represented by $q_n(k) = \delta(\mathbf{k} - \mathbf{k}_0)$. Equation (37) is valid in one or three dimensions. The position operator X is represented by

$$i\frac{d}{dk}\,\delta_{nn'} + X_{nn'}(k), \tag{38}$$

$$X_{nn'}(k) \equiv i\int u_{nk}^*(r)\frac{d}{dk}\,u_{n'k}(r)\,d\mathbf{r}, \tag{39}$$

where $u_{nk}(r)$ is the cell periodic part of the Bloch function and the integration is over the unit cell.

In the case of our one-dimensional model in Eqs. (6)–(11) we can write the eigenfunctions in zero field in the form

$$\begin{aligned} \psi_+ &= \alpha(k)\exp[i(k+\tfrac{1}{2}K)x] + \beta(k)\exp[i(k-\tfrac{1}{2}K)x] \\ \psi_- &= \beta(k)\exp[i(k+\tfrac{1}{2}K)x] - \alpha(k)\exp[i(k-\tfrac{1}{2}K)x]. \end{aligned} \tag{40}$$

In this case the X operator becomes

$$X_{+-} = -\,isV_0/2[\mathscr{E}(k)]^2, \tag{41}$$

where $\mathscr{E}$ is given by Eq. (11). A similar result may be obtained in the three-dimensional case.

If the terms $X_{nn'}$ in Eq. (37) are neglected, the electric field eigenfunctions may be written

$$q_n(k) = A_n \exp\left\{(i/F)\int_0^k (\mathscr{E}_n(k') - E)\,dk'\right\}. \tag{42}$$

Equation (42) can be put in a more intuitive form by making a change of variables suggested by the WKB viewpoint. Define the variable x' by

$$E \equiv \mathscr{E}_n(k') + Fx'. \tag{43}$$

Equation (43) is conservation of energy in the WKB approximation, where a k vector and a position x can be simultaneously defined. However, as we are using it, Eq. (43) is not an approximation but merely a change of variables. Substituting (43) in (42) and integrating by parts gives

$$q_n(k) = A_n \exp i\left(\int_{x_n}^{x} k'\,dx' - kx\right), \tag{44}$$

where x_n is the solution of Eq. (43) with $k = 0$. Since the phase of the basis states is e^{ikx}, Eq. (44) shows that the proper phase is not kx but $\int_{x_n}^{x} k'\,dx'$, as would be expected from a WKB viewpoint. The absolute phase determined in Eq. (42) by taking $k = 0$ as the lower limit of integration is, of course, completely arbitrary.

If one thinks of the energy band function $\mathscr{E}_n(k)$ as a function of the complex variable k, it turns out that one does not have many functions corresponding to the band index n but one single analytic function defined on a many-sheeted Riemann surface, one sheet for each band. Krieger has pointed out, on the basis of an n-band secular equation, that *every* band meets every other band at a branch point somewhere in complex k space.

In the case of the two-band model we have been discussing the $\mathscr{E}(k)$ relation is given by Eq. (11). The two bands meet in a pair of branch points k_B given by

$$k_B = \pm iV_0/s. \tag{45}$$

Since these branch points are of the square root type, any contour which goes around either branch point once brings us to the other sheet of the two-sheeted Riemann surface. Taking the analytic-function point of view, Eq. (42) also defines an analytic function on a many-sheeted Riemann surface. The only singularities of q are the branch point singularities of $\mathscr{E}_n(k)$. Clearly, if one goes around the branch point connecting bands 1 and 2, say, one will go from the function q_1 to the function q_2. However, the multiplicative factors A_1 and A_2 are not independent but are related to each other by the analytic continuation process.

The relation is

$$A_2 = A_1 \exp\left\{(i/F) \int_{C_{12}} [\mathscr{E}(k') - E]\, dk'\right\}, \tag{46}$$

where the contour C_{12} begins at $k = 0$ in band 1 and ends at $k = 0$ in band 2, having circumnavigated one of the branch points. Suppose we use Eq. (11) of the two-band model and we go around

$$k_{B+} \equiv i\varkappa_B = iV_0/s. \tag{47}$$

Take the contour $k = i\varkappa$ along the imaginary axis in band 1 from $k = 0$ to $i\varkappa_B$, go around the branch point, and return to $k = 0$ along the imaginary axis in band 2. This gives

$$\begin{aligned} A_2 &= A_1 \exp\left\{-(1/F) \int_0^{\varkappa_B} [\mathscr{E}_+(\varkappa) - \mathscr{E}_-(\varkappa)]\, d\varkappa\right\} \\ &= A_1 \exp\left[-(2/F) \int_0^{\varkappa_B} (V_0^2 - s^2\varkappa^2)^{1/2}\, d\varkappa\right] \\ &= A_1 \exp[-\pi\varkappa_B^2 s/(2F)] = A_1 e^{-\pi\mu}, \end{aligned} \tag{48}$$

where we have used Eqs. (22) and (45). Comparing Eq. (48) with the barrier attenuation in Eq. (33), we note that $| A_2/A_1 |^2$ as given by analytic continuation is exactly the same as the barrier transmission computed by solving the two-band equation. This indicates that Eq. (42), the solution of Eq. (37) given by neglecting the interband coupling terms $FX_{nn'}(k)$ is, in fact, an adequate solution of the tunneling problem when looked on as a complete analytic function. The interband coupling is provided by the branch point structure of $\mathscr{E}_n(k)$, and the explicit coupling $FX_{nn'}$ can be ignored in the limit of small F.

If the process of analytic continuation seems rather unphysical, we note that we can use Eq. (44) to cast the result in a very intuitive form. Equation (48) becomes

$$A_2 = A_1 \exp -\left(\int_{x_1}^{x_B} \varkappa_1 \, dx + \int_{x_B}^{x_2} \varkappa_2 \, dx\right), \tag{49}$$

where $i\varkappa_1$ and $i\varkappa_2$ are the solutions of Eq. (43) for $n = 1$ and 2, respectively, and x_1 and x_2 are the classical turning points where $k = 0$. This result agrees exactly with the semiclassical picture of an electron tunneling from band 1 to band 2 through the forbidden gap. The total attenuation is determined by the integral of $\varkappa \, dx$, where $\varkappa$ is given by the energy conservation relation, Eq. (43). For $E - Fx > \mathscr{E}(k_B)$ the electron is in band 1 with the attenuation constant $\varkappa_1$, and for $E - Fx < \mathscr{E}(k_B)$ it is in band 2 with attenuation constant $\varkappa_2$. There is a "smooth" transition from band 1 to band 2 at the branch point. Thus the result is just what one would expect on the basis of the attenuating character of Bloch states in the forbidden band together with the semiclassical assumption that $\varkappa(x)$ is sufficiently well defined that the interval dx contributes attenuation $e^{-\varkappa(x)\,dx}$.

There are also some slight complications of the analytic-continuation viewpoint. Equation (42) shows that for a given band all values of k are equally probable. Thus there is no possibility of considering a purely "outgoing wave" in this approach. This is connected with the fact that in a constant field an electron in a given band will undergo Bragg reflection when it reaches the zone boundary and thus will periodically traverse the zone from $-K/2$ to $K/2$. (Choosing the field in the direction of the reciprocal lattice vector K.) To remedy this defect one must go to a finite junction model such as that of Fredkin and Wannier, of Shuey, or of Takeuti.

If we return to Eq. (46) and choose the coutour C_{12} to circumnavigate the branch point k_{B-},

$$k_{B-} \equiv -i\varkappa_B = -iV_0/s. \tag{50}$$

The analog of Eq. (48) is

$$A_2 = A_1 e^{+\pi\mu}. \tag{51}$$

Thus this is a different sheet of the function $q(k)$ from that obtained by going around k_{B+}, although both sheets correspond to band 2. The k_{B-} branch point corresponds physically to an electron in band 2 tunneling into band 1. This inverse action of the branch points k_{B+} and k_{B-} produces an infinite number of sheets of the $q(k)$ function. There are only two types of functional dependences corresponding to bands 1 and 2, but there are an infinite number of prefactors given by

$$\begin{aligned} A_1{}^n &= A_1 e^{2n\pi\mu}, \\ A_2{}^n &= A_1 e^{(2n-1)\pi\mu}, \end{aligned} \tag{52}$$

where n is any integer, positive or negative. Circling either branch point one merely alternates between two sheets, but circling both branch points clockwise, n decreases, while circling counterclockwise, n increases.

We now wish to discuss the generality of our results. As we have already remarked, the two-band model generalizes to three dimensions. Blount and Krieger have shown that in general two bands meet in a branch point of the square root type, so that Eq. (46) should be generally valid.

In giving the WKB interpretation implied in Eq. (49) it was assumed that one could circle the branch point by a contour on which $\mathscr{E}(k)$ is real so that one could interpret Eq. (43) in WKB terms with a real value of x. Branch points can occur for complex values of $\mathscr{E}$, and in this case it is not immediately obvious that a WKB interpretation will be applicable.

We study one such example, namely a model for the valence band of a semiconductor involving the light-mass band and the spin-orbit split-off band, which have an indirect mutual interaction via the conduction band. Krieger has studied this model in considerable detail. Defining the energy at $k = 0$ to be 0 and $-\Delta$ for the light mass band and split-off band, respectively, the secular equation is

$$\mathscr{E}(\mathscr{E} + \Delta) + p^2k^2(\mathscr{E} + \tfrac{2}{3}\Delta) = 0, \tag{53}$$

where p is a parameter determining the mass. The solution of this equation for real $\mathscr{E}$ is most easily seen by plotting k^2 as a function of $\mathscr{E}$ as shown in Fig. 3. The two bands are not connected along real $\mathscr{E}$ because the coefficient of k^2 goes to zero at $\mathscr{E} = \frac{2}{3}\Delta$. Hence k^2 goes to $+$ and $-$ infinity on the opposite sides of this point. The general solution of Eq. (53) can be written

$$\mathscr{E} = [-\Delta - p^2k^2 \pm (\Delta^2 - \tfrac{2}{3}\Delta p^2k^2 + p^4k^4)^{1/2}]/2, \tag{54}$$

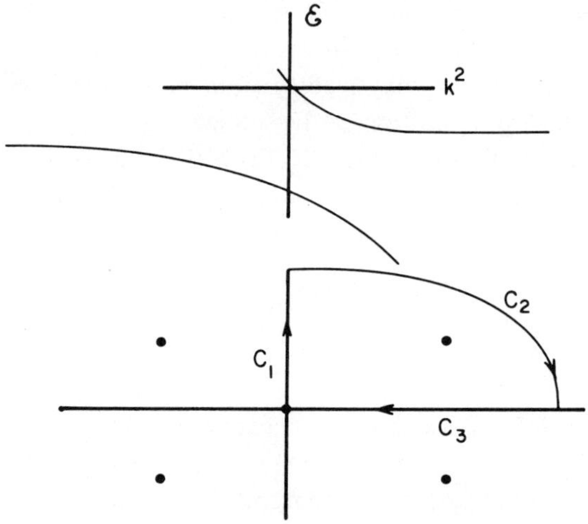

Fig. 3.

which has a branch point at

$$k_B{}^2 = (\Delta/3p^2)(+1 \pm i2\sqrt{2}), \tag{55}$$

with a branch point energy

$$\mathscr{E}_B = (\Delta/3)(-2 \mp i\sqrt{2}). \tag{56}$$

The existence of these branch points with complex $\mathscr{E}$ was first pointed out by Krieger. We wish to demonstrate that in spite of this apparent complication the analytic connection between these two bands can still be described completely in WKB terms along the real $\mathscr{E}$ line.

We take a contour in the complex k plane as shown in Fig. 3. This contour surrounds one branch point, so if we start out in band 2 ($\mathscr{E} = -\Delta$ at $k = 0$), we will end up in band 1 ($\mathscr{E} = 0$ at $k = 0$). Starting in band 2 along the positive imaginary axis, $\mathscr{E}_2(k)$ is a monotone function of k, so we can make the WKB transformation from Eq. (42) to (44) and write

$$(i/F) \int_{C_1} \mathscr{E}_2(k')\, dk' = -\int_{x_1}^{x_\infty} \varkappa_2\, dx + \varkappa_\infty x_\infty . \tag{57}$$

We have used Eq. (43) to define x in terms of $\varkappa = ik$ and we have set $E = 0$ for simplicity. Then $\mathscr{E}_2(\varkappa_\infty) = -\frac{2}{3}\Delta \equiv -Fx_\infty$, as can be seen from Eq. (54) using the $-\sqrt{}$ branch at $k = 0$.

For large $|\, k \,|$ on the arc C_2, $\mathscr{E}_2(k)$ maintains the constant asymptotic value $-\frac{2}{3}\Delta = -Fx_\infty$. Hence this integral can be written as

$$(i/F)\int_{C_2} \mathscr{E}_2(k')\,dk' = -x_\infty i(k_\infty - i\varkappa_\infty). \tag{58}$$

On the contour C_3, $\mathscr{E}(k)$ is again monotone, only now in band 1, and hence we can write

$$(i/F)\int_{C_3} \mathscr{E}(k')\,dk' = i\int_{x_\infty}^{x_1} k_1\,dx + ik_\infty x_\infty. \tag{59}$$

Adding the three pieces together, we find

$$\begin{aligned} A_1 &= A_2 \exp\left[(i/F)\int_C \mathscr{E}(k')\,dk'\right] \\ &= A_2 \exp\left(-\int_{x_2}^{x_\infty} \varkappa_2\,dx + i\int_{x_\infty}^{x_1} k_1\,dx\right) \end{aligned} \tag{60}$$

Hence the connection between the two bands can be described entirely in terms of WKB integrals along the lines where $\mathscr{E}(k)$ is real, as the naive WKB approach would suggest, in spite of the complex character of the energy branch point.

REFERENCES

1. W. Franz, in: Encyclopedia of Physics, Springer-Verlag, Berlin, 1956, p. 155.
2. E. O. Kane, J. Phys. Chem. Solids 1:249 (1957); Semiconductors and Semimetals, Vol. I, Physics of III–V Compounds (R. K. Willardson and A. C. Beer, eds.), Academic Press, New York, 1967, p. 75.
3. A. G. Aronov and G. E. Pikus, Soviet Phys.—JETP 24:188 (1967).
4. P. M. Morse and H. Feshbach, Methods of Theoretical Physics, Vol. II, McGraw-Hill Book Co. New York, 1953, p. 1565.
5. E. T. Whittaker and G. N. Watson, Modern Analysis, Cambridge University Press, 1947.
6. L. V. Keldysh, Soviet Phys.—JETP 6:763 (1958).
7. E. O. Kane, J. Phys. Chem. Solids 12:181 (1959).
8. E. I. Blount, Solid State Physics, Vol. 13, Academic Press, New York, 1962, p. 305.
9. J. B. Krieger, Ann. Phys. (N.Y.) 36:1 (1966).
10. R. T. Shuey, Phys. Rev. 137:A1268 (1965); D. R. Fredkin and G. H. Wannier, Phys. Rev. 128:2054 (1962); Y. Takeuti and H. Funada, J. Phys. Soc. Japan 20:1854 (1965).

Chapter 7

Interband Tunneling–Theory

R. T. Shuey*

Laboratory of Solid State Physics
Swiss Federal Institute of Technology
Zürich, Switzerland

INTRODUCTION

This chapter discusses two theoretical models for interband tunneling in semiconducting *pn* junctions, the Zener field-emission model and the Fredkin–Wannier junction potential model. In Chapter 5 of this volume Esaki reviewed the main experimental facts and mentioned these two types of theory. I want to further contrast the two theoretical pictures and to emphasize two mathematical effects where the models seem to differ. These are the so-called Stark ladder and the band-edge breakdown of the WKB connection formula. Both are still somewhat controversial and of current research interest. I do not propose to explain specific calculations or to consider tunneling "assisted" by entities such as phonons, photons, or localized states.

BAND-THEORETIC PRELIMINARIES

Before considering the two models individually we recall several results of general band theory. In both pictures of interband tunneling a slowly varying external potential V is superposed on the periodic crystal potential. When the change in V over a lattice constant is small compared to the amplitude of the periodic potential an electron in the combined potentials can be approximately treated as a particle affected by only the external potential V but responding according to "effective" dynamical laws. If the

* USAFOSR Postdoctoral Research Fellow. Present address: Geophysics Department, University of Utah, Salt Lake City, Utah.

effective Hamiltonian is considered classically, the electron at any instant has a band index n, a crystal momentum k, and a position x. The quasi-classical equations of motion for k and x are

$$\dot{x} = \frac{1}{\hbar}\frac{dE_n}{dk}, \qquad \hbar\dot{k} = -\frac{dV}{dx}, \tag{1}$$

where $E_n(k)$ is a quantum-mechanical energy eigenvalue for an electron in the periodic crystal potential only. The sum U of band energy $E_n(k)$ and external energy $V(x)$ is a constant of the motion:

$$\dot{U} = \dot{k}\frac{dE_n}{dk} + \dot{x}\frac{dV}{dx} = 0. \tag{2}$$

Using this fact, a deformed-band diagram may be used to trace trajectories is position space. Such a diagram shows on a coordinate plane of x versus U those allowed regions where some n and real k exist satisfying

$$U = E_n(k) + V(x).$$

This effective dynamics represents a quantum treatment of the crystal potential and a classical treatment of the external potential. With a full quantum method there arises the theoretical possibility of a change in band index n by interband tunneling. Most commonly, the tunneling probability P is evaluated by an adaptation of the WKB method familiar in the theory of free-particle tunneling:

$$P_{\text{WKB}} = |\exp(-2\int k\,dx)|. \tag{3}$$

The path of integration in Eq. (3) was discussed by Kane and Blount in Chapter 6. One first makes an analytic continuation of the Brillouin zone to complex wave vectors and then finds a trajectory leading from one band into complex momentum space, around a branch point connecting the bands, back to the real Brillouin zone, and into another band. The value of the integral depends on the band structure and the potential barrier shape.

THE UNIFORM FIELD AND STARK LADDER

The older picture of interband tunneling is the field-emission theory initiated by Zener over thirty years ago. (A brief history appears in Ref. ([1]).) Here the external potential is taken to be $V = -Fx$, representing a constant field F. For simplification we suppose a fictional one-dimensional semiconductor of lattice constant a. This is not unrealistic, since tunneling is largely a one-dimensional phenomenon.

To understand the theoretical behavior of an electron in combined periodic and uniform fields, it is helpful to consider first the quasiclassical result. In momentum space the motion is periodic: the electron cycles through the Brillouin zone at a constant rate $\hbar\dot{k} = F$. Since the length of the one-dimensional Brillouin zone is $2\pi/a$, the period is

$$T = \frac{2\pi}{a}\frac{\hbar}{F} = \frac{h}{aF}. \tag{4}$$

The motion in position space is also periodic, the electron oscillating back and forth between two points where its band energy becomes forbidden [see Fig. 1 (a)]. The period is of course the same as the period in momentum space. For a typical bandwidth of 10 eV and a large field of 10^5 V/cm the width of the oscillations is 1 μ.

Suppose we now treat the external field quantum mechanically. The Bohr–Wilson quantization rule says [([2]), p. 165] that when the classical

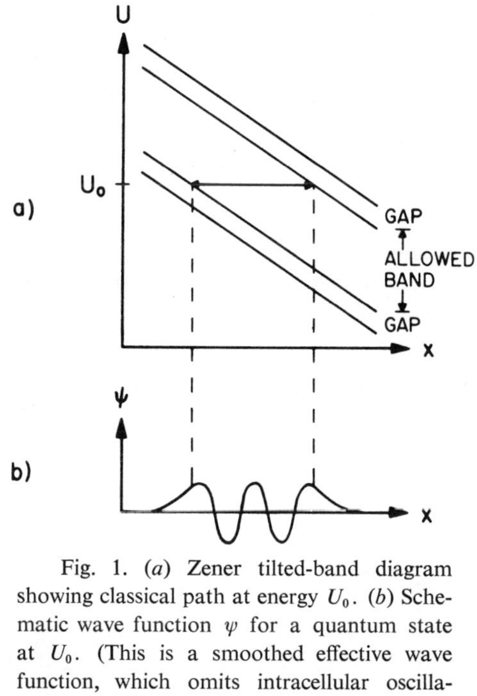

Fig. 1. (a) Zener tilted-band diagram showing classical path at energy U_0. (b) Schematic wave function ψ for a quantum state at U_0. (This is a smoothed effective wave function, which omits intracellular oscillations.)

motion is periodic the quantum spectrum is discrete, with a level spacing

$$\Delta U = k/T = Fa. \tag{5}$$

This is indeed the quantum spectrum provided one removes from the Hamiltonian the interband matrix elements of the field. In this artificial situation one can find the eigenfunctions explicitly in the crystal momentum representation ([3]). In the position representation they oscillate in the allowed interval and die off exponentially in the forbidden interval [Fig. 1 (*b*)]. The eigenfunctions for one band are all identical in shape and displaced from each other in space by multiples of the lattice constant a and in energy by multiples of Fa. This array of artificial eigenstates has been dubbed the Stark ladder, for it represents a sort of Stark splitting of the band into a ladder-like array.

The question of the physical reality of the Stark ladder remains somewhat controversial. It is frequently pointed out that the period [Eq. (4)] is of the order of electron lifetime due to phonon or impurity scattering. Furthermore, in three dimensions the classical motion is not generally periodic unless the field is aligned along a reciprocal lattice vector. Either effect might destroy the periodic motion and hence the discrete ladder-like spectrum. On the other hand, it was once reported ([4]) that the ladder spectrum had been observed in the current–voltage characteristic of a diode. [A plausible alternative explanation by Tiemann of these data is reported in ([5]).]

Another point is that when the interband matrix elements of the uniform field are retained in the Hamiltonian the spectrum is apparently a continuum. No complete discussion seems to be in the literature, but the following argument is offered: Suppose the periodic potential is bounded. Then far enough from the classical turning point x_0 *for the uniform field alone* the periodic potential can be neglected in comparison, and the eigenfunction looks like an Airy function. The latter is the eigenfunction of a free particle in a uniform field [([2]), p. 71]. If at energy U we make a linear substitution

$$z = \left(x + \frac{U}{F}\right)\left(\frac{2mF}{\hbar^2}\right)^{1/3}$$

to measure distance from $x_0 = -U/F$, then the asymptotic form of the Airy function A is:

$$\begin{aligned} A &\sim z^{-1/4} \sin(\tfrac{2}{3}z^{3/2} + \tfrac{1}{4}\pi), \qquad & z \gg 0 \\ A &\sim \tfrac{1}{2} \mid z \mid^{-1/4} \exp - \tfrac{2}{3} \mid z \mid^{3/2}, \qquad & z \ll 0. \end{aligned} \tag{6}$$

We argue that sufficiently far from x_0 the addition of the crystal potential has a negligible effect on the oscillation frequency or decay rate, respectively, in Eq. (6). The only effect in the asymptotic regions would be a change in the phase of the standing wave or the amplitude of the exponential tail. These changes exhibit ladder-like periodicity, since it is still true that eigenfunctions differing in energy by $\Delta U = Fa$ have the same shape, being simply displaced by $\Delta x = a$.

The usual way of finishing a Zener calculation is as follows: One supposes that at the initial instant $t = 0$ there is a full valence band and an adjacent empty conduction band. The motion is allowed to develop according to the full Hamiltonian. Phase coherence between repeated passes on an electron near the band gap, and hence any Stark ladder effect, is ignored. A transmission probability such as Eq. (3) is calculated and integrated over the Brillouin zone. The process is not allowed to continue long enough for the distribution of electrons in band states to change significantly, i.e., only an initial rate is calculated. The final result is expressed as the number of electron–hole pairs generated per unit volume per unit time by the applied field F.

It was once thought the Zener theory applied to the breakdown of a lightly doped *pn* diode at large reverse bias. It is now generally agreed, however, that in this case a more accurate understanding comes from analogy with the gas avalanche. The Zener theory may be adapted to the rather nonuniform field of an Esaki diode by integrating over the volume in the junction where the field is high and the Zener rate therefore large. However, the Esaki diode led to the second type of interband tunneling theory, which we now consider.

THE FREDKIN-WANNIER JUNCTION POTENTIAL

The major theoretical implication of Esaki's discovery of the tunnel diode was that the proper criterion for interband tunneling is not a high field but instead a close proximity (~ 100 Å) of p and n semiconducting regions. This is obtained by doping heavily, so the space charge and hence the field gradient are high in the junction depletion region. Fredkin and Wannier [6] therefore considered the motion of Bloch electrons in an external potential which changes over a short distance from one constant value to another [Fig. 2 (*b*)]. The total value of the step in potential is the sum of four terms: the energy gap, the Fermi energies in the p and n bulk, and the applied voltage.

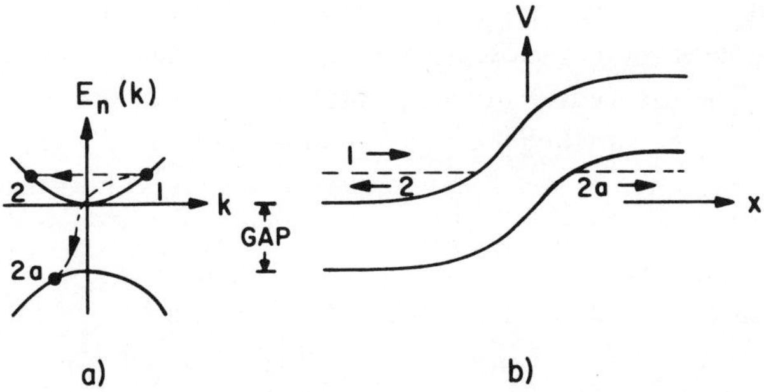

Fig. 2. (*a*) Energy band structure of a one-dimensional semiconductor, showing initial (1) and final (2 and 2*a*) wave vectors for a junction collision. (*b*) Deformed-band diagram for Fredkin–Wannier problem.

Let us again find the quasiclassical motion first. Suppose an electron initially has crystal momentum k_1 in the conduction band and is traveling toward the potential step at constant velocity [Eq. (1)]. When it collides with the barrier the crystal momentum begins to decrease, and at the classical turning point ($\dot{x} = 0$), k passes through an energy extremum. Then the electron retreats from the barrier with a terminal velocity opposite to its initial velocity. [We continue to suppose a one-dimensional semiconductor, for which the symmetry of $E_c(k)$ and hence altisymmetry of velocity about the band edge is ensured by time reversal. In three dimensions the kinematics can be more complicated.]

The quasiclassical motion just described is unbound rather than periodic. Therefore the quantum energy spectrum is continuous, with no Stark ladder. This is an important difference between the Zener and Fredkin-Wannier models.

The quantum eigenstate corresponding to a given classical trajectory is a standing Bloch wave in the semiinfinite region which the quasi classical electron traversed with constant initial or final velocity. If tunneling is energetically possible [Fig. 2 (*b*)], the miniumum amplitude of the standing wave is not quite zero, so there is a net probability current toward the junction. On the classically unattainable far side of the junction this current is carried by an outgoing Bloch wave of small amplitude. A wave packet following the classical path toward the junction would split up into two packets during the collision. The larger packet would rebound with crystal momentum k_2 like the classical electron, while the smaller packet would move away on the far side with crystal momentum k_{2a} in the valence band.

According to conservation of total energy U, the change of crystal energy $E_c(k_1) - E_v(k_{2a})$ is equal to the change of potential energy V.

The tunneling probability P is by definition the ratio of transmitted to incident current. Since the velocities of incidence and transmission are generally different, P is not simply the ratio of the wave packet norms. To a first approximation P is given by a WKB expression [Eq. (3)], but a possibly significant correction is discussed in the next section.

The technical intricacies of superposed periodic and step potentials have been discussed by several authors ([1, 6, 7]). In addition to the methods of formal band theory, some results of collision theory ([8]) are applied. This requires that the left and right sides of the junction be considered as separate *channels* [p. 81 and 112 of ([8])]. The separation of a step-like potential perturbation from the total energy is performed differently for the two channels; V vanishes asymptotically to the left in one case and to the right in the other. This point is mentioned explicitly in ([7]) and is handled in ([1]) by choosing different integration contours in the complex crystal momentum plane.

In the Fredkin–Wannier model the answer is no longer the *volume rate* of electron–hole pair generation in a high field, but the *cross section* for scattering across to the other side when an electron collides with the junction barrier. This cross section is then used in a kinetic calculation of the net current transferred between a gas of electrons on one side and a gas of holes on the other.

BREAKDOWN OF THE WKB APPROXIMATION

The essential step in many interband tunneling calculations is an approximation of WKB type. When is this step justified? Certainly the WKB exponent should be large, so that the tunneling probability is small. This indicates that the decay of the wave function in the forbidden region is accurately given by WKB. There is no problem here for typical tunnel diode parameters. For example, with a two-band model and a uniform field, the WKB factor [Eq. (3)] is

$$P_{\mathrm{WKB}} = \exp\left(- \frac{\pi}{2\sqrt{2}} \frac{\sqrt{m^* E_g}}{\hbar} W \right), \tag{7}$$

with m^* the effective mass, E_g the band gap, and W the width of the classically forbidden region. For the typical values: $m^* = 0.1 \times$ free electron mass, $E_g = 1$ eV, and $W = 100$ Å we have $P_{\mathrm{WKB}} = \exp(-12.6)$.

In the Zener model there is no way to distinguish between the electrons, so that any error in $P_{\rm WKB}$ is the same for all electrons. (In three dimensions this means all electrons with a given component of crystal momentum parallel to the junction plane). In the Fredkin–Wannier model, on the other hand, electrons are distinguished by their different initial or final wave vectors. This means that inaccuracy of $P_{\rm WKB}$ can affect not just the normalization but also the shape of a calculated current–voltage characteristic.

It happens that in a collision model of tunneling there is a systematic error in $P_{\rm WKB}$ when the asymptotic wavelength in the classically allowed region becomes large, i.e., when an initial or final wave vector is near a band edge. [In three dimensions "band edge" as used here means an extremum in the one-dimensional plot of $E_n(k)$ for k along a line perpendicular to the junction plane.] In this situation the WKB connection formula fails, with the consequence that the exact P drops linearly to zero at the band edge while $P_{\rm WKB}$ does not (Fig. 3). This effect was emphasized in ([1]) and ([7]), and was implicitly recognized in ([6]). In this chapter I will neglect the complications due to the periodic crystal potential and discuss the corresponding effect in free-particle tunneling.

Consider first the rectangular barrier, which is easy to solve exactly. If W is the barrier width, k the propagation wave vector outside the barrier, and K the attenuation constant inside the barrier, the tunneling probability is

$$P_{\rm exact} = \left[\cosh^2 KW + \frac{1}{4}\left(\frac{k}{K} - \frac{K}{k}\right)^2 \sinh^2 KW\right]^{-1}. \tag{8a}$$

When the WKB exponent $2KW$ is large this becomes

$$P_{\rm exact} \approx 4\left[1 + \frac{1}{4}\left(\frac{k}{K} - \frac{K}{k}\right)^2\right]^{-1} \exp(-2KW), \tag{8b}$$

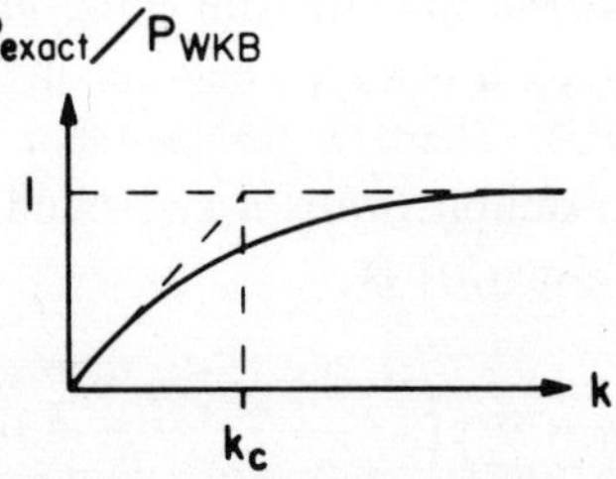

Fig. 3. Schematic error in $P_{\rm WKB}$ as a function of initial or final wave vector k measured from the one-dimensional band edge. The critical wave vector k_c is the reciprocal of the initial slope.

so that

$$\frac{P_{\text{exact}}}{P_{\text{WKB}}} \approx 4\left[1 + \frac{1}{4}\left(\frac{k}{K} - \frac{K}{k}\right)^2\right]^{-1}. \tag{8c}$$

This illustrates the effect mentioned above and in Fig. 3: When k is small, corresponding to a wave vector near the band edge, $P_{\text{exact}}/P_{\text{WKB}} \approx 16k^2/K^2$. The dependence at small k is quadratic, rather than linear as in Fig. 3, because both the initial and final wavelengths are becoming long, so a linear correction factor is contributed by both connection regions.

That the correction factor Eq. (8c) is not unity even for k large is a peculiarity of the abrupt sides of the rectangular barrier. Since the width of the transition region between classically allowed and forbidden regions is zero, the asympotic wavelength in the allowed region is always large in some sense, so that the WKB connection formula is never valid. This property of abrupt barriers was noted in Harrison's ([9]) review of the WKB approximation in solids. [See also p. 175 of ([2]).] For nonabrupt barriers the general picture in Fig. 3 obtains.

The tunneling probability can be evaluated analytically for a number of barrier shapes. Although complete analytic results are generally not as simple as for the rectangular barrier, the band-edge breakdown of WKB can be characterized by a single parameter, the slope of $P_{\text{exact}}/P_{\text{WKB}}$ at $k = 0$. The reciprocal of this slope can be considered as a critical wave vector k_c (Fig. 3), the minimum separation of an initial or final wave vector from an energy extremum permitted by the WKB connection formula. In Table I we give this parameter k_c for three further analytic shapes. As long as the WKB exponent is large, the validity of the connection formula may be judged independently at each side of the barrier. The example of the rectangular barrier Eqs. (8a)–(8c) shows how the corrections from two such independent turning points multiply together.

To close this chapter we consider when the effect just described will affect computed tunneling current–voltage characteristics. Some specific numerical studies have been made by Conley and Mahan ([10]). In general a reasonable criterion for the use of WKB is that the radius k_F of the Fermi surface in a degenerate semiconductor is greater than k_c. To show that in particular cases this may or may not be satisfied, we give an example where $k_F \approx k_c$: With a spherical Fermi surface k_F is related to carrier density N by

$$N = k_F{}^3/3\pi^2. \tag{9}$$

If we suppose $N = 10^{18}\ \text{cm}^{-3}$, then $k_F = 3.1 \times 10^6\ \text{cm}^{-1}$. The parabolic

TABLE I

Breakdown of the WKB Connection between Classically Allowed ($x < 0$) and Forbidden ($x > 0$) Regions*

Barrier	Analytic form		Critical wave vector
Abrupt	$V = 0;$	$x < 0$	$k_c = \frac{1}{4} K$
	$V = \dfrac{\hbar^2 K^2}{2m};$	$x > 0$	
Linear	$V = 0;$	$x < 0$	$k_c = \pi 3^{-2/3}[\Gamma(\frac{1}{3})]^{-2} K$ $\approx 0.21\ K$
	$V = \dfrac{\hbar^2 K^3}{2m} x;$	$x > 0$	
Parabolic	$V = 0;$	$x < 0$	$k_c = 4\pi[\Gamma(\frac{1}{4})]^{-2} K$ $\approx 1.0\ K$
	$V = \dfrac{\hbar^2 K^4}{2m} x^2;$	$x > 0$	
Exponential	$V \propto \exp Kx$		$k_c = K/4\pi$

* One side of a potential barrier $V(x)$ is characterized by a parameter K; the other side is unspecified. The WKB connection is invalid when the wave vector for propagation in the allowed region is less than the parameter k_c defined in the text and in Fig. 3. The solution of the exponential barrier is due to G. D. Mahan ([10]), Appendix C.

barrier in Table I is associated by Poisson's equation with a uniform space charge N' according to

$$\hbar^2 K^4/m = 4\pi N' e^2/\varepsilon, \tag{10}$$

where ε is the semiconductor dielectric constant and e the electronic charge. If we also take N' as 10^{18} cm^{-3}, $\varepsilon = 10$, and $m = 0.1$ free-electron mass, then $K = 2.2 \times 10^6$ cm^{-1}. Since $K \approx k_c$ for the parabolic barrier (Table I), these typical tunnel-diode parameters give $k_c \approx k_F$, and hence indicate a questionable applicability of WKB.

REFERENCES

1. R. T. Shuey, Phys. Rev. 137:A1268 (1965).
2. L. D. Landau and E. M. Lifshitz, Quantum Mechanics, Addison-Wesley, Mass., Reading, 1958.
3. E. O. Kane, J. Phys. Chem. Solids 12:181 (1959).
4. A. G. Chynoweth, G. H. Wannier, R. A. Logan, and D. E. Thomas, Phys. Rev. Letters 5:57 (1960).

5. E. O. Kane, "Report on the Symposium on Electronic Tunneling in Solids, Philadelphia, 1961," J. Phys. Chem. Solids 23:173 (1962).
6. D. R. Fredkin and G. H. Wannier, Phys. Rev. 128:2054 (1962).
7. Y. Takeuti and H. Funada, J. Phys. Japan 20:1854 (1965).
8. M. L. Goldberger and K. M. Watson, Collision Theory, John Wiley and Sons, New York, 1964.
9. W. A. Harrison, Phys. Rev. 123:85 (1961).
10. J. W. Conley and G. D. Mahan, Phys. Rev., 161:681 (1967); also, private communication.

Chapter 8

Tunneling in Schottky Barrier Rectifiers

R. Stratton

Texas Instruments Inc.,
Dallas, Texas

When a metal is brought into intimate contact with a semiconductor there sometimes arises a potential barrier within the semiconductor which impedes the flow of electrons between it and the metal. The formation of such Schottky barriers will be described by Professor Mead and this chapter will be mainly concerned with the tunneling of electrons through them. Clearly, if a bias is applied between a metal and a semiconductor, current will flow. If the barriers are sufficiently thin and the temperature is sufficiently low, then the current flow will be primarily by electron tunneling. Conversely, for thicker barriers and higher temperatures the current flow will be due to thermionic emission over the barrier. We will also consider intermediate ranges of thickness and temperature, where the current is primarily due to tunneling of thermally excited carriers. This is the so-called TF emission first introduced by Dolan and Dyke ([1]) in connection with field emission from a metal tip.

Duke has developed an equation to describe the tunneling of electrons from a conducting region 1 through a potential barrier into a conduction region 2 at zero temperature. This equation, generalized to a finite temperature, will be briefly rederived and then applied to the special case of a Schottky barrier. Figure 1 shows the potential energy diagram and serves to define the symbols that will be used for the tunnel equation. This can be written as ([2])

$$j = \frac{2q}{h^3} \int_0^\infty [f_1(E_1) - f_2(E_2)] v_{1x} P \, d^3p_1, \tag{1}$$

where $f_1(E_1)$ and $f_2(E_2)$ are the Fermi functions or occupation probabilities for electrons of energy E_1 and E_2 in conductors 1 and 2, respectively,

$$f_1(E_1) = \{1 + \exp[-(\zeta_1 - E_1)/kT]\}^{-1} \tag{2}$$

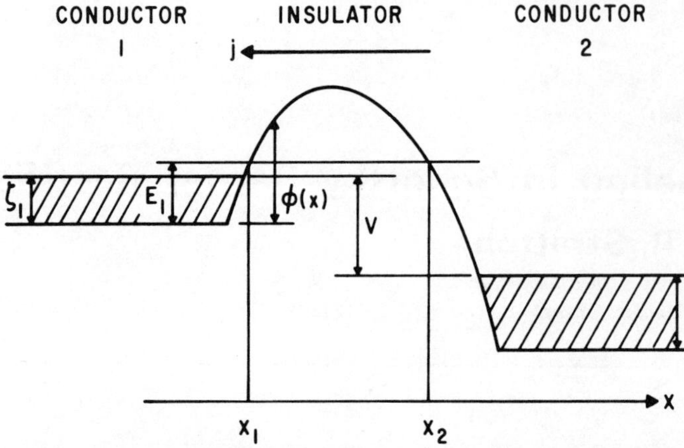

Fig. 1. Potential energy diagram for a tunneling structure. The insulating region represents either an insulating film sandwiched between two conductors or a space charge barrier formed at the contact between a metal and semiconductor.

$$f_2(E_1) = \{1 + \exp[-(\zeta_1 - V - E)/kT]\}^{-1}; \tag{3}$$

the group velocity v_{1x} is given by

$$v_{1x} = \partial E_1/\partial p_{1x}; \tag{4}$$

and P is the ratio of the transmitted to the incident current, or transition probability, for electrons of energy E_1. The integration is over all the wave vector or quasimomentum for conductor 1. Substituting from Eq. (4) into Eq. (1) leads to

$$j = \frac{2q}{h^3} \int_0^\infty [f_1(E_1) - f_2(E_1)]\, dE_1 \int P(E_1, p_{1y}, p_{1z})\, dp_{1y}\, dp_{1z}, \tag{5}$$

which reduces to the equation previously derived by Duke at zero temperature. If the functional form of the transition probability is known, then it is a simple matter to substitute this expression into Eq. (5) and compute the voltage–current characteristic for any temperature. However, since neither the exact shape of the barrier nor the relationship between energy and wave vector in each of the regions are ever well known, it is often preferable to simplify Eq. (5) as far as possible without making any specific assumptions about any of these relations. In this way it becomes possible to derive much simpler equations for various tunneling structures which give some idea of the important factors that dominate the volt–current characteristic. By comparing the voltage–current characteristic thus derived with experiment, it may then be possible to deduce either the barrier shape or the relationship between energy and wave vector in the tunneling region.

It will be assumed that the WKB approximation is applicable. This has been discussed in great detail by Kane and Blount (Chapter 6), Shuey (Chapter 7), and others; its application to the specfic case of the Schottky barrier has been discussed in ([11]). In addition, for the present the free-electron energy–momentum relation will be assumed in each of the three regions. The latter restriction will be removed later. Thus the negative energy of the electrons in the insulating region is given by

$$E_1 - \varphi(x) = \frac{p_i^2}{2m} = \frac{p_{ix}^2}{2m} + \frac{p_\perp^2}{2m}, \tag{6}$$

where $\varphi(x)$ is the potential energy of the electron in the barrier region and p_i will be imaginary, as will p_{ix}, the component of momentum in the x direction. It will be assumed that the boundary conditions correspond to specular transmission* so that the tangential electron momentum is conserved and is a real quantity. The WKB approximation can then be written in the form

$$\ln P = -\frac{2i}{\hbar}\int_{x_1}^{x_2} p_{ix}\,dx = -\frac{2}{\hbar}\int_{x_1}^{x_2} (p_\perp^2 - p_i^2)^{1/2}\,dx. \tag{7}$$

Then, substituting for the momentum from Eq. (6), this can be rewritten in the more familiar form

$$\begin{aligned}\ln P &= -\frac{2(2m)^{1/2}}{\hbar}\int_{x_1}^{x_2} [E_\perp - (E_1 - \varphi)]^{1/2}\,dx \\ &= -\frac{2(2m)^{1/2}}{\hbar}\int_{x_1}^{x_2} [\varphi - E_{1x}]^{1/2}\,dx.\end{aligned} \tag{8}$$

In these equations $p_\perp$ is related to the momentum component perpendicular to the x direction and E_{1x} is the portion of the energy corresponding to motion in the x direction. From Eq. (7) we see immediately that for a given energy of the electron the tunneling probability decreases as the perpendicular momentum increases. Thus for a given energy the tunneling probability is greatest when the electron is moving perpendicular to the

* More general boundary conditions involving diffuse scattering have also been considered ([3]).

barrier. The reason for writing the WKB approximation in the form of Eq. (7) is that Harrison ([2]) has shown that this equation still applies if the free-electron approximation does not hold provided that the band structure changes slowly with position across the interface between the conductors and the insulators. This is similar to the condition for the WKB approximation; namely, that the electron wavelength be small compared to the region over which appreciable variations in the potential of the electron occur. In this more general situation the relationship between the electron momentum and the energy given by Eq. (6) no longer applies.

We now develop Eq. (5) further without making any specific assumption about the relation between energy and electron momentum in any of the three regions. Since the tunneling probability given by Eq. (7) decreases rapidly as $p_\perp$ increases from 0, we can make a Taylor expansion for the exponent of the transition probability; namely,

$$-\ln P = b + (p_\perp^2/p_0^2) + O(p_\perp^4), \tag{9}$$

where the first two coefficients of the expansion are given by

$$b = \frac{2}{\hbar}\int_{x_1}^{x_2} \bar{p}\, dx, \qquad \frac{1}{p_0^2} = \frac{1}{\hbar}\int_{x_1}^{x_2} \frac{dx}{\bar{p}} \tag{10}$$

and where the complex momentum is replaced by a real quantity $\bar{p}$ given by

$$\bar{p}^2 \equiv -p_i^2. \tag{11}$$

It can be shown that the term of order $p_\perp^4$ in Eq. (9) is of the order b^{-1}, which is usually small compared to unity and can thus be neglected. Substituting this expression for P into the integral over p_{1y} and p_{1z} leads to

$$\int P\, dp_{1y}\, dp_{1z} = \pi p_{01}^2 e^{-b_1}\left[1 - \frac{1}{2\pi}\int_0^{2\pi} \exp(-p_m^2/p_0^2)\, d\varphi\right], \tag{12}$$

where the variables of integration have been changed to $p_\perp$ and the angle φ in the plane of the junction. The maximum value of $p_\perp$ is defined by p_m, which will be a function both of the energy of the incoming electron E_1 and the angular coordinate φ. The p_m is given by the shadow of the energy surface in conductor 1 on a plane parallel to the junction, as illustrated in Fig. 2. The range of integration will actually be given by the common area of the shadows on this plane for the energy surfaces in conductor 1 and 2, but it will be assumed that the energy surface in 1 will usually be

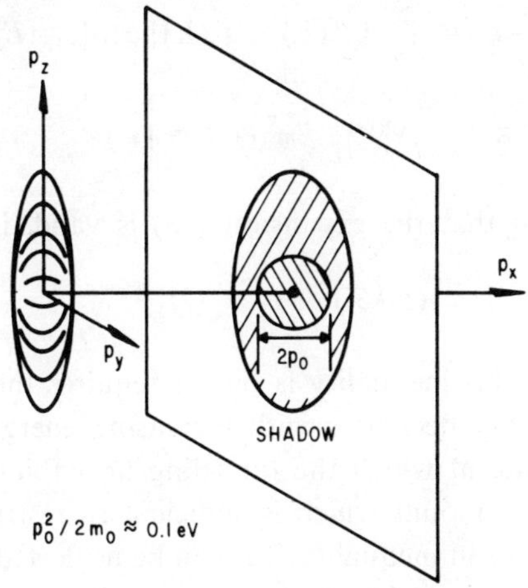

Fig. 2. The construction for shadows of energy surfaces in the plane of the junction.

the smaller one and its shadow will lie completely inside the shadow of the energy surface 2. (There are more complicated situations where the two energy surfaces are centered on different points of momentum space and tunneling can only take place by the assistance of a phonon which supplies the needed change in tangential quasimomentum.) Equation (12) shows that there is a circular region of radius p_0 within the shadow for which tunneling is appreciable. Typically, the magnitude of p_0 is given by $p_0^2/2m_0 \approx 0.1$ eV. This means that unless the energy surface corresponds to quite small energies there will be no effect on the integral in Eq. (12) due to the shape of the energy surface in conductor 1.

We now expand the quantity b_1 by recognizing that the tunneling probability will be greatest for electrons near the Fermi level of the conductor 1 at sufficiently low temperatures. Thus

$$b_1 = b_{1F} + c_{1F}(\zeta_1 - E_1) + f_{1F}(\zeta_1 - E_1)^2 + \dots, \tag{13}$$

where the first of these coefficients is given by

$$c_{1F} = -\frac{2}{\hbar}\int_{x_1}^{x_2}\left(\frac{d\bar{p}}{dE}\right)_{\zeta_1} dx. \tag{14}$$

If we now substitute from Eq. (13) into Eq. (12) and then into the expression for the tunnel current given by Eq. (5), we find

$$j = \frac{2\pi q p_{01F}}{h^3} e^{-b_{1F}} \int_0^\infty [f_1(E) - f_2(E)] \exp[c_{1F}(E_1 - \zeta_1)]\, dE$$
$$\times \left[1 - \frac{1}{2\pi} \int_0^{2\pi} \exp(-p_m{}^2/p_0{}^2)\, d\varphi\right], \tag{15}$$

under the condition that the expansion (13) is valid, i.e.,

$$(1/kT) - c_1 > (2f_1)^{1/2}. \tag{16}$$

The significance of this inequality is that it requires the rate at which the occupation probability decreases with increasing energy (this is given by $1/kT$) exceed the rate at which the tunneling probability increases (this is given by c_{1F}) by an amount which is sufficient to restrict energies so that the higher order terms in inequality (13) can be neglected. There is a further condition for Eq. (15) to hold which is a generalization of equations given by Murphy and Good ([4]) and relates to the fact that the tunneling probability for electrons near the top of the barrier must be small. [Equation (15) has been used to analyze voltage–current characteristics for metal–insulator–metal structures ([5–7]). We will not dwell on this point further, since it has been covered in other chapters of this volume.]

We now consider the special case of a Schottky barrier, which is illustrated in Fig. 3. We identify the degenerate semiconductor (Fermi energy ζ_1) with the conducting region 1. The metal (Fermi energy ζ_2) is conducting region 2, and the Schottky potential barrier formed in the semiconductor, near the surface of metal 2, is identified with the insulating region of Fig. 1. The shape of this potential barrier is determined by the distribution of impurities within it, assumed to be uniform for a Schottky barrier, and by the surface states which occur between the metal and semiconductor. The formation of such barriers will be described in detail by Mead. For my purpose I will assume that the quantity E_b and the barrier height remain fixed and that the potential barrier shape is parabolic. This is the ideal Schottky barrier. Changes in the potential barrier shape near the foot of the barrier due to the space charge of the electrons will be neglected. Effects due to this change of barrier shape are generally quite small. Initially, for simplicity we will assume that the energy shadows are large, so that the last integral in Eq. (15) can be neglected, and that the Fermi energy of the semiconductor is sufficiently large to satisfy the inequality.

$$\zeta_1/kT > c_1\zeta \gg 1. \tag{17}$$

It then follows that the voltage–current characteristic Eq. (15) is approxi-

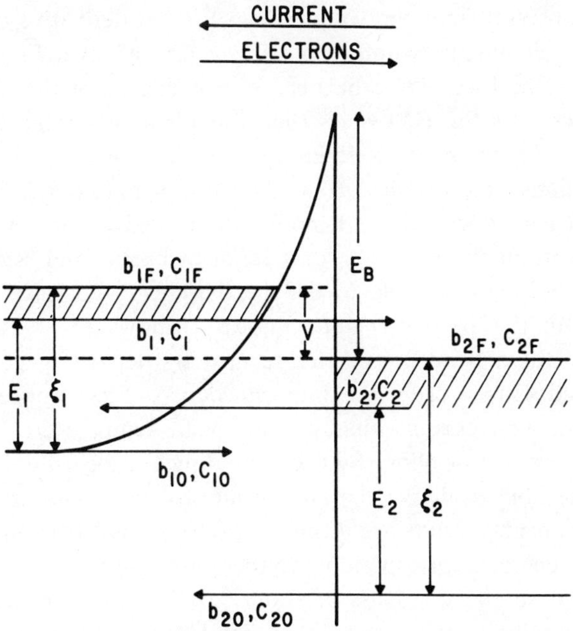

Fig. 3. Energy diagram of a Schottky barrier under forward-bias condition. The letters b and c refer to the first coefficients of an expansion of the transparency of the barrier around the energy level indicated by their position on the diagram.

mately given by the simple relation

$$j \propto \exp(-\, b_{1F}) \qquad \text{if} \quad c_1 V > 1 \tag{18}$$

(neglecting a slowly varying preexponential factor), which implies that the tunneling current is dominated by the probability of tunneling for electrons at the Fermi level of conductor 1. The quantity b_{1F} is given by

$$b_{1F} = \frac{2}{\hbar} \int_0^{E_B - V} \frac{\bar{p}\; d\psi_F}{d\psi_F/dx} = \frac{1}{\hbar} \left(\frac{2\varepsilon}{Nq^2} \right)^{1/2} \int_0^{E_B - V} \frac{d\psi_F\; \bar{p}}{(\psi_F + \zeta_1)^{1/2}}, \tag{19}$$

where the energy of the electrons in the barrier is given by

$$\psi_F = (Nq^2 x^2 / 2\varepsilon) - \zeta_1. \tag{20}$$

Differentiating the expression for b_{1F} leads to

$$\frac{db_{1F}}{dV} = \frac{1}{\hbar} \left(\frac{2\varepsilon}{Nq^2} \right)^{1/2} \frac{\bar{p}(E_B - V)}{(E_B + \zeta_1 - V)^{1/2}}. \tag{21}$$

It must be observed that the simple form for the derivative of b_{1F} arises because the field in the Schottky barrier, expressed as a function of the distance from the foot of the barrier, is independent of the applied bias. Thus differentiating Eq. (19) yields the value of the integrand at the upper limit, and we do not have to differentiate under the integral sign. Image force corrections or other deviations from the simple parabolic shape of the potential barrier would yield more complicated results. By measuring the voltage–current characteristic for a Schottky barrier and using Eqs. (18) and (21), it becomes possible to deduce the energy–complex momentum relationship in the barrier region. Figure 4 shows the voltage–current characteristic for a Schottky barrier formed by depositing gold films on heavily doped GaAs (2×10^{18} impurities/cm^3). First notice the small change in the volt–current characteristic with temperature, which is a standard feature of tunneling. By differentiating the logarithmic characteristic, we were able to derive the energy momentum relationship given in Fig. 5. Experimental results are given by the horizontal bars and the extrapolation is to energies corresponding to the bottom of the conduction band, which leads to an effective mass of about 0.07 free electron mass, close to the experimentally known conduction band mass of GaAs. The figure shows the result of measurements on one diode, but similar results have

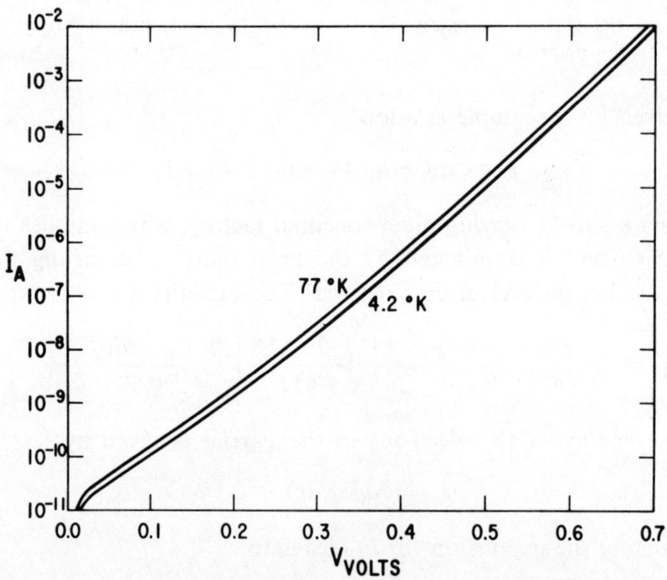

Fig. 4. Voltage–current characteristics for a gold GaAs diode (2×10^{18} donors/cm^3) at 77 and 4.2 °K. From Padovani and Stratton [8].

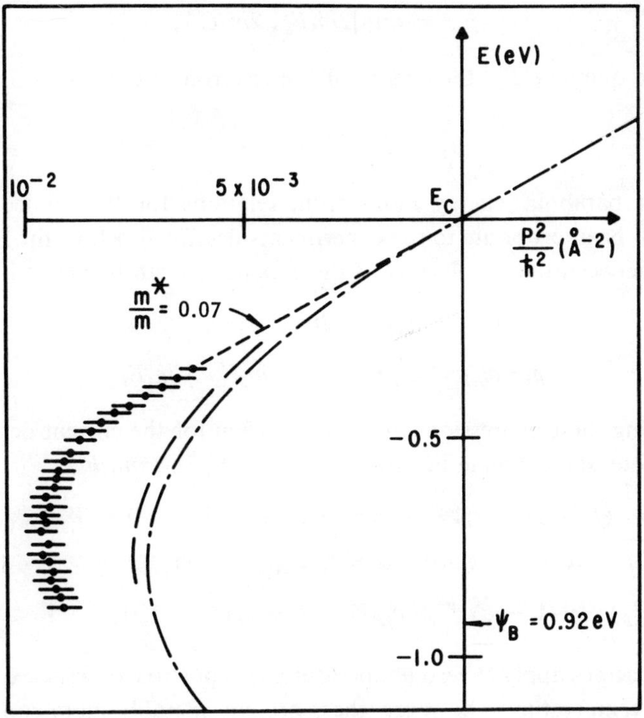

Fig. 5. Experimentally derived values of electron energy E versus wave vector $(p/\hbar)$. Dash-dot line: Kane's band structure; dashed line: approximate result obtained by equating $d(\ln I)/dV$ to db_1/dV. Dots with bars through them: experimental points including error bars.

been obtained for six other diodes which confirm the experimental results shown. The dashed curve is based on the **k·p** band structure calculation by Kane ([9]) for indium antimonide and applied to other semiconducting compounds by various investigators. The curve involves the so-called two-band approximation and is not in good agreement with the experimental results obtained by us. More recent calculation of the forbidden energy gap band structure in GaAs carried out by Chaves *et al.* ([10]) yields results that are not very different from this simple two-band calculation, although the discrepancy between theory and the experiments is slightly reduced.*

We now remove the restriction of large energy shadows and consider the situation where there are different masses in the conduction band of the semiconductor and in the metal. The shadow region will then be given by

* See note **A** added in proof.

$$p_m{}^2 = \min[2m_1E_1\,,\ 2m_2E_2]\,, \tag{22}$$

where the quantity E_2, the energy of the electron in conductor 2, is given by

$$E_2 = E_1 + V + \zeta_2 - \zeta_1. \tag{23}$$

Assuming parabolic energy momentum relations for the semiconductor, which will be appropriate to the experiments described below, the following relation between the coefficients of the expansions can be derived

$$p_{01F}^2 = 2m_1/c_{1F}\,, \tag{24}$$

$$p_m{}^2/p_{01F}^2 = c_{1F}E_1 \qquad \text{if} \quad m_2E_2 > m_1E_1. \tag{25}$$

Substituting these expressions into the equation for the current density and carrying out some simple integrations gives ($K_1 = 4\pi m_1/h^3$)

$$j = (K_1/c_{1F}^2)e^{-b_{1F}}[1 - e^{-c_{1F}V} - c_1Ve^{-c_1\zeta_1}]\,, \qquad 0 \le V \le \zeta_1, \tag{26}$$

$$\begin{aligned} j &= (K_1/c_{1F}^2)e^{-b_{1F}}[1 - e^{-c_{1F}\zeta_1} - c_1\zeta_1e^{-c_1\zeta_1}]\,, \qquad \zeta_1 \le V; \\ j &= (K_1/c_{2F}^2)e^{-b_{2F}}\{1 - e^{c_{2F}V} + c_{2F}V\exp[-c_{2F}(\zeta_1 - V)]\}\,, \qquad V < 0. \end{aligned} \tag{27}$$

These equations apply at zero temperature. It is possible to write expressions for finite temperatures; however, these are considerably more complicated and will involve the use of series expansions as discussed in ([11]). For most purposes the expressions relating to zero temperature will be sufficient, since the deviations of the current from the zero-temperature value are quite small, although these deviations have been the subject of both experiment and analysis. Figure 6 shows the result of plotting the differential resistance of Schottky barrier diodes as a function of bias using Eq. (26) and evaluating the coefficients which occur in this equation for parameters appropriate to germanium with the doping levels given in the figure. The dashed curves are the results of our calculations and the solid curves are the results of a calculation by Conley *et al.* ([12]) using an exact transition probability in the sense that the Schrödinger equation has been solved for a Schottky potential barrier, without, however, taking into account the detailed band structure of germanium.* It will be seen that their results and ours are fairly similar and predict a peak in the conductivity in the forward direction at an energy which is close to the Fermi energy of the degenerate semiconductor. In our case the peaks occur naturally because of the change of the functional form of the characteristic at the Fermi level which led to

* See note **B** added in proof.

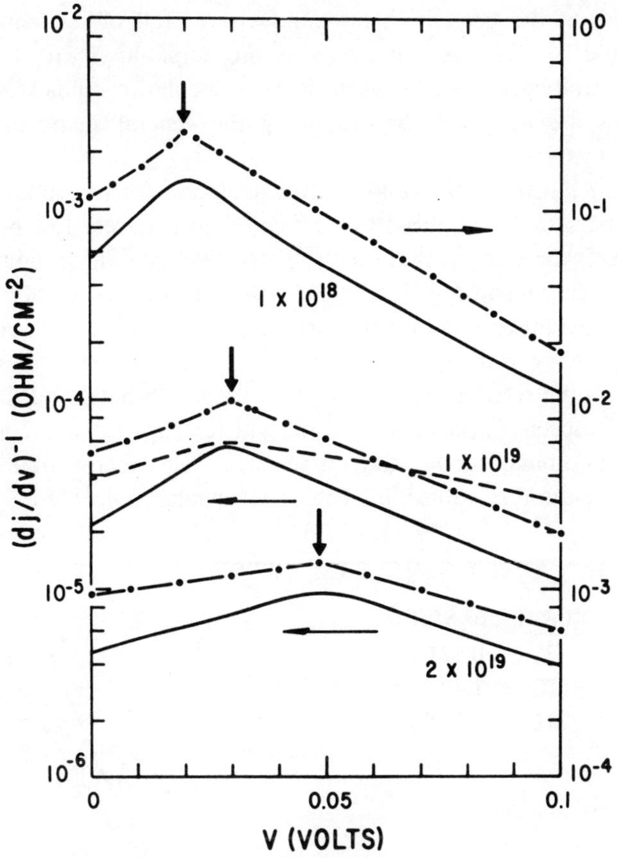

Fig. 6. Plot of $(dj/dV)^{-1}$ versus V for a Ge Schottky barrier. Solid curve: Theoretical results of Conley *et al.* Dashed curve: Experimental results of Conley *et al.* shifted vertically to make the maximum coincide. Dash-dot curve: Our results using a WKB approximation.

a cusp. This is a result of the simplification we have adopted using the WKB approximation for electrons of all energies. Now, it was shown by Shuey (Chapter 7) that the WKB approximation certainly breaks down when the energy of the electron in the semiconductor approaches zero. Then the true tunneling probability also tends to zero and is very approximately proportional to the WKB approximation for tunneling multiplied by the square root of the energy of the electron in the semiconductor. It turns out that the discontinuity in the second derivative of the current implied by the cusps in Fig. 6 is directly proportional to the tunneling probability for electrons at the bottom of the conduction band of semiconductor 1, which should be zero but is nonzero in our simple approximation. However, the overall agreement between the two calculations indicates that our assumptions are adeguate for describing the general shape of voltage–current characteristics.

Figure 7 illustrates the results of a calculation for the voltage–current characteristic of a GaAs Schottky barrier rectifier at very low biases. The doping level is the same as that used to determine the nonparabolic energy–momentum relations in Fig. 5. There are three theoretical curves in Fig. 7, each for a different assumption concerning the relation between energy and momentum vector in the forbidden gap. Curve I corresponds to extrapolating the quadratic relation, valid near the conduction band and the valence band edge, respectively, toward the center of the gap. Curve 2 corresponds to Kane's two-band model and curve 3 to our experimentally derived energy–momentum relationship. The experimental points are shown by

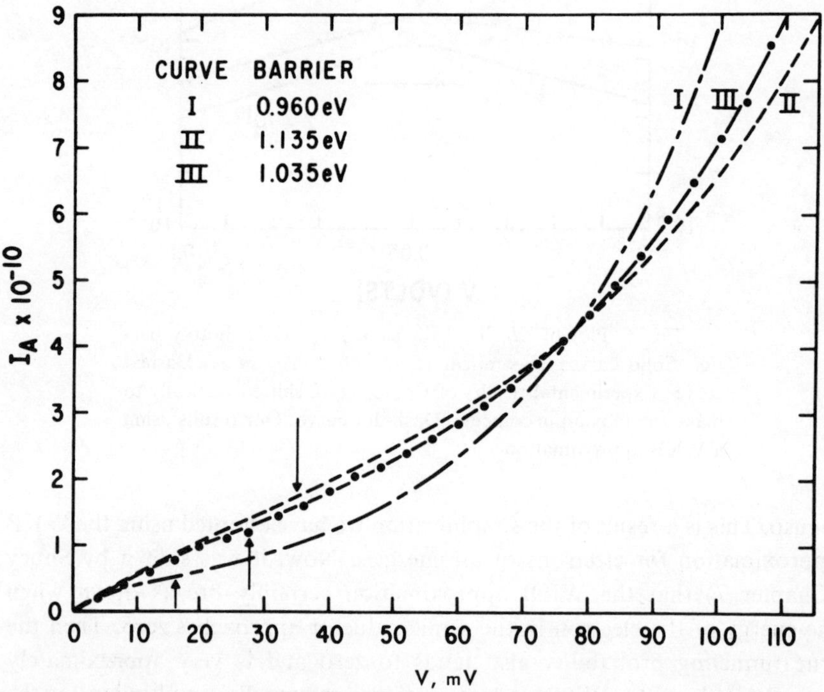

Fig. 7. Low-bias forward characteristic of a gold *n*-type GaAs Schottky barrier (2×10^{18} donors/cm³). Computed curves: — - — - Parabolic bands. - - - - - - Kane's band. Solid curve: Experimentally deduced bands. Dots indicate experimental points and arrows inflection points.

dots and all the theoretical curves have been fitted to coincide with the experimental point at the Fermi energy, which is 79 meV. It should be observed that there is a point of inflection predicted by each calculated curve, which in turn means that there is a resistivity maximum in the forward direction at a bias which is now less than the Fermi energy, as is verified by out experimental data. In fitting the theory and experiment, we have an adjustable parameter: the barrier height. The best agreement between theory and experiment is for curve 3, which uses our empirically determined energy–momentum relationship, and it turns out that the corresponding height is also close to that deduced from other measurements on the Schottky barrier diode.

We now consider the voltage–current characteristic in more detail and write down explicit expressions which apply for sufficiently large voltages applied in either the forward or reverse direction. The coefficient b_1 for a Schottky barrier can be shown to be given by

$$b_1 = \frac{1}{E_{00}} \Big\{ (E_B - V + \zeta_1)^{1/2} (E_B - V + \zeta_1 - E_1)^{1/2} - E_1 \ln\Big[\frac{(E_B - V + \zeta_1)^{1/2} + (E_B - V + \zeta_1 - E_1)^{1/2}}{\sqrt{\zeta_1}} \Big] \Big\}, \quad (28)$$

where

$$E_{00} = q\hbar \sqrt{N} / 2(m_1 \varepsilon)^{1/2}. \quad (29)$$

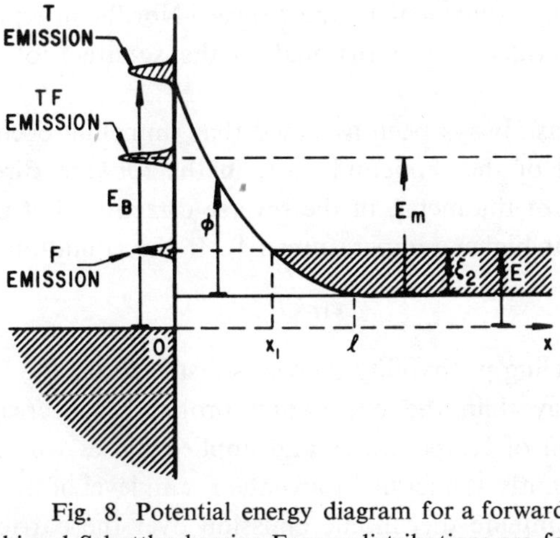

Fig. 8. Potential energy diagram for a forward-biased Schottky barrier. Energy distributions are for thermionic (T), field (F), or thermally assisted field (TF) emission.

If we now assume that the Fermi energy is small compared to the effective barrier height, i.e., if

$$\zeta_1 \ll E_B - V, \tag{30}$$

then

$$b_{1F} \approx (E_B - V)/E_{00}, \tag{31}$$

$$c_{1F} \approx \ln[4(E_B - V)/\zeta_1]/2E_{00}, \tag{32}$$

so that the voltage–current characteristic in the forward direction will be given by an expression of the form

$$j \propto \exp(V/E_{00}), \qquad V \geq \zeta_1 \gg 1/c_{1F}, \tag{33}$$

where we neglect the voltage dependence of the preexponential factors in the general expression for the current density.

In the reverse directions the voltage–current characteristics are rather complicated for low biases, but if the voltage is sufficiently large, i.e., if $-V > E_B$, then

$$b_{2F} \approx 2E_B^{3/2}/3E_{00}(E_B - V)^{1/2} \tag{34}$$

$$c_{2F} \approx \sqrt{E_B}/E_{00}(E_B - V)^{1/2}, \tag{35}$$

so that the voltage–current characteristic is given by

$$j \propto \exp\left[-\frac{2E_B^{3/2}}{3E_{00}(E_B - V)^{1/2}}\right]. \tag{36}$$

This is essentially equivalent to the Fowler–Nordheim equation, since the field at the interface is proportional to the square root of the effective barrier height.

So far it has always been assumed that tunneling occurs mainly from the Fermi level of the semiconductor, in the forward direction, or from the Fermi level of the metal, in the reverse direction. Let us now consider what happens at higher temperatures ([13]). If the condition

$$c_{1F}kT > 1 \tag{37}$$

holds, the tunneling probability increases faster above the Fermi level with increasing energy than the occupation probability decreases. There will then be a region of temperatures and applied biases where tunneling will occur predominantly at energies above the Fermi level of the semiconductor, but will still dominate thermionic emission over the barrier. If this is the case, the Fermi function in the general expression for the current density can be replaced by an exponential

$$\exp[(E_1 - \zeta_1)/kT] \gg 1. \tag{38}$$

We now expand the quantity b_1, the exponent in the tunneling probability, about the energy E_m, which corresponds to the maximum of the exponent of the integrand in the expression for the current density. This leads to the following expression for the current density ([13]):

$$j = \frac{2\pi q p_{01m}^2}{h^3} e^{-b_{1m}} \int_0^\infty \exp\left[\frac{\zeta - E_1}{kT} - c_{1m}(E_m - E_1) - f_{1m}(E_m - E_1)^2\right] dE_1, \quad \text{if} \quad p_m{}^2 \gg p_0{}^2, \tag{39}$$

where the energy E_m, for which the exponent of the integrand has a maximum, is given by

$$c_{1m}(E_m)kT = 1. \tag{40}$$

Substituting from Eq. (39) into (40) leads to

$$j = \frac{2\pi q p_{01m}^2}{h^3} \exp\left\{-\left[b_{1m} + \frac{(E_m - \zeta_1)}{kT}\right]\right\} \int_0^\infty \exp[-f_{1m}(E_m - E_1)^2]\, dE_1, \tag{41}$$

which shows that the energy distribution has a Gaussian-shape peaked at the energy E_m. Substituting specific values for the various coefficients in Eq. (41) and performing the integrations leads to the simple functional form (neglecting the preexponential factor)

$$j \propto \exp(V/E_0), \tag{42}$$

where the coefficient E_0 is given by

$$E_0 = E_{00} \coth(E_{00}/kT). \tag{43}$$

Thus the logarithm of the current is linear in the voltage for sufficiently high voltages applied in the forward direction, both for tunneling and for TF emission, and the slope of the curve will depend on temperature. Figure 9 gives the measured voltage–current characteristic for a gold GaAs Schottky barrier rectifier with a doping level of 5×10^{17} impurities/cm^3. As predicted by our simple theory, a semilogarithmic plot yields a series of straight lines with varying slopes depending on the temperature. The deviations from the straight line, previously used to deduce the energy band structure for more heavily doped Schottky barrier diodes, are not seen in this figure. This is because measureable currents require forward biases which are sufficiently high for most of the tunneling to occur at energies which are in the parabolic energy momentum relation range close to the bottom of the conduction band edge in the semiconductor. Figure 10 shows the experimentally determined slopes of the voltage–current characteristic as a function of temperature in comparison with the theoretically calculated expression for E_0 using Eq. (43). The theoretical curve has only one parameter, which is completely determined by the effective mass, the doping level, and the dielectric constant, all of which are known. The excellent agreement between theory and experiment, without any adjustable parameters, confirms our general concepts used in deriving

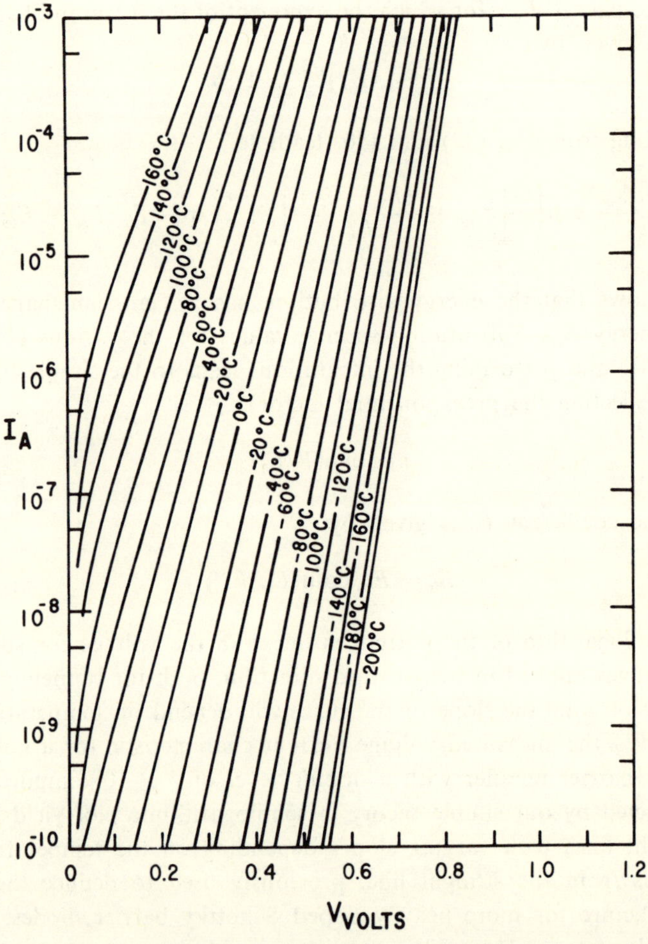

Fig. 9. Forward *I–V* characteristic of a gold GaAs Schottky barrier (5×10^{17} donors/cm^3)

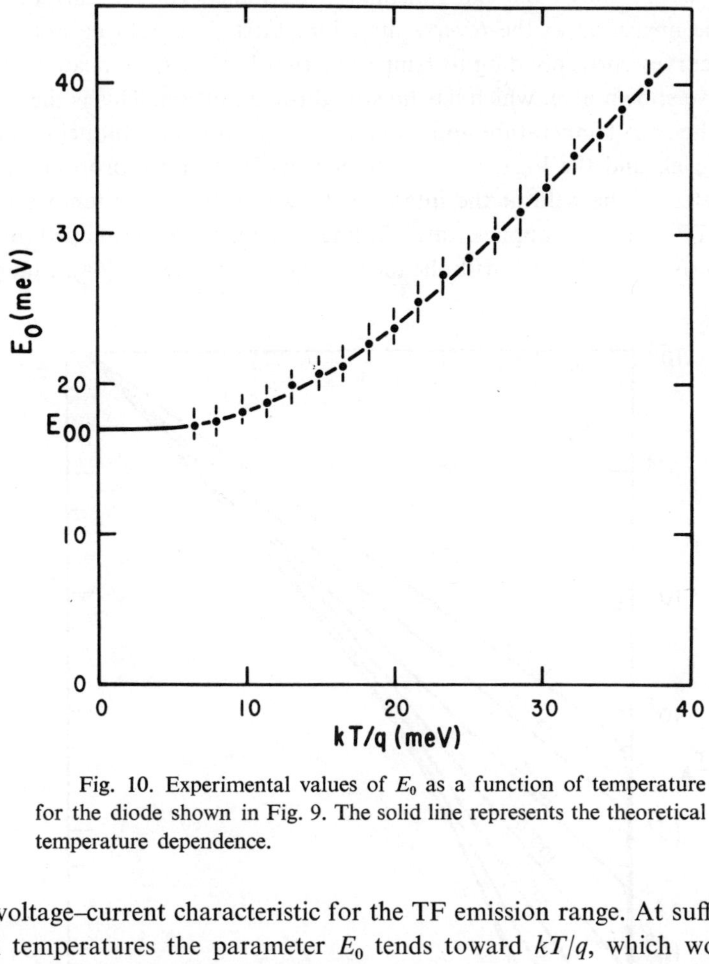

Fig. 10. Experimental values of E_0 as a function of temperature for the diode shown in Fig. 9. The solid line represents the theoretical temperature dependence.

the voltage–current characteristic for the TF emission range. At sufficiently high temperatures the parameter E_0 tends toward kT/q, which would be expected for pure thermionic emission. At sufficiently low temperatures E_0 tends to the constant value E_{00}, which corresponds to the pure field emission case. Thus Eq. (41) interpolates between the three types of emission.

In the reverse direction it can be shown [13] that the characteristic is again given by a simple expression in the TF emission range, namely,

$$j \propto \exp(-V/\mathscr{E}'), \tag{44}$$

where the parameter which measures the slope of the log j–voltage plot is given by

$$\mathscr{E}' = \frac{E_{00}}{(E_{00}/kT) - \tanh(E_{00}/kT)}. \tag{45}$$

Figure 11 shows the experimentally determined voltage–current characteristic measured in the reverse direction. First, it should be noted that all the curves corresponding to temperatures below 20°C coincide with the one curve shown here, which has no straight-line portion. This is the region where the zero-temperature approximation holds for pure tunneling at the Fermi level, and the logarithm of the current is inversely proportional to the inverse of the field at the interface. However, at higher temperatures, where TF emission applies, straight-line portions are developed in the characteristic which agree with the theoretical expression Eq. (44). Figure 12

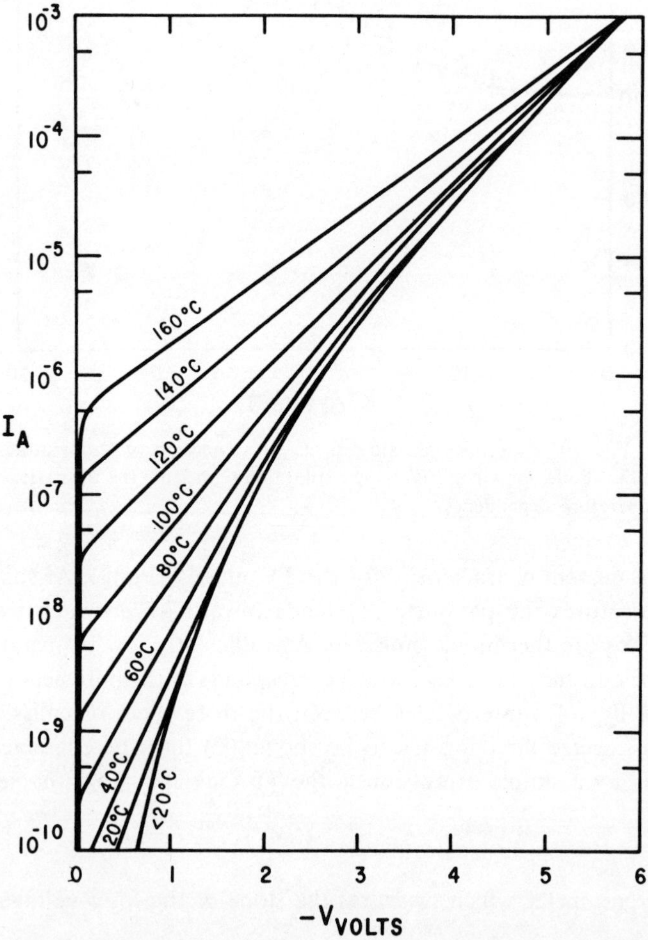

Fig. 11. Reverse I–V characteristic of a gold GaAs Schottky barrier (5×10^{17} donors/cm^3).

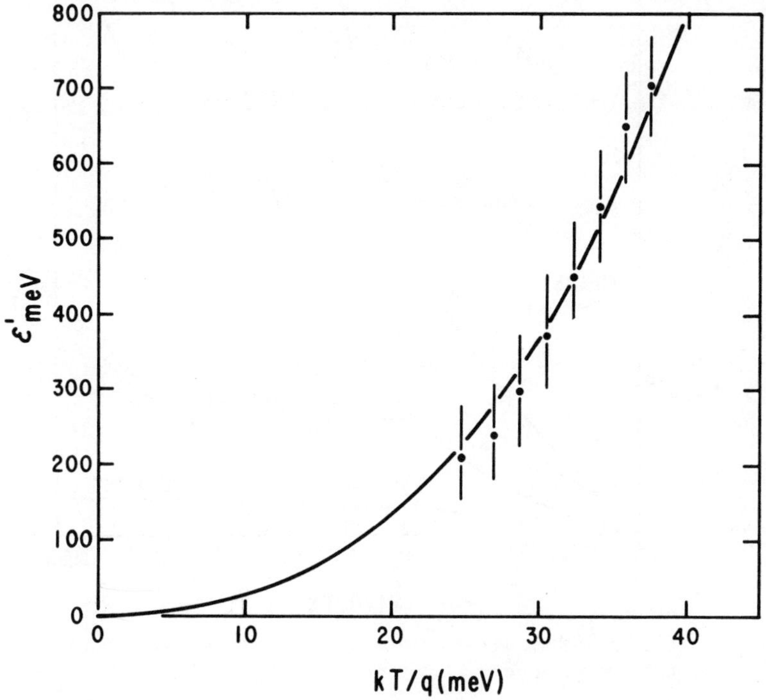

Fig. 12. Experimental values of $\mathscr{E}'$ as a function of temperature for the diode shown in Fig. 11. The solid line represents the theoretical temperature dependence.

compares the measured slopes as functions of temperature with the theoretical curve derived from Eq. (45), and again shows excellent agreement between theory and experiment.

Figure 13 is of interest in connection with the history of rectifier theories. The thermionic emission and diffusion theories predict that the direction of easy current flow for an *n*-type Schottky barrier rectifier is in the "forward" direction; that is, when electrons go from the semiconductor to the metal and the direction of high resistance is in the "reverse" direction as indicated in the figure. This is in general agreement with most Schottky barrier rectifiers that have been constructed and gives the right direction of rectification. However, one of the first theories of rectification was due to Wilson ([14]), who considered the transfer of electrons between the semiconductor and metal due to tunneling. It then turns out that at low biases, at least, the direction of rectification is reversed. Indeed, if we plot on the same figure the characteristic in the low-bias range for a heavily doped GaAs metal rectifier, we do get a reversal in the direction of rectification.

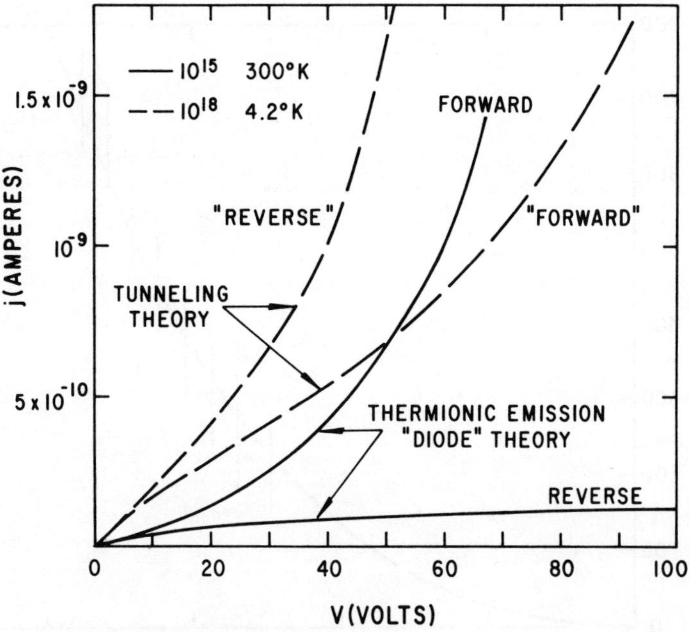

Fig. 13. Voltage–current characteristics for gold GaAs Schottky barriers illustrating reversal of rectification.

It is interesting that one of the first theories of rectification is qualitatively correct provided one uses the right experiments to verify it.

REFERENCES

1. W. W. Dolan and W. P. Dyke, Phys. Rev. 61:339 (1942).
2. W. A. Harrison, Phys. Rev. 123:67 (1962).
3. R. Stratton, Phys. Rev. 136:A837 (1964).
4. E. L. Murphy and R. H. Good, Jr., Phys. Rev. 102:1464 (1956).
5. R. Stratton, J. Phys. Chem. Solids 23:1177 (1962).
6. R. Stratton, G. Lewicke, and C. A. Mead, J. Phys. Chem. Solids 27:1599 (1966).
7. G. Lewiski and C. A. Mead, Phys. Rev. Letters 16:939 (1966).
8. F. A. Padovani and R. Stratton, Phys. Rev. Letters 16:1202 (1966).
9. E. O. Kane, J. Phys. Chem. Solids 1:249 (1957).
10. C. M. Chaves, N. Majlis, and M. Cardona, Solid State Commun. 4:631 (1966).
11. R. Stratton and F. A. Padovani, Solid-State Electron. 10:813 (1967).
12. J. W. Conley, C. B. Duke, G. D. Mahan and J. J. Tieman, Phys. Rev. 150:466 (1966).
13. F. A. Padovani and R. Stratton, Solid-State Electron. 9:695 (1966).
14. A. H. Wilson, Proc. Roy. Soc. (London) A136:487 (1932).

NOTES ADDED IN PROOF

A. Recently G. H. Parker and C. A. Mead [Phys. Rev. Letters 21:605 (1968)] used the same technique to derive the forbidden gap, i.e., energy versus wave vector relation in InAs, and find excellent agreement with the calculated band structure based on a simple two-band $\mathbf{k} \cdot \mathbf{p}$ approximation. In addition, experiments by J. W. Conley and G. D. Mahon [Phys. Rev. 161:681 (1967)] for GaAs give much better agreement with Kane's band structure than those reported in ([8]). We now believe that our analysis of the experimental data could have been in error due to our use of an incorrect value of the carrier density deduced from capacitance measurements and from the device area.

B. More recent calculations, by Robert Stratton and F. A. Padovani in Phys. Rev. (to be published), based on many-valley conduction band structure, reveal only a minimal correction to the calculated curves in Fig. 6, and still leave a discrepancy between theory and experiment.

Chapter 9

Some Properties of Exponentially Damped Wave Functions

C. A. Mead

California Institute of Technology
Pasadena, California

We know from Bloch's theorem that the wave function ψ of an electron in a semiconducting or insulating crystal can be written

$$\psi = e^{ikx}u(x)\,, \tag{1}$$

where k is the wave vector of the state and $u(x)$ is a periodic function with the lattice period of the crystal. Over reasonable distances the $u(x)$ part of the function averages out, and we can view the propagating part e^{ikx} as a quasi-particle behaving in much the same way as a particle in free space. The dynamics of this particle are given by the dispersion relation between the wave vector k and the energy E. For a semiconductor or insulator with a forbidden gap the dispersion relation is something like the one shown in Fig. 1. Note that we have plotted E versus k^2 rather than the more conventional E versus k. The reason is as follows: It can be shown rather generally that near one of the band edges (E_c or E_v) the energy is quadratic in k, and is usually written $E = \hbar^2k^2/2m^*$ by analogy with the free electron. Here m^* is the effective mass of the quasi-particle. This form is shown by the straight line in Fig. 1. For $E > E_c$ and $E < E_v$, k^2 is positive; therefore k is real, and propagating solutions result. For energies in the forbidden gap $E_c < E < E_v$, k^2 is negative, k is imaginary, and exponentially damped solutions result. It is these damped solutions with which we are concerned in tunneling problems. A typical experiment is to measure the tunneling current in a structure consisting of two metal layers separated by a thin insulating layer. The energy diagram for such a structure is shown in Fig. 2. The tunneling current through such a structure at any particular applied voltage V is proportional to the tunneling probability P, which in

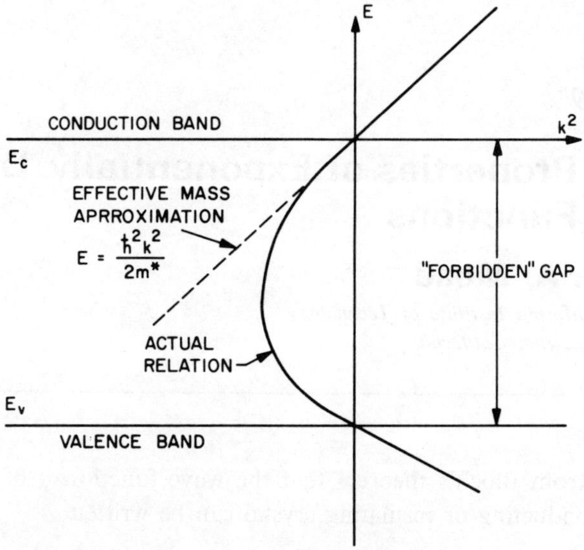

Fig. 1. Energy–momentum relationship in a semiconductor or insulator.

turn is given by a WKB type integral

$$P = \exp - 2 \int_0^t k(x)\, dx, \tag{2}$$

where t is the insulator thickness and k is here the imaginary part of the propagation vector. The dependence of k upon x arises from the dependence of k on E and the electric field in the insulator, which causes E to depend on x. For a uniform field E is proportional to x plus a constant, and the probability can be rewritten

$$\log P = \frac{-2t}{\varphi_1 - \varphi_2 + qV} \int_{E_1}^{E_2} k(E)\, dE, \tag{3}$$

where E_1 and E_2 are, respectively, the initial and final state energies of the particular electron under consideration relative to the insulator conduction band.

Since normally we are considering energies in the upper half of the forbidden gap, k generally increases as E decreases. Since k appears in the exponent of the tunneling probability, electrons with higher energies contribute much more to the tunnel current than those with lower energies. However, above the Fermi level E_{f1} we rapidly run out of electrons. Hence

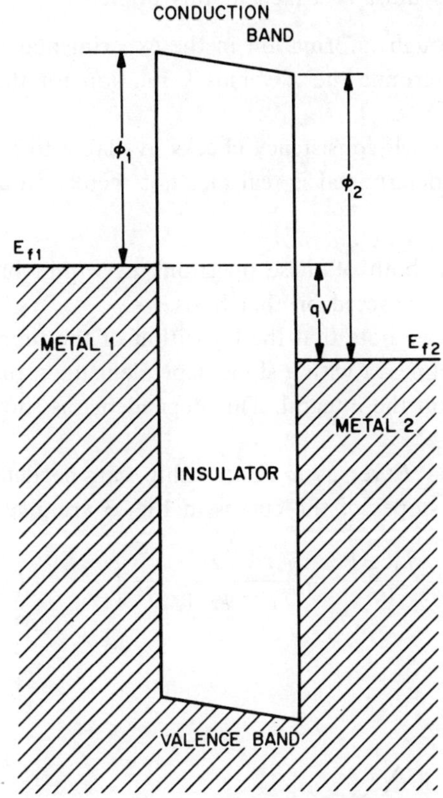

Fig. 2. Energy diagram of metal–insulator–metal tunneling structure.

for most experimental conditions the tunneling current is a very sharply peaked function at E_{f1}. Referring to Fig. 2 and defining our energy zero at the insulator conduction band edge E_c, we can write the tunneling probability

$$\log P = \frac{2t}{qV + \varphi_1 - \varphi_2} \int_{-\varphi_1}^{-\varphi_2+qV} k(E)\, dE. \tag{4}$$

At this point we can go no further without a knowledge of the dependence of k on E. It is clear from Fig. 1 that any analysis based upon the effective-mass approximation is doomed to dismal failure if the energies involved are an appreciable fraction of the forbidden gap, and even great quantities of mathematical manipulation do not serve to alleviate the problem. This point is the basis of a large portion of the diasgreement between theory and experiment in the tunneling literature.

Two questions must be asked at this point:

1. Is there enough information in the experimental tunneling data to unambiguously determine the E versus k relation for the insulator being used?

2. Are enough self-consistency checks available to be sure that the E versus k curve so determined is real and not merely an agglomerated uncertainty factor?

The answer to both of these questions is, within broad limits, affirmative, and we will proceed on that basis.

From Eq. (4) we note that the logarithm of the tunneling probability is linear in the thickness t with a slope dependent upon the applied voltage through the limit on the integral. This dependence is shown schematically in Fig. 3.

From here the E versus k relationship can be determined directly. From Eq. (4) the slopes of the curves in Fig. 3 are given by

$$\text{slope} = \frac{\partial \log P}{\partial t} = \frac{2}{\varphi_1 - \varphi_2 + qV} \int_{-\varphi_1}^{-\varphi_2+qV} k(E)\, dE; \tag{5}$$

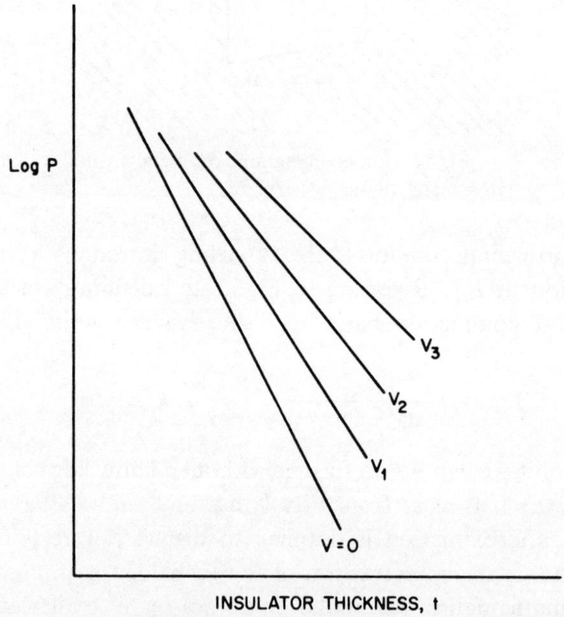

Fig. 3. Dependence of tunneling probability on insulator thickness for different applied bias V.

$k(E)$ itself may now be evaluated, as it is just the derivative of the integral with respect to its upper limit:

$$\frac{\Delta(\frac{1}{2}(\varphi_1 - \varphi_2 + qV) \times \text{slope})}{\Delta(qV)} = k(-\varphi_2 + qV). \tag{6}$$

Thus the k versus E relationship is given directly by the differences in slope for curves of different voltages. The tunneling probability is subject to direct experimental measurement. At zero voltage the small-signal tunnel resistance R of the sample and its capacitance C can be measured. It can be shown that ([1])

$$P \propto 1/RC. \tag{7}$$

At applied biases larger than typically a few tenths of a volt the tunneling probability is, to a good approximation, proportional to the current times the square of the insulator thickness ([2]). Thus curves similar to Fig. 3 are accessible by direct electrical measurement.

The relationship given by Eqs. (5) and (7) is particularly direct and is demonstrated experimentally in Fig. 4. The data are from Al–AlN–Mg structures reported by Lewicki ([3]) and Lewicki and Mead ([4]). The k versus E curve for these samples resulting from an analysis similar to that described above is shown in Fig. 5. The k versus E curve is very much similar to what we sketched in Fig. 1, and its slope changes by more than a factor of two over the energy range involved in the tunneling. Thus it is not surprising that calculations based upon an effective-mass approximation give thoroughly erroneous results.

We have now come to the point where we have (within the limits of the technique) a complete description of the system and can check the self-consistency of the results.

1. The E versus k relation should be a property of the insulating material and therefore not dependent on the electrode barrier energy φ. In Fig. 5 the results of the analysis are shown for two bias polarities where the φ's were different, and the same E versus k relation holds for both. Similar results obtain for other electrode materials where the barrier energies differ by even a larger amount. In the latter case the volt–ampere curves are very different, and would not be expected to yield the same results unless the technique were self-consistent. This is an important point and should always be checked in each experimental situation.

2. The technique yields values for the barrier energies φ_1 and φ_2. These may be checked independently by two methods: (*a*) the cusps in

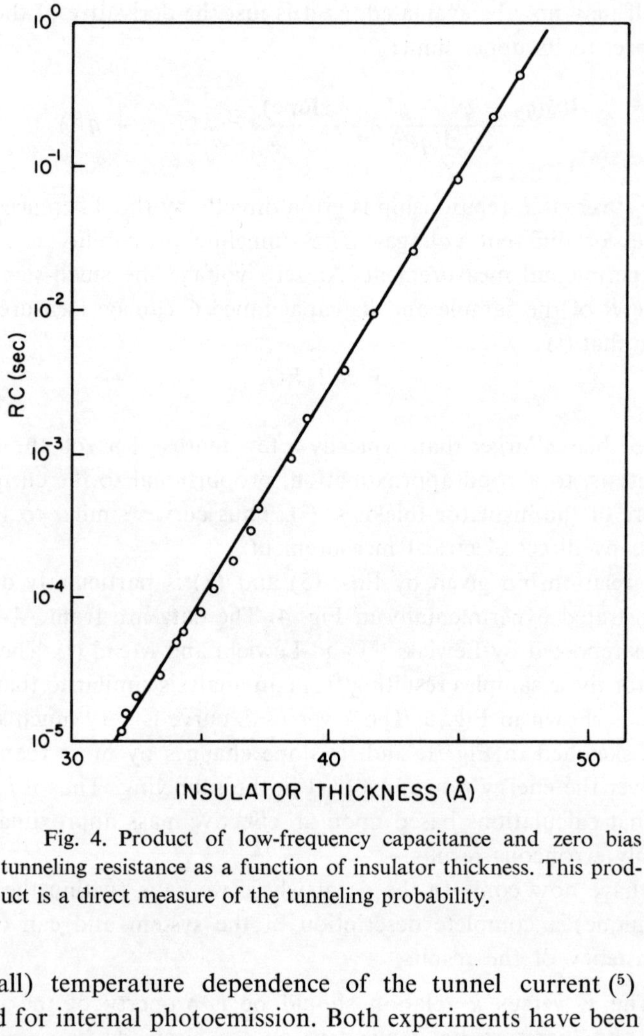

Fig. 4. Product of low-frequency capacitance and zero bias tunneling resistance as a function of insulator thickness. This product is a direct measure of the tunneling probability.

the (small) temperature dependence of the tunnel current ([5]) and (*b*) threshold for internal photoemission. Both experiments have been done on the AlN samples and the results agree with the *E*–*k* method within experimental uncertainty.

3. Using the *E*–*k* relation shown, it is possible to compute not only the form of the *I*–*V* characteristics, but the absolute magnitude as well. When this is done the *I*–*V* curves *of both polarities* agree with those observed, and the magnitude is within a factor of two of that computed from the usual theoretical models.

Thus we may conclude that within the limitation of the somewhat rough and ready method tunneling in thin films can be characterized in a

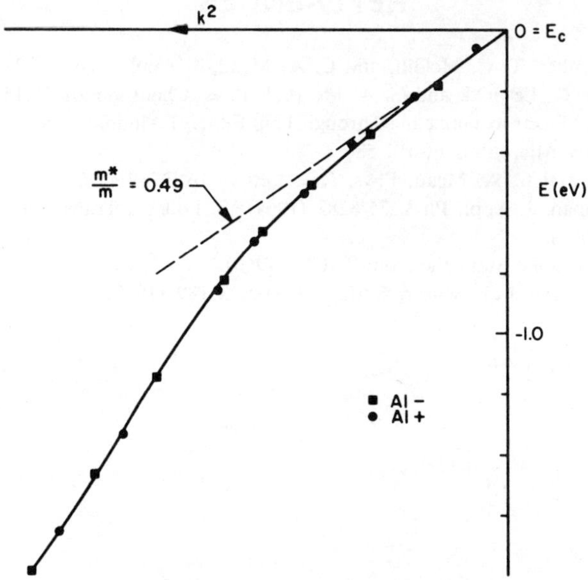

Fig. 5. The E–k relationship as derived by Lewicki on Al–AlN–Mg structures using the method outlined in the text. Note strong peviation from effective-mass approximation.

simple and demonstrably self-consistent manner. Undoubtedly, details will accumulate and refinements of experiment and theory will ensue, but it is important to have the first-order ideas correct before such additions are attempted.

Exponentially attenuated wave functions are also important in other experimental situations. In particular, these states, like any other states, are either occupied or empty, and the charge stored in them can be appreciable. As an example, on the free surface of a semiconductor there can be one such state for every surface atom, or approximately 10^{14} per cm^2. If every one were occupied with one electronic charge, a field of 10^8 V/cm would result. Turning the argument around, a field of 10^6 V/cm terminating on the semiconductor surface would induce charge in only 1% of the available states. This serves to illustrate the powerful effect these states can have on the electronic nature of surfaces. A review of the interaction of these states at metal–semiconductor interfaces has been given previously and will not be repeated here ([6]). In addition, an excellent description of this effect on the semiconductor–vacuum surface is given by Van Laar and Scheer ([7]).

REFERENCES

1. K. K. Thornber, T. C. McGill, and C. A. Mead, J. Appl. Phys. 38:2384 (1967).
2. R. Stratton, G. Lewicki, and C. A. Mead, J. Phys. Chem. Solids 27:1599 (1966).
3. G. Lewicki, "Electron Tunneling Through Thin Films of Aluminum Nitride," available as University Microfilms 66-10, 586.
4. G. Lewicki and C. A. Mead, Phys. Rev. Letters 16:D939 (1966).
5. T. E. Hartman, J. Appl. Phys. 35:3283 (1964); S. Pollack, Trans. Met. Soc. AIME 233:497 (1965).
6. C. A. Mead, Solid-State Electron. 9:1023 (1966).
7. J. van Laar, and J. J. Scheer, Surface Science 3:189 (1965).

Chapter 10

Image Force in Metal–Oxide–Metal Tunnel Junctions

J. G. Simmons

Physics Dept., University of Lancaster
Lancaster, England

IMAGE FORCE AT A SIMPLE METAL SURFACE

Since it is necessary to supply energy to a metal in order to observe electron emission from a metal surface, it is evident that the potential energy of an electron within a metal is lower than that of an electron at rest outside the metal. Other theoretical arguments support this conclusion ([1]). It follows then that the electron must undergo a change in potential as it crosses the metal–vacuum interface. Figure 1 is the energy diagram of the model envisaged by Sommerfeld ([2]). The metal is represented by a potential energy well of depth V_0; the vacuum level represents the energy of an electron at rest within the metal. At 0°K the electrons, which exist in pairs in discrete energy levels, fill up the well to an energy η, the Fermi level. The distance $\psi = V_0 - \eta$ from the Fermi level to the vacuum level is known as the work function of the metal, and represents the mini-

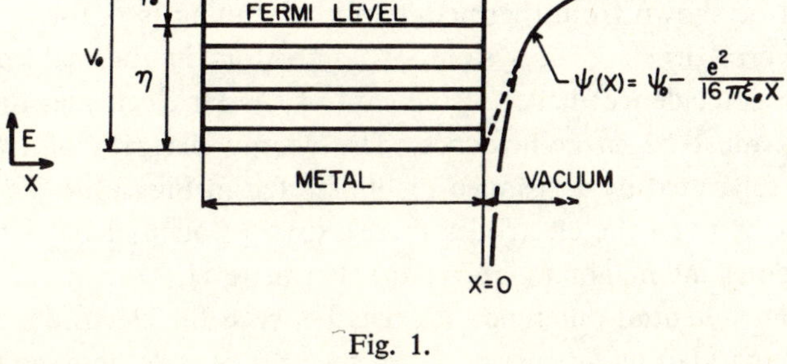

Fig. 1.

mum energy required to free electrons from the interior of the metal at 0°K.

The abrupt change in the potential barrier at the metal–vacuum interface is not only an ideal model, but is also physically unrealizable, since abrupt changes in potential imply infinite fields. In actual fact the barrier changes smoothly as a result of the image force. To understand the meaning of image force, consider an electron approaching a flat metal surface. The electron polarizes the metal surface, which in turn exerts an attractive force P_i (the image force) on the electron of magnitude $(1/4\pi\varepsilon_0)(e^2/4x^2)$. The potential V_i of the electron, measured with respect to the vacuum level, when it is a distance x from the metal surface is

$$V_i = \int_\infty^x P_i \, dx$$

$$V_i = -(1/4\pi\varepsilon_0)(e^2/4x).$$

Thus the height of the potential step for $x > 0$ is no longer the constant value ψ_0, but instead is a function of the distance x from the metal surface:

$$\psi(x) = \psi_0 + V_i(x). \tag{1}$$

However, Eq. (1) implies that the potential energy of the electron at the metal surface is equal to $-\infty$. Schottky ([3]) circumvents this anomaly by assuming that the image potential holds only for values of x greater than some critical value x_0, approximately 1 Å; thereafter he assumes it decreases linearly until at $x = 0$ it coincides with the energy of the bottom of the potential well.

POTENTIAL BARRIER BETWEEN SIMILAR ELECTRODE

Without Image Forces

It can be shown from thermodynamic reasoning [see, for example ([4])] that the Fermi levels of a system of conductors in thermal equilibrium must be in coincidence; hence in the case of similar electrodes the vacuum levels also must be in coincidence. The energy diagram of two similar electrodes separated by a distance s is illustrated in Fig. 2 (*a*). The diagram shows that between the electrodes there exists a potential barrier of width s and of constant height ψ_0 above the Fermi level.

When a potential difference V exists between the electrodes the Fermi levels are separated by an energy eV on the energy scale, moving the Fermi

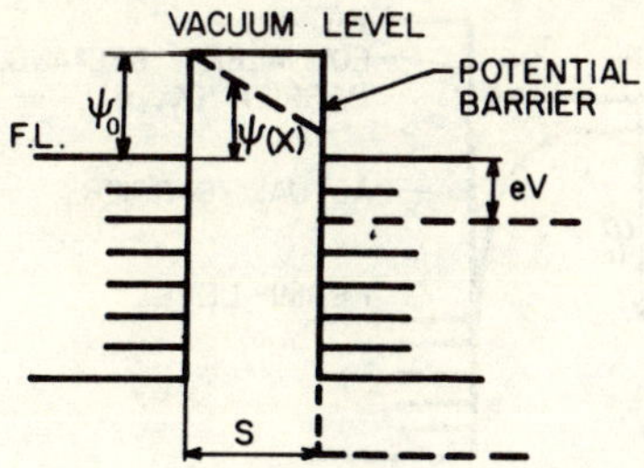

(a)

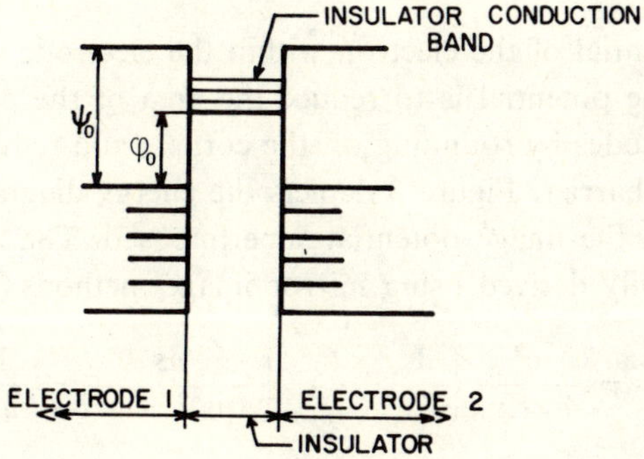

(b)

Fig. 2.

level of the negative-biased electrode vertically upward. The height of the barrier $\psi(x)$ above the Fermi level that exists between the electrodes is now trapezoidal in shape; i.e., it is a linear function of the distance within the barrier:

$$\psi(x) = \psi_0 - (eVx/s). \tag{2}$$

When two metallic electrodes are separated by an insulating film of thickness s the equilibrium conditions require that the bottom of the conduction band of the insulator be positioned above the Fermi level and below the vacuum level of the electrodes. Thus the action of the insulator is to lower the potential barrier existing between the two electrodes by an amount $\psi_0 - \varphi_0$, where φ_0 is the distance between the bottom of the insulator conduction band and the Fermi level of the metal electrodes [Fig. 2 (*b*)].

With Image Forces—Zero Potential

When an electron exists between two parallel and closely spaced electrodes it polarizes both of the electrodes. As a result, both of the electrodes

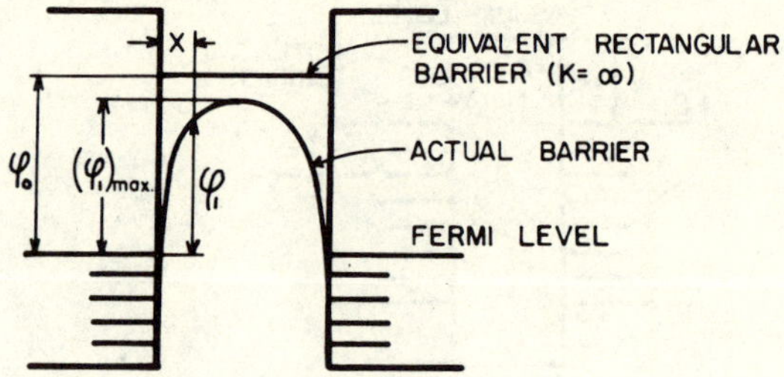

Fig. 3.

influence the potential of the electron within the electrode separation. The effect of the image potential is to reduce the area of the potential barrier between the electrodes by rounding off the corners and reducing the height and width of the barrier. Figure 3 depicts the energy diagram of a rectangular barrier with the image potential superimposed. The image potential $V_i(x)$ can be readily derived using mirror image methods ([5]), and is

$$V_i(x) = -\frac{e^2}{8\pi K\varepsilon_0}\left\{\frac{1}{2x} + \sum_{n=1}^{n-\infty}\left[\frac{ns}{(ns)^2 - x^2} - \frac{1}{ns}\right]\right\}, \tag{3}$$

where e is the unit of electronic charge, x is distance in the insulator measured from the electrode 1 – insulator interface, ε_0 in the permittivity of free space, and K is the dielectric constant of the insulating film.

When the image potential is superposed on the rectangular barrier the modified barrier height φ_i is a function of distance in the insulator:

$$\begin{aligned}\varphi_i &= \varphi_0 + V_i(x)\\ &= \varphi_0 - \frac{e^2}{8\pi K\varepsilon_0}\left\{\frac{1}{2x} + \sum_{n=1}^{n=\infty}\left[\frac{ns}{(ns)^2 - x^2} - \frac{1}{ns}\right]\right\}.\end{aligned} \tag{4}$$

Equation (4) written in the form shown is interpreted as follows: if we let $s \to \infty$, Eq. (4) reduces to Eq. (1); thus the summation term in Eq. (4) is simply the contribution of the second electrode to the image potential.

A normalized energy diagram ([6]) of a rectangular barrier of height φ_0 and width s, with the image potential included, is shown by the dotted line in Fig. 4. In this diagram s is expressed in angstroms and φ_0 in electron volts. For example, if it is assumed that $s = 20$ Å and $K = 5$, then the Fermi level of a $\frac{1}{2}$-eV barrier will be located

$$50/Ks = 50/5\times 20 = \tfrac{1}{2}\text{ eV}$$

below the top of the barrier. Thus in Fig. 4 abcd represents a rectangular

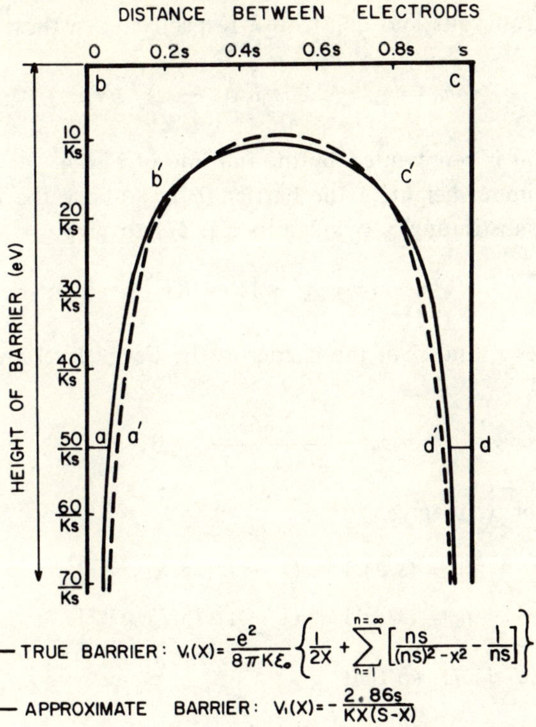

Fig. 4.

barrier wherein $\varphi_0 = \frac{1}{2}$ eV and $s = 20$ Å; the size of the true barrier (with image forces included and $K = 5$) is illustrated by the dotted curve a′b′c′d′ inside this rectangle.

Image Force Approximation

The unwieldy expression for the image force, Eq. (4), can be expressed in a simpler and more convenient approximate form as follows ([6]):

$$V_i = -\frac{0.795se^2}{16\pi K\varepsilon_0 x(s-x)}, \tag{5}$$

thus Eq. (5) reduces to

$$\varphi_i = \varphi_0 - \frac{0.795se^2}{16\pi K\varepsilon_0 x(s-x)}. \tag{6}$$

Henceforth it will be more convenient to use as the units of energy and measure electron volts and angstroms, respectively; in these units Eq. (6) becomes

$$\varphi_i = \varphi_0 - 2.86s/Kx(s-x) \quad \text{eV}. \tag{7}$$

This expression is represented by the full line in Fig. 4.

The maximum height of the barrier $(\varphi_i)_{\max}$ above the Fermi level is obtained by substituting $x = s/2$ into Eq. (7) to give

$$(\varphi_i)_{\max} - 11.44/Ks.$$

The extremities s_1 and s_2 of the barrier at the Fermi level are obtained by setting $\varphi_i = 0$,

$$\varphi_0 - \frac{2.86s}{Kx(s-x)} = 0, \tag{8}$$

and solving for x, to give

$$s_1 = (s/2)[1 - (1 - 11.75/K\varphi_0 s)^{1/2}],$$
$$s_2 = (s/2)[1 + (1 - 11.75/K\varphi_0 s)^{1/2}].$$

Usually, $s\varphi_0 K \gg 11.75$, so that

$$s_1 \approx 3/K\varphi_0, \tag{9}$$
$$s_2 \approx s - (3/K\varphi_0).$$

Thus the height and width of a rectangular barrier are reduced, respectively, by an amount $11.44/Ks$ and $6/K\varphi_0$ due to image forces.

Figure 5 illustrates the manner in which a rectangular barrier of height 2 eV and width 20 Å is modified by the image forces and the dielectric constant of the insulating film. It will be observed that the greater the value of the dielectric constant, the greater is the area of the potential barrier; at $K = \infty$ the true barrier coincides with the ractangular barrier, since from Eq. (8) $\varphi = \varphi_0$, and from Eq. (9) $s_1 = 0$ and $s_2 = s$.

Similar Electrodes at Different Potentials

The foregoing discussion can be readily extended to similar electrodes held at a potential difference V. To do this, it is necessary to superpose the effect of the applied field, $F = V/s$, upon the potential barrier:

$$\varphi_i = \varphi_0 - [2.86s/Kx(s-x)] - sFx, \tag{10}$$

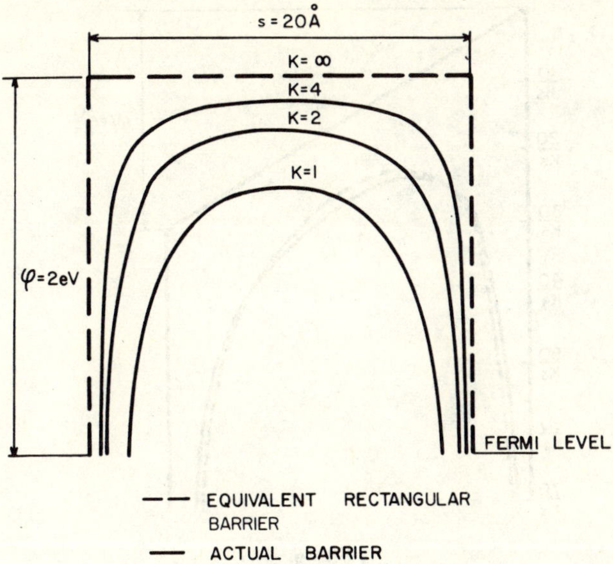

Fig. 5.

where V is the applied voltage. The effect of the applied field is to depress and move the barrier maximum closer to the negatively biased electrode and to reduce the width of the barrier, as shown by the normalized energy diagram in Fig. 6. This diagram illustrates the case for a voltage bias of magnitude $30/Ks$. The extremities s_1 and s_2 of the barriers at an energy corresponding to the Fermi level of the negatively-biased electrode, which are required when we come to study the effect of image forces on metal–insulator–metal tunneling, are obtained by setting φ_i in Eq. (10) equal to zero and solving for x. This procedure, however, results in a cubic equation for x, and s_1 and s_2 cannot be obtained in a convenient tractable form. It has been shown ([6]), however, that s_1 and s_2 can be written to a good approximation as:

For $V \lesssim \varphi_0$

$$s_1 = 3/K\varphi_0,$$
$$s_2 = s\{1 - (23/Ks)[3\varphi_0 + (10/Ks) - 2V]\} + s_1. \tag{11}$$

For $V > \varphi_0$

$$s_1 = 3/K\varphi_0,$$
$$s_2 = [\varphi_0 - (14/Kx)](s/V). \tag{12}$$

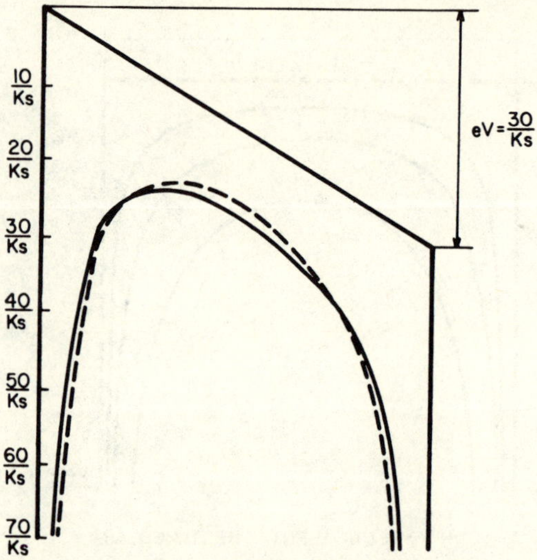

Fig. 6.

POTENTIAL BARRIERS BETWEEN DISSIMILAR ELECTRODES

Zero Bias

It is well known that a contact potential exists between two closely spaced dissimilar metal electrodes. As a result of this contact potential, extremely high intrinsic fields exist in asymmetric tunnel junctions (7), in which the electrode separation is normally less than 100 Å. The magnitude of the intrinsic field F_{in} (V/Å) is

$$F_{\text{in}} = (\psi_2 - \psi_1)/s\,, \tag{13}$$

where s is the thickness of the insulating film in angstroms and ψ_1 and ψ_2 are the work functions of the electrodes (in eV). Figure 7 is the energy diagram for two dissimilar electrodes separated by an insulating film. The height of the barrier between the electrodes is no longer constant, as shown in Fig. 2 (*b*), but is now trapezoidal. As a result, the barrier heights at the metal–insulator interfaces are no longer equal, but are related as follows:

$$\begin{aligned} \varphi_2 &= \varphi_1 + F_{\text{in}}s \\ &= \varphi_1 + \psi_2 - \psi_2 \\ &= \varphi_1 + \Delta\psi\,, \end{aligned} \tag{14}$$

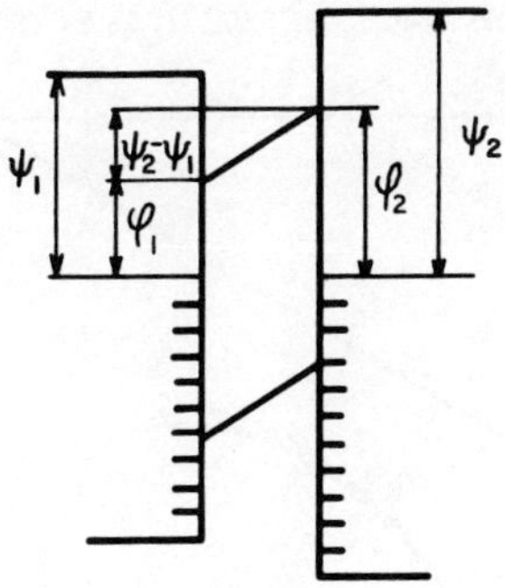

Fig. 7.

where $\Delta\psi = \psi_2 - \psi_1$. The height of the barrier $\varphi(x)$ between the electrodes is readily deduced as

$$\varphi(x) = \varphi_1 + (\Delta\psi x/s). \tag{15}$$

Superposing the image potential on the trapezoidal barrier gives the true barrier φ_i as a function of distance within the electrodes:

$$\varphi_i = \varphi_1 + \Delta\psi(x/s) - [2.86s/Kx(s-x)] \text{ eV}. \tag{16}$$

A normalized energy diagram of an asymmetric barrier ([8]), in which $\Delta\varphi = 30/Ks$, is shown in Fig. 8. For a barrier of $s = 15$ Å and $K = 4$ the contour abcd of Fig. 8 represents an ideal barrier of $\varphi_1 = 30/Ks = 30/60 = \frac{1}{2}$ eV, and $\varphi_2 = 60/60 = 1$ eV; the dotted curve a′b′c′d′ represents the practical barrier.

With Bias

When a voltage bias is applied to an asymmetric junction the field in the insulator is augmented by the intrinsic field when the electrode of greater work function is negatively biased, and decreased when the electrode of lower work function is negatively biased. Thus the relevant equations for the potential barrier when the electrode of lower work function is:

Negatively biased,

$$\varphi_i = \varphi_1 - (V - \Delta\psi)(x/s) - [2.86s/Kx(s-x)].$$

Positively biased,

$$\varphi_i = \varphi_1 + (V + \Delta\psi)(x/s) - [2.86s/Kx(s-x)].$$

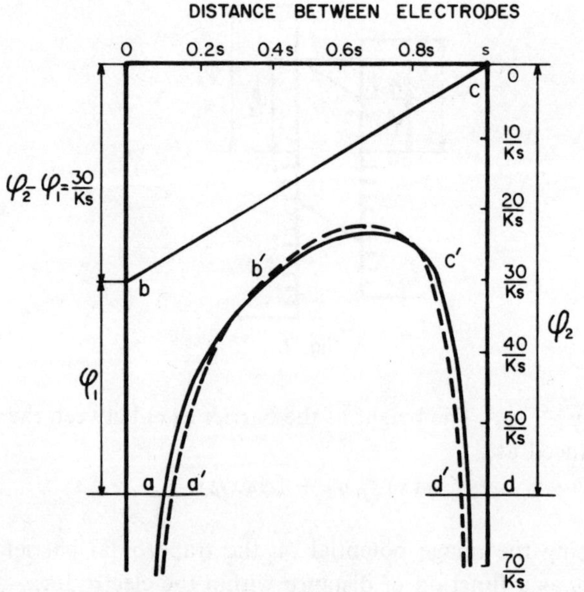

Fig. 8.

In the above equations the distance x is measured from the surface of the electrode of lower work function.

TUNNEL I-V CHARACTERISTICS

Generalized Tunneling Formula

The tunnel current flowing between electrodes separated by a thin insulating film is of the form (see Chapter 4 is this volume)

$$J = \frac{2e}{h} \int dE\, [f_R(E) - f_L(E + eV)] \int \frac{d^2K_{\parallel}}{(2\pi)^2} D(E, K_{\parallel}). \tag{17}$$

The integral in Eq. (17) can normally be solved only for the very simplest form of potential barrier. Thus in order to incorporate image force effects into the theory, Sommerfeld and Bethe and subsequent workers found it necessary to approximate the image potential by a parabolic function. This method has been shown to be unsatisfactory, particularly for other than extremely small voltage biases.

We now introduce the generalized formula for the tunnel effect between metal–insulator–metal tunnel junctions. This formulation is readily applicable to potential barriers of *arbitrary* shape and to *all* voltage ranges. The generalized formula gives the relationship connecting the tunnel current density with the applied voltage for a barrier of arbitrary shape as

$$j = j_0\{\bar{\varphi}\exp(-A\bar{\varphi}^{1/2}) - (\bar{\varphi} + eV)\exp[-A(\bar{\varphi} + eV)^{1/2}]\}, \tag{18}$$

where $j_0 = e/2\pi h(\beta\,\Delta s)^2$, $A = 4\pi\beta\,\Delta s(2m)^{1/2}/h$, $\Delta s = s_2 - s_1$ is the width of the barrier at the Fermi level of the negatively biased electrode, $\bar{\varphi}$ is the mean barrier height above the Fermi level of the negatively biased electrode, and β is a function of barrier shape and is usually approximately equal to unity, a condition we will assume throughout. When $\bar{\varphi}$ and Δs are expressed, respectively, in electron volts and angstroms, j in amps/cm^2, and m is taken to be the free electron mass, Eq. (18) becomes

$$j = \frac{6.2 \times 10^{10}}{(\Delta s)^2}\{\bar{\varphi}\exp(-1.025\,\Delta s\,\bar{\varphi}^{1/2}) - (\bar{\varphi} + V) \times \exp[-1.025\,\Delta s(\bar{\varphi} + V)^{1/2}]\}. \tag{19}$$

Determination of the Junction Parameters

Clearly, if $\bar{\varphi}$ and Δs can be determined for all V for the barrier under consideration, the I–V characteristics of the junction can be obtained from Eq. (19), and this will be our next objective. For conciseness the following discussion will be confined to the symmetrical tunnel junction, and the reader is referred to the literature for the asymmetrical junction results ([8]).

From the definition of $\bar{\varphi}$ it follows that

$$\bar{\varphi} = (1/\Delta s)\int_{s_1}^{s_2}\varphi_i\,dX. \tag{20}$$

Substituting Eq. (10) into Eq. (20), we get

$$\bar{\varphi} = \frac{1}{\Delta s}\int_{s_1}^{s_2}\left[\varphi_0 - \frac{Vx}{s} - \frac{2.86s}{Kx(s-x)}\right]dx,$$

which on integration yields

$$\bar{\varphi} = \varphi_0 - \frac{V}{2s}(s_1 + s_2) - \frac{2.86}{K(s_2 - s_1)}\ln\frac{s_2(s - s_1)}{s_1(s - s_2)}, \tag{21}$$

where s_1 and s_2 are given by Eqs. (11) and (12).

Substituting Eq. (21) into Eq. (19) gives the I–V characteristic for a tunnel junction with image force effects included. The effect of image

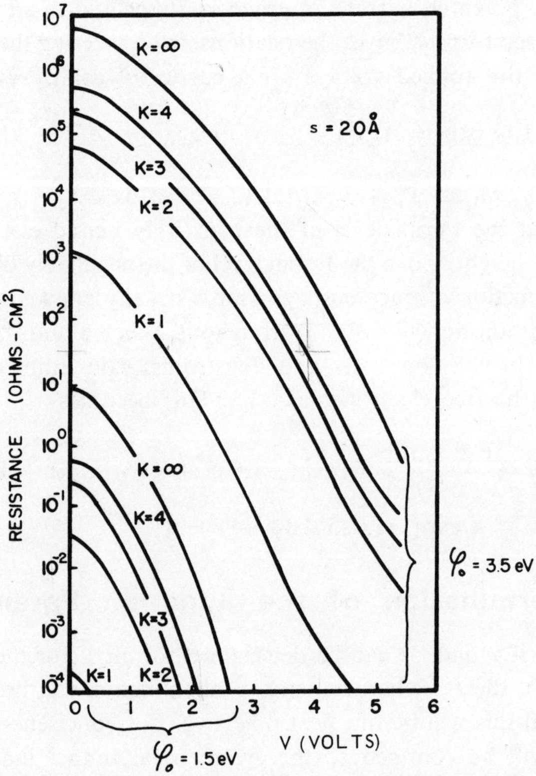

Fig. 9.

force on a rectangular barrier for various values of K, φ_0, and V is shown in Fig. 9.

Effect of Image Force on Low-Voltage (Ohmic) Characteristic

It is not clear from the complicated expression of Eq. (21) how the effect of the image force on the I–V characteristic varies with the parameters K, φ_0, and s. If, however, we confine our attention to very low voltages, where the tunnel resistance is ohmic, Eq. (21) reduces to a much simpler form and the variation of image force effects with the parameters K, φ_0, and s becomes much clearer. At low voltages Eq. (21) reduces to

$$j = \frac{3.16 \times 10^{10}}{\Delta s} \bar{\varphi}^{1/2} V \exp(-1.025\, \Delta s \bar{\varphi}^{1/2}), \tag{22}$$

which shows that the junction is ohmic. We need only consider the effect on the *I–V* characteristics of changes in Δs and $\bar{\varphi}^{1/2}$ due to the image forces, where these parameters appear in the exponential factor; similar changes in these parameters in the preexponential factor play relatively unimportant roles in modifying the *I–V* characteristic.

Without image forces then

$$\Delta s\bar{\varphi}^{1/2} = s\varphi_0^{1/2}$$

at low voltages. With image forces present we have from Eqs. (21) and (11)

$$\begin{aligned}\Delta s\,\bar{\varphi}^{1/2} &= \left[s - \frac{23}{K[3\varphi_0 + (10K/s)]}\right]\left[\varphi_0 - \frac{2.86}{K(s_2 - s_1)}\ln\left(\frac{s_2}{s_1}\right)^2\right]^{1/2} \\ &\simeq s\varphi_0^{1/2} - \frac{1}{K\varphi_0^{1/2}}\left[\frac{23}{[3 + (10/K\varphi_0 s)]} + 6.58\log_{10}\left(\frac{s_2}{s_1}\right)\right]. \qquad (23)\end{aligned}$$

Thus the image force reduces the low-voltage (ohmic) junction *resistance* by the multiplying factor

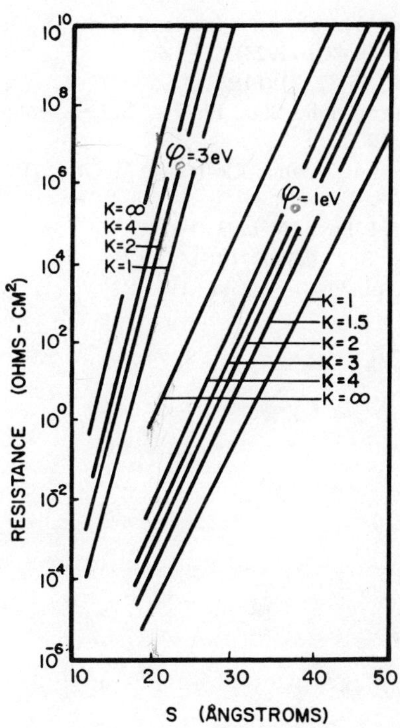

Fig. 10.

$$\exp\left(-\frac{1}{K\varphi_0^{1/2}}\left[\frac{23}{[3+(10/K\varphi_0 s)]}+6.58\log_{10}\left(\frac{s_2}{s_1}\right)\right]\right). \tag{24}$$

Equation (24) is clearly influenced most strongly by the dielectric constant and to a lesser extent by the barrier height; the thickness of the dielectric, which is implicit in s_2, appears in the logarithmic function and, as such, only slightly influences Eq. (24). Increasing the values of K and φ_0 reduces the effect of image forces on the junction resistance, while increasing s increases the effect. These conclusions are quantitatively illustrated in Fig. 10, in which the low-voltage junction resistance is plotted as a function of K, φ_0 and s. It is worth noting here that for a barrier of $\varphi_0 = 1$ eV and $K = 1$ the junction resistance with image forces included is over *five* orders of magnitude lower than for the ideal ($K = \infty$) barrier; for $K = 4$ it is approximately two orders of magnitude lower; the high-frequency dielectric constant of most insulators usually lies within these two values of K.

REFERENCES

1. H. Frolich, Elektronen Theorie der Metalle, J. Springer, Berlin, 1936, Chapter 11.
2. A. Sommerfeld, Z. Physik 47:1 (1928).
3. W. Schottky, Physik Z. 15:872 (1914).
4. C. Kittel, Introduction to Solid State Physics, 2nd ed., John Wiley and Sons, New York, 1956, Chapter 10.
5. W. R. Smythe, Static and Dynamic Electricity, McGraw-Hill Book Co., New York, 1950, Chapter IV.
6. J. G. Simmons, J. Appl. Phys. 34:1793 (1963).
7. J. G. Simmons, Phys. Rev. Letters 10:10 (1963).
8. J. G. Simmons, J. Appl. Phys. 34:2581 (1963).

Chapter 11

Phonon-Assisted Semiconductor Tunneling

R. A. Logan

Bell Telephone Laboratories, Inc.
Murray Hill, New Jersey

Studies of tunneling currents in semiconductor junctions have been a useful method of measuring phonon energies in many semiconductors. A review will be given of (1) junction fabrication and measurement of impurity concentrations in tunnel junctions, (2) the current–voltage characteristic and the parameters which control the tunnel current, (3) the techniques of measuring phonon energies in various semiconductors, and (4) the effects of phonons upon the tunnel current in indirect energy gap semiconductors. In addition, the anomalous changes observed in the diode conductance near zero bias will be described.

When the impurity doping level in a semiconductor changes abruptly from p to n type the width W of the p–n junction space charge region is given by the relation ([1])

$$W^2 = \frac{KA^2}{2\pi q}\left(\frac{1}{N_A} + \frac{1}{N_D}\right)(V_a + V_i),$$

where K is the dielectric constant, q is the electronic charge, V_a is the applied voltage, V_i is the built-in voltage, and N_A and N_D are the acceptor and donor concentrations on the p and n sides of the junction, respectively. If the current flow across the junction is to be dominated by tunneling processes, then the junction width W must be $\lesssim 100$ Å, so that N_A and N_D must be increased to values where both sides of the junction are degenerately doped. In practice this is easily achieved in junction fabrication by alloying, where the dopant metal is alloyed to a degenerate semiconductor using a brief heating cycle. The short heating cycle is necessary to ensure that impurity diffusion is negligible. The characteristic impurity decrement ($\sqrt{Dt} \approx 1000$ Å, where D is the diffusion constant and t the diffusion time) prevents the formation of abrupt junctions by substitutional diffusion mechanisms. With the exception of interstitial Zn diffusion in InSb [2] and GaSb [3] tunnel junctions cannot in general be made by diffusion processes.

The alloying process is in general quite simple. To make a tunnel junction in silicon, for example, one selects wafer of degenerately doped *n*-type silicon and, by heating to $\sim 700°C$ on a strip heater in an inert ambient, alloys a gold wire to it to serve as an ohmic contact. The *p*-type region can be formed by reheating the sample to $\sim 725°C$ and alloying an aluminum wire to the wafer. The solution of aluminum and silicon that is formed at the alloying temperature regrows *p*-type silicon onto the *n*-type wafer during the cooling cycle. For such two component systems the solid solubility of aluminum in the regrowth can be estimated from solubility data [4]; thus the *p*-type regrowth would contain $\sim 9 \times 10^{18}$ Al atoms/cm^3. Higher *p*-type doping can be achieved by using boron as the acceptor. In this case wire is made from Al + 1% B and is alloyed to form the *p*-type contact. It is generally not possible to estimate the doping level in such three-component systems from published data. A direct measurement is possible by alloying a length of wire onto the semiconductor using the same heating procedure. When the excess aluminum is dissolved in caustic solution the resistivity 0.005 ohm-cm of the regrowth can be directly measured, and corresponds to an acceptor concentration of 2.3×10^{19} cm^{-3}. The correlation of resistivity with impurity concentration can be established by measurement of Hall effect. The results of these measurements [5] on *n*-type silicon doped, respectively, with As, P, and Sb exhibited an anomaly which has a counterpart in tunneling studies. For a given carrier density the resistivity was lowest in Sb-doped crystals and highest in As-doped crystals. Measurement of carrier mobility in these crystals shows that this was due to the result that at a given impurity level $\mu_{Sb} > \mu_P > \mu_{As}$, where μ_{Sb} is the mobility in Sb-doped crystals, etc. These mobility differences are not predicted by the theory of impurity scattering, where only the density of impurities determines the mobility. Since the mobility increases in the same order as the impurity ionization energy decreases, the results suggest that the scattered electron interacts at close range with the scatterer and that Sb has the smallest central cell potential for such interaction. As discussed below, tunneling via impurity states gives rise to an excess current, and this current is lowest in Sb-doped diodes, in accord with the mobility results. It should also be noted that the melting point of the compound of group III acceptor impurities such as Al and B with Sb is much lower than those with As or P, so that Sb + acceptor compounds would be less stable than those with As or P. Thus diodes with Sb doping would also contain relatively fewer donor–acceptor pairs in the junction, and these

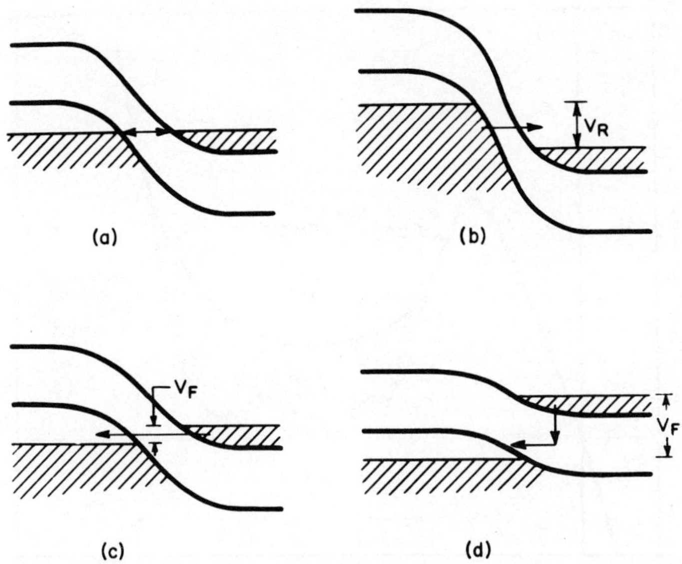

Fig. 1. The potential energy diagram of a tunnel junction, (*a*) at zero bias, (*b*) at reverse bias V_R, (*c*) at small forward bias V_F, and (*d*) at large forward bias V_F. In each case the arrow indicates the tunneling path.

may also account for the relatively low excess current in Sb-doped diodes.

The potential energy diagram of a tunnel junction at zero bias is shown in Fig. 1 (*a*). The dominant term in the expression for tunnel current is the tunneling probability, given by ([6])

$$I \approx \exp[-\lambda m^{*1/2} E_B{}^{3/2}/F], \tag{2}$$

where λ is a constant, m^* is the reduced carrier mass, E_B is the barrier height, and F is the electric field. From Fig. 1 (*a*) it is evident that

$$E_B = E_G + E_N + E_P, \tag{3}$$

where E_G is the energy gap and E_N and E_P are the energy differences between the Fermi level and the conduction and valence bands, respectively. In reverse bias, Fig. 1 (*b*), valence electrons tunnel to the empty states in the conduction band. At low forward biases, Fig. 1 (*c*), electrons in the conduction band tunnel into empty states in the valence band, and this process would cease at a bias where the bottom of the conduction band is opposite to the top of the valence band. At higher forward biases one would expect the dominant current flow to be injection current, where electrons and holes

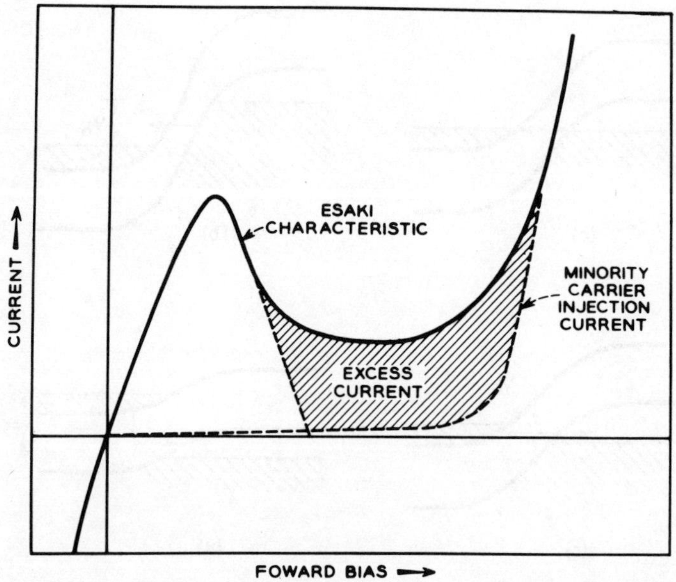

Fig. 2. A typical forward current–voltage characteristic of an Esaki diode.

would be excited over the potential barrier at the junction. However, the actual current observed is much larger than the injection current and has been shown (7) to be tunneling via states in the forbidden gap, as indicated schematically in Fig. 1 (*d*). Figure 2 shows a typical current–voltage characteristic and indicates the three current components discussed above, the band-to-band tunneling (or Esaki characteristic), the excess current, and the minority carrier injection current.

Experimental studies (8) of junctions made in Ge, Si, and InSb have confirmed that the tunneling probability given in Eq. (2) does account quantitatively for the variation of the tunneling current with bias, with impurity doping level (through change in the junction width and the field, F), and with temperature (through change in the energy gap). Energy gap E_G may also be varied with pressure and effectively varied by application of a magnetic field through the creation of magnetic sub-bands of both the conduction and valence band.

Using the samples of uniformly doped n-type silicon on which Hall effect data had been obtained a series of diodes was made using identical procedures to alloy Al + 1% B wire to the silicon chips obtained from the Hall effect samples. The logarithm of the current density at a forward bias of 0.025 V is plotted in Fig. 3 against $[(1/n) + (1/p)]^{1/2}$, the latter being

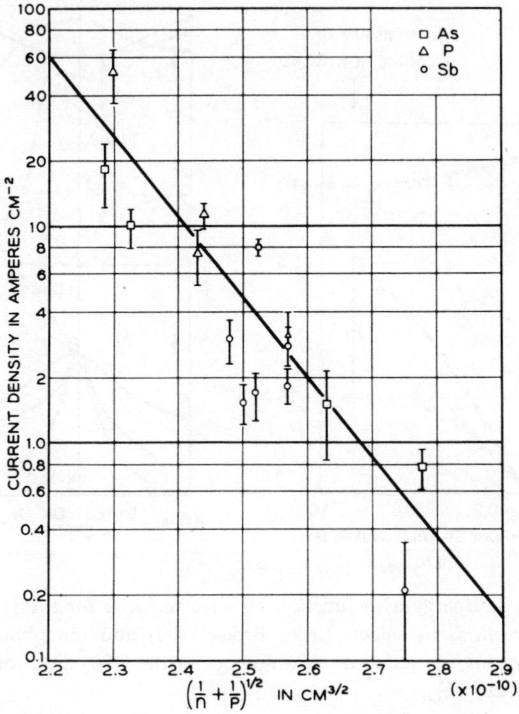

Fig. 3. The logarithm of the current density plotted against $[(1/n) + (1/p)]^{\frac{1}{2}}$ for a series of Si diodes. From Logan *et al.* ([5]).

proportional to the junction width, as shown in Eq. (1). The mean deviation of the current densities at each doping level is also indicated. While the current densities are independent of the specific n-type dopant used, it is apparent that the deviations encountered at each doping level are quite large and indicate inhomogeneities in the Al + 1% B wire or in the doping level of the crystals. If one assumes that the latter is responsible for the scatter in current density, then using the average of the data of Fig. 3 as a reference, the observed current density at 0.025 V may be used as a useful estimate of n in the diodes made in this way.

The current–voltage characteristics are measured with the conventional four-terminal connections (two current leads and two potential leads) to the diodes. The first derivative is obtained by applying a 500-cps signal through a high impedance to the junction and measuring the in-phase voltage developed with a lock-in amplifier. The output of the lock-in amplifier then gives dV/dI directly. The second derivate, d^2I/dV^2, is obtained

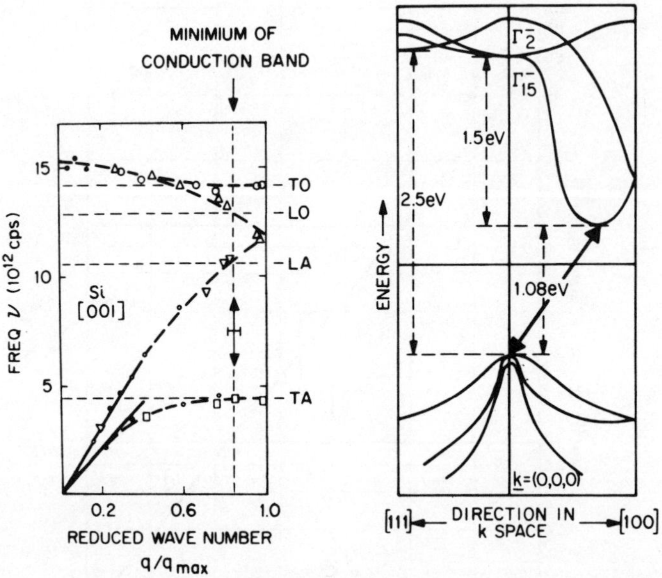

Fig. 4. Energy as a function of wave vector k for the [111] and [100] directions in silicon [from Brooks ([16])] and the phonon frequency versus the reduced wave number in the [100] direction [from Broockhouse ([17])].

by measuring the second harmonic of the applied signal using a circuit developed by Thomas and Rowell ([9]).

The role played by phonons in tunneling in semiconductor junctions is summarized in Fig. 4, which shows the conductance and valence bands in silicon as a function of wave vector in k space in the [100] and [111] directions. It is evident that the minimum energy gap is 1.08 eV and occurs at 0.85 of the edge of the Brillouin zone. The direct energy gap at $k = 0$ is 1.5 eV higher in energy. It is evident that tunneling will occur across the indirect energy gap, and since tunneling transitions must occur between states of equal energy and momentum, the electron momentum must change during the tunneling process. This may occur by emission or absorption of a phonon. The phonon energies as a function of position in the Brillouin zone are also shown in Fig. 4, where the conduction band minimum at $q = 0.85_{\text{Max}}$ is indicated. The two lower branches are acoustic phonons and the two higher branches are optical phonons. The values of the fundamental phonon energies in silicon that would conserve momentum in a tunneling transition are the phonon energies at the indirect band edge, and are the transverse acoustic (TA) at 17.9 mV, the longitudinal acoustic (LA)

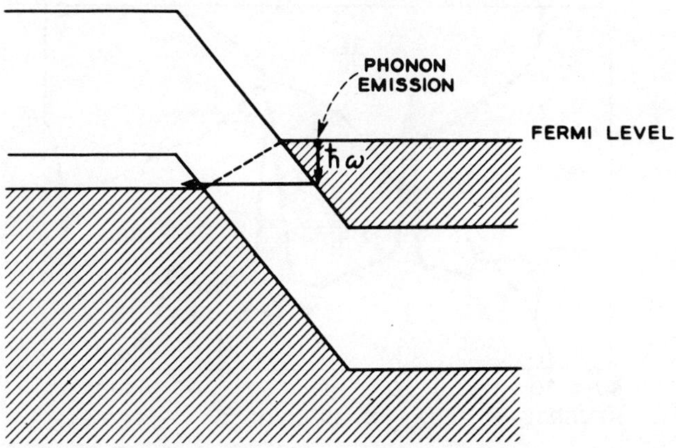

Fig. 5. The potential energy diagram of an indirect tunnel junction showing the relationship between the forward bias and the phonon energy $\hbar\omega$ that assists the tunneling.

at 43.7 mV, the longitudinal optic (LO) at 53.2 mV and the transverse optic (TO) at 58.5 mV. At low temperatures, where the density of phonons is low, tunneling occurs across the indirect gap with phonon emission. Figure 5 shows the potential energy diagram of a tunnel junction, and it is evident that no tunneling can occur until the bias equals the lowest phonon energy, $\hbar\omega$. At this bias the electron may emit a phonon and tunnel across the gap, so that tunneling will occur between states of equal energy and momentum. It is noted that in indirect tunneling described here the energy barrier through which the electron tunnels is reduced by the phonon energy in forward-bias tunneling (and is increased by the phonon energy in reverse-bias tunneling). No current flow in forward bias should thus occur until the bias is equal to $\hbar\omega$. In practice some current flow occurs, and this current correlates in magnitude with the excess current, suggesting that the momentum unbalance in pre-phonon tunneling is compensated by scattering with states in forbidden gap. The current–voltage characteristic of a silicon junction at 4.2°K is shown in Fig. 6, where the prominent phonon-induced inflections of the TA and TO phonons are evident ([10]). The conductance dI/dV, is also shown, and rapid changes in dI/dV occur near the inflections in the I–V curve. For vanishingly small interrogation signals and in the absence of thermal smearing of the Fermi energies the inflections would appear as sharp corners, giving rise to an abrupt discontinuity in the conductance. Because of the smearing effects, the increase takes place over a finite voltage range, and to a first approximation the phonon energy corresponds to the bias at the midpoint of the upward swing in dI/dV. This

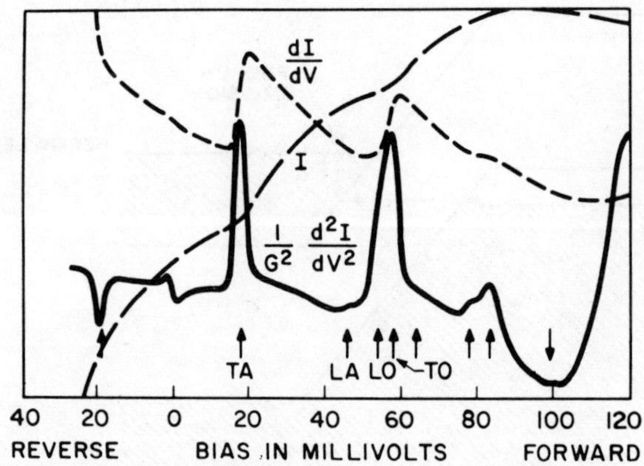

Fig. 6. The characteristic I versus V, dI/dV versus V, and d^2I/dV^2 of a silicon diode at 4 °K.

point corresponds to the peak in d^2I/dV^2, and the plot of the latter in Fig. 6 gives a precise determination of the phonon energies. Figure 7 shows the second-derivative plot of this junction in greater detail, in particular, phonon combinations at energies higher than the four fundamental phonons. These phonon combinations are combinations of the primary phonons with zero wave number optical phonons, O, or acoustic phonons S_1 and S_2.

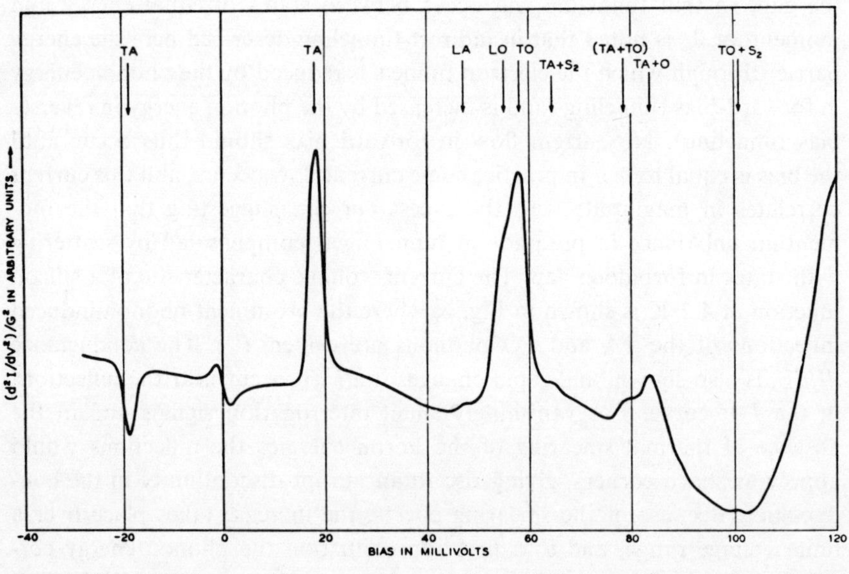

Fig. 7. The d^2I/dV^2 versus bias for a silicon diode at 4 °K. From Chynoweth *et al.* ([10]).

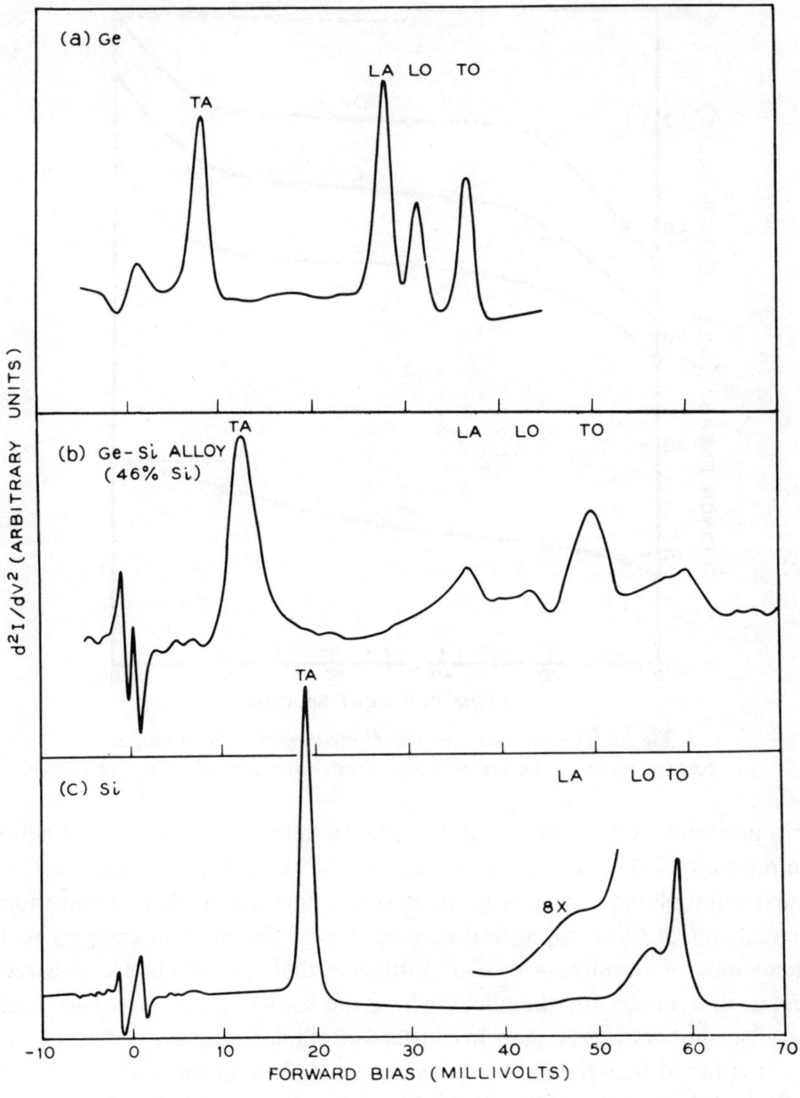

Fig. 8. The d^2I/dV^2 versus bias for junctions of (*a*) Ge, (*b*) Ge–Si alloy containing 46% Si, and (*c*) Si at 0.8 °K. From Logan *et al.* ([11]).

Figure 8 shows similar data ([11]) for junctions made by alloying aluminum to *n*-type Ge, where the phonons are well resolved because of their greater separation in energy. This data is compared to data for a silicon junction and for a junction made on a Ge–Si single-crystal alloy containing 46% Si. In the alloy the four fundamental phonons are easily resolved, and

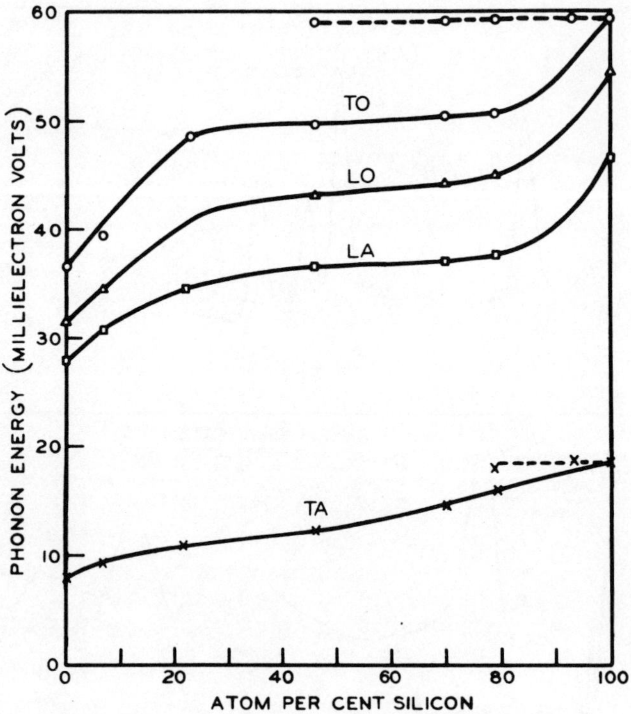

Fig. 9. The phonon energies plotted against the atom percentage of Si in the Ge–Si alloy. From Logan *et al.* ([11]).

their positions were determined for several alloys as functions of alloy composition of Ge and Si. The results are shown in Fig. 9, where the four fundamental phonon energies are plotted as a function of alloy composition. Comparison of the composition dependence of the phonon energies with calculations of Braunstein *et al.* ([12]) indicate that the results are in agreement with a model for the alloys where the similar atoms occur as next-nearest-neighbors rather than in clusters of similar atoms.

It is noted that the indirect energy gap in Si is in the $\langle 100 \rangle$ direction in k space but in the $\langle 111 \rangle$ direction in Ge. One might therefore expect some discontinuity in the tunneling currents in the alloy junctions as one changed from Si-like to Ge-like transitions, but no unusual effects were observed ([11]).

In addition to the phonon effects in Figs. 6–8, an additional structure is observed near $V = 0$. In the Ge junction of Fig. 8 (*a*) this corresponds to a dip in conductance at $V = 0$. In the case of the alloy and the Si junction in Fig. 8 superimposed upon the characteristic is that expected for tunneling from a metal or semiconductor into a superconductor with an energy gap of about 0.3 mV. While the conductance dip or peak grows slowly as the temperature is decreased from 20° to 1°K, the superconductor tunneling characteristic appears abruptly in many diodes at temperatures in the range 1.6°–2°K. It was found that alloys of Al and/or Ge and Si undergo superconducting transitions in this range, and it is assumed that the new characteristic corresponds to tunneling between the Al alloy and the *p*-type regrowth.

Let us now examine the effect of phonons upon the *I–V* characteristics of Si tunnel junctions. As the temperature is lowered from 300°K to 4°K, inflections in the *I–V* characteristic due to phonon emission become prominent, and the tunneling currents at a given bias decrease due to the temperature dependence of the energy gap and the phonon density. At low temperatures the semilogarithmic plot of the *I–V* characteristic exhibits an extensive linear portion in the bias range where the excess current is dominant ([7]). If this linear excess current region is extrapolated to lower biases, it will be seen that at biases less than V_p, the peak voltage, the excess current is negligible compared to the band-to-band tunneling, and the position of V_p will not be influenced to any significant extent by the background excess current. Figure 10 shows the initial portion of the *I–V* characteristic at 4°K, and clearly demonstrates that for forward biases $\leq V_p$ the forward tunnel current is the sum of three components. The first component, i_0, is the tunnel current that flows without phonon participation, and this component accounts for all the observed current for biases less than approximately 18 mV, corresponding to emission of the TA

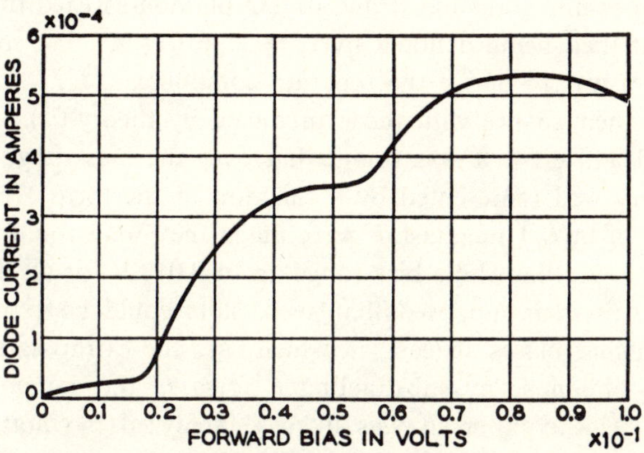

Fig. 10. The initial segment of the *I–V* characteristic of a Si tunnel diode at 4.2 °K. From Logan and Chynoweth ([13]).

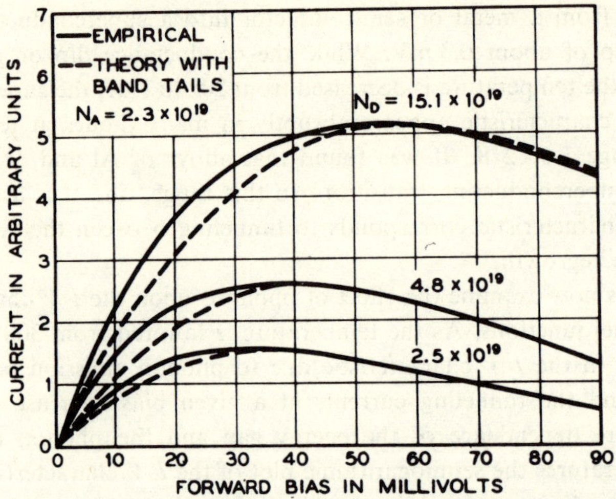

Fig. 11. Comparison between the empirically derived current–voltage characteristics and theoretical curves computed by Kane ([14]). The pairs of curves are scaled to similar peak currents but there is no significance in the relative magnitudes of the curves. From Logan and Chynoweth ([13]).

phonon. If it were not for the phonon-assisted current components, i_0 would be seen to peak at a bias somewhat greater than 18 mV and thereafter pass into a negative resistance region. At 18 mV the TA phonon-assisted component, i_1, starts, so that the total observed current is then $i_0 + i_1 = i_1'$, say. The i_1 component alone would also be expected to follow a hump-shaped curve somewhat similar to that of i_0. However, at 58 mV a third component, i_2, is added due to TO phonon-assisted tunneling, the total current then being denoted by $i_2' = i_0 + i_1 + i_2$. The major aim in this work ([13]) was to derive the separate components, i_0, i_1, and i_2, and to compare their shapes with those predicted by theory ([14]).

By trial and error it was found that over the bias range 0–18 mV i_0 could be very well represented by a function of the form $V \exp(-B'V)$. When plots of $\ln(i_0/V)$ against V were made they were found to be good straight lines over the whole bias range up to 0.018 V for all the junctions studied. It was then supposed that such plots could be extrapolated to somewhat higher biases at least, in which case the extrapolated value for i_0 could be obtained; by substracting i_0 form i_1' the component i_1 was determined. This component was likewise analyzed by plotting $\ln[i_1/(V - 0.018)]$ against V. Again a good straight line was always obtained, and its slope, again, supposedly, B', was generally in excellent agreement with the value found from analyzing i_0. The whole procedure could be repeated again so as to derive component i_2, and it too was found to be a good straight line when $\ln[i_2/(V - 0.058))$ was plotted against V.

The three separate components so derived all have a very similar voltage dependence. This similarity in shape confirms that the effect of phonons is simply to generate a set of components with displaced origins. Figure 11 compares the I–V band-to-band tunneling characteristic obtained from junctions covering a range of donor concentrations of 2.5×10^{19}–$1.5 \times 10^{20}\ \text{cm}^{-3}$ to curves calculated from Kane's theory ([14]). It is clear that agreement of position of peak voltage and of the general shape of the curves is very good. The empirical expression $V \exp(-BV)$ predicts a peak voltage at $V = B^{-1}$, and this relationship was confirmed in most diodes. It is noted that while the empirical expression appears to describe the observed I–V characteristic in a simple analytical form, the theoretical expression for the I–V characteristic does not appear to be reducible to the empirical form.

In direct-energy-gap semiconductors phonon emission is not essential to the tunneling mechanism, but inflections in the current–voltage characteristic are observed which correspond to the emission of $k = 0$ optical pho-

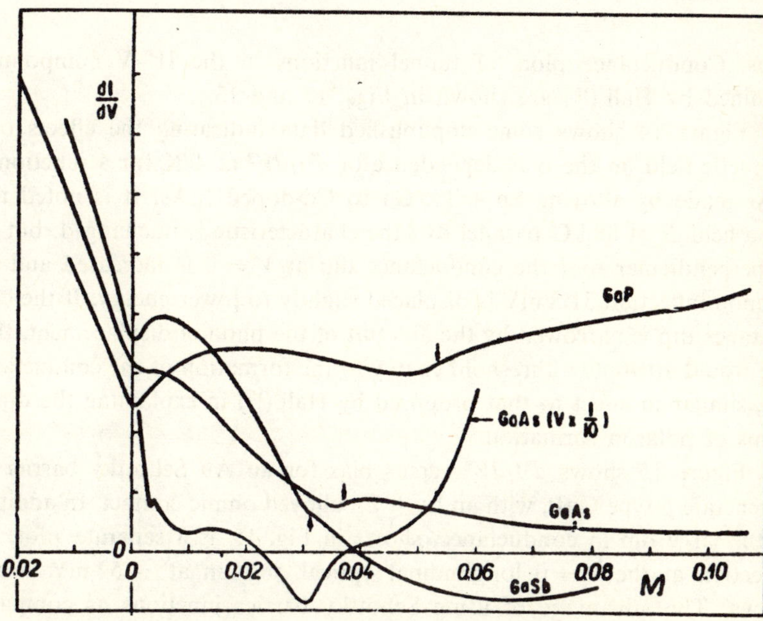

Fig. 12. The conductance of GaSb, GaAs, and Ga P junctions at 4.2 °K. From Hall and co-workers ([15]).

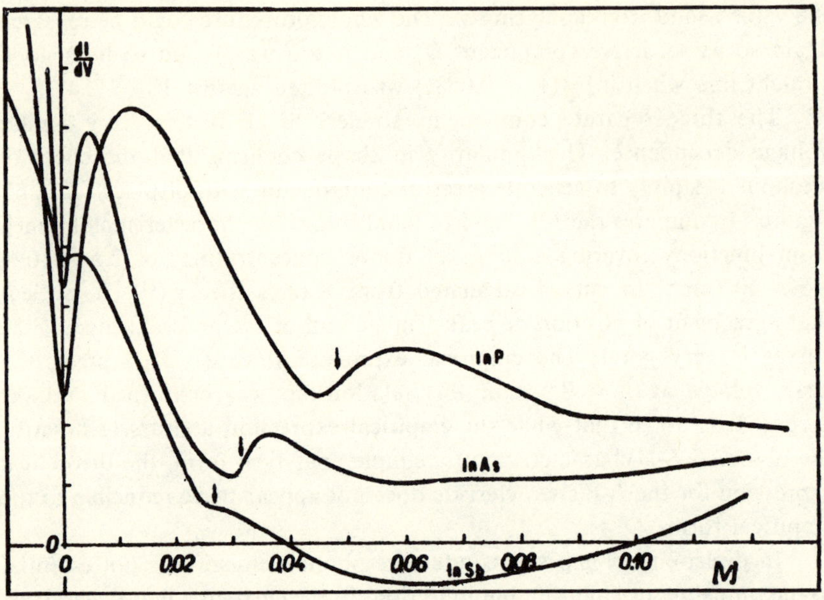

Fig. 13. The conductance of InSb, InAs, and In P junctions at 4.2 °K. From Hall and co-workers ([15]).

nons. Conductance plots of tunnel junctions in the III–V compounds, obtained by Hall ([15]) are shown in Figs. 12 and 13.

Figure 14 shows some unpublished data indicating the effects of a magnetic field on the bias dependence of d^2I/dV^2 at 4°K for a junction of InAs made by alloying Sn + 1% Ga to Cd-doped InAs. It is noted that for a field H of 88 kG parallel to I the characteristic is unchanged, but for H perpendicular to I the conductance dip at $V = 0$ is increased and the phonon inflection 31.8 mV is displaced slightly to lower energy. If the conductance dip is narrowed by the amount of the phonon displacement, then one would attribute a threshold energy to the formation of the conductance dip, similar in effect to that proposed by Hall ([15]) in explaining the dip in terms of polaron formation.

Figure 15 shows d^2I/dV^2 versus bias for an Au Schottky barrier on degenerate p-type GaP, with an In + Zn alloyed ohmic contact. In addition to the large dip in conductance, shows in Fig. 15 as a separate plot, the inflection at the $k = 0$ longitudinal optical phonon at ~ 52 mV is also shown. The advantage of using Schottky barrier junctions as compared to p–n junctions is that the doping levels are more accurately determined, since a single Hall effect measurement on the bulk crystal before forming the

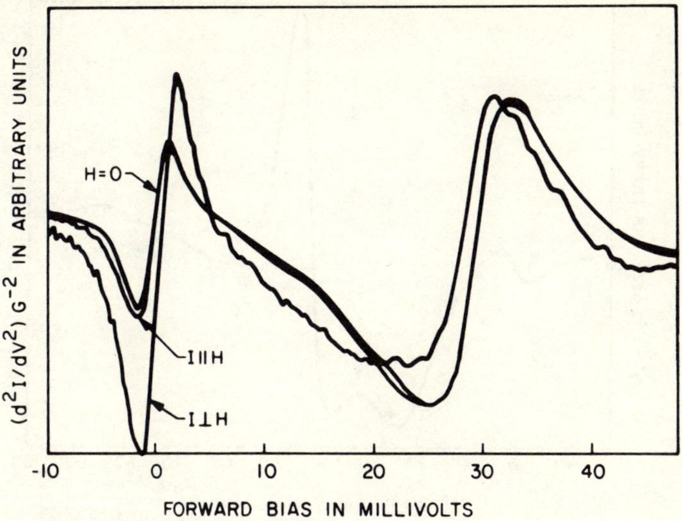

Fig. 14. The second derivative characteristic of an InAs tunnel diode in zero field and in fields of 88 kG parallel and perpendicular to the direction of current flow at 4.2 °K.

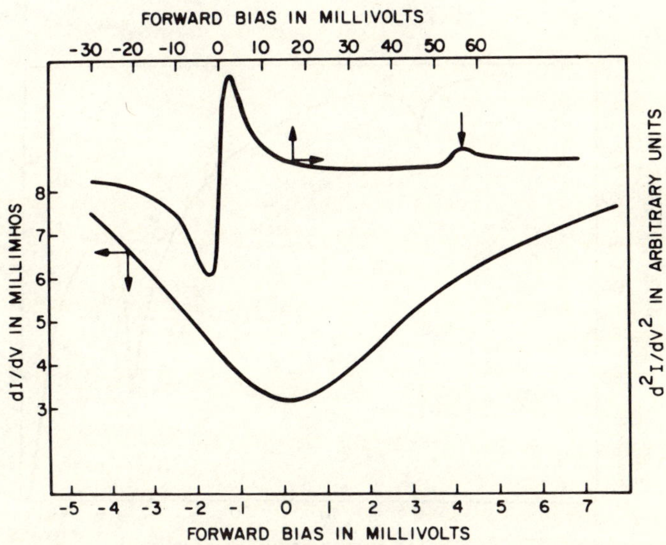

Fig. 15. The characteristics of an Au Schottky barrier junction made on *p*-type GaP showing dI/dV near zero bias and d^2I/dV^2 versus V at 4.2°K. The arrow indicates an inflection observed at the optical phonon energy.

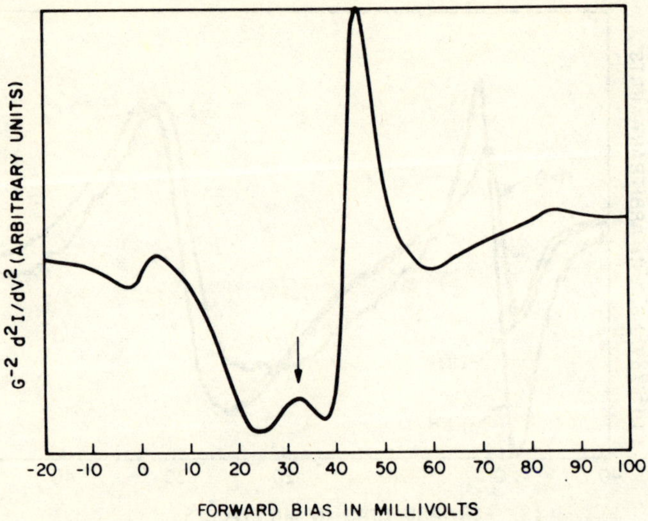

Fig. 16. The second derivative characteristic at 0.9 °K of an Au Schottky barrier junction made on *p*-type InAs. The arrow indicates an inflection at the phonon energy which is distinct from the large peak at slightly higher bias.

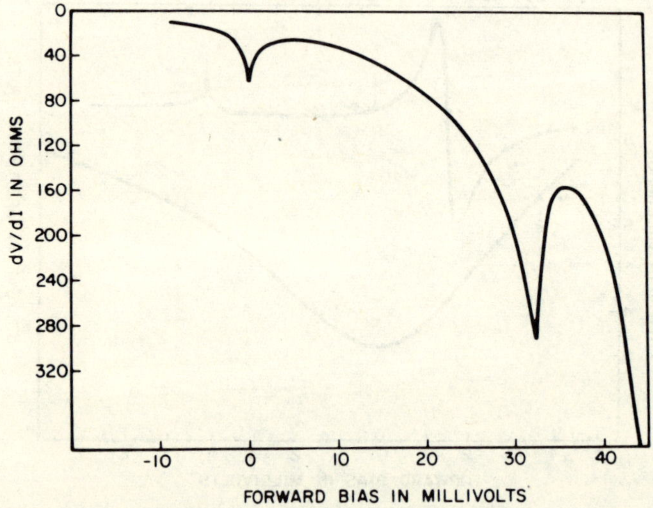

Fig. 17. The first derivative characteristic at 0.9 °K of an Au Schottky barrier junction made on *p*-type InAs more lightly doped than that in Fig. 16.

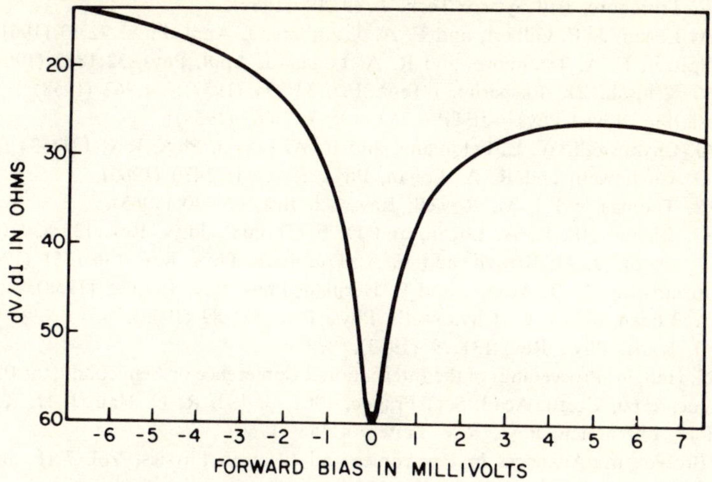

Fig. 18. The conductance dip near zero bias at 0.9 °K of the junction of Fig. 17.

barrier is all that is needed. On the other hand, surface effects have not been studied in detail, so that in many instances the detailed knowledge required of inversion layers, oxide films, etc. is a handicap in utilizing this technique.

Figure 16 shows similar data for an Au Schottky barrier on *p*-type InAs doped to an acceptor concentration of 3×10^{18} cm^{-3}. In addition to the conductance dip at $V = 0$ and the phonon inflection at 31.8 mV, a large peak is observed at 43 mV. The origin of this large peak is not certain. As the acceptor concentration increases, this peak occurs at higher biases. Figure 17 shows a conductance plot of a similar barrier junction made on *p*-type InAs containing 2×10^{17} acceptors/cm^3, where the peak occurs at approximately the phonon energy. While the new peak is close to that in energy of a plasma frequency, ω_p, where $\omega_p = (4\pi ne^2/mE_0)^{1/2}$, with n the carrier density, e the electronic charge, m the carrier mass, and E_0 the dielectric constant, the dependence of ω_p upon doping did not give a sufficiently precise correlation to establish the origin of the new peak. Figure 18 shows the conductance dip at $V = 0$ of the junction of Fig. 17 and demonstrates the value of Schottky junctions in studying this aspect of tunneling.

REFERENCES

1. W. Shockley, Bell System Tech. J. 28:435 (1959).
2. K. T. Rogers, A. J. Dawes, and R. L. Bell, J. Appl. Phys. 36:1493 (1965).
3. W. L. Feldmann and A. G. Chynoweth, private communication.

4. F. A. Trumbore, Bell System Tech. J. 39:205 (1960).
5. R. A. Logan, J. F. Gilbert, and F. A. Trumbore, J. Appl. Phys. 32:13 (1961); W. G. Spitzer, F. A. Trumbore, and R. A. Logan, J. Appl. Phys. 32:1882 (1961).
6. L. V. Keldysh, Zh. Eksperim. i Teor. Fiz. 33:994 (1957); 34:962 (1958) [English translation: Soviet Phys.—JETP 6:763 (1958); 7:665 (1958)].
7. A. G. Chynoweth, W. L. Feldmann, and R. A. Logan, Phys. Rev. 121:684 (1961).
8. A. G. Chynoweth and R. A. Logan, Phys. Rev. 118:1470 (1960).
9. D. E. Thomas and J. M. Rowell, Rev. Sci. Inst. 36:130 (1965).
10. A. G. Chynoweth, R. A. Logan, and D. E. Thomas, Phys. Rev. 125:877 (1962).
11. R. A. Logan, J. M. Rowell, and F. A. Trumbore, Phys. Rev. 136:1751 (1964).
12. R. Braunstein, A. R. Moore, and F. Herman, Phys. Rev. 109:695 (1958).
13. R. A. Logan and A. G. Chynoweth, Phys. Rev. 131:89 (1963).
14. E. O. Kane, Phys. Rev. 131:79 (1963).
15. R. N. Hall, in: Proceedings of the International Conference on Semiconductor Physics, Prague, 1960, Czech. Acad. Sci., Prague, 1961, p. 193. R. N. Hall, J. H. Rocette, and H. Ehrenreich, Phys. Rev. Letters 4:456 (1960).
16. H. Brooks, in: Advances in Electronics and Electron Physics, Vol. 7 (L. Martin, ed.), Academic Press, New York, 1955.
17. B. N. Brockhouse, Phys. Rev. Letters 2:256 (1959).

Chapter 12

Effect of Stress on Interband Tunneling in Semiconductors*

H. Fritzsche

The James Franck Institute and Department of Physics
The University of Chicago
Chicago, Illinois

The study of effects of stress on interband tunneling in semiconductors has added to our understanding of both the tunneling process and the semiconductor materials. The same degree of success is not to be expected for the case of metal–insulator–metal tunneling. In order to appreciate the difference between metal and semiconductor tunneling, a few important factors governing interband tunneling in semiconductors should be pointed out.

First, in semiconductors both the initial and final tunneling states are confined to small regions in k space. These regions contain not more than a few per cent of the Brillouin zone volume. They are situated around points or axes of high symmetry. As a consequence, the tunneling process requires some interaction with the lattice if there is a crystal momentum difference between the initial and final tunneling states. The nature of the interaction allows one to distinguish a (*a*) phonon-assisted tunneling, (*b*) impurity-assisted tunneling, and (*c*) direct tunneling.

Second, the energy eigenstates and electron wave functions of the semiconductor are noticeably altered if a small fraction (as small as 10^{-5} in Ge and 10^{-6} in InSb) of host atoms are replaced by impurity atoms. The perturbed wave function for an energy E may be written as a sum of Bloch waves of the bands n and wave number $\mathbf{k}$ as

$$\Psi_E(\mathbf{r}) = \sum_{n,k} A_n(\mathbf{k}) u_{n,\mathbf{k}}(\mathbf{r}) \exp(i\mathbf{k} \cdot \mathbf{r}). \quad (1)$$

* Work supported by the U.S. Air Force Office of Scientific Research, AFOSR 49(638)-1653.

As a consequence one observes the following:

1. States associated with impurities and defects are introduced into the forbidden band gap, giving rise to the excess tunneling current. This excess current arises from tunneling between initial band states and such rather localized gap states, with subsequent recombination to final band states.

2. The sharpness of the band edges disappears at larger impurity concentrations, and so-called band tails extend into the forbidden gap [see the review paper by Pankove ([1])].

3. Some impurities—for example, As and P in germanium and most donor impurities in silicon—are known to produce strongly localized central cell potentials ([2]) which introduce components $A_n(\mathbf{k})$ into the wavefunction Ψ_E from large parts of the Brillouin zone including all of the conduction band minima and the region around $k = 0$. This allows tunneling into any region of k space permitted by energy conservation and is the basis for impurity-assisted tunneling.

Finally, interband tunneling in semiconductors occurs in one single-crystal region with a potential barrier for charge carriers but not for phonons. Perhaps the elastic constants and deformation potentials are slightly different on the n-type and p-type sides because of the differences in carrier concentrations and their screening effects. As a consequence, shear stress as well as pressure experiments can be performed on oriented p–n tunnel junctions. Most effects of elastic stress on tunneling have been observed with Esaki-type, heavily-doped, germanium p–n junctions. A few remarks about the preparation of tunnel junctions and the stress arrangements are deferred to the Appendix. The band structure $E_n(\mathbf{k})$ of pure Ge is shown in Fig. 1 along a few important symmetry directions ([3]). Although the large impurity concentrations will decrease the gap and cause the modifications mentioned above, this band structure forms the basis for explaining the experimental results. Figure 2 shows the relevant band extrema for the n-type side (left) and the p-type side (right) with the Fermi levels ξ_n and ξ_p as measured from the conduction band minima at L and the valence band at Γ, respectively, and separated by the applied potential eV. Before we discuss the stress effects and write down an expression for the tunneling current let us identify all the features of the tunneling characteristic given in Fig. 3. This figure shows the I–V curve of a Ge tunnel diode at 4°K doped with Ga acceptors and Sb donors on the respective sides. These impurities cause a particularly small disturbance of the host lattice, so that the mixing expressed by Eq. (1) is small, and tunneling proceeds via

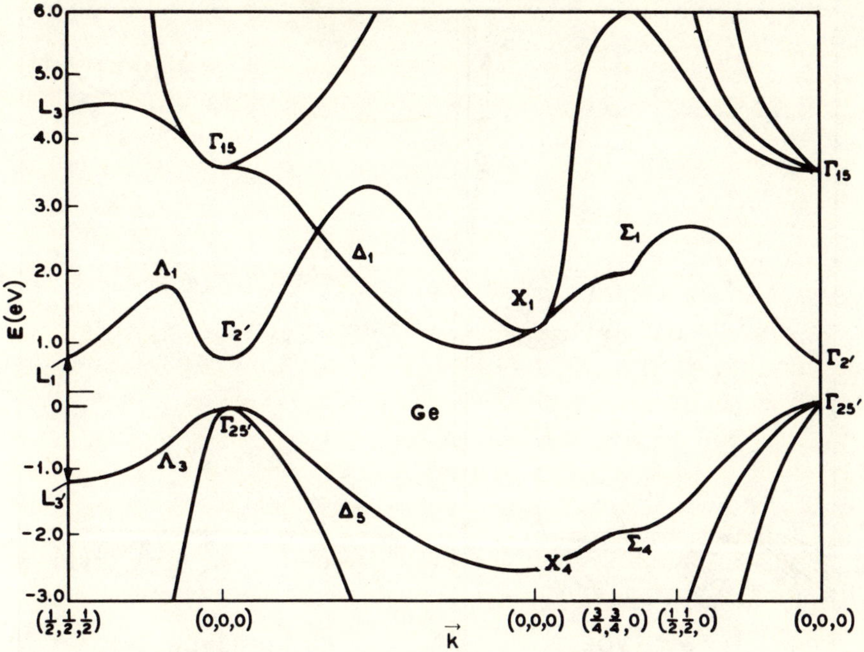

Fig. 1. Band structure of Ge (neglecting spin orbit interactions), along certain symmetry directions in k space. From Phillips ([3]).

phonon emission (absorption is negligible at 4°K), supplying the wave number difference between the states near $\Gamma_{25'}$ and those near L_1. At small V the energy difference between initial and final states is insufficient to create a phonon of this wave number, $\mathbf{k} = (2\pi/a)(\frac{1}{2}\ \frac{1}{2}\ \frac{1}{2})$, and the small current observed represents impurity-induced tunneling. For $|eV| > 7.8$ meV, TA (transverse acoustic) phonons and for $|eV| > 27.5$ meV also LA (longitudinal acoustic) phonons can be emitted, and a sudden rise in tunnel current indicates the onset of these channels ([4]). At negative biases of about $-$ 0.14 V the Fermi level on the p-type side has reached the energy of the $\Gamma_{2'}$ minimum on the n-type side and direct tunneling can set in ([5,6]). Since the direct tunneling process proceeds with a much larger probability than phonon-assisted tunneling, the current scale of the I–V curve showing this onset is greatly reduced, dwarfing the phonon-assisted current.

Since there are four minima at L and six along Δ and a light- and heavy-hole valence band near $\Gamma_{25'}$, the tunnel current consists of several contributions. Each of the phonon-assisted contributions consists in turn of several channels according to the type of phonons [TA, LA, LO (longitudinal optical), and TO (transverse optical)] emitted. These channels and

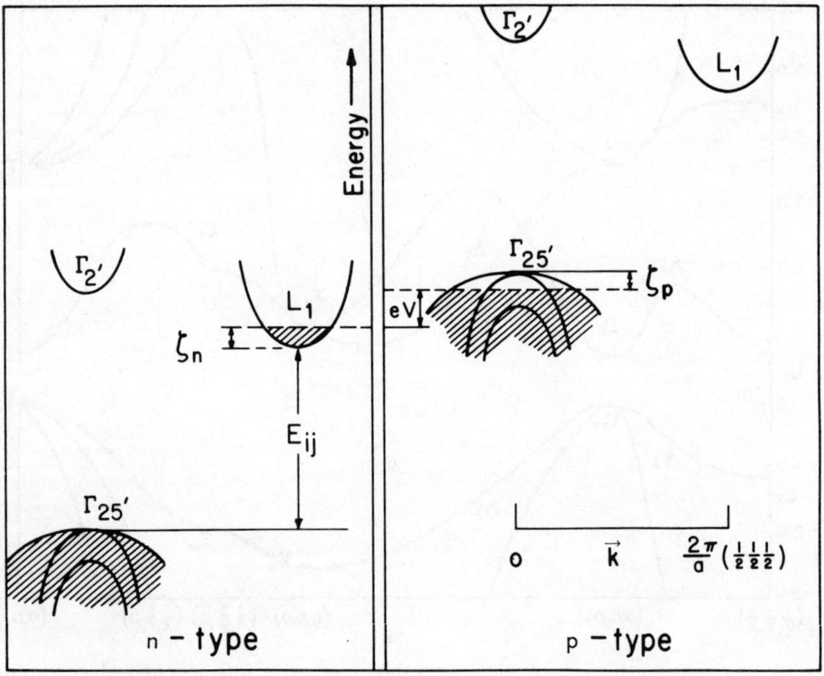

Fig. 2. Sketch of $E(k)$ for Ge near Γ and L for the n-type and p-type sides of the tunneling junction. As drawn, the Fermi levels on the two sides are separated by a negative bias near the onset of direct electron tunneling from the valance band on the p side to $\Gamma_{2'}$ on the n side. Only one of the E_{ij} is shown.

contributions set in with their specific tunneling probabilities at definite threshold voltages which are a measure of the phonon energies or of the energy separations of band extrema.

One of the interesting effects of both pressure and shear stress on tunneling is caused by the change of these energies. Hydrostatic pressure leaves the symmetry, and hence the degeneracy of the critical band edge points, intact, but changes their energies. A shear causing a trigonal distortion of the cubic lattice (stress along the body diagonal) lifts the degeneracy of the L points and along Λ without disturbing the degeneracy along the cubic axis. A tetragonal distortion (stress along one cubic axis), on the other hand, lifts the degeneracy along the cubic axes but preserves the degeneracy at L and along Λ.

A simple expression for the interband tunneling current was given by Kane ([7]). For direct transitions it is

$$I_d = C_d D_d \exp(-\alpha) \qquad \text{with} \qquad \alpha = \lambda_d E_d^{3/2} m^{*1/2}/F. \tag{2}$$

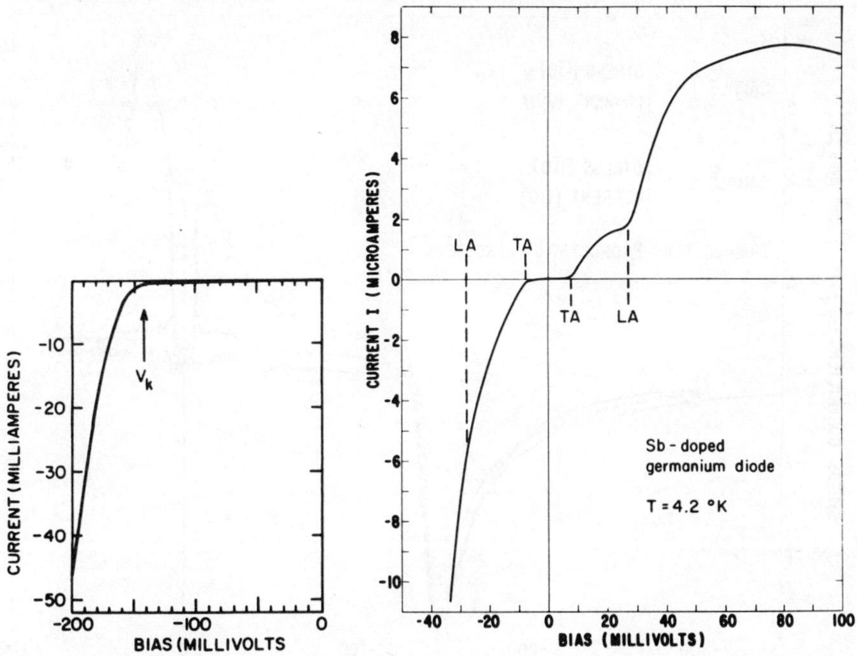

Fig. 3. Current–voltage characteristic of an Sb-doped Ge tunnel junction at 4.2 °K. Note the onset voltages for phonon-assisted tunneling and V_k for the direct tunneling to $\Gamma_{2'}$ (on the left-hand side with condensed current scale).

For transitions with emission of phonons of energy $\hbar\omega$ it is

$$I = \sum_{ij} CD_{ij} (|eV| > \hbar\omega) \exp(-\beta) \qquad \text{with} \qquad \beta = \lambda(E_{ij} \pm \hbar\omega)^{3/2}(m_{ij}^*)^{1/2}/F. \tag{3}$$

The sum includes all combinations of contributing conduction and valence band extrema, labeled by i, j, separated by the energy E_{ij}; C includes the electron–phonon coupling strength, which is different for different types of phonons; F is the average junction field; and λ and λ_d are constants. The upper sign in Eq. (3) is for p to n tunneling (reverse bias) and the lower sign for n to p tunneling (forward bias). The factor D involves an integral over the initial and final tunneling states. Energy conservation causes D_{ij} to vanish for biases $|eV|$ smaller than $\hbar\omega$. For direct tunneling D vanishes below the threshold voltage. The mass m_{ij}^* is obtained from the effective-mass components of the initial and final state extrema along the tunneling direction x,

$$m_{ij}^* = [(1/m_{ix}) + (1/m_{jx})]^{-1}. \tag{4}$$

The different tunneling theories developed later attempt to treat the spatial

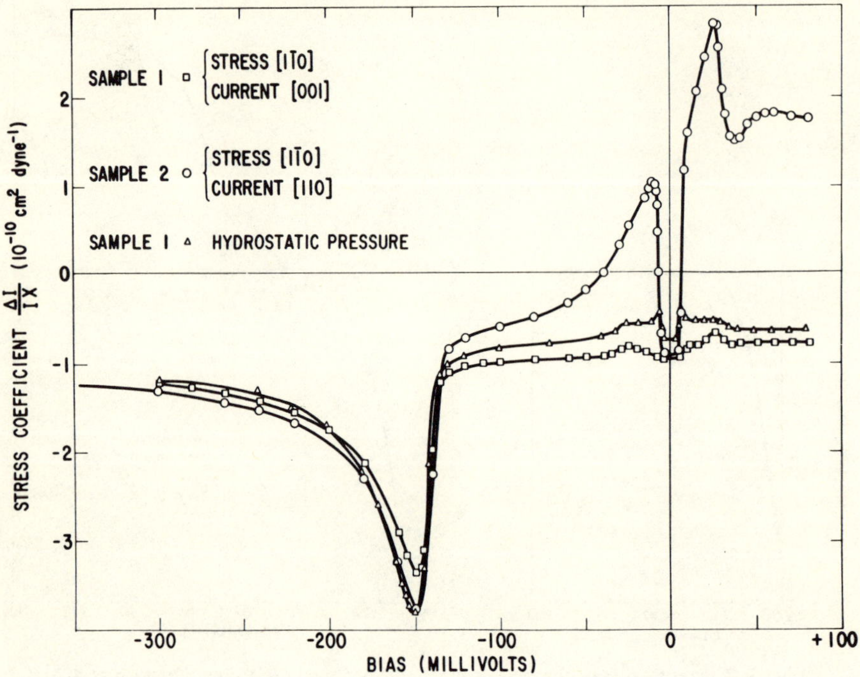

Fig. 4. Stress coefficients $\Delta I/IX$ and $\Delta I/I3p$ as a function of bias at 4.2 °K for Sb-doped Ge tunnel junctions. The stress coefficient for pure shear is the difference $(\Delta I/IX) - (\Delta I/I3p)$.

dependence of the junction field more realistically ([8]) and to account for the selection rules in the electron–phonon interactions imposed by the symmetry properties of the phonons connecting the different symmetry points in the Brillouin zone ([9]). Some of these refinements change the tunneling exponent of Eq. (3) and it is expected that stress experiments which change the gap energies in a predictable manner will be able to test the validity of the various theories.

The effect of both uniaxial compression and hydrostatic pressure on the tunneling characteristic ([10]) of Fig. 3 is shown in Fig. 4, where the relative change of tunnel current per unit stress is plotted against bias voltage. Pronounced structure is observed at the threshold voltages for phonon-assisted and direct tunneling. The sharp negative at $V = 0.14$ V results from the pressure-induced increase of the energy separation between $\Gamma_{2'}$ and L_1. The almost constant negative pressure coefficient at lower voltages is due to the increase of the indirect gap E_{ij} and, to a smaller extent, of the mass m_{ij}^* with pressure [see Eq. (3)].

Since uniaxial compression X produces not only a shear but also volume change equivalent to that of a hydrostatic pressure $p = X/3$, the pressure coefficient has been divided by a factor three. The difference $\Delta I/IX - \Delta I/I3p$ is then entirely due to pure shear. The energy motion of the conduction band valleys at L depends on the dilatation and shear deformation potentials E_1 and E_2 and the strain tensor $\mathbf{u}$ [notation of Brooks ([11])] as

$$\Delta E^{(i)} = \hat{\mathbf{n}}^{(i)}[E_1\mathbf{1} + E_2(\mathbf{u} - \tfrac{1}{3}\,\mathrm{tr}\,\mathbf{u}\mathbf{1})]\hat{\mathbf{n}}^{(i)}\,, \tag{5}$$

where $\hat{\mathbf{n}}^{(i)}$ is the unit vector pointing to the ith valley. The pure shear part of the stress causes the valleys to shift with respect to one another in such a way that their average energy is unchanged. Shear due to X along $[1\bar{1}0]$ moves the [111] and $[11\bar{1}]$ valleys up and the $[1\bar{1}1]$ and $[1\bar{1}\bar{1}]$ valleys down. Since the shifts are the same,* however, a large first-order effect of the shear is observed only for a tunneling direction for which the individual valley currents are different. Because of Eq. (4) and the effective mass anisotropy of the L_1 valleys, m_{ij}^* is different for the two pairs of valleys (25% larger for the [111] and $[11\bar{1}]$ valleys than for the other two) when the current I is along [110]. This results in the large positive shear contribution seen in Fig. 4 for this orientation. The much smaller shear contribution for I along [001] must arise from other effects, such as a shear deformation of the mass ellipsoids and the valence bands, because the mass components along [001] are the same for all four valleys.

The large effect of shear on the tunneling characteristic for some orientations of I and X and its absence for other orientations can be used as a tool for determining whether or not the electron tunnels from a state localized in a particular valley.

One can apply this argument immediately to the low-bias region, $|V| < 7.8$ meV, of Fig. 4, where impurity-induced tunneling is the only tunneling process. The large positive shear effect is absent, confirming our earlier description of this type of tunneling as arising from contributions to Eq. (1) from regions near $k = 0$. This contribution is so large for arsenic

* A more general statement is the following: For a cubic crystal pure shear causes energy shifts $\Delta E(\mathbf{k}_m)$ of the band energies at an arbitrary point in $\mathbf{k}$ space and at the equivalent points $\mathbf{k}_m$ generated by the symmetry elements α_m of the cubic point group in such a way that the sum of the shifts $\sum_m \Delta E(\mathbf{k}_m) = 0$ to first order in strain. Proof. Let the first order shift in strain $\mathbf{u}$ be described by a deformation potential tensor Ξ_{k_m} as $\Delta E(\mathbf{k}_m) = \Xi_{k_m}\cdot\mathbf{u}$. Then $\sum_m \Delta E(\mathbf{k}_m) = \sum_m \Xi_{k_m}\cdot\mathbf{u} = \mathbf{D}\cdot\mathbf{u}$. Since the symmetry elements α_m change only the ordering of the terms in the sum, the second rank tensor $\mathbf{D}$ is invariant to α_m and hence is a scalar. This scalar is the energy shift per unit volume change, which is zero for pure shear.

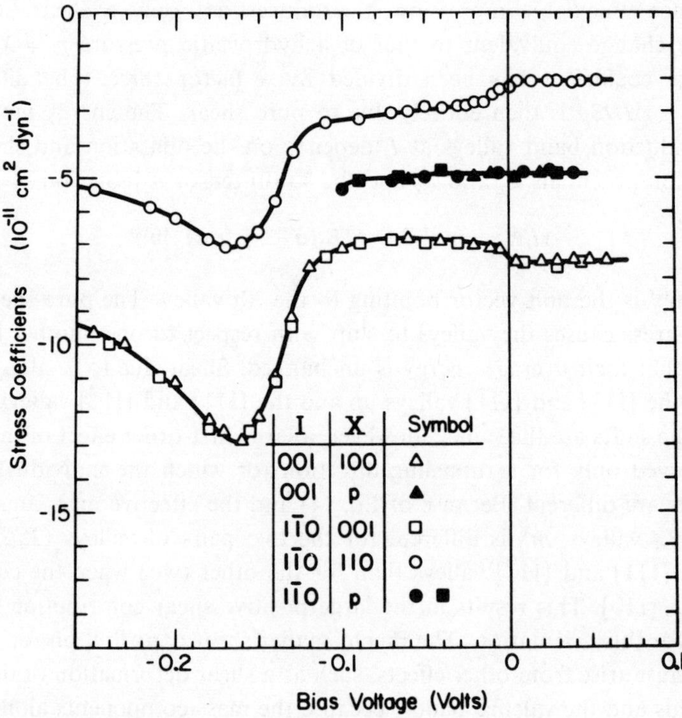

Fig. 5. Stress coefficients as in Fig. 4 for As-doped Ge tunnel junctions in which impurity-induced tunnelling dominates. Note the absence of the large positive shear coefficient seen in the previous figure.

donors in germanium (because of their large central cell potential) that impurity-induced tunneling dominates phonon-assisted tunneling at all biases and no shear effect due to the anisotropic tunnel current from the [111] valleys is observed ([12,13]). The results on As-doped Ge tunnel junctions are shown in Fig. 5. One observes that the pure shear effect of the $I[1\bar{1}0]$, $X[110]$ orientation is identical in magnitude and of opposite sign to the shear effect of the $I[1\bar{1}0]$, $X[001]$ and $I[010]$, $X[001]$ orientations.

These shear effects are much smaller than that expected for tunneling from individual [111] valleys and more nearly equal to shear-induced changes of effective masses. A quantitative analysis of these effects have to wait until a tractable theory of impurity-induced tunneling has been worked out.* Fig. 5 shows, however, that impurity-induced tunneling is quite distinct from direct tunneling and of considerably smaller probability,

* As far as I know, the only paper treating this problem is ([14]).

since a clear break of the I–V curve, and hence of the stress coefficients for tunneling, is observed near $V = -0.14$ V, where direct tunneling to the $\Gamma_{2'}$ conduction band minimum sets in.

In the excess current region stress experiments on Sb-doped Ge junctions did show a large positive effect for the $I[1\bar{1}0]$, $X[110]$ orientation ([15]). This indicates that the excess current involves tunneling from the individual [111] valleys to gap states on the p-type side of the junction followed by recombination with a hole of the valence band.

These examples show how the location in k space of initial or final tunneling states can be identified by using the appropriate current and shear stress orientation. For such an identification only the relative sizes of the effects need be interpreted. A quantitative analysis requires a knowledge of the energy shifts and deformations of the energy bands with stress as well as an accurate expression for the tunneling current. The threshold voltages for the onset of the various tunneling processes, on the other hand, are obtained by simple physical arguments and their shift under stress can

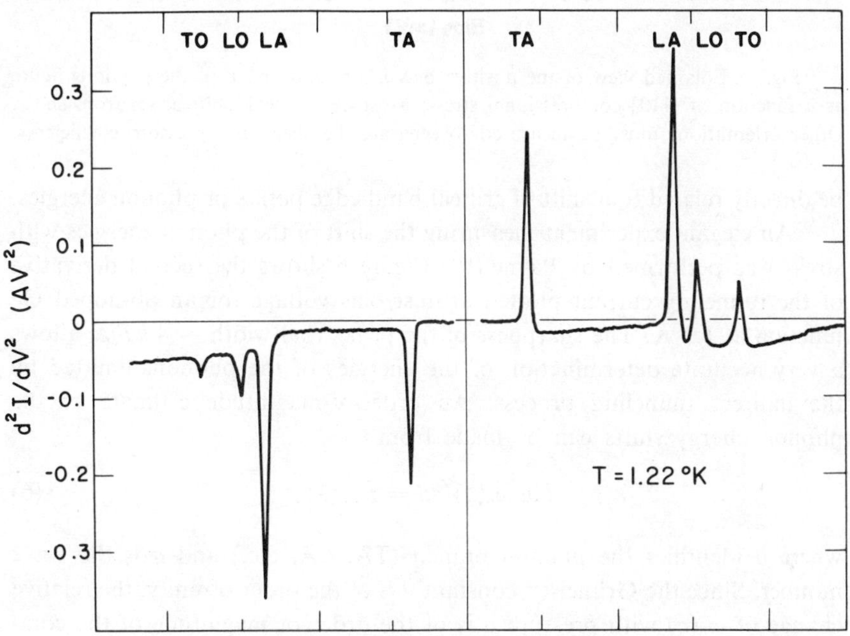

Fig. 6. The curvature of the I–V characteristic of Fig. 3 at 1.2 °K as a function of bias. All four phonon thresholds are resolved.

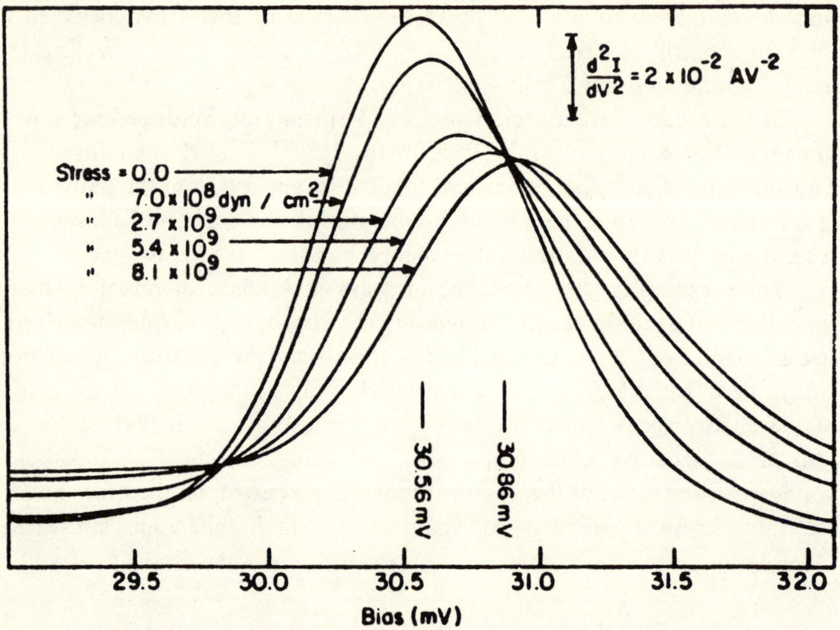

Fig. 7. Enlarged view of the positive bias LO phonon peak of the previous figure as a function of [110] compressional stress. Most of the peak shift arises from shear. Other orientations must be measured to separate the shear and pressure coefficients.

be directly related to a shift of critical band edge points or phonon energies.

An elegant experiment measuring the shift of the phonon energies with stress was performed by Payne ([16]). Figure 6 shows the second derivative of the tunneling current plotted against bias voltage for an Sb-doped Ge junction at 1.2°K. The sharpness of the peaks (halfwidth $\sim 4\,kT/e$) allows a very accurate determination of the energies of the phonons emitted by the indirect tunneling process. An order-of-magnitude estimate of the phonon energy shifts can be made from

$$d \ln \omega_b(q)/dp = K\gamma_b(q), \tag{6}$$

where b identifies the phonon branch (TA, LA, etc.) and q is the wave number. Since the Grüneisen constant γ is of the order of unity, the relative change of $\omega_b(q)$ with pressure p is of the order of magnitude of the compressibility K. This is about 20–50 times smaller than the relative changes of typical band edge energies with pressure, and therefore requires rather large stresses. In order to avoid the difficulties involved in high-pressure experiments at He temperatures, Payne used uniaxial compression and

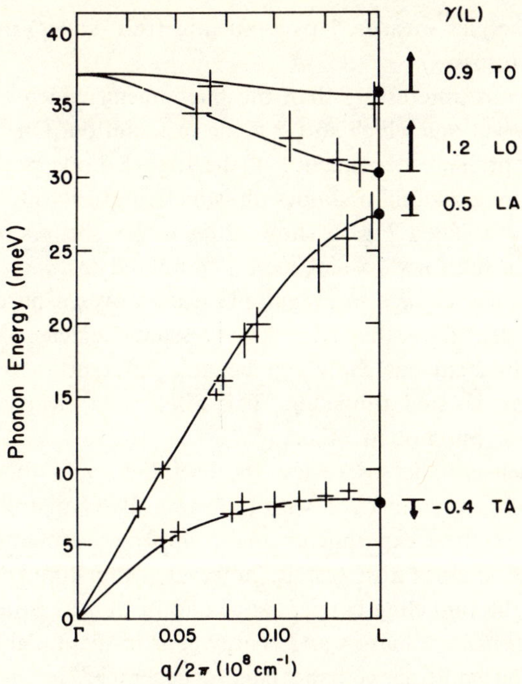

Fig. 8. Phonon dispersion curve for q along [111]. Not only the phonons at L but those from a certain volume around L assist the tunneling process because of the finite size of the Fermi surfaces on the n and p sides. The size of the crosses represent the uncertainties in the neutron scattering data and the points at L were obtained from tunneling data. On the right-hand side the magnitudes of the Grüneisen constants are listed.

separated the effects due to the shear and volume change by varying the crystallographic orientations of the stress. Figure 7 shows as an example an enlarged view of the LO phonon peak of the previous figure as a function of [110] compressional stress. The Grüneisen constants [as defined by Eq. (6)] obtained in this experiment are shown on the right-hand side of Fig. 8.

The most interesting result is the negative sign of $\gamma(L)$ of the TA branch. At low temperatures only these low-energy phonons are excited, yielding a negative average Grüneisen constant and hence an explanation of the anomalous negative coefficient of thermal expansion observed in diamond-like semiconductors ([17]). Figure 8 also shows the good agreement between the phonon energies measured by tunneling (full circles) and by neutron scattering ([18]) (crosses).

Perhaps more interesting than the phenomena which are well understood are observations which so far elude explanation. One of these is the change of the pressure dependence of the tunnel diode peak current near 8000 kG/cm^2 in germanium diodes in which impurity-induced tunneling predominates ([19]). Then Fig. 5 shows 20% larger shear stress coefficients for tunneling in the forward-bias region compared to those of the reverse-bias direction. The change in magnitude occurs over a narrow bias range which is symmetric about the origin. The pressure coefficients of the phonon-assisted tunneling current show significantly different values for TA and for LA phonon-assisted tunneling. This effect was clearly seen in some experiments ([20]), but not in others ([21]).

As a conclusion, let us review the usefulness and the limitations of stress experiments in connection with tunneling. By changing critical energy band parameters, these experiments might supply a useful test of the tunneling theories. A quantitative test is, however, often limited by a lack of knowledge of the tunneling barrier shape and of the stress-induced changes of the relevant effective masses and energy gaps in these highly doped semiconductors. The splitting of band edge degeneracies in multivalley semiconductors with shear enables one to identify the location in k space of some initial or final tunneling states. The stress-induced shifts of the onset voltages of the different tunneling processes are most accessible to interpretation. They might be useful in identifying the structure recently observed in Bi–I–Al and in lead salt junctions.

APPENDIX: EXPERIMENTAL CONSIDERATIONS

Interband tunneling is observed when the depletion region of the p–n barrier is sufficiently narrow, less than about 500 Å, and when the p- and n-type regions of the semiconductor are sufficiently doped to have the Fermi level in the valence and conduction bands, respectively. For these stress experiments the tunnel diodes were prepared by alloying indium dots which were doped with 0.7% gallium in a protective hydrogen atmosphere onto the appropriate faces of freshly etched Ge bars which contained more than 4×10^{18} cm^{-3} antimony or arsenic donors. The alloying cycle reaches a temperature of about 540°C in less than 1 min, after which the bars are quickly cooled by means of a gas stream. This short time assures a rather sharp concentration gradient of the readily diffusing gallium atoms. The

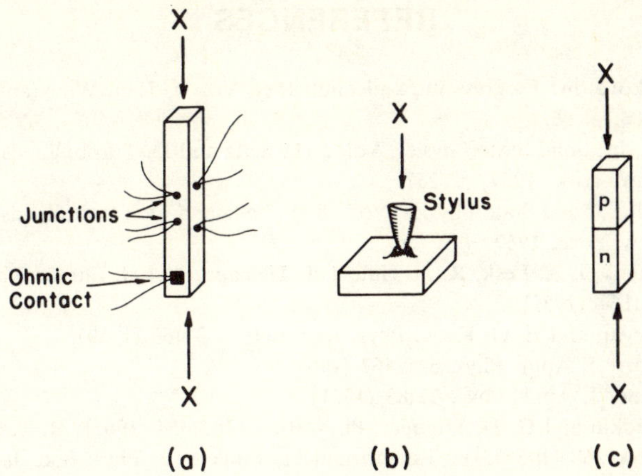

Fig. 9. Various experimental arrangements used in stress experiments.

concentration of gallium acceptors can only be estimated from the solubility curves. One obtains a Fermi level on the *p*-type side of about 0.14 eV below the top of the valence band.

Double contact wires were attached to the dots and to the common ohmic contact, so that four-probe measurements avoiding errors due to the lead resistances can be made. The samples were briefly etched to remove the perimeter of the junctions, which has the wrong crystallographic orientation. The dot diameter was about 0.05 cm.

For the stress experiments the arrangement of Fig. 9 (*a*) was used because it allows the placement of several dots on the various crystal faces in the same alloying procedure. Measurements made on opposite junctions were averaged to correct for effects introduced by a possible buckling of the bar under uniaxial compression.

Figure 9 (*b*) shows a diamond stylus pressing onto a *p–n* junction ([22]). This arrangement produces an inhomogeneous stress distribution which cannot be readily analyzed. The configuration of Fig. 9 (*c*) avoids this difficulty ([23]), but does not allow for measurements along different current directions on the same bar. Furthermore, if one restricts oneself to first-order stress effects on the tunneling current, then the stress coefficient for current along the stress direction is a linear combination of the stress coefficients for transverse uniaxial stress and for hydrostatic pressure ([12,13]).

Much higher accuracy and ease of measurement might be obtained by employing alternating stresses and synchronous detection of the resulting AC component in the tunnel current, very similar to the techniques recently developed for measuring the stress-optical constants of solids ([24]).

REFERENCES

1. J. I. Pankove, in: Progress in Semiconductors, Vol. 9, John Wiley and Sons, New York, 1965, p. 46.
2. W. Kohn, in: Solid State Physics, Vol. 5 (F. Seitz and D. Turnbull, eds.), Academic Press, New York, 1957, p. 257.
3. J. C. Phillips, Solid State Physics, Vol. 18 (F. Seitz and D. Turnbull, eds.), Academic Press, New York, 1967.
4. H. Holonyak, I. A. Lesk, R. N. Hall, J. J. Tiemann, and H. Ehrenreich, Phys. Rev. Letters 3:167 (1959).
5. J. V. Morgan and E. O. Kane, Phys. Rev. Letters 3:466 (1959).
6. W. N. Carr, J. Appl. Phys. 34:2467 (1963).
7. E. O. Kane, J. Appl. Phys. 32:83 (1961).
8. D. R. Fredkin and G. H. Wannier, Phys. Rev. 128:2054 (1962); R. T. Shuey, Phys. Rev. 137:A1268 (1965); Y. Takeuti and H. Funada, J. Phys. Soc. Japan 20:1854 (1965).
9. J. J. Tiemann and H. Fritzsche, Phys. Rev. 137:A1910 (1965); L. Kleinman, Phys. Rev. 140:A637 (1965).
10. H. Fritzsche and J. J. Tiemann, Phys. Rev. 130:617 (1963).
11. H. Brooks, in: Advances in Electronics and Electron Physics (L. Marton, ed.), Vol. 7, Academic Press, New York, 1955, p. 85.
12. H. Fritzsche and J. J. Tiemann, Phys. Rev. 139:A920 (1965).
13. A. E. Long and K. F. Hulme, Brit. J. Appl. Phys. 16:147 (1965).
14. Kan Tzuchao, Acta Physica Sinica 19:25, 49 (1963).
15. S. Fujita, H. Fritzsche, and J. J. Tiemann, J. Phys. Soc. Japan 20:1443 (1965); D. Meyerhofer, G. A. Brown, and H. S. Sommers, Phys. Rev. 126:1329 (1962).
16. R. T. Payne, Phys. Rev. Letters 13:53 (1964); Phys. Rev. 139:A570 (1965).
17. R. D. McCammon and G. K. White, Phys. Rev. Letters 10:234 (1963); S. I. Novikova, Soviet Phys.—Solid State 7:2170 (1966).
18. B. N. Brockhouse and P. K. Iyengar, Phys. Rev. 111:747 (1958).
19. S. L. Miller, M. I. Nathan, and A. C. Smith, Phys. Rev. Letters 4:60 (1960).
20. H. Fritzsche and J. J. Tiemann, Proceedings of the 7th International Conference on the Physics of Semiconductors, Paris, 1964, Academic Press, New York, 1965, p. 599.
21. R. T. Payne, Phys. Rev. 154:730 (1967).
22. W. Rindner, J. Appl. Phys. 33:2479 (1962); W. Rindner and I. Braun, J. Appl. Phys. 34:1958 (1963); T. Imai, Japan. J. Appl. Phys. 2:463 (1963).
23. T. Imai, M. Uchida, H. Sato, and A. Kobayashi, Japan. J. Appl. Phys. 4:102 (1965); T. Imai and M. Uchida, Japan. J. Appl. Phys. 4:409 (1965).
24. W. E. Engeler, H. Fritzsche, M. Garfinkel, and J. J. Tiemann, Phys. Rev. Letters 14:1069 (1965); G. W. Gobeli and E. O. Kane, Phys. Rev. Letters 15:142 (1965); W. E. Engeler, M. Garfinkel, and J. J. Tiemann, Phys. Rev. 155:693 (1967); U. Gerhardt, D. Beaglehole, and R. Sandrock, Phys. Rev. Letters (August 7, 1967).

Chapter 13

Phonon-Assisted Tunneling in Semiconductors

L. Kleinman

Department of Physics
University of Southern California

The experimental data we shall try to explain are shown in the Figs. 1 and 2. In Fig. 1 we see the current voltage for a *p–n* junction heavily doped with Sb donors. There is seen to be no current until a voltage is reached such that $eV = \hbar\omega$, where ω is the frequency of a particular phonon needed to conserve momentum. For forward voltage four such onsets, corresponding to four different phonons, can be seen; for back voltage the current due to the first phonon becomes so large that the onset due to the other phonons is not seen. In Fig. 2 we see dI/dV versus V, which gives us peaks for those voltages at which a new phonon process becomes possible. Note that these peaks are completely smeared out for those junctions with donors other than Sb. This is because when Sb is substituted for a Ge atom in the crystal the effect is just to add a single positive charge on the Ge site, whereas the other impurities give a drastic change in the central cell potential as well. Why Sb does not have these central cell corrections is not known, but it is an experimental fact that of all the impurities only Sb has bound electronic energy levels obtained by assuming the impurity potential is equal to the potential Ge plus one proton. The central cell potentials, being localized in $\mathbf{r}$ space, contain high-$\mathbf{k}$ components which can scatter the tunneling electrons and supply the momentum which otherwise would have to be supplied by the phonons. The acceptors do not seem to contribute appreciably to the scattering, which can be understood as follows. Conduction band electrons are repelled by the negative acceptor impurities and hence are not drawn into the central cell to be strongly scattered. On the other hand, the holes (or negative-mass) valence electrons are attracted by the acceptors, but have *p*-like wave functions which vanish at the center of the central cell, and so again are only slightly scattered.

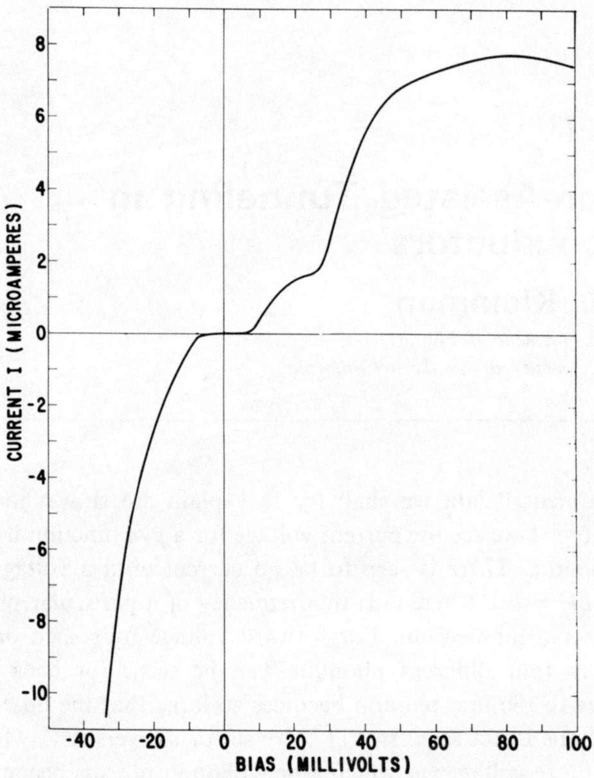

Fig. 1. Current versus voltage for *p–n* junction heavily doped with Sb donors. From Fritzsche and Tiemann ([1]).

From now on we shall limit ourselves to Ge Esaki diodes where the donor is Sb and the momentum for an indirect tunneling must be supplied by a phonon. In Fig. 3 we show the energy bands on the two sides of the junction. In the absence of an applied voltage the Fermi levels on the *n* and *p* sides of the junction must be equal, so that if we assume a junction region of thickness *l* with constant electric field we must have

$$Eel = \Delta_L + \zeta_h + \zeta_e - eV, \tag{1}$$

where ζ_h and ζ_e are the hole and electron Fermi levels, respectively, Δ_Γ and Δ_L the direct and indirect gaps, respectively, and V the applied voltage. Without any calculations we can qualitatively understand the experimental results of Figs. 1 and 2 if we accept the fact that the interband term in the Hamiltonian only connects states with the same **k** vector. This approxima-

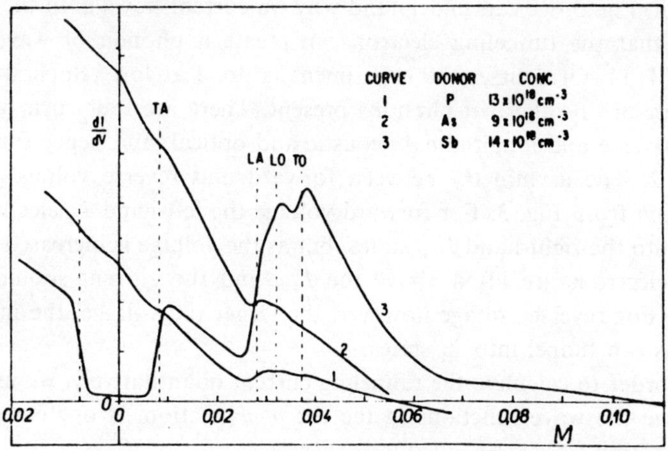

Fig. 2. Plot of dI/dV versus V. From Hall ([2]).

tely true statement is a consequence of the fact that the weak electric field in the junction only slightly destroys the perfect periodicity of the crystal, unlike oxide barrier junctions in which the momentum (or **k**-vector) component normal to the barrier is not even approximately conserved on tunneling. Given this momentum conservation and the fact that the Fermi surface on the p side is near $\mathbf{k} = 0$ and on the n side near $\mathbf{k} = (\pi/a)(1, 1, 1)$,

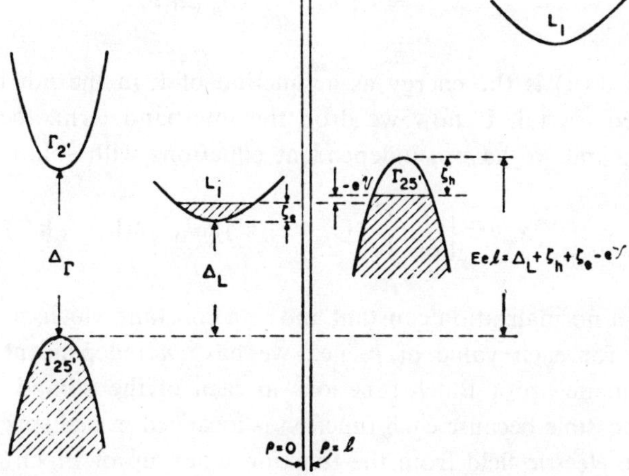

Fig. 3.

as seen in Fig. 3, we can understand why no current flows until the voltage is such that the tunneling electron can create a phonon of wave vector $(\pi/a)(1, 1, 1)$. Of course, the experiment is done at low temperatures, so that there are no thermal phonons present. There are four such phonons, the transverse and longitudinal acoustic and optical, and hence four peaks in Fig. 2. The asymmetry between forward and reverse voltages is also easily seen from Fig. 3. For forward voltage the left-hand L_1 electrons can tunnel into the right-hand $\Gamma_{25'}$ states, but as the voltage is increased further, the L_1 electrons are lifted above the $\Gamma_{25'}$ and the current should return to zero. For reverse voltage however, the larger the voltage, the more $\Gamma_{25'}$ electrons can tunnel into L_1 states.

In order to calculate the tunneling current quantitatively, we must first determine the wave functions in the the p–n junction. If one expands the wave function

$$\Psi = \sum_n \sum_{\mathbf{k}} A_n(\mathbf{k}) b_n(\mathbf{r}, \mathbf{k}), \tag{2}$$

where the $b_n(\mathbf{r}, \mathbf{k}) = (\exp i\mathbf{k}\cdot\mathbf{r})u_{n\mathbf{k}}$ are Bloch functions of wave vector $\mathbf{k}$ and band n, it can be shown that ([3]) in the presence of an electric field E the expansion coefficients $A_n(\mathbf{k})$ obey the Schrödinger equation

$$\left[\varepsilon_n(\mathbf{k}) - ieE\frac{\partial}{\partial k_{\parallel}} - \varepsilon\right] A_n(\mathbf{k}) - \sum_{n'} eEX_{nn'}A_{n'}(\mathbf{k}) = 0, \tag{3}$$

where the interband operator is given by

$$X_{nn'} = i\int u_{nk}^{*}\frac{\partial}{\partial k_{\parallel}} u_{n'k}\, d^3\mathbf{r}, \tag{4}$$

and where $\varepsilon_n(\mathbf{k})$ is the energy as a function of $\mathbf{k}$ in the nth band of the unperturbed crystal. If now we drop the interband term, the A_n are all uncoupled, and we have n independent equations with solutions

$$A_n(\mathbf{k}) = \varkappa^{-1/2}\alpha_n \exp\left\{\frac{i}{eE}\int_0^{k_{\parallel}} [\varepsilon - \varepsilon_n(\mathbf{k}')]\, dk_{\parallel}'\right\}\delta(\mathbf{k}_{\perp} - \mathbf{k}_{\perp 0}), \tag{5}$$

where $\varkappa$ is a normalization constant and α_n a constant which may be chosen differently for each value of n, i.e., we have n independent degenerate functions made up of Bloch functions in each of the n bands. This degeneracy is possible because each function is localized in real space a distance l down the electric field from the function made up of Bloch functions in the zeroth band, such that $Eel = \Delta_{on}$, where Δ_{on} is the energy gap between the bands o and n. Now, if we have n degenerate eigenfunctions, we may combine them in an infinite variety of ways to obtain n different degenerate eigenfunctions. However, in the p–n junction problem we want to take our functions to be the ones we have defined, i.e., each function made up of Bloch functions within a single band. This is because, physically, when an electron goes from the wire lead, say on the right, it goes into a definite band ($\Gamma_{25'}$ in Fig. 3); it then is scattered by some mechanism, either an electron–phonon scattering or the interband term we have dropped from Eq. (3) or a combination of both, into some definite band on the left side of the junction (L_1 in Fig. 3). Thus, in order to calculate the current in a p–n junction, we need only to calculate the scattering rate between right-hand states and left-hand states.

In Chapter 6 Kane and Blount explain how to evaluate Eqs. (2) and 5 in the two-band model by changing the sum over $\mathbf{k}$ to an integral in the complex plane. Because the n independent (on the real $\mathbf{k}$ axis) energy bands $\varepsilon_n(\mathbf{k})$ form a single n-sheeted function in the complex plane, they obtain a function Ψ in which both α_0 and α_1 have finite values. (The in-band functions I use have $\alpha_0 = 1$, $\alpha_1 = 0$, and $\alpha_0 = 0$, $\alpha_1 = 1$.) They can get either of two functions by integrating about either of two branch points in the complex plane, so that by combining their functions we can get back to the in-band functions and vice versa. They consider the amplitude of α_1 (in that function in which α_1 is small) to give the amplitude for direct inter-band tunneling and get a result in agreement (to within factors of $\pi/3$) with the usual result without ever considering the interband term $X_{nn'}$. Although it is difficult to argue with success, I object to this procedure on several grounds: (1) All the functions we are talking about are standing waves and therefore carry no current. (2) As I have pointed out, one can take any combination of these standing waves one wants; it appears to be pure coincidence that by integrating about a branch point one gets exactly the right combination. Perhaps this is due to the fact that the branch point is also a pole for the interband matrix element. Should one try to extend the procedure of Kane and Blount to phonon-assisted or combination direct plus phonon-assisted tunneling it is not at all clear that this coincidence would hold. (3) As I previously pointed out, the physical problem involves an electron which starts out in one band and ends up in another, not an electron which is in a state consisting of several bands.

In order to more easily match the boundary conditions between the field-free region and the junction region, we shall work in the Wannier representation rather than in the crystal momentum representation (CMR). The Wannier functions are defined by

$$W_n(\mathbf{r}-\mathbf{R}_i) = N^{-1/2}V(2\pi)^{-3}\int_z \exp(-i\mathbf{k}\cdot\mathbf{R}_i)\, b_n(\mathbf{r},\mathbf{k})\, d^3\mathbf{k}, \tag{6}$$

where the integration is over all $\mathbf{k}$ in the Brillouin zone and $\mathbf{R}_i$ is a lattice vector. This can be inverted to give

$$b_n(\mathbf{r},\mathbf{k}) = N^{-1/2}\sum_{\mathbf{R}_i}\exp(i\mathbf{k}\cdot\mathbf{R}_i)\, W(\mathbf{r}-\mathbf{R}_i). \tag{7}$$

The Wannier functions are made up out of Bloch functions in a single band and are localized in real space about the lattice points $\mathbf{R}_i$. The in-band eigenfunctions are then given by

$$\Psi_n(\mathbf{r}) = \sum_{\mathbf{R}_i}\beta_n(\mathbf{R}_i)W_n(\mathbf{r}-\mathbf{R}_i). \tag{8}$$

The prescription ([3]) for obtaining the coefficients of the Wannier functions β is to take the formulas for the coefficients A of the Bloch functions in the CMR and substitute $-i\,\partial/\partial\varrho_r$ for $k_\parallel$. Thus we have

$$[\varepsilon_n(\mathbf{k}_\perp, -\partial/\partial\varrho_\parallel) - eE\varrho_\parallel - \varepsilon]\beta(\boldsymbol{\rho}) = 0. \tag{9}$$

$\beta(\boldsymbol{\rho})$ is strictly meaningful only when $\boldsymbol{\rho}$ takes the discrete values $\mathbf{R}_i$. Note, however, that if only Bloch functions from the bottom of the band contribute to $\Psi_n(\mathbf{r})$, we may take $u_{n\mathbf{k}}(\mathbf{r}) \approx u_{no}(\mathbf{r})$ and write [substituting Eq. (6) into Eq. (8) and dropping factors of N, V, π, etc.]

$$\begin{aligned}\Psi_n &= \sum_{\mathbf{R}_i}\int_z \beta(\mathbf{R}_i)\exp(-i\mathbf{k}\cdot\mathbf{R}_i)\exp(i\mathbf{k}\cdot\mathbf{r})\, u_{no}(\mathbf{r})\, d^3\mathbf{k} \\ &= \sum_{\mathbf{R}_i}\beta(\mathbf{R}_i)\delta(\mathbf{r}-\mathbf{R}_i)\, u_{no}(\mathbf{r}) = \beta(\mathbf{r})u_{no}(\mathbf{r}),\end{aligned} \tag{10}$$

so that our wave function is approximated by an envelope function $\beta(\mathbf{r})$, defined for all $\mathbf{r}$, times the Bloch function at the bottom of the band. This approximation is not too bad for the p–n junction, where eEl is much less than the band width, but probably cannot be used for the constant-field case, where l can be infinite and all the Bloch functions in a band are mixed into Ψ_n.

Applying Eq. (9) to the L_1 band on the left side of the junction, one has in the junction region

$$\begin{aligned}&-\tfrac{1}{2}\hbar^2[\mathbf{k}\cdot(\tilde{m}_{L_1}^{-1})\cdot\mathbf{k} - (m_{L_1}^{-1})_\parallel\, \partial^2/\partial\varrho_\parallel{}^2]\beta_{L_1}(\varrho_\parallel) + eE\varrho_\parallel\beta_{L_1}(\varrho_\parallel) \\ &\quad = \varepsilon_{L_1}\beta_{L_1}(\varrho_\parallel),\end{aligned} \tag{11}$$

where we have used the effective-mass approximation for $\varepsilon_n(\mathbf{k}, -i\,\partial/\partial\varrho)$.

Following Fredkin and Wannier [4] we introduce the classical turning point

$$\varrho_{L_1} = \hbar^2 k_{\parallel L_1}^2 (m_{L_1}^{-1})_{\parallel} / 2eE\,, \tag{12}$$

where $\mathbf{k}_{L_1}$ is the wave vector of the electron in the region to the left of the junction, where there is no electric field, and $\mathbf{k}$ is an exact quantum number. Making the substitution

$$\xi = \alpha_{L_1}[1 - (\varrho/\varrho_{L_1})], \qquad \alpha_{L_1} = (\varrho_{L_1} k_{\parallel})^{2/3}, \tag{13}$$

Eq. (11) becomes

$$d^2\beta_{L_1}/d\xi^2 + \xi\beta_{L_1} = 0\,, \tag{14}$$

which has the solution

$$\beta_{L_1}(\varrho_n) = N_{L_1} A(\xi) = N_{L_1} A\{\alpha_{L_1}[1 - (\varrho_n/\varrho_{L_1})]\}\,, \qquad 0 < \varrho_{\parallel} < l, \quad (15$$

where $A(\xi) = \int_{-\infty}^{\infty} \exp[-\, i(\frac{1}{3}z^3 - \xi z)]\, dz$ is the Airy function. Using the asymptotic value of A for $\varrho_{\parallel} \ll \varrho_{L_1}$, we have

$$\beta_{L_1}(\varrho_{\parallel}) \sim 2iN_{L_1}\pi^{1/2}\alpha_{L_1}^{-1/4}[1 - (\varrho_{\parallel}/\varrho_{L_1})]^{-1/4} \sin\{\tfrac{2}{3}\alpha_{L_1}^{3/2}[1 - (\varrho_{\parallel}/\varrho_{L_1})]^{3/2} + \tfrac{1}{4}\pi\}\,,$$
$$0 < \varrho_{\parallel} \ll \varrho_{L_1}. \tag{16}$$

In the region to the left of the junction, where $E = 0$, the solution to Eq. (11) is a sine wave whose amplitude and phase are determined by matching the solution (16) at $\varrho_{\parallel} = 0$:

$$\beta_{L_1}(\varrho_{\parallel}) = 2iN_{L_1}\pi^{1/2}\alpha_{L_1}^{-1/4} \sin[-\, k_{\parallel}\varrho_{\parallel} + \tfrac{1}{4}\pi], \qquad \varrho_{\parallel} < 0. \tag{17}$$

The normalization constant N_{L_1} is determined by requiring the amplitude of the incoming plane wave to be unity:

$$N_{L_1} = -\, (V_l\pi)^{-1/2}\alpha_{L_1}^{1/4} \exp[i(\tfrac{2}{3}\alpha_{L_i}^{3/2} + \tfrac{1}{4}\pi)].$$

Note that for $\varrho_{\parallel} \gg \varrho_{L_1}$ the Airy function falls off asymptotically like $\exp\{-\frac{2}{3}[(\varrho_{\parallel}/\varrho_{L_1}) - 1]^{3/2}\}$, so that Eqs. (17) and (15) give us $\beta_{L_1}(\varrho_{\parallel})$ in every region where it is appreciably different from zero. In Fig. 4 we schematically show β_{L_1} as well as $\beta_{\Gamma_{25'}}$ and $\beta_{\Gamma_{2'}}$, which are calculated in an identical manner.

Now that we have our wave functions we merely apply the golden rule to calculate the transition rate from a state on one side of the junction to a state of the other side. That times the electric charge e is the current. Let us first consider the current due to scattering by transverse acoustic

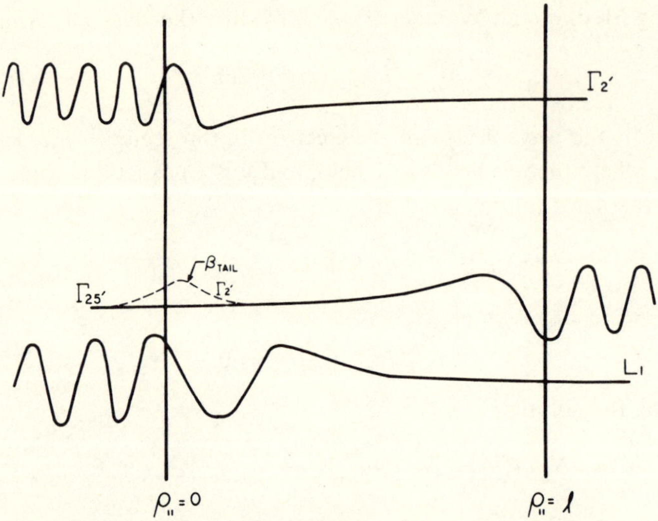

Fig. 4.

phonons between the $\Gamma_{25'}$ and L_1 states of Fig. 4. We have

$$J=\frac{2\pi e}{\hbar}\sum_{\mathbf{k}_{\Gamma_{25'}}}\sum_{\mathbf{k}_{L_1}}\sum_{\text{phonon}} |\langle\Psi_L(\mathbf{k}_{L_1}) \mid H_{\text{TA}} \mid \Psi_{\Gamma_{25'}}(\mathbf{k}_{\Gamma_{25'}})\rangle|^2$$
$$\times (f_{L_1}-f_{\Gamma_{25'}})\,\delta(\varepsilon_L-\varepsilon_{\Gamma_{25'}}\mp\hbar\omega), \tag{18}$$

where the $-$ $(+)$ $\hbar\omega$ holds for positive (negative) voltage in the energy-conserving δ-function. The f_{L_1} and $f_{\Gamma_{25'}}$ are Fermi functions which may be taken to be step functions at the low temperatures at which the tunneling experiments are done. Because the wave functions in the matrix element may be written in the form Eq. (10), where $\beta(\mathbf{r})$ is slowly varying compared to $u_{no}(\mathbf{r})$, the matrix element may be factored

$$\langle\Psi_{L_1} \mid H_{\text{TA}} \mid \Psi_{\Gamma_{25'}}\rangle = M_{\text{TA}}V_{\text{ph}}^{-1/2}\langle\beta_{L_1}(k_{\parallel L_1}) \mid \beta_{\Gamma_{25'}}(k_{\parallel\Gamma_{25'}})\rangle, \tag{19}$$

where M_{TA} is the ordinary TA phonon–electron matrix element and V_{ph} the normalizing volume for the phonons. The β overlap integral and the integrations over $\mathbf{k}_{\Gamma_{25'}}$, $\mathbf{k}_{L_1}$, and phonons are done in detail in ([5]). The calculated current versus voltage is shown in Fig. 5. Kane ([6]), using the constant electric field model and the crystal momentum representation, but otherwise the same process, got essentially the same result, differing only by a prefactor whose numerical value is $\frac{2}{3}$. This current is about 250 times

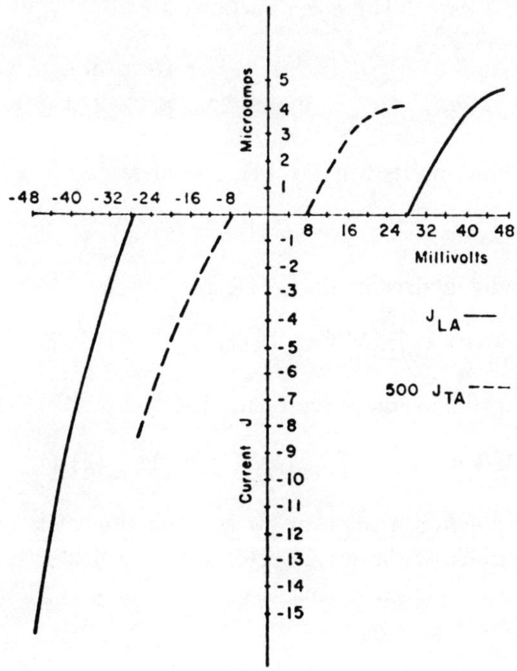

Fig. 5.

smaller than the experimental current and this process applied to the longitudinal-acoustic-phonon assisted tunneling yields a current 1000 times too small.

We therefore consider a second-order process in which the $\Gamma_{25'}$ electron tunnels into the $\Gamma_{2'}$ state via the interband term. This step does not conserve energy, as may be seen from Fig. 3, so a phonon is then emitted and the electron scattered into the L_1 state. It has been shown by Kane [7] that in the two-band model the Hamiltonian is

$$H = \begin{pmatrix} \Delta_\Gamma + (\hbar^2k^2/2m) & (\hbar/m)\mathbf{k} \cdot \boldsymbol{\rho} \\ (\hbar/m)\mathbf{k} \cdot \boldsymbol{\rho} & \hbar^2k^2/2m \end{pmatrix}, \tag{20}$$

which has eigenvalues

$$E = \tfrac{1}{2}\,\Delta_\Gamma + [(\hbar^2k^2/2m) \pm \tfrac{1}{2}\eta], \tag{21}$$

where

$$\eta = [\Delta_\Gamma{}^2 + (\Delta_\Gamma\hbar^2k^2/m_r)]^{1/2} \tag{22}$$

and $m_r^{-1} = m_{\Gamma_{25'}}^{-1} + m_{\Gamma_{2'}}^{-1}$. The eigenfunctions are

$$u_k^{\Gamma_{2'}} = (2\eta)^{-1/2}\{(\eta + \Delta_\Gamma)^{1/2} u_0^{\Gamma_{2'}} + (\eta - \Delta_\Gamma)^{1/2} u_0^{\Gamma_{25'}}\}$$
$$u_k^{\Gamma_{25'}} = (2\eta)^{-1/2}\{(\eta - \Delta_\Gamma)^{1/2} u_0^{\Gamma_{2'}} - (\eta + \Delta_\Gamma)^{1/2} u\}_0^{\Gamma_{25'}}. \tag{23}$$

If these u's are now inserted in Eq. (4) ,one finds

$$X_{\Gamma_{2'}\Gamma_{25'}} = i\hbar\, \Delta_\Gamma^{3/2}/2m_r^{1/2}\eta^2. \tag{24}$$

Thus the interband matrix in the CMR is

$$M(\mathbf{k}) = eE\mathscr{A} \int A_{\Gamma_{2'}}(\mathbf{k}) X_{\Gamma_{2'}\Gamma_{25'}} A_{\Gamma_{25'}}(\mathbf{k})\, d^3\mathbf{k}\,, \tag{25}$$

which becomes in the Wannier representation

$$M(\mathbf{k}) = eE\mathscr{A} \int \beta_{\Gamma_{2'}}(\boldsymbol{\rho}) X(i\,\partial/\partial\varrho_\parallel) \beta_{\Gamma_{25'}}(\boldsymbol{\rho})\, d\varrho_\parallel\,, \tag{26}$$

where $\mathscr{A}$ is the junction area. Here M is a function of $\mathbf{k}$ because the β's are explicit functions of the wave vector $\mathbf{k}$ in the field-free region. If we define the Green's function g such that $X\beta = \int g(\varrho_\parallel, \varrho_\parallel')\beta(\varrho_\parallel')\, d\varrho_\parallel'$, then $X^{-1}(\varrho_\parallel) g(\varrho_\parallel, \varrho_1') = \delta(\varrho_\parallel - \varrho_\parallel')$, or

$$[2m_r^{1/2}/i\hbar\, \Delta_\Gamma^{3/2}][\Delta_\Gamma^2 + \hbar^2 m_r^{-1} \Delta_\Gamma (k_\perp^2 + \partial^2/\partial\varrho_\parallel^2)] g(\varrho_\parallel, \varrho_\parallel') = \delta(\varrho_\parallel - \varrho_\parallel')\,, \tag{27}$$

which has the solution

$$g = -\tfrac{1}{4} i(\Delta_\Gamma m_r)^{1/2}(m_r\Delta_\Gamma + \hbar^2 k_\perp^2)^{-1/2} \exp\{-|\,\varrho_\parallel - \varrho_\parallel'\,|\,(\Delta_\Gamma m_r \hbar^{-2} + k_\perp^2)^{1/2}\} \tag{28}$$

and

$$M(\mathbf{k}) = eE\mathscr{A} \int \beta_{\Gamma_{2'}}(\varrho_\parallel) g(\varrho_\parallel - \varrho_\parallel') \beta_{\Gamma_{25'}}(\varrho')\, d\varrho_\parallel'\, d\varrho_\parallel\,. \tag{29}$$

Once $M(\mathbf{k})$ is evaluated [see ([5])] we can proceed to calculate the admixture of the left-hand $\Gamma_{2'}$ states into the right-hand $\Gamma_{25'}$ states. By first-order perturbation theory this is just the matrix element times the $\Gamma_{2'}$ wave function divided by the energy difference between the $\Gamma_{25'}$ and $\Gamma_{2'}$ states summed over all $\Gamma_{2'}$ states, i.e. we have a $\Gamma_{2'}$ "tail" on the $\Gamma_{25'}$ function

$$\beta_{\text{tail}}(\varrho_\parallel) = \frac{V_l^{1/3}}{2\pi} \int_0^\infty \frac{M(\mathbf{k})\beta_{\Gamma_{2'}}(\varrho_\parallel)\, dk_{\parallel\Gamma_{2'}}}{\Delta_0^2 + \frac{1}{2}\hbar^2[(k_{\parallel\Gamma_{2'}}^2/m_{\Gamma_{2'}}) + (k_\perp^2/m_r) + (k_{\parallel\Gamma_{25'}}^2/m_{\Gamma_{25'}})]}\,, \tag{30}$$

where $\Delta_0 = \Delta_\Gamma - eEl = \Delta_\Gamma - \Delta_L - \zeta_h - \zeta_e + eV$. We have been able to evaluate this integral ([5]) only in the region $\varrho_\parallel < 0$, where $\beta_{\Gamma_{2'}}$ is given by Eq. (17) and in the region $\varrho_{\|} \gg \varrho_{\Gamma_{2'}}$ (where $\varrho_{\Gamma_{2'}}$ is the classical turning point). In both regions β_{tail} has the same exponential fall off (as a function of $|\varrho_{\|}|$). We show β_{tail} as the envelope function for the $\Gamma_{2'}$ admixture into the $\Gamma_{25'}$ state in Fig. 4.

We have estimated the longitudinal-acoustic-phonon assisted tunneling as twice that due to the scattering from the $\varrho_{\|} > 0$ region of β_{tail} into the L_1 state. Equation (18) is again valid (with H_{TA} replaced by H_{LA}), but now

$$\langle \Psi_{L_1} \mid H_{\text{LA}} \mid \Psi_{\Gamma_{25'}} \rangle = M_{\text{LA}} V_{\text{ph}}^{-1/2} \langle V_l^{-1/2} \mid \beta_{\text{tail}}(\varrho_{\|}) \rangle \,, \tag{31}$$

where V_l is the normalizing volume on the left side of the junction. The calculated current shown in Fig. 5 is in almost exact agreement with the experimental current of Fig. 1, which is highly fortuitous because quantities like the junction area are not known to within better than a factor of two. This second-order process can also be applied to the TA case. A symmetry-imposed selection rule makes M_{TA} vanish between exact $\Gamma_{2'}$ and L_1 states; however, for $\mathbf{k} \neq 0$ we see from Eq. (23) that the $\Gamma_{2'}$ states contain both the symmetry of $\Gamma_{2'}$ and $\Gamma_{25'}$ states at $\mathbf{k} = 0$. One can estimate what $\mathbf{k}$'s are needed to construct a wave packet which has the exponential fall off of β_{tail}, and hence from Eq. (23) the percentage of $\Gamma_{25'}$ symmetry in β_{tail}. Similarly, the L_1 electrons are at the Fermi surface near, but not exactly at, the point L_1, and hence contain some $L_{3'}$ symmetry. The total symmetry breaking is about 13% effective; since there are two TA phonons, the second-order process should yield a J_{TA} about $\frac{1}{4}$ of J_{LA}, which Fig. 1 shows to be the case experimentally.

Besides accounting for the magnitude of the current, our second-order process accounts for the independence from junction orientation found experimentally ([1]) for the current. Because the tunneling from one side of the junction to the other occurs between $\Gamma_{25'}$ and $\Gamma_{2'}$ states, both of which have essentially spherical masses, the exponential tunneling factors are independent of orientation. On the other hand, the first-order process between L_1 and $\Gamma_{25'}$ contains $m_{Lr}^{1/2} = [m_{\Gamma_{25'}}^{-1} + (m_{L_1}^{-1})_{\|}]^{-1/2}$ in the tunneling exponential. The large asymetry of m_{L_1} leads to about a factor of two dependence of J on junction orientation for the first-order process. Finally, the pressure coefficients $\pi = J^{-1}\, d\varrho/dP$ for forward and reverse voltages have been measured ([8]). In our formulas for the current, energy gaps, effective masses, and Fermi levels are pressure dependent. The latter two, however, may be related to the pressure dependence of the energy gaps, which is known experimentally. Thus our current formulas also predict the pressure coefficients. The results are shown in Table I, which compares

TABLE I

Comparison of Experimental and Theoretical Pressure Coefficients $\pi = J^{-1}\, dJ/dP$*

	Experiment $\pi \times 10^{-10}$ cm²/dyn	Theory $\pi \times 10^{-10}$ cm²/dyn
π_{LA^+}	− 2.07	− 2.62
π_{LA^-}	− 2.46	− 3.13
π_{TA^+}	− 1.47	− 2.38
π_{TA^-}	− 1.56	− 2.47

experiment with the second-order theory. The theory is seen to predict very well the LA–TA differences as well as the forward–reverse voltage asymmetries.

REFERENCES

1. H. Fritzsche and J. J. Tiemann, Phys. Rev. 130:617 (1963).
2. R. N. Hall, in: Proceedings of the International Conference on Semiconductor Physics, Prague, 1960, Czechoslovak Academy of Sciences, Prague, 1960, p. 193.
3. E. N. Adams, J. Chem. Phys. 21:2013 (1953).
4. D. R. Fredkin and G. H. Wannier, Phys. Rev. 128:2054 (1962).
5. L. Kleinman, Phys. Rev. 140:A637 (1965).
6. E. O. Kane, J. Appl. Phys. 32:83 (1961).
7. E. O. Kane, J. Phys. Chem. Solids 12:181 (1959).
8. H. Fritzsche and J. J. Tiemann, in: Proceedings of the International Conference on the Physics of Semiconductors, Paris, 1964, Academic Press, New York, 1965, p. 599.

Chapter 14

Excess Currents in Semiconductor Tunneling

C. T. Sah*

Department of Electrical Engineering
Department of Physics
and Materials Research Laboratory
University of Illinois, Urbana, Illinois

Those components of the junction current in semiconductor tunnel diodes which cannot be explained in terms of simple tunnel transition across the junction barrier are known as excess currents. The origin and physical mechanisms of these excess currents have been stipulated and investigated in detail by many workers since Esaki first observed and interpretated the negative resistance phenomena in degenerate germanium *p–n* junctions ([1]).

The observed excess currents can be classified into the two types of voltage dependences illustrated schematically in Fig. 1. The solid curve represents what is normally observed in a germanium or silicon tunnel *p–n* junction at room temperature. The curves given by the longer dashes are the currents predicted by the simple tunneling theory based on electron and hole tunneling transition across the potential barrier of the *p–n* junction with the assistance of phonons to conserve momentum, and the minority carrier injection current based on the simple diffusion model of Shockley ([2]). The difference between the experimental curve and the two calculated components, shown in Fig. 1 as cross-hatched area, is one component of the excess current. This is the most commonly observed component. It is relatively temperature insensitive compared with the injection-diffusion component, which has an activation energy of the thermal band gap. The magnitude of this excess component depends on the type of semiconductor used and to an appreciable extent on the fabrication processes. In practical

* Supported in part by the Advanced Research Projects Agency and Air Force Office of Scientific Research.

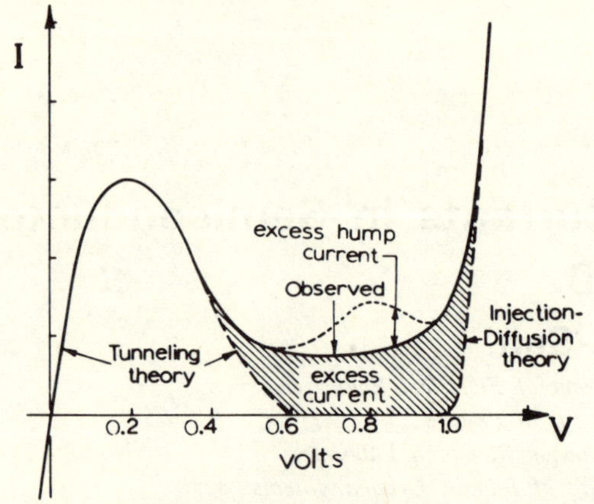

Fig. 1. The current–voltage characteristics of a tunnel diode, showing the excess current components. The theory is indicated by the long dashed curves, the solid curve is that normally observed, where the cross-hatched area is the humpless component of the excess current, and the short dash corresponds to the presence of an additional hump component of the excess current which may be produced by radiation damage or impurity doping.

applications one of the figures of merit used to characterize the tunnel diodes is the peak to valley current ratio, which serves as a measure of this component of the excess current and its dependence on the fabrication processes in one type of semiconductor material. However, it is not a good gauge for the study of excess current mechanisms in different materials, since the peak current (see Fig. 1) is strongly dependent on the band structure of the material. For example, the peak to valley current ratio in silicon tunnel diodes has not exceeded about 4.3 at room temperature after considerable effort was made to control and improve the fabrication techniques, while in germanium and gallium arsenide tunnel junctions ratios of 10 or more can be readily obtained. This large difference reflects the difference between the direct and indirect (phonon-assisted) tunneling rates due to the magnitude of the band gap and the location of the band edges more than the difference in excess currents.

The second excess current component, shown as a curve of shorter dashes superimposed onto the observed I–V curve in Fig. 1, has structures or current peaks. More than one peak has been observed, although in the figure only one peak is illustrated. This excess current component, which has been referred to as the hump current, is usually observed in tunnel

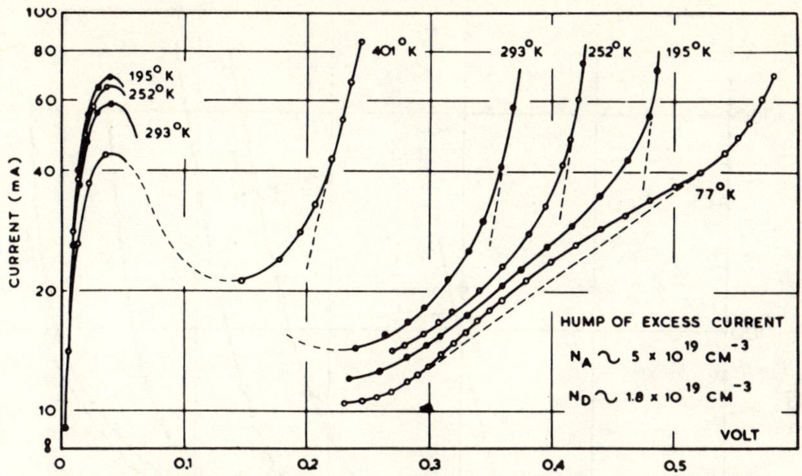

Fig. 2. Temperature dependence of the I–V characteristics of a germanium tunnel diode showing the hump component of the excess current at 77 °K. From Esaki (4).

diodes doped with deep level impurity centers or irradiated with nuclear particles. As a consequence, the hump component has been subjected to a number of detailed studies under controlled experimental and fabrication conditions.

In this chapter we shall first make a survey of the experimental observations on the excess current components under various conditions and in several materials, and then present a brief review of the various models put forth to account for the excess currents.

SURVEY OF EXPERIMENTAL OBSERVATIONS

The two excess current components, the humpless and the hump, have been observed in many semiconductor tunnel diodes, including Ge and Si and GaAs, InP, and other compound semiconductors. The earliest observation and discussion of possible mechanisms was made by Yajima and Esaki in 1958 in their study of the correlation between the excess current and low-frequency noise (3). Both of the excess current components were observed in germanium tunnel junctions. The hump component became evident at 77°K. The noise power spectrum in the valley region displayed a $1/f$ frequency dependence, indicating a distribution of time constants associated with the fluctuation. An example of these two excess current components in germanium p–n junction is shown in Fig. 2.

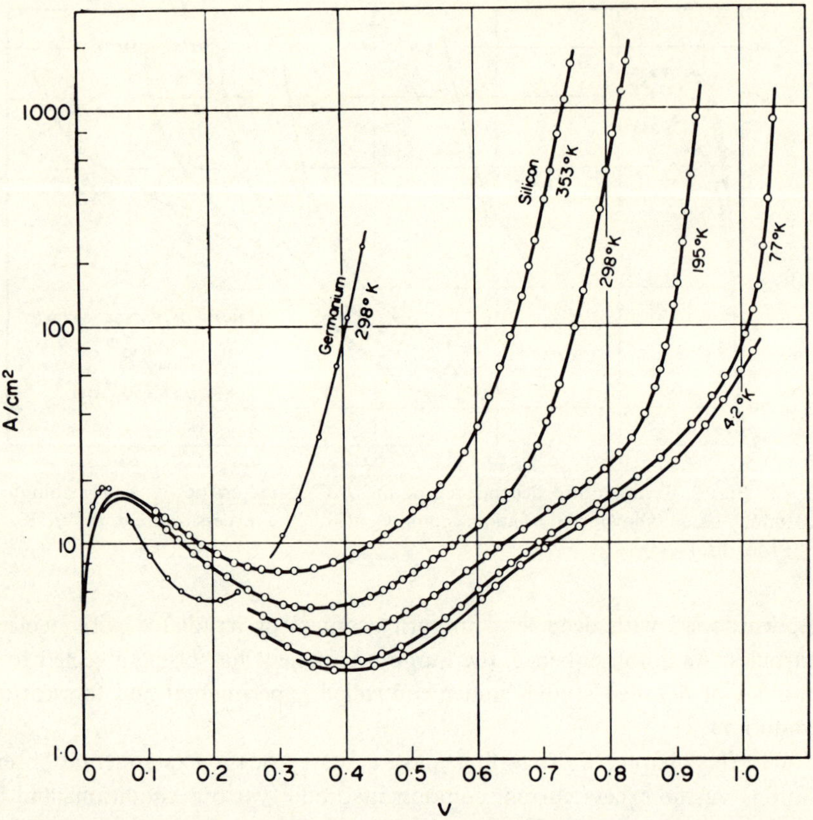

Fig. 3. Excess hump current in a silicon tunnel diode which becomes increasingly evident at low temperature. From Esaki and Miyahara (5).

Several mechanisms were proposed as possible origins of these excess currents (3). These included crystal imperfections in the bulk due to impurity aggregates from heavy impurity doping, surface leakage due to the very narrow junction transition region, and gap states causing additional tunneling current through these localized states in the transition region of the junction, which ceases at a considerably higher forward bias voltage than that required to uncross the band edges on the two sides of the junction transition region.

The two excess current components were also observed in silicon by Esaki and Miyahara (5) (Fig. 3). The hump components in both germanium and silicon (Figs. 2 and 3) diodes appear to set in at about $E_G/2$ (Ge: 0.3 eV, Si: 0.55 eV), which suggests they are due to transition through gap states, as pointed out by Esaki and co-workers, since it is well known that

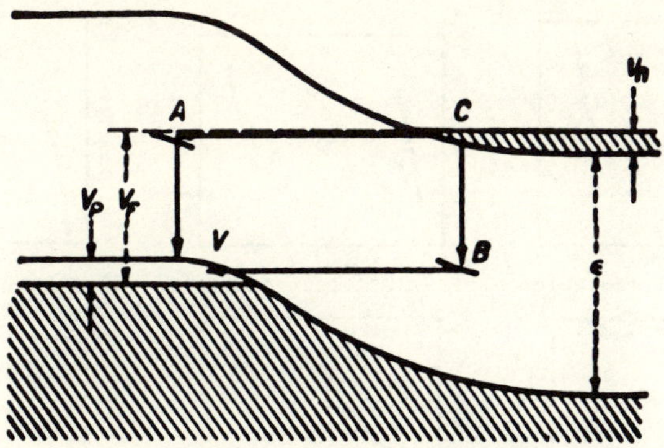

Fig. 4. A detailed discrete gap state model of two-step tunneling processes for the excess currents. For a given discrete level there are two alternative and simultaneous paths for electronic transition from the *n*-type degenerate conduction band to the *p*-type degenerate valence band: CAV and CBV. From Cheynoweth and Logan ([6]).

there are many defect and impurity levels located near the midgap position of these materials.

A detailed and specific model of tunneling transition through the gap states was analyzed in detail by Chynoweth and Logan in 1960 ([6]). This model made use of a discrete-level gap state. As illustrated in Fig. 4, the excess tunnel current comes from the two-step transition processes between the valence band on one side of the junction transition region and the conduction band on the other (*n*-type) side of the junction. It is evident that for a single discrete level and a forward-bias condition two possible paths of current-carrying transition are possible. In both of these cases one of the two steps is the tunneling step. For example, an electron from the conduction band of the degenerate *n*-type side makes a tunneling transition, horizontally in energy, to the trap state located at A (see Fig. 4) as a first step. In the second step of this tunneling path the trapped electron at A makes a vertical transition and drops into a hole state in the valence band of the *p*-type degenerate side. This is the well-known step of hole capture in the Shockley–Read–Hall (SRH) model of electron–hole recombination and generation in semiconductors ([7]). It has been shown by Lax ([8]) that in these vertical SRH-type of transitions the energy involved may be supplied and carried away by cascade absorption or emission of many phonons.

The discrete-gap-state model employed by Chynoweth and Logan to account for the smooth or humpless excess current component in silicon

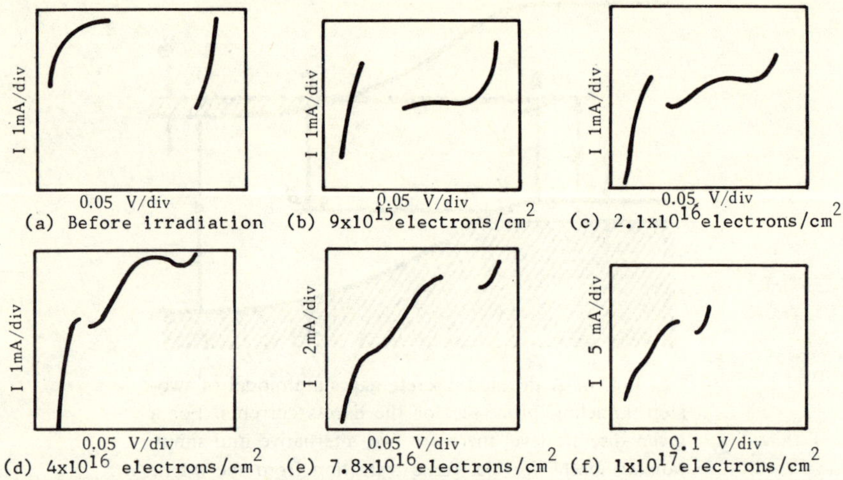

Fig. 5. The effect of increasing electron-irradiation dosage on the excess current in germanium tunnel diode. From Classen ([11]).

tunnel diodes was later shown to be inadequate, and more extensive correlation of experiments with the model by Chynoweth *et al.* ([9]) showed that a special distribution of the gap states as a function of energy located in the *p–n* junction transition region was required to interpret the observed voltage and temperature dependences.

Several experiments were designed to test the hypothesis of the gap state model of excess current. One type of experiment made use of radiation damage to generate a set of controlled gap states in tunnel diodes already fabricated and then carried out measurements to correlate the observed excess currents and the amount of radiation dosage or damage. A second type of experiment made use of deep level impurity doping to generate gap states and increase excess current. A third type of experiment made use of controlled doping of the shallow level impurities to generate relatively shallow impurity bands in the gap and thereby promote the excess current components. A number of these experiments are discussed below and some of the results are illustrated.

The radiation damage experiments were first performed by Longo in germanium tunnel diodes where structures and humps were observed in the valley current range ([10]). Extensive studies of excess current due to electron irradiation damage were made by Classen in germanium, silicon, and gallium arsenide tunnel diodes. A dramatic demonstration of the direct correlation of excess current with electron radiation damage is illustrated in Fig. 5, where the *I–V* characteristics of a germanium tunnel diode are

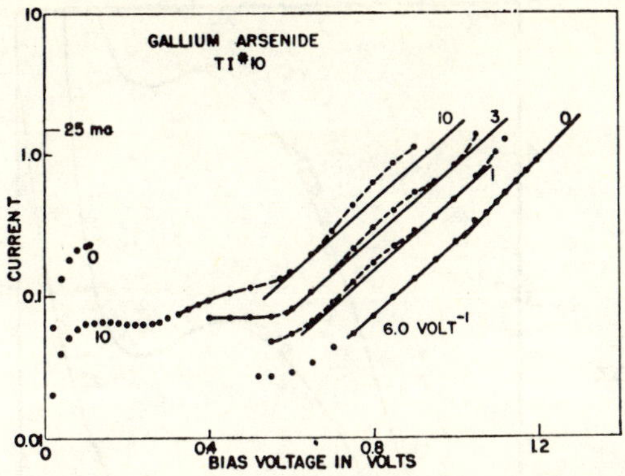

Fig. 6. The effect of increasing electron-irradiation dosage on the excess current in GaAs tunnel diode. From Classen ([11]).

shown for increasing radiation dosages. A similar set of characteristics for electron-irradiated GaAs tunnel diodes are shown in Fig. 6. In this figure curves labeled 0, 1, 3, and 10 correspond, respectively, to 0, 0.25, 0.74, and 3.2×10^{18} electrons/cm^2 dosage from a 2-MeV electron beam. It is evident in Fig. 5 that the excess hump current sets in about 0.1 eV and would be associated with a gap level at about this energy due to radiation damage in germanium. In the case of GaAs (Fig. 6) two distinct humps set in at about 0.3 and 0.6 eV, corresponding to electron-damage gap states at these energy levels.

Detailed excess current experiments concerning electron-irradiation-induced defects in silicon and their subsequent annealing have also been carried out by Roth *et al.* ([12]). These results showed a general increase of excess current after irradiation, with several distinct humps. Subsequent annealing at high temperature (433°K) with or without junction bias showed considerable removal of the radiation damage and corresponding decrease of the irradiation-induced excess current.

The second type of experiments designed to make a detailed study of the excess currents were performed by doping the tunnel junctions with deep level impurities ([13]). A number of metals were used in silicon diodes by diffusion-doping prior to alloy junction formation. A number of humps and fine structures were observed. A detailed analysis of the impurity-induced excess current was made and correlated quantitatively with the two-step tunneling model through the gap states using the theoretical tunneling rate calculation due to Price ([14]). A typical result of a family of silicon tunnel diodes with increasing amount of gold doping is shown in

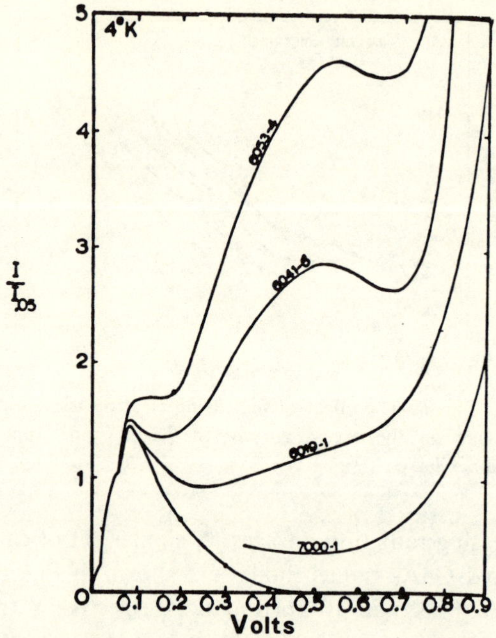

Fig. 7. The effect of gold doping on the excess current in silicon tunnel diode. The gold concentrations are 0, 1.9, 4.1, and 5.3 $\times$ 10^{16} atoms/cm^3 for the samples 7000-1, 6019-1, 6041-6, and 6053-4, respectively. The curve passing through the dots corresponds to the result of extrapolation to zero gold concentration using the top three curves and additional data. From Sah ([13]).

Fig. 7. The fine structures both in the I–V curves and in the conductance (dI/dV)–voltage curves shown in Fig. 8 are correlated with the two-step processes associated with the two gold levels in the band gap of silicon ([13]), one located at $E_C - E_T = 0.54$ eV and the other located at $E_T - E_V = 0.35$ eV.

A demonstration of impurity gap state induced excess current in germanium tunnel diode was recently made using gold impurity ([14]). Figure 9 illustrates the I–V curve of germanium junctions, doped and undoped with gold, showing considerable excess currents due to gold gap states at $E_C - E_T = 0.2$ eV and $E_T - E_V = 0.16$ eV. A second hump current associated with the second route illustrated in Fig. 4 should become evident at about a forward bias of $E_G - V$ (onset voltage of the first hump) $\doteq 0.6 - 0.15$

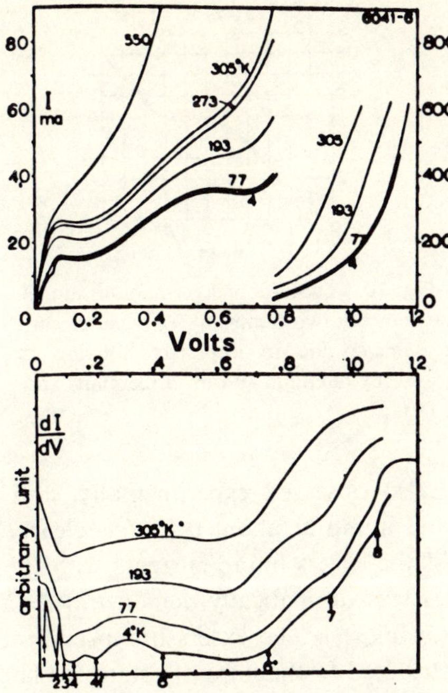

Fig. 8. The hump currents and the structures in the conductance–voltage characteristics of a gold-doped (4.1×10^{16} atoms/cm^3) silicon tunnel diode. From Sah ([13]).

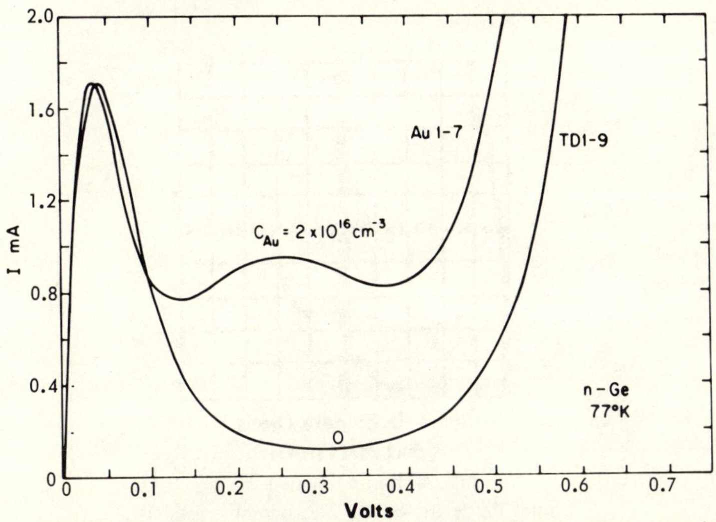

Fig. 9. The hump excess current in gold-doped germanium tunnel diode at 77 °K. Streetman and Sah ([14]).

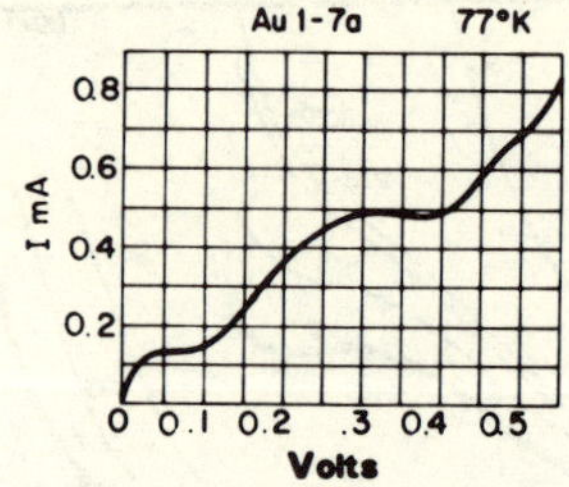

Fig. 10. A gold-doped germanium tunnel diode showing two humps at 77 °K after surface damage due to imperfect alloying is removed by etching. From Streetman and Sah ([14]).

$= 0.45$ eV. This was also observed experimentally, and is illustrated in Fig. 10, where a second hump at about 0.45 V is clearly evident.

The association of the excess hump currents with the shallow-energy-level doping impurities was dramatically demonstrated in GaAs and InP tunnel diodes by Holonyak ([15]). The excess hump current in GaAs diode shown in Fig. 11 is attributed to the impurity states or band of the Cd–Zn p-type dopant in the degenerate starting material, since only an appropriate amount of Cd or Zn impurity (sufficiently high to cause impurity band formation but not so high that the impurity band is merged into the valence band) will cause an appreciable excess hump component. This hump

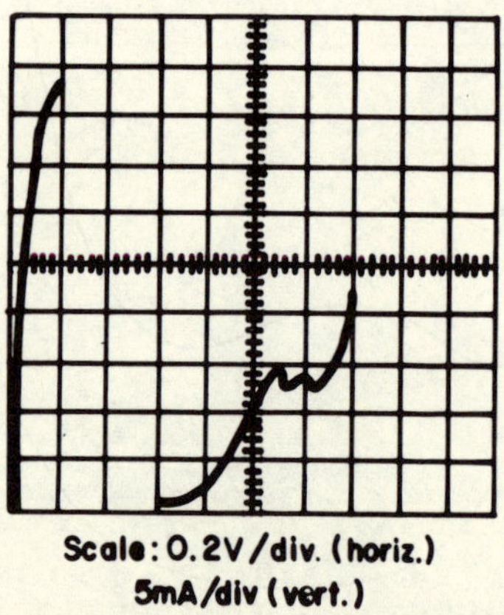

Fig. 11. A hump current in a GaAs tunnel diode at 78 °K. The parent crystal is doped with Cd–Zn, while the alloy region is doped with Sn.

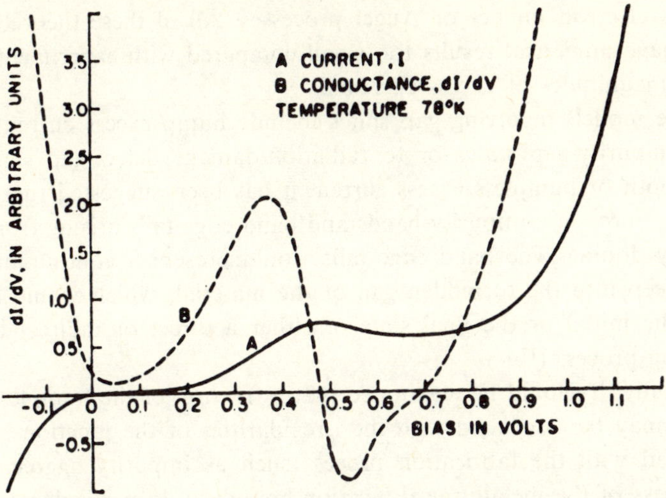

Fig. 12. A hump current in an InP tunnel diode at 78 °K which has little or no direct or one-step tunnel current component.

component may again be interpreted in terms of the two-step process illustrated in Fig. 4 (path CBV), where the acceptor state is located at approximately 0.8 eV from the conduction band edge. The alternative two-step component of hump current (path CAV) would then have an onset of 0.2 eV and perhaps much smaller magnitude due to the large barrier height and long forbidden path for tunneling, so that it is obscured in the negative resistance region.

The hump-type excess current associated with dopant impurity was also and more dramatically observed in InP tunnel diodes (Fig. 12). This excess is again associated with the *p*-type dopant in the substrate material, which gives rise to a gap state or band at about 0.2 eV from the valence band edge and provides the two-step tunneling path.

SURVEY OF THEORETICAL CALCULATIONS AND MODELS

A summary of the various possible excess current models was given by Kane ([16]), who divided them into two catagories: those involving gap states and those involging direct interactions of electrons and holes and no gap states. Calculations have been made for the magnitude of the excess current without gap states, where energy and momentum conservation are achieved by incorporation of photons, phonons, and multiphonons or by electron–electron impact or Auger processes. All of these theoretical estimates gave numerical results too small compared with experimentally observed magnitudes of excess currents.

The models involving gap states include hump excess current due to either impurity-gap states or to radiation-damage defect-gap states. For the smooth or humpless excess current it has been suggested that the important sources are impurity bands and band edge tails arising from heavy impurity doping. The band edge tails would present a continuum of gap states deep into the forbidden gap of the material, which would serve as either the initial or the final state of either a direct or indirect two-step tunneling process ([17,18]).

Finally, it should be mentioned that a fraction of the humpless excess current may be associated with the irregularities of the junction structure associated with the fabrication process such as impurity aggregates and ruggedness of the metallurgical junction boundary. In particular, the irregular thickness of the very thin regrowth layer of alloyed tunnel diodes not only would provide shorter tunneling paths and high-field regions for higher tunneling rates, but would also enhance the injection current component. Thus a critical fabrication step is the use of long annealing and slow cooling to obtain uniform junction boundary at as low a temperature as possible so that negligible diffusion would occur during the alloying cycle. A systematic study of the fabrication problem associated with the excess current has not appeared in the literature.

REFERENCES

1. L. Esaki, "New Phenomenon in Narrow Germanium *p–n* Junctions," Phys. Rev. 109:603–604 (1958).
2. W. Shockley, "Theory of *p–n* Junctions and Transistors," Bell System Tech. J. 28:435–489 (1949).
3. T. Yajima and L. Esaki, "Excess Noise in Narrow Germanium *p–n* Junctions," J. Phys. Soc. Japan 13:1281–1287 (1958).
4. L. Esaki, "Properties of Heavily-Doped Germanium and Silicon Narrow *p–n* Junctions," in: Solid State Physics in Electronics and Telecommunication, Academic Press, New York, 1960, Vol. I, pp. 514–523.
5. L. Esaki and Y. Miyahara, "A New Device Using the Tunneling Process in Narrow *p–n* Junctions," J. Solid State Electronics, 1:13–21 (1960).
6. A. G. Chynoweth and R. A. Logan, "On the Excess Current in Silicon Esaki Junctions," in: Proceedings of the International Conference on Semiconductor Physics, Prague, 1960, Academic Press, New York, 1961, pp. 201–203.
7. W. Shockley and W. T. Read, Jr., "Statistics of Recombinations of Holes and Electrons," Phys. Rev. 87:835–842 (1952); R. N. Hall, "Germanium Rectifier Charac-

teristics," Phys. Rev. 83:228 (1951); R. N. Hall, "Electron–Hole Recombination in Germanium," Phys. Rev. 87:387 (1952); R. N. Hall, "Recombination Processes in Semiconductors," Proc. IEE Vol. 106B, Suppl. 17, pp. 923–931, 1959.

8. M. Lax, "Cascade Capture of Electrons in Solids," Phys. Rev. 119:1502–1523 (1960).
9. A. G. Chynoweth, W. L. Feldmann, and R. A. Logan, "Excess Tunnel Current in Silicon Esaki Junctions," Phys. Rev. 121:684–694 (1961).
10. T. A. Longo, Bull. Am. Phys. Soc. 5:160 (1960).
11. R. S. Classen, "Excess and Hump Current in Esaki Diodes," J. Appl. Phys. 32:2372–2378 (1961).
12. H. Roth, W. Bernard, A. P. Schmid, and P. Zeldes, "Voltage Anneal of Radiation-Induced Excess Current in Tunnel Diodes," J. Appl. Phys. 34:669–671 (1963).
13. C. T. Sah, "Electronic Processes and Excess Currents in Gold-Doped Narrow Silicon Junctions," Phys. Rev. 123:1594–1612 (1961).
14. B. Streetman and C. T. Sah, "Excess Currents in Au-Doped Ge Tunneling Junctions," Proc. IEEE, Vol. 55 (1967).
15. N. Holonyak, Jr., "Evidence for States-Bands in the Forbidden Gap of Degenerate GaAs and InP — Secondary Tunnel Currents and Negative Resistances," J. Appl. Phys. 32:130–131 (1961).
16. E. O. Kane, "Theory of Tunneling," J. Appl. Phys. 32:83–91 (1961).
17. E. O. Kane, "Thomas–Fermi Approach to Impurity Semiconductor Band Structure," Phys. Rev. 131:79–88 (1963).
18. R. A. Logan and A. G. Chynoweth, "Effect of Degenerate Semiconductor Band Structure on Current–Voltage Characteristics of Silicon Tunnel Diodes," Phys. Rev. 131:89–95 (1963).

Chapter 15

Photon-Assisted Tunneling (Franz–Keldysh Effect)

W. Franz

Institute of Theoretical Physics
University of Münster

Before we treat photon-assisted tunneling between two bands we will review optical absorption in the absence of fields. Then we will give a rough theory of photon-assisted tunneling using the WKB method, and follow that with an improved theory. Next we will treat a simple model for the exponential absorption tail, and finally we will mention some further theoretical developments and some experiments.

OPTICAL ABSORPTION WITHOUT AN ELECTRIC FIELD

Optical absorption occurs in a pure semiconductor or in a insulator crystal if the energy of an incident photon $\hbar\omega$ is high enough to lift an electron from the valence band into an empty state in the conduction band. There are two possibilities. The absorption may be direct, with conservation of electron momentum, or indirect, with a net change of electron momentum through the emission or absorption of a phonon. The momentum of the photon is negligible in both cases. In contrast, the phonon *energy*, is much smaller than the photon energy. For crystals, which have a direct absorption threshold at ω_0, the absorption constant is very small for energies below $\hbar\omega_0$ and rises very rapidly for energies above $\hbar\omega_0$. It is proportional to the matrix element for the optical transition and to the density of states, which is $\alpha(\omega - \omega_0)^{1/2}$. There is a further sharp increase near energies which can lift an electron from the valance band to the conduction band at so-called critical points, where

$$\nabla_p[E_c(p) - E_v(p)] = 0.$$

For an indirect threshold the absorption constant is much smaller than for a direct one because of the phonon process involved. Thus in most cases it requires considerable refinement to measure the absorption for energies above the first absorption threshold.

The absorption is connected with reflectivity through the Kramers–Kronig dispersion relation. Therefore reflectivity measurements were used in the experimental determination of thresholds above the first edge, but by means of the usual method it was not possible to resolve structures smaller than 0.1 eV. Only a derivative technique based on the photon-assisted tunneling is capable of determining energies of interband transitions with errors of about 1 meV or better.

If there are impurity levels in the gap, absorption also occurs for energies below the first threshold. In the absence of other absorption processes

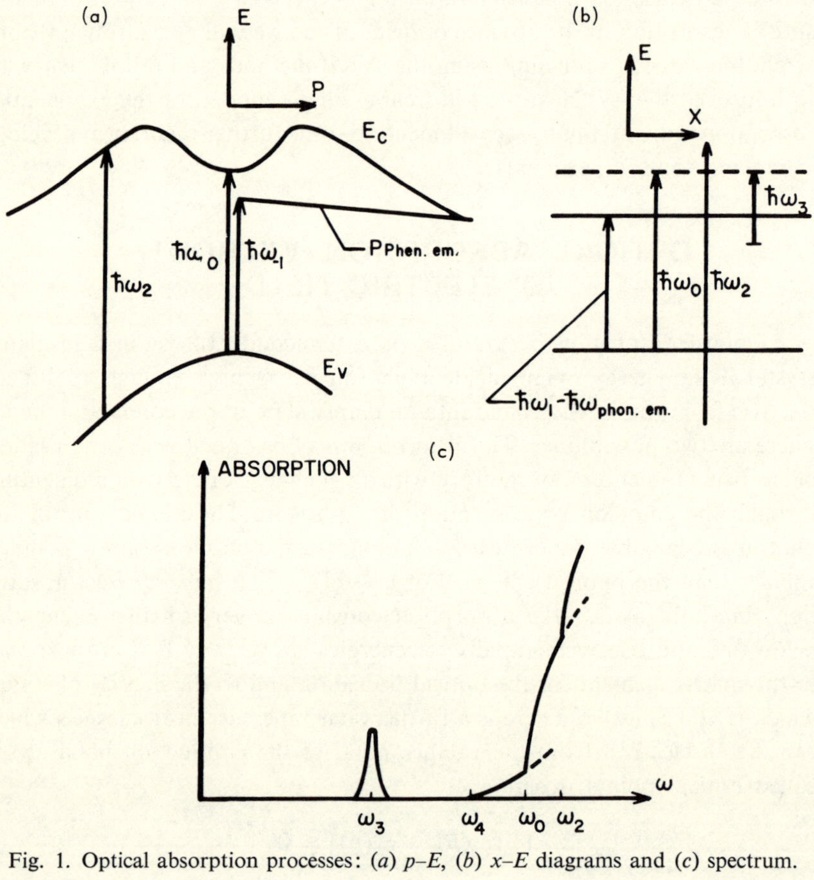

Fig. 1. Optical absorption processes: (*a*) *p*–*E*, (*b*) *x*–*E* diagrams and (*c*) spectrum.

they should give rise to an absorption band due to transitions of electrons from the impurity levels to the conduction band. All these cases are shown in Fig. 1.

PHOTON-ASSISTED TUNNELING BETWEEN BANDS. THEORY USING THE WKB METHOD

We will now treat the optical absorption when a field is applied. We first discuss a rough theory using damped electron waves in the forbidden gap and then a more abstract theory using Houston functions.

In an external constant electric field the bands in the E–x diagram are tilted. One finds an electron of fixed total energy confined to an interval of the x axis where its kinetic energy is within the valence band. For band electrons tunneling into the forbidden gap we have a probability dropping like e^{2ipx}, where p is the imaginary momentum.

In contrast, an impurity level is not tilted by the electric field and, having the simple model of a potential well for the impurity as shown in Fig. 2, we have an unsymmetrically damped wave outside of this potential well. Therefore we get an altered absorption band which is more smeared out than in the free-field case, but the maximum of absorption is essentially unshifted.

The photon-assisted interband tunneling process is now thought to take place in the following manner (see Fig. 3): An electron tunnels from the top of the valence band at x_1 into the forbidden gap to $\bar{x}$, is lifted there in energy by absorption of a photon $\hbar\omega$, and then tunnels further to x_2, where it reaches the bottom of the conduction band. The $\bar{x}$ is determined by the energy $\hbar\omega$ and conservation of imaginary momentum. We expect now the optical transition rate to be proportional to the squares of the wave amplitudes at the point of transition, which are calculated by the WKB method:

$$A \propto \left[\exp \frac{i}{\hbar} \int_{x_1}^{\bar{x}} p_v \, dx\right]^2 \cdot \left[\exp \frac{i}{\hbar} \int_{\bar{x}}^{x_2} p_c \, dx\right]^2. \tag{1}$$

We integrate by parts to get

$$A \propto \exp\left\{\frac{2i}{\hbar}\left[\int_{\bar{p}}^{0} (x_v - \bar{x}) \, dp + \int_0^{\bar{p}} (x_c - \bar{x}) \, dp\right]\right\}$$

with $\bar{p} = p(\bar{x})$. From Fig. 3 we have for the energy of the electron

$$E(p) = \text{const} + eFx,$$

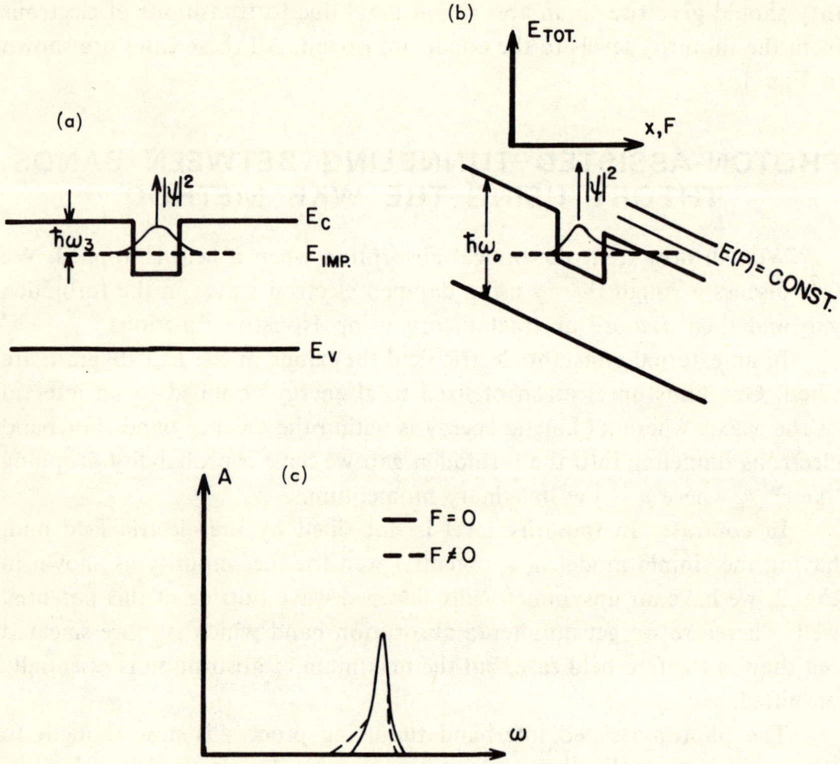

Fig. 2. Field effect on impurity absorption. (*a*) Band scheme without field. (*b*) Band scheme, field applied. (*c*) Spectrum.

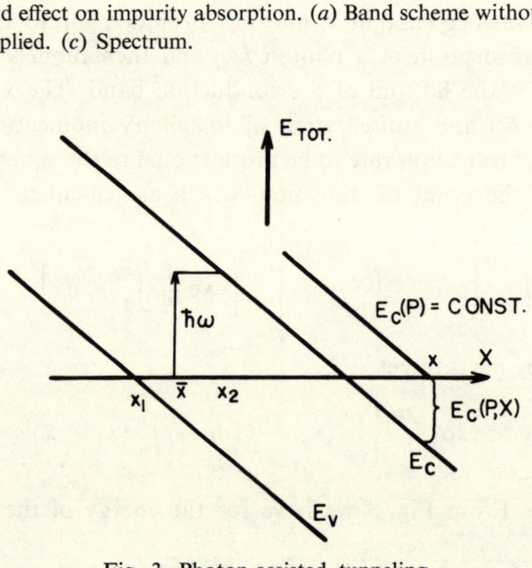

Fig. 3. Photon-assisted tunneling.

and therefore

$$x_v - \bar{x} = [E_v(p) - E_v(\bar{p})]eF.$$

The energy difference $E_c(\bar{p}) - E_v(\bar{p})$ is the photon energy $\hbar\omega$, and we get

$$A \propto \exp\left[\frac{2i}{\hbar eF}\int_0^{\bar{p}} [E_c(p) - E_v(p) - \hbar\omega]\, dp\right]. \tag{2}$$

For photon energies near the gap energy I we may use the approximation $E_c - E_v = p^2/m^*$, with the average effective mass near the band edge

$$1/m^* = (1/2m_v) + (1/2m_c).$$

Then we get

$$A \propto \exp\left\{-\frac{4\sqrt{m^*}\,(I - \hbar\omega)^{3/2}}{3\hbar eF}\right\}.$$

From this formula we see that A decreases exponentially below the gap.

IMPROVED THEORY

The theory as developed above is rather rough and only one dimensional. Therefore we will treat the problem again with an improved theory without using any "state" in the forbidden gap. Figure 4 shows an $E(p)$ diagram for a very simple band structure. The momentum of an electron in a band changes with time due to the electric field in a way similar to the free-electron problem:

$$dp/dt = eF, \qquad p(t) = p_0 + eFt,$$

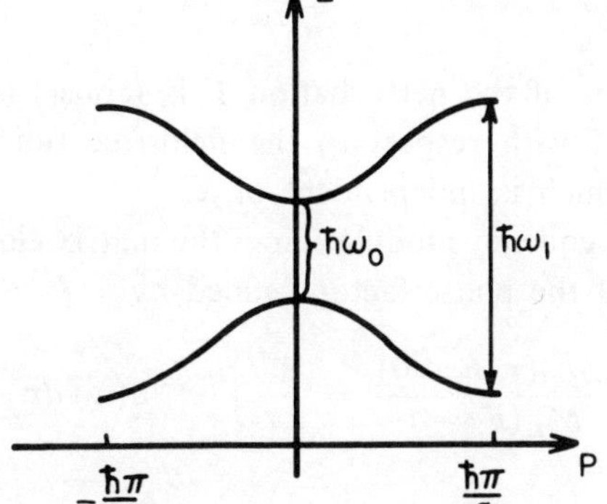

Fig. 4. Simple band structure.

where e is the charge of the electron ($e < 0$). Because $(\hbar\pi/a)$ and $-(\hbar\pi/a)$ give the same state, the electron's state will oscillate in energy between the top and the bottom of the band. In the theory of the many-electron problem the wave function Ψ of the system is represented by a Slater determinant:

$$\Psi = \frac{1}{\sqrt{N!}} \begin{vmatrix} \psi_1(x_1) & \cdots & \psi_N(x_1) \\ \psi_1(x_2) & \cdots & \\ \vdots & & \vdots \\ \psi_1(x_N) & \cdots & \psi_N(x_N) \end{vmatrix} \exp(-i/\hbar)Et,$$

where the one-electron wave functions may be chosen to be, e.g., Bloch, Wannier, or Houston wave functions. In this case the Houston functions are most powerful. We get them from the Bloch functions $\psi_B = u(p, x) \times \exp\{(i/\hbar)px\}$ by replacing the momentum by the time-dependent momentum

$$p(t) = p_0 + eFt.$$

Thus we get

$$\psi_H = u(p, x) \exp\left\{(i/\hbar)\left[p(t)x - \int^t E[p(\tau)]\, d\tau\right]\right\},$$

where the $E(p)$ remain the same functions as before. One can show that these functions are an approximate solution of the Schrödinger equation for a crystal in an electric field.

If we now calculate the matrix element for the optical transition from a state ψ_1 in the valance band to a state ψ_2 in the conduction band, we get

$$\begin{aligned} M_{21}(t, F \neq 0) &= \langle \psi_2 \mid V \mid \psi_1 \rangle \\ &= M_{21}(F = 0) \exp\left[(i/\hbar) \int^t (E_1 - E_2)\, d\tau\right] \end{aligned}$$

because the operator of the perturbation V is (const) $(\partial/\partial x)$ and ψ_1 and ψ_2 are only altered with respect to the field-free functions by a factor $\exp[(i/\hbar) \int^\tau E\, d\tau$, which is independent of x.

Thus we get a frequency modulation of the matrix element. The Fourier components $a(\omega)$ of the phase factor defined by

$$\frac{M_{21}(\tau, F \neq 0)}{M_{21}(F = 0)} = \frac{1}{2\pi} \int^t e^{i\omega\tau} a(\omega)\, d\tau$$

are

$$a(\omega) = \int_{-\infty}^{\infty} \exp\left\{i \int_0^t [\Omega(t') - \omega]\, dt'\right\} dt,$$

where

$$\Omega(t) = (1/\hbar)\{E_c[p(t)] - E_v[p(t)]\}.$$

For $\hbar\omega > I$ the integrand has a point of stationary phase for $t = t_s$, where $\Omega(t_s) = \omega$. For $\hbar\omega < I$ this equation has no real solution; instead it gives a complex saddlepoint t_s of the integrand. In both cases the value of the integrand at the saddlepoint determines an approximate value of $a(\omega)$. For t_s real this is just the ω contribution to the spectrum, which we put in by modulation. For t_s imaginary we get an exponentially small contribution, as ω is outside the modulation range.

Thus we get for imaginary t_s

$$a(\omega) \approx \exp\left[(i/\hbar)\int_0^{t_s} [E_c(p) - E_v(p) - \hbar\omega]\, d\tau\right],$$

or, using $dp/dt = eF$, we have for the square of the matrix element

$$A \propto \exp\left[\frac{2i}{\hbar eF}\int_0^{\bar{p}} [E_c(p) - E_v(p) - \hbar\omega]\, dp\right], \tag{3}$$

which is exactly expression (2), but now obtained from an improved theory.

So far we have discussed the optical absorption due to a single electron. In order to get the contribution of all the electrons in the valence band one has to integrate over a plane in p space $d^2p_\perp$ perpendicular to the direction of the field. This calculation yields as result

$$A(\omega, F) = \frac{1}{4\sqrt{\pi i}}\int_{-i\infty+\varepsilon}^{+i\infty} dt\, t^{-3/2}\exp[\omega_F{}^3t^3 + (\omega - \omega_0)t] \tag{4}$$

with $\omega_F = (e^2F^2/12\hbar m^*)^{1/3}$. Here $A(\omega, F)$ is the normalized absorption constant of the crystal using parabolic bands. For the field-free case one gets

$$A(\omega, F = 0) = (\omega - \omega_0)^{1/2} \quad \text{for} \quad \omega \geq \omega_0$$
$$A = 0 \quad \text{for} \quad \omega \leq \omega_0.$$

Far below the edge the following expansion for frequencies smaller than ω_0 is valid:

$$A(\omega, F) \approx \sqrt{3\omega_F}\,\frac{\omega_F}{4(\omega_0 - \omega)}\left(1 - \left[\frac{3\omega_F}{\omega_0 - \omega}\right]^{3/2}\cdots\right)$$
$$\times \exp\left\{-2\left(\frac{\omega_0 - \omega}{3\omega_F}\right)^{3/2}\right\}.$$

This means that the edge has an exponential tail due to the field (see Fig. 5).

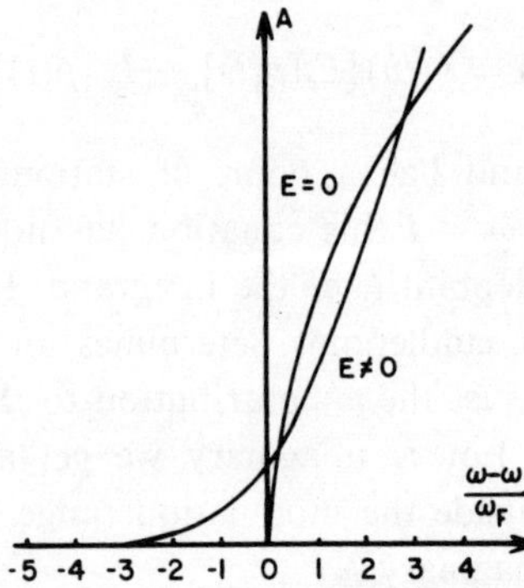

Fig. 5. Ideal band edge without and with field applied.

THE EXPONENTIAL ABSORPTION EDGE

Experimentally, usually one already finds in zero field an optical absorption constant increasing exponentially below the gap frequency of the direct transition. It is known that many effects may contribute to this tail, including excitons, temperature broadening, the influence of impurities on the conduction and valence bands, absorption into and out of impurity levels, and electric field effects on the absorption constant due to the local fields of charged impurities.

For simplicity we will treat the exponential absorption edge as a sum of ideal edges $(\omega - \omega_1)^{1/2}$ due to processes with different gap frequencies ω_1. We assume that the probability of finding a process with gap frequency ω_1 decreases exponentially below ω_0, where $\hbar\omega_0$ is the width of the forbidden band.

Therefore we have to alter Eq. (4) to

$$A(\omega) = \frac{\lambda}{4\sqrt{\pi i}} \int_0^{\omega_0} d\omega_1 \int_{-i\infty}^{i\infty} dt\, t^{-3/2} \exp[\omega_F{}^3 t^3 + (\omega - \omega_1)t + (\omega_1 - \omega_0)\lambda].$$

The integration leads to

$$A(\omega, F) = \frac{\lambda}{4\sqrt{\pi i}} \int_{-i\infty}^{i\infty} \frac{dt}{\lambda - t}\, t^{-3/2} \exp[\omega_F{}^3 t^3 + (\omega - \omega_0)t],$$

which is the result in a closed form.

From this equation we get for the field-free case

$$A(\omega, F=0) = \tfrac{1}{2}\sqrt{(\pi/\lambda)}\exp[\lambda(\omega-\omega_0)] \qquad \text{for} \quad \omega \leq \omega_0.$$

and

$$A(\omega, F=0) = \frac{1}{2}\sqrt{\frac{\pi}{\lambda}}\exp[\lambda(\omega-\omega_0)] - \lambda(\omega-\omega_0)^{3/2}\sum_{n=0}^{\infty}\frac{\Gamma(\frac{3}{2})}{\Gamma(n+\frac{5}{2})} \times [\lambda(\omega-\omega_0)]^n \qquad \text{for} \quad \omega \geq \omega_0.$$

We have obtained the exponential tail below the gap, as expected. For $F \neq 0$ we can make two expansions, one valid over the whole frequency range:

$$A(\omega, F\neq 0) = \frac{1}{2}\sqrt{\frac{\pi}{\lambda}}\exp[\lambda^3\omega_F{}^3 + \lambda(\omega-\omega_0)] - \lambda\omega_F^{2/3}\sum_{n=0}^{\infty}\frac{1}{3n!}\left(\frac{\omega-\omega_0}{\omega_F}\right)^n\sum_{l=0}^{\infty}\frac{\Gamma(\frac{3}{2})}{\Gamma[\frac{3}{2}+(l-n)/3]}(\lambda\omega_F)^l,$$

and a second one for very low frequencies:

$$A(\omega, F\neq 0) \sim \frac{1}{2}\sqrt{\frac{\pi}{\lambda}}\exp[\lambda^3\omega_F{}^3 + \lambda(\omega-\omega_0)] \qquad \text{for} \quad \left(\frac{\omega_0-\omega}{\omega_F}\right)^{1/4}\left[\sqrt{\frac{\omega_0-\omega}{3\omega_F}} - \lambda\omega_F\right] \gg 1. \tag{5}$$

This second expansion, which is only valid far below the gap and not in the maximum of the change in absorption, gives a "shift" in the absorption constant to lower frequencies of $\Delta\omega = \lambda^2\omega_F{}^3$ or, more correctly, a factor to the absorption constant of $\exp(\lambda^3\omega_F{}^3)$. Because ω_F is connected with the combined effective masses this formula gives a method to measure these if one is careful enough in using the model of the exponential tail given here and in using the expansion (5).

FURTHER THEORETICAL DEVELOPMENTS AND EXPERIMENTS

This theory of the photon-assisted tunneling, which was published in more mathematical detail in 1958 ([1]), was also treated by Keldysh ([2]) for the ideal edge at the same time. The theory is extended to indirect absorption by Penchina ([3]) and Chester and Fritzsche ([4]) and was recast in the effective-mass approximation by Aspnes ([5]) for all kinds of direct transitions.

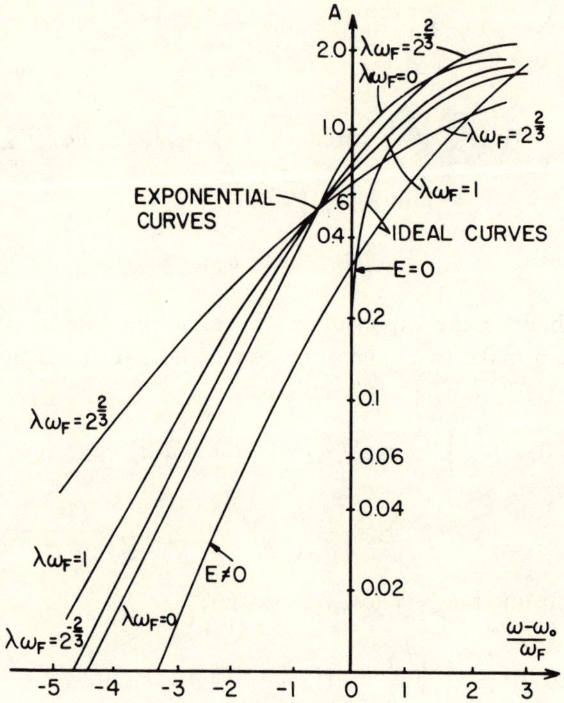

Fig. 6. Exponential band edge for different applied field strengths.

A theoretical work by Krieg ([6]) includes the effect of internal fields due to charged impurities. Duke and Alferieff ([7]) have given a hydrogen-like model with which it should be possible to calculate the field effect on absorption due to excitons. Callaway ([8]) predicts Stark levels to be seen in electroabsorption, but they certainly cannot be observed, since this would only be possible if the electron could be accelerated throughout the conduction band within its impact lifetime (because of the uncertainty principle). But this is not possible for fields below the avalanche breakdown field.

The broadening effect of an absorption edge due to electric field also takes place for higher interband transitions, i.e., for critical points with photon energies higher than $\hbar\omega_0$ as described above. Because the main effect is near the onset of an edge, measurements with a field-modulation technique give curves like a derivative of the absorption spectrum. Due to the Kramers–Kronig relation this is also valid for reflectivity measurements. This electroreflectance method, which it seems is becoming a tool in band structure analysis, is treated extensively both experimentally and theoreticaly by Seraphin ([9]), Seraphin and Bottka ([10]), Phillips and Seraphin ([11]), and other workers.

Experimental results are reported for several materials, especially for Si, Ge, and GaAs. Moss ([12]) compares the "shift" of the edge in GaAs with known effective masses. Stuke and Weiser ([13]) determine the effective masses of Se by means of this method. Frova and Handler ([14]) measured the change of absorption in Ge *p–n* junctions for both direct and indirect transitions. A more recent paper about electroabsorption in Si is by Frova *et al.* ([15]). They believe to have measured oscillations in ΔA, which follow from all theories for frequencies greater than ω_0. [This is due to the fact that Eq. (4) may be written as an integral over Airy functions.] Similar oscillations in GaAs are found by Paige and Rees ([16]). Seraphin and coworkers have measured the effect of changing the potential barrier inside a reflecting surface of Ge ([17]), GaAs ([18]), and Si ([19]). Recently Cardona and *et al.* ([19]) have obtained results on several other materials by means of changing the potential barrier in a semiconductor–electrolyte interface.

Rees ([20]) has measured electroabsorption with a field modulation frequency of 10 GHz. He did not find a difference from measurements with low frequencies. This may open a method for high-frequency modulation of light beams.

REFERENCES

1. W. Franz, Z. Naturforsch. 13a:484 (1958).
2. L. V. Keldysh, Soviet Phys.—JETP 7:788 (1958).
3. C. Penchina, Phys. Rev. 138:A924 (1965).
4. M. Chester and L. Fritsche, Phys. Rev. 139:A578 (1965).
5. D. E. Aspnes, Phys. Rev. 147:554 (1966); Phys. Rev. 153:972 (1957).
6. J. G. Krieg, Z. Physik, to be published.
7. C. B. Duke and M. E. Alferieff, Phys. Rev. 145:583 (1966).
8. J. Callaway, Phys. Rev. 130:549 (1963) and Phys. Rev. 134:A998 (1964).
9. B. O. Seraphin, Phys. Rev. 140:A1716 (1965).
10. B. O. Seraphin and N. Bottka, Phys. Rev. 145:628 (1966).
11. J. C. Phillips and B. O. Seraphin, Phys. Rev. Letters 15:107 (1965).
12. T. S. Moss, J. Appl. Phys. Suppl. 32:2136 (1961).
13. J. Stuke and G. Weiser, Phys. Stat. Sol. 17:343 (1966).
14. A. Frova and P. Handler, Phys. Rev. 137:A1852 (1965).
15. A Frova, P. Handler, A. Germane, D. E. Aspnes, Phys. Rev. 145:575 (1966).
16. E. G. S. Paige and H. D. Rees, Phys. Rev. Letters 16:444 (1966).
17. B. O. Seraphin, R. B. Hess, N. Bottka, J. Appl. Phys. 36:2242 (1965).
18. B. O. Seraphin, J. Appl. Phys. 37:721 (1966).
19. M. Cardona *et al.*, Phys. Rev. 154:696 (1967).
20. H. D. Rees, Phys. Letters 27:629 (1966).

Chapter 16

Magnetotunneling Effects in Semiconductors

W. Zawadzki*

National Magnet Laboratory†
Massachusetts Institute of Technology
Cambridge, Massachusetts

INTRODUCTION

The subject of this chapter is the interband (Zener) tunneling and the photon-assisted tunneling (Franz–Keldysh effect) in semiconductors in the presence of a transverse magnetic field. First, electrons in uniform, constant, crossed fields are treated using the one-band effective-mass approximation (EMA). It is shown that in this approximation, which is applicable to the case of low electric fields, there are no tunneling effects. The limitation of this approach is discussed. It is then shown that for high electric fields a model of two interacting energy bands has to be applied. This model is used to calculate the tunneling and photon-assisted tunneling in crossed fields. Finally, the experimental results on the Franz–Keldysh effect in crossed fields are presented and compared with the theory.

THE ONE-BAND EFFECTIVE-MASS APPROXIMATION

We begin with the use of the one-band effective-mass approximation ([1]). In other words, we shall treat electrons in the conduction and valence bands as free electrons with the free mass replaced by an effective mass for the band in question. We shall subsequently discuss the limitations of this approach.

* Present address: Institute of Electron Technology, Polish Academy of Sciences, Warsaw, Poland.

† Supported by the U.S. Air Force Office of Scientific Research.

The EMA Schrödinger equation for conduction electrons in the presence of electric and magnetic fields is

$$\left[\frac{1}{2m_2}\left(\mathbf{p}+\frac{e}{c}\mathbf{A}\right)^2+e\mathbf{E}\mathbf{r}\right]f_2=\mathscr{E}_2 f_2. \tag{1}$$

We put $\mathbf{E}=[-E,0,0]$; the magnetic field is introduced by means of the vector potential: curl $\mathbf{A}=\mathbf{H}$; m_2 is the effective mass of the conduction electrons and $\mathscr{E}_2$ their energy; e denotes the absolute value of the electron charge; and c is the velocity of light.

Equation (1) determines f_2, the envelope functions of electrons in the conduction band. According to the effective-mass approximation theory, the complete solutions are given by $\Psi_2=f_2u_{20}$, where u_{20} is the periodic part of the Bloch function taken at the bottom of the conduction band, i.e., for $k=0$.

For the crossed field case we choose the gauge $A=[0,Hx,0]$, i.e., the magnetic field is along the z axis. The coordinates in Eq. (1) are separable, and it can be seen that k_y and k_z are the constants of motion. The differential equation in x is just the well-known harmonic oscillator problem, and its solutions are given by the harmonic oscillator functions. The eigenenergies are

$$\mathscr{E}_2=\hbar\omega_{c2}\left(n_2+\frac{1}{2}\right)+\frac{\hbar^2k_{z2}^2}{2m_2}+eEL^2k_{y2}-\frac{m_2c^2E^2}{2H^2}, \tag{2}$$

where $\omega_{c2}=eH/m_2c$ is the cyclotron frequency and $L=(\hbar c/eH)^{1/2}$ is called the magnetic radius. For $E=0$ we obtain the well-known Landau formula, with the motion quantized in the xy plane, and free along the magnetic field direction. Thus the electric field does not affect the motion dramatically. It simply shifts the center of oscillations and brings down the $k_y=0$ energies of the Landau levels by the amount $m_2c^2E^2/2H^2$.

In order to interpret the above quantum mechanical results, we now consider the motion of a free electron in crossed fields according to classical nonrelativistic mechanics (see Fig. 1).

The Newtonian equation of motion for our problem is

$$m_0\, d^2\mathbf{r}/dt^2=-e\mathbf{E}-(e/c)(\mathbf{v}\times\mathbf{H}). \tag{3}$$

This can easily be solved. With the initial conditions $\mathbf{r}_0=\mathbf{v}_0=0$ the resulting motion can be described as follows. The electron describes a cycloid which is a superposition of the cyclotron motion with cyclotron frequency $\omega_{c0}=eH/m_0c$ and a transverse motion with a constant velocity $v_{\text{tr}}=eE/H$.

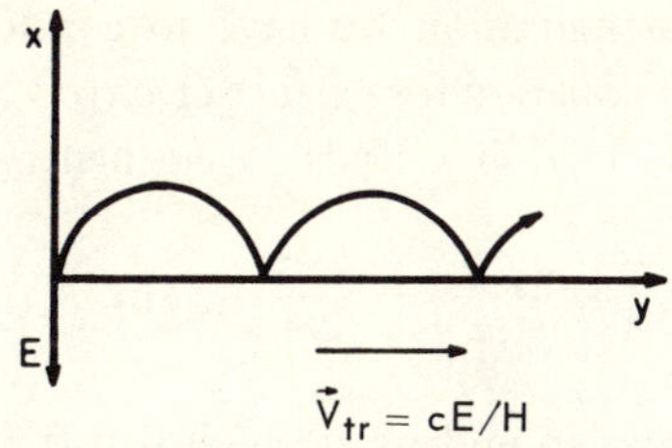

Fig. 1. Motion of a free electron in crossed electric and magnetic fields according to classical nonrelativistic mechanics, with initial conditions $r_0 = v_0 = 0$. The magnetic field is in the z direction. The motion is a superposition of the cyclotron motion with the frequency $\omega_c = eH/m_0c$ and the transverse motion with constant velocity $v_{tr} = cE/H$.

It should be observed that the constant velocity is in the direction perpendicular to both the electric and magnetic fields. The kinetic energy of the transverse motion, $E_{\text{kin}} = \frac{1}{2}m_0v_{\text{tr}}^2 = m_0c^2E/2H^2$, is developed at the expense of potential energy due to the shift of the center of oscillations along the electric field by the amount $eEL^2/\hbar\omega_{c0}$. Since the center of the cyclotron motion in the absence of the electric field is $x_0 = L^2k_y$, an interpretation of the electric field terms in the expression (2) for the eigenenergies is very simple. The term linear in E represents the energy of the center in the uniform electric field, whereas the term quadratic in E is the energy due to the shift of the center in the electric field.

Two important points should be made here. First, the motion is essentially of the magnetic field type: the oscillations with the cyclotron frequency are still the dominant feature, and there is no average acceleration in the direction of the electric field. It seems that the motion in crossed fields is always of the magnetic field type, even for large electric and small magnetic fields. This clearly does not make much sense physically. It should also be noted that one gets into trouble with Eq. (2) for large values of the E/H ratio. The clue to both difficulties is given by the transverse velocity. As long as this velocity is small compared to the velocity of light, i.e., $v_{\text{tr}} \ll c$, our description is correct. This, however, leads to the limitation that $E/H \ll 1$ for the free electrons. If this restriction is not satisfied, a relativistic treatment has to be used, both in classical and quantum theories. The scale of values is different for electrons in a solid, but the qualitative arguments remain the same. We shall return to this point later.

In order to see what kind of interband transitions can occur in the above crossed field configuration we have to consider the valence band. The EMA Schrödinger equation for valence electrons has the form of Eq. (1) with m_2 replaced by $-m_1$. This leads to eigenenergies

$$\mathscr{E}_1 = -\mathscr{E}_g - \hbar\omega_{c1}\left(n_1 + \frac{1}{2}\right) - \frac{\hbar^2 k_{z1}^2}{2m_1} + eEL^2 k_{y1} + \frac{m_1 c^2}{2}\frac{E^2}{H^2}, \quad (4)$$

where $\mathscr{E}_g$ is the energy gap and we have chosen the zero of energy at the bottom of the conduction band.

Thus the electric field shifts the *valence* Landau levels *up* by the amount $m_1 c^2 E^2/2H^2$, bringing them closer to the levels of the conduction band. The energies for direct interband transitions are

$$\mathscr{E}_2 - \mathscr{E}_1 = \mathscr{E}_g + \hbar\omega_{c1}(n_1 + \tfrac{1}{2}) + \hbar\omega_{c2}(n_2 + \tfrac{1}{2}) - [(m_1 + m_2)c^2 E^2/2H^2]. \quad (5)$$

We have omitted the k_z terms, since, as follows from the theory of magneto-optical phenomena, they determinate the shape of absorption peaks rather than their energies.

The energy differences as given by Eq. (5) can be and were observed in the light absorption experiments in germanium ([2,3]). Whenever the photon energy is equal to the separation between the levels in both bands a strong peak in absorption is observed, just as in normal magnetoabsorption.

For magnetic fields of $H = 100$ kG and electric fields up to $4 \cdot 10^6$ V/cm in germanium the lowest conduction levels and the highest valence levels are already in the energy gap, so that the transitions for photon energies smaller than that of the gap can be observed. However, the absorption has a resonant, oscillatory behavior and it lacks the tunneling characteristics. It is also clear from Eq. (5) that as long as $\mathscr{E}_2 - \mathscr{E}_1 > 0$ no Zener tunneling can occur, since the allowed energies in both bands do not overlap.

Thus we conclude that in the one-band effective-mass approximation described above there is neither tunneling nor photon assisted-tunneling.*

However, it should be clear by now that the one-band EMA breaks down for higher electric fields. It is obvious that the one-band picture cannot be used if the energies under consideration become comparable to the energy gap, since then the exact interaction between the bands has to be considered. This suggests the following criteria for the EMA to hold:

$$\hbar\omega_c(n + 1) \ll \mathscr{E}_g; \qquad \frac{\hbar^2 k_z^2}{2m} \ll \mathscr{E}_g; \qquad \frac{mc^2 E^2}{2H^2} \ll \mathscr{E}_g. \quad (6)$$

* It was pointed out in discussion by Dr. L. Kleinman that the energies in the two bands can overlap if the change in k_y is taken into account (indirect interband transition).

The third criterion restricts the E/H ratio. An exact treatment of the EMA validity for the crossed field problem ([4]) confirms the above intuitive result. The mass used in Eq. (6) stands for the effective mass of a band in question.

To treat our crossed field problem properly for all values of E/H ratio the interaction between the bands has to be treated exactly. The simplest of such treatments is represented by the two-band model.

THE TWO-BAND MODEL

First, in order to gain an intuitive understanding of the problem we again recourse to classical mechanics. It has been shown by Kane ([5]) (see also Chapter 6 by Kane and Blount in this volume) that the energy–momentum relation in an energy band interacting with another band is of the form

$$\mathscr{E}(\mathbf{k}) = \pm\left[\left(\frac{\mathscr{E}_g}{2}\right)^2 + \mathscr{E}_g\frac{\hbar^2k^2}{2m}\right]^{1/2}, \tag{7}$$

where the zero of energy has now been chosen in the middle of the forbidden gap for symmetry purposes. Now, the relation (7) is exactly of the relativistic type with the following correspondence:

$$2m_0c^2 \to \mathscr{E}_g; \qquad m_0 \to m. \tag{8}$$

This suggests that we can treat our problem classically using the relativistic equation of motion. The treatment of the problem of motion in crossed fields according to classical relativistic mechanics is given in standard electrodynamics textbooks ([6]). For small values of the E/H ratio the motion is, of course, of the type described in the first part of this chapter. It turns out that, as long as $cE/H < c$, the motion is essentially of the magnetic field type. For $cE/H > c$ the motion is of the electric field type: there are no oscillations and there is an average acceleration along the electric field direction. Thus, according to the relativistic mechanics, there is a transition from the magnetic to the electric type of motion as the electric field is increased, which is the only reasonable physical result. It is easy to see what the limiting velocity is in our two-band system:

$$c = (2m_{0c}^2/2m_0)^{1/2} \to (\mathscr{E}_g/2m)^{1/2} = v_{\max}. \tag{9}$$

Thus the critical point between two types of motion should occur at

$$cE/H = (\mathscr{E}_g/2m)^{1/2} \qquad \text{or} \qquad \omega_c^2 = 2e^2E^2/m\mathscr{E}_g. \tag{10}$$

We expect that the tunneling phenomena should occur for electric fields higher than that needed to reach the critical point. The above intuitive arguments are confirmed by the rigorous treatment ([7]) which we shall now present in brief.

We start with the Hamiltonian for an electron in a periodic potential in the presence of uniform electric and magnetic fields

$$\mathscr{H} = \frac{1}{2m_0}\left(\mathbf{p} + \frac{e}{c}\mathbf{A}\right)^2 + e\mathbf{E}\mathbf{r} + V(\mathbf{r}). \tag{11}$$

We now represent the Hamiltonian using the Kohn–Luttinger representation $x_{nk} = \exp(i\mathbf{kr})u_{n0}(\mathbf{r})$. After returning to the coordinate representation the set of equations for the envelope functions is obtained in the following form

$$\sum_{n'} [(\mathbf{P}^2/2m_0 + e\mathbf{E}\mathbf{r} + \mathscr{E}_n - \mathscr{E})\delta_{nn'} + \mathbf{P}\boldsymbol{\pi}_{n'n}]f_{n'}(\mathbf{r}) = 0. \tag{12}$$

Here the $\boldsymbol{\pi}_{n'n} = (1/m_0) < u_{n'} \mid \mathbf{p} \mid u_n >$ are determined by the interband matrix elements of momentum, $\mathscr{E}_n$ is the energy at the bottom of the nth band (at $k = 0$), and $\mathbf{P} = \mathbf{p} + (e/c)\mathbf{A}$ is the kinetic momentum. To simplify the problem, we neglect the free-electron term in Eq. (12), which is equivalent to neglecting the free mass in the effective-mass formula

$$\frac{1}{m_n} = \frac{1}{m_0} + 2\sum_{n \neq n'} \frac{|\pi_{nn'}|^2}{\mathscr{E}_n - \mathscr{E}_{n'}}, \tag{13}$$

and consider only two spherical nondegenerate bands. Taking as before $\mathbf{E} = [-E, 0, 0]$, we obtain the Schrödinger equation in the following matrix form:

$$\begin{bmatrix} -eEx + \frac{1}{2}\mathscr{E}_g & \boldsymbol{\pi}_{21}\mathbf{P} \\ \boldsymbol{\pi}_{12}\mathbf{P} & -eEx - \frac{1}{2}\mathscr{E}_g \end{bmatrix} \begin{bmatrix} f_1 \\ f_2 \end{bmatrix} = \mathscr{E} \begin{bmatrix} f_1 \\ f_2 \end{bmatrix}. \tag{14}$$

With spin included this equation can be written in terms which display its similarity to the Dirac equations in the diagonalizing coordinate system $\pi_{12}^{\alpha}\pi_{21}^{\beta} = \delta_{\alpha\beta}\mathscr{E}g/2m$, defining the effective mass for both bands. We can solve the system of equations (14) by substitution. For the crossed field configuration, taking as before $\mathbf{A} = [0, Hx, 0]$, we obtain

$$\begin{aligned} \Big\{ -\frac{\hbar^2}{2m}\frac{\partial^2}{\partial x^2} - \alpha x + \frac{m}{2}\left[\left(\frac{eH}{mc}\right)^2 - \frac{2e^2E^2}{m\mathscr{E}_g}\right]x^2 \\ + \frac{3\hbar^2}{8m}\frac{eE}{eEx + \mathscr{E} + \frac{1}{2}\mathscr{E}_g}\Big\}\varphi(x) = \lambda\varphi(x), \end{aligned} \tag{15}$$

where

$$\alpha = 2e\left(\frac{\mathscr{E}}{\mathscr{E}_g}E - \frac{\hbar k_y}{2mc}H\right)$$

and

$$\lambda = \frac{1}{\mathscr{E}_g}\left(\mathscr{E}^2 - \frac{\mathscr{E}_g^{\,2}}{4}\right) - \frac{\hbar^2}{2m}(k_y^{\,2} + k_z^{\,2});$$

the solution f_1 is given in terms of $\varphi(x)$ as

$$f_1(\mathbf{r}) = | \, eEx + \tfrac{1}{2}\mathscr{E}_g + \mathscr{E} \, |^{1/2}\varphi(x)\exp(ik_y y + ik_z z)$$

and f_2 can be obtained from f_1 using the second of equations (14).

The last term on the left-hand side of Eq. (15) is identically equal to zero for $E = 0$. The detailed evaluation shows that this term is always small for practically achievable electric fields, and it can be treated as a small perturbation.

A closer examination of Eq. (15) is very instructive. First, we notice that for small electric fields, when $(eH/me)^2 \gg 2e^2E^2/m\mathscr{E}_g$, it reduces to the one-band EMA Hamiltonian (we also have to put $\mathscr{E} \ll \mathscr{E}_g$), but those are exactly the criteria (6) for the EMA to hold. Second, we observe that as long as $\omega_c^2 > 2e^2E^2/m\mathscr{E}_g$ Eq. (15) is the harmonic oscillator equation with energy quantized in terms of $\omega_{\text{eff}} = [\omega_c^2 - (2e^2E^2/m\mathscr{E}_g)]^{1/2}$. This means that in this range the cyclotron frequency is decreased by the electric field. It can be shown that the eigenenergies for the conduction and valence bands resulting from Eq. (15) do not overlap as long as we are in the above "magnetic field range." In this region there is no Zener tunneling, and the interband optical transitions have oscillatory nontunneling behavior. For still higher electric fields, when $\omega_c^2 < 2e^2E^2/\mathscr{E}_g m$, the eigenvalues of Eq. (15) are not quantized and we deal with the continuous energy spectrum, characteristic of the electric field-type solutions. The eigenfunctions of Eq. (15) (with the small term omitted) are given now by the Weber functions (also called the parabolic cylinder functions). In this region we expect tunneling behavior for interband transitions, both with and without photon assistance.

In order to show this explicitly, we solve Eq. (15) using the semiclassical WKBJ approximation, i.e., we treat p_x as a number rather than an operator. The same result could be obtained by using the asymptotic expansion of the Weber function. We can now solve Eq. (15) for energy as a function of x and p_x, with the following result

$$\mathscr{E}_{1,2} = -eEx \pm \left\{\left[\frac{\mathscr{E}_g^{\,2}}{4} + \mathscr{E}_g\right]\left[\frac{p_x^{\,2}}{2m} + \frac{m\omega_c^{\,2}}{2}(x - k_y L^2)^2 + \frac{\hbar^2 k_z^{\,2}}{2m}\right]\right\}^{1/2}, \quad (16)$$

where the plus and minus signs refer to the conduction and valence bands, respectively. Values $p_x = \hbar k_y = \hbar k_z = 0$ determine the turning points of the motion in both bands.

The plots of the turning points as functions of x for both bands are shown in Fig. 2 as solid lines. The shaded areas represent classically accessible regions in the two bands (real values of momentum p_x). It can be seen that the motion in each band is bound only on one side, so that the energy spectrum must be continuous, in agreement with the conclusion from Eq. (15). The unshaded area represents the classically forbidden region (imaginary values of p_x), in which the quantum mechanical solutions decay exponentially. Thus, due to the presence of the electric field, there is a finite probability of finding an electron in the forbidden gap, although

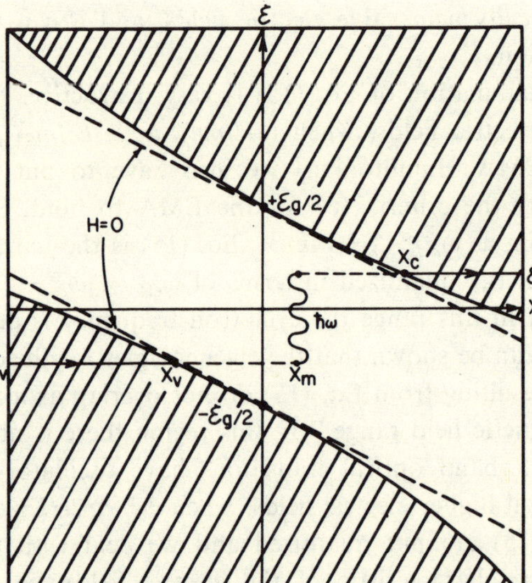

Fig. 2. Photon-assisted tunneling in crossed electric and magnetic fields for $E > H(\mathscr{E}_g/2me^2)^{1/2}$. The dashed lines represent the valence and conduction band edges (or turning points) for $H = 0$. The shaded regions represent the valence and conduction band regions with both fields present. A valence electron of energy $\mathscr{E}_v$ tunnels from its turning point x_v to x_m, absorbs a photon, and tunnels to x_c, the turning point for a conduction electron of energy $\mathscr{E}_c = \mathscr{E}_v + \hbar\omega$. The magnetic field acts to reduce the absorption by curving the band edges away from each other, thus enlarging the tunneling region. From Weiler *et al.* (9).

it is appreciable only in the vicinity of the allowed band edges. This is, of course, the reason that the electron can tunnel through the gap, both with and without photon assistance. Now we can also see the effect of the magnetic field. For $H = 0$ Eq. (16) yields the usual tilted band picture, shown in Fig. 2 by two parallel dashed lines. The magnetic field curves the allowed regions away from each other, enlarging the tunneling region. Thus the magnetic field clearly acts to decrease the tunneling currents. (If the magnetic field is increased further still, the curving becomes so strong that the motion in a band has two turning points, i.e., becomes quantized. This corresponds to the condition $\omega_c^2 > 2e^2E^2/\mathscr{E}_g m$. At the same H value the energies in the two bands cease to overlap, so that the Zener tunneling becomes impossible.) The photon-assisted tunneling can now be viewed as follows. A valence electron of energy $\mathscr{E}_v$ tunnels from its turning point x_v to x_m, absorbs a photon, and tunnels to $x_e = \mathscr{E}_v + \hbar\omega$. Thus the transition can also occur for photon energies smaller than the energy gap $\mathscr{E}_g$ (Franz–Keldysh effect). It can be seen that at $x = 0$ the curves with and without magnetic field coincide, so that for photon energy $\hbar\omega = \mathscr{E}_g$ the magnetic field has no effect on the photon-assisted tunneling current. This is confirmed by the exact calculation.

The tunneling current has been computed with the use of the golden rule for the transition rate between two states. The states in the valence and conduction bands are quasi-stationary, since we have neglected one term in Eq. (15). This term is responsible for the Zener tunneling, and it is used as the perturbation in the golden rule procedure. For the photon-assisted tunneling the perturbation due to the incident radiation is introduced in the usual way by putting $\mathbf{P} \to \mathbf{P} + (e\mathbf{A}_\omega/c)$ in the Hamiltonian (14), where $\mathbf{A}_\omega$ is the vector potential of the light wave. The solutions of Eq. (15) for initial and final states have been obtained using WKBJ approximation and the integrals carried out by the method of steepest descent. The details of the calculation can be found in Weiler *et al.* [9]. It should be mentioned that practically identical results were obtained by Aronov and Pikus [10] using methods based on an analogy to the relativistic Dirac electrons.

The final results are as follows. The tunneling (Zener) current is

$$j = \frac{e^2E^2}{3\pi^2\hbar^2}\left(\frac{2m}{\mathscr{E}_g}\right)^{1/2} \exp\left[-\frac{\pi}{2\hbar e E_{\text{eff}}}\left(\frac{m}{2}\right)^{1/2}\mathscr{E}^{3/2}\right], \tag{17}$$

where $E_{\text{eff}}^2 = E^2 - (H^2\mathscr{E}_g/2mc^2)$. It can be seen that the magnetic field decreases the tunneling current. At the crytical point $\omega_c^2 = 2e^2E^2/\mathscr{E}_g m$ the effective field $E_{\text{eff}}^2 = 0$, and the tunneling current vanishes. The magnetic field dependence predicted by formula (17) has been confirmed experimentally [see ([10])].

For photon-assisted transitions the light absorption serves as a measure of tunneling current, since to every absorbed photon there corresponds one electron undergoing the interband transition. For photon energies smaller than that of the gap the final result reads

$$\frac{\alpha(E, H)}{\alpha(E, 0)} \approx \frac{E_{\text{eff}}}{E} \exp\left[- \frac{8(\mathscr{E}_g - \hbar\omega)^{5/2}}{15\hbar e \mathscr{E}_g} m^{1/2} \frac{H_{\text{eff}}^2}{E^3} \right], \tag{18}$$

where $\alpha(E, H)$ is the light absorption constant in the presence of both fields and $H_{\text{eff}}^2 = H^2\mathscr{E}_g/2mc^2$. Two features of this result should be noticed. First, for $\hbar\omega = \mathscr{E}_g$ we get $\alpha(E, H) = \alpha(E, 0)$, i.e., the magnetic field has no effect on the photon-assisted tunneling current, as we have concluded from Fig. 2. Second, for magnetic fields such that $\omega_c^2 \geq 2e^2E^2/\mathscr{E}_g m$, i.e., $E_{\text{eff}}^2 \leq 0$, there is no photon-assisted tunneling.

EXPERIMENTAL RESULTS

The above theoretical conclusions have been confirmed in the experiments on photon-assisted tunneling in crossed electric and magnetic fields carried out by Reine *et al.* ([11]) in Ge *p–n* junction. Figure 3 presents the absorption constant $\alpha(E, H)$ plotted as a function of photon energies below that of the direct optical gap of Ge. For $H = 0$ the exponential tail of absorption below the gap is observed, which is just the ordinary Franz–Keldysh effect. The magnetic field decreases the photon-assisted tunneling current, but its effect becomes smaller as the photon energies approach that of the gap, in agreement with the above predictions. We can now determine a shift of energy as a function of magnetic field for a fixed value of the absorption constant and electric field. This is plotted in Fig. 4. A theoretical relation for this shift obtained from the above theory is

$$\Delta\mathscr{E} = \frac{2}{15} \frac{(\mathscr{E}_g - \hbar\omega)^2}{mc^2} \frac{H^2}{E^2}. \tag{19}$$

We can see from Fig. 4 that the quadratic dependence on H is well observed. This is a rather interesting result. As we have shown before [Eq. (5)], for small electric and large magnetic fields the energy shift depends on fields as E^2/H^2, whereas for large electric and small magnetic fields it behaves as H^2/E^2. This result, which is confirmed experimentally, can be obtained only with the use of the two-band model.

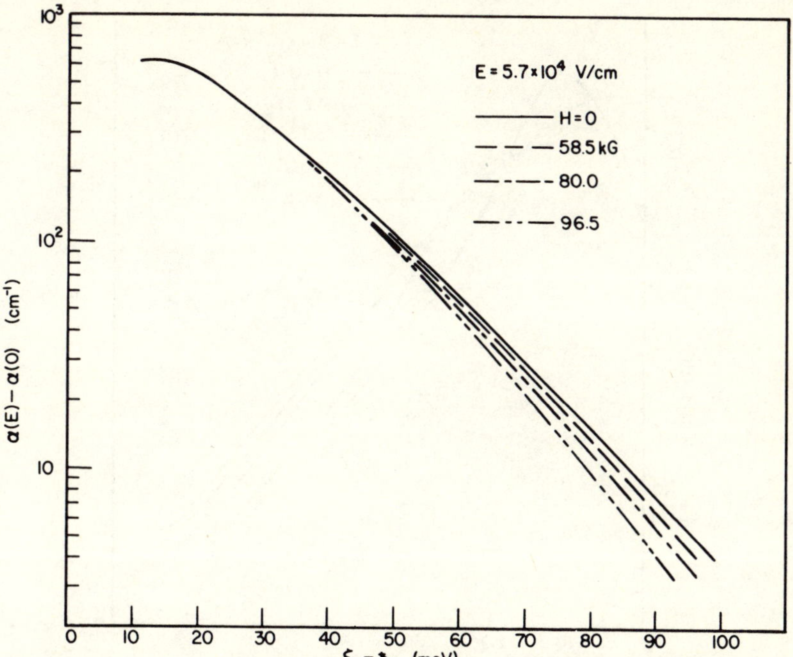

Fig. 3. The change, as a function of $\mathscr{E}_g - \hbar\omega$, in the absorption coefficient below the direct gap of Ge due to an electric field E in the presence of a transverse magnetic field H for $E = 5.7 \times 10^4$ V/cm and for various values of H. The effect of the magnetic field is greater for photon energies away from the gap value, in agreement with the theoretical predictions. From Reine *et al.* ([11]).

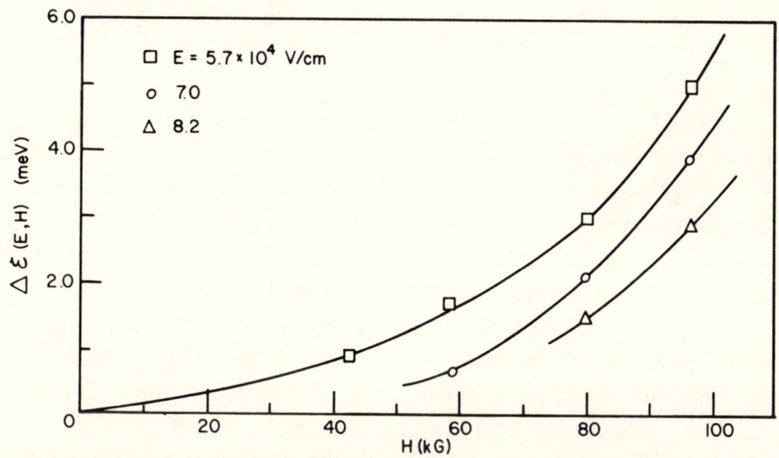

Fig. 4. The magnetic shift $\Delta\mathscr{E}$ as determined from Fig. 3 for a chosen value of the absorption coefficient $[\alpha(E) - \alpha(0) = 20\ \text{cm}^{-1}]$ at a constant electric field. The shift is quadratic in the magnetic field, in agreement with the theory. From Reine *et al.* ([11]).

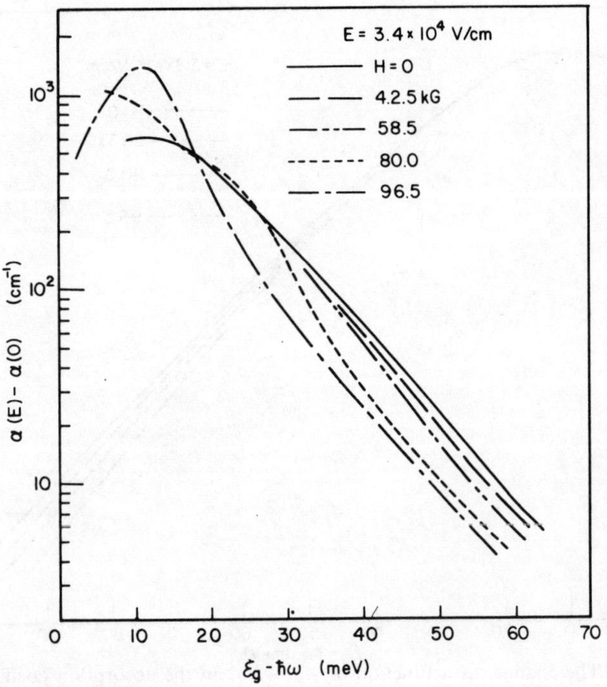

Fig. 5. The change, as a function of $\mathscr{E}_g - \hbar\omega$, in the absorption coefficient of Ge due to an electric field E in the presence of a transverse magnetic field H. The continuous transition from electric-field-type interband photon-assisted tunneling (at zero magnetic field) to magnetic-field-type resonant interband absorption (at high magnetic fields) can be seen. From Reine *et al.* ([11]).

Finally, Fig. 5 shows the transition from electric- to magnetic-type of behavior as the magnetic field is increased in the crossed field configuration. Again the optical absorption of Ge is plotted as a function of photon energies below the direct gap. For $H = 0$ the Franz–Keldysh exponential tail is observed, which is an indication of the electric-field-type behavior. As the magnetic field is increased, we first observe the decrease of absorption below the gap, and then for magnetic fields sufficiently high the well-pronounced resonance in absorption below the gap. This is characteristic of magnetic-field-type quantized behavior shifted below the zero-field gap by the transverse electric field. The experimental value of the E/H ratio at the critical point agrees rather well with the theoretical prediction for the actual values of $\mathscr{E}_g$ and m of Ge. The transition range is fairly broad, since, as was indicated before, for magnetic fields just above the critical point the quantization energy $\hbar\omega_{\mathrm{eff}}$ is very small and cannot be resolved experimentally.

The problem of interband tunneling and photon-assisted tunneling in parallel electric and magnetic fields has also been recently investigated both theoretically and experimentally, and results can be found in original publications ([9,11]).

REFERENCES

1. A. G. Aronov, Fiz. Tverd. Tela 5:552 (1963) [English translation; Soviet Phys.–Solid State 5:402 (1963)].
2. Q. H. F. Vrehen, Phys. Rev. Letters 14:558 (1965).
3. Q. H. F. Vrehen, W. Zawadzki, and M. Reine, Phys. Rev. 158:702 (1967).
4. J. Zak and W. Zawadzki, Phys. Rev. 145:536 (1966).
5. E. O. Kane, J. Phys. Chem. Solids 1:249 (1957).
6. L. D. Landau and E. M. Lifshitz, Classical Theory of Fields, Addison-Wesley, Reading, Mass., 1959, p. 59.
7. W. Zawadzki and B. Lax, Phys. Rev. Letters 16:1001 (1966).
8. J. M. Luttinger and W. Kohn, Phys. Rev. 97:869 (1955).
9. M. H. Weiler, W. Zawadzki, and B. Lax, Phys. Rev. 163:733 (1967).
10. A. G. Aronov and G. E. Pikus, Zh. Experim. i Teor. Fiz. 51:281 (1966); 51:505 (1966) [English translation Soviet Phys.—JETP 24:188 (1967); 24:399 (1967)].
11. M. Reine, Q. H. F. Vrehen, and B. Lax, Phys. Rev. 163:726 (1967).

Chapter 17

Molecular Excitations in Barriers I

J. Lambe
and R. C. Jaklevic

Scientific Laboratory, Ford Motor Co.
Dearborn, Michigan

INTRODUCTION

Molecular excitations ([1]) in barriers are one aspect of the more general topic of inelastic tunneling. By this latter term we mean the various processes whereby an electron loses energy as it tunnels through a barrier. In the case of molecular excitations the energy loss is to the vibrational modes of molecular species in the barrier. This inelastic process provides an additional path or channel for current flow, and so one sees breaks in the *I–V* curve as shown in Fig. 1.

We will see a break in the *I–V* curve at threshold V_0, and that $eV_0 = h\nu$, where ν is the vibrational frequency which we will be exciting in the inelastic process. In general, of course, molecules will have many such frequencies, so we will see many breaks in the *I–V* curve. This will then yield the vibrational spectrum of the molecule in question, and we would expect this "inelastic" spectrum to be related to the infrared spectrum and/or Raman spectrum of the molecule.

In this chapter we will discuss some simple physical analogs to this situation and then present the experimental methods and results.

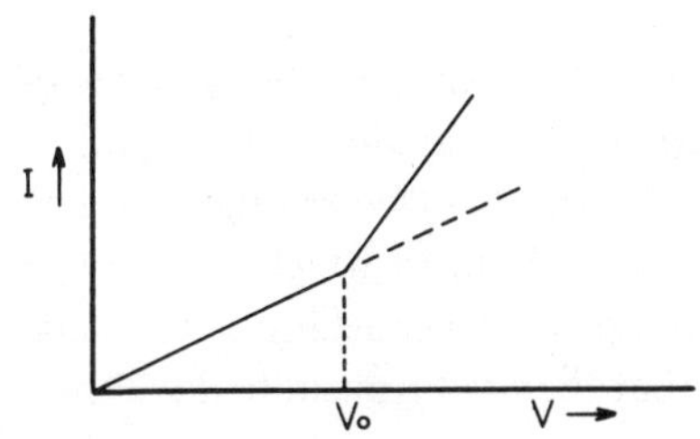

Fig. 1.

Fig. 2.

VIBRATIONAL SPECTRA

The usual method for obtaining vibrational spectra is by the absorption of light in the infrared region, which we call infrared spectra, or by the inelastic scattering of visible light, which we call Raman spectra.

Infrared spectra are the result of a coupling between the electric field of the infrared radiation and the dipole moment of the molecule. The interaction energy will be of the form $\bar{\mu} \cdot E$, where $\bar{\mu}$ represents the dipole moment of the molecules. This means that transitions will be related to the dipole matrix element.

Raman spectra occur as the result of the interaction between the optical electric field and the polarizability of the molecule. The interaction is of the form αE^2, where α, the polarizability of the molecule, will be dependent on the vibrational motion of the molecule, and this is the basis of Raman spectra. In the scattered beam one finds frequencies $\nu \pm \nu_{\text{vib}}$ in addition to the primary optical frequency ν.

It is helpful to note which vibrational modes can give rise to infrared spectra and which to Raman spectra. Consider, for example, the CO_2 molecule. The mode illustrated in Fig. 2(*a*) gives rise to an oscillating electric dipole, and hence will be infrared-active. It is not Raman-active, however. The mode in Fig. 2(*b*) is Raman-active, since α can change linearly with the displacement of the oxygens. Since mode (*b*) does not give rise to an oscillating dipole, it is not infrared-active.

AN OPTICAL ANALOG TO INELASTIC TUNNELING

To gain some physical insight into the process of inelastic tunneling, we can consider a rather crude analog using light.

In Fig. 3 we show a situation where there is a bump in the value of the refractive index of a medium in which we are transmitting a beam of light. Moreover, this bump is not a constant, but its height varies sinusoidally with time. We expect that the light transmitted will be given by

$$E_t = (E_0 \cos 2\pi\nu_1 t)(1 + \delta \cos 2\pi\nu_2 t),$$

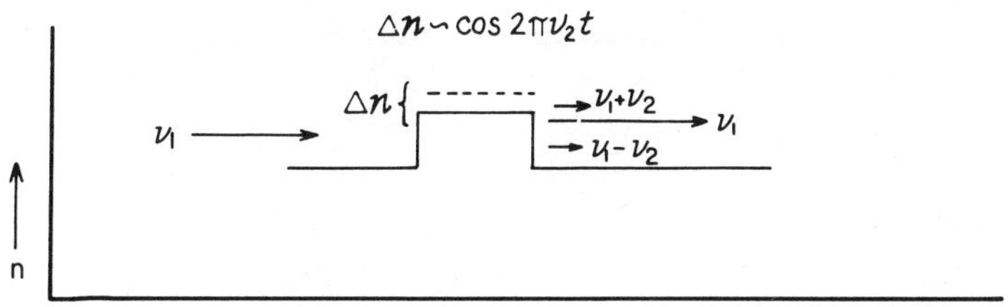

Fig. 3.

where E_0 is the transmitted amplitude in the absence of the modulation. Then

$$E_t = E_0 \cos 2\pi\nu_1 t + \tfrac{1}{2}E_0\delta \cos 2\pi(\nu_1 + \nu_2)t + \tfrac{1}{2}E_0\delta \cos 2\pi(\nu_1 - \nu_2)t.$$

The transmitted beam will contain frequencies $\nu_1 \pm \nu_2$. Thus the time dependence of the refractive index results in frequency changes in the light beam. This is rougly similiar to the inelastic tunneling effect. In our case it is the potential barrier we deal with rather than refractive index. The time variation of the potential barrier can then be related to the electric fields of the molecules which vary at the vibrational frequencies of the molecules. When we carry out the detailed calculation we will see that at low temperatures we in effect get only the frequency $\nu_1 - \nu_2$ since the barrier will not be able to add energy to the primary frequency.

It is worth noting that the electron which is tunneling can couple to the dipole moment of the molecule (infrared-like spectra) or to the polarizability of the molecule (Raman-like spectra).

EXPERIMENTAL TECHNIQUES

Sample Preparation

Films

The tunneling junctions were of the usual metal film–insulating oxide–metal film structure, and with few exceptions these were Al–Al oxide–Pb. Preparation was done in bell-jar-type vacuum chambers. Two types of vacuum system were used; one an oil-diffusion-pumped ($\sim 10^{-7}$ Torr) with liquid-nitrogen trapping and a zeolite trap on the foreline. The other was an ion-pumped oil-free system ($\sim 10^{-9}$ Torr ultimate). Both systems used viton gasket seals and could be given a moderate degassing by heating to 100°C by internal operation of a filament. All fixtures were either of stainless steel, copper, aluminum, Teflon, or ceramic. Both systems were equipped with a gas-discharge electrode made of aluminum which could be operated with O_2 gas immediately after initial pumpdown to promote cleanup of residual impurities. This discarge was also used for promoting oxide growth later in the process.

Sample substrates were of fused silica of dimension $\frac{1}{2} \times 1\frac{1}{2}$ in. $\times$ 1 mm with prefired silver paste electrodes. They were first scrubbed in detergent, rinsed in deionized water for $\frac{1}{2}$ hr or more, and finally dried in propyl alcohol vapor. The samples were then mounted in the vacuum chamber in a rotatable mask changer which allowed the entire fabrication to take place without exposure to atmosphere. The mask provided for initial evaporation of an aluminum strip 1–2 mm wide ($\sim$ 2000-Å thickness), the length of the substrate. Masks could then be moved aside for oxidation of the aluminum. Three Pb films were then evaporated crosswise to form a junction 1×1 mm^2. This area was found to be large enough for easy masking and yet to produce a large enough yield of defect-free samples. The area was large enough that with the impedances used ($\sim$ 100 ohms), current densities were not so high as to cause serious breakdown and noise problems.

Oxidation

Oxidation of the aluminum film was done by one of two methods. In the first the sample was removed from the vacuum system and exposed to room air for 18–24 hr. In this case the edges of the aluminum film were usually masked by painting over the edges with Formvar paint after it was suspected that shorts and other defects occurred there. This method suffers from unavoidable contamination of the junction by organic impurities from both the paint and the atmosphere. On the other hand, it was found that simple exposure of the aluminum to pure oxygen at temperatures as high as 200°C would not result in appreciable tunneling oxide layers. In this case resistances of less than 0.1 ohm were obtained, which indicated oxides $<$ 10-Å thickness. In such cases the junctions showed a good Josephson effect when both metals were superconducting. Water and organic films are probably responsible for the thicker insulators ($\sim$ 30 Å) formed by exposure to air.

Therefore the gas-discharge oxidation technique described by Miles and Smith ([2]) was used. An electrical discharge was set up between an aluminum cathode at $-$ 500 V DC and the rest of the system while maintaining a pressure of approximately 50 μ of either high-purity or standard tank oxygen. Care was taken to keep the aluminum electrode close enough (3 in.) for efficient oxidation and yet to remain shielded from a direct optical path to the samples. Without shielding the results were very erratic. Exposure of the samples to the discharge for 20 min or more produced a 10–20-Å insulator. One can also apply a small DC oxidizing potential (1 V) to the aluminum films to promote slightly thicker oxide growth closer to 30 Å. Samples prepared this way showed no organic contamination. When following this procedure no significant differences were found between use of either type of vacuum system. Apparently, with proper trapping contamination from pump oil is not appreciable. As will be seen, we have not been successful in eliminating effects caused by water vapor, present either as residuals in the chamber or water liberated from the system walls during operation of the discharge.

Molecular Additions

Deliberate doping of junctions by impurities was done by exposing the aluminum oxide surface to a vapor of the substance to be deposited. For the case of the liquids (acetic and propionic acids and methyl and ethyl alcohols; water) a drop or two was sealed in a small capillary and broken inside the vacuum chamber after oxidation. A resistance heater was provided to aid evaporation of less volatile materials (cyanoacetic acid). The pressure quickly rose to 10^{-2}–10^{-1} Torr as measured by a thermocouple gauge with subsequent slow decrease due to absorption of the vapor on the chamber surfaces and sample. After 5–10 min exposure the system is again evacuated and the top Pb film deposited. For the case of the hydrocarbon (Convoil 20 pump oil) the vapor pressure was low enough to simply place a tiny amount ($1 \sim 10^{-4}$ gm) in a tantalum boat approximately 10 in. from the sample and evaporate the material at the appropriate time by heating slowly to several hundred degrees centigrade. In each case it is estimated that approximately one monolayer or less is deposited. This estimate is based on the measured pressure drop after exposure or, in the case of pump oil, actual amount evaporated. In particular, the acids are known to absorb to a single monolayer. None of the organics used presented a contamination problem, and they could be successfully purged from the system with a draft of warm air and routine pumping. Although the majority of junctions were Al–Al oxide–Pb combination, a number of metals were successfully used. Au, Sn, Al, Mg, and Cr in place of the Pb all showed inelastic tunneling, as did the combination Ta–Ta oxide–Pb.

Junctions made according to the procedure outlined had nominal tunneling resistances of 30–300 k-ohms, although the most convenient resistances were in the 30–200 ohm range. Usually, for the doped samples several attempts were necessary before the right combination of time of exposure to anodization and organic vapor was achieved. Experience indicates that the presence of the organic layer increases the resistance of the tunnel junction, in some cases amounting to an equivalent Al oxide dielectric thickness of about 10–20 Å. It was usually necessary to decrease the amount of oxidation to accommodate the presence of the organic film and obtain a workable junction resistance. All of the samples were reasonably stable for a period of hours up to 1 day. However, those doped with organics were more unstable and tended to deteriorate, the resistance dropping, and becoming noisy over a period of weeks, probably due to decomposition or reaction of the organics. The pure oxide junctions were the most stable structures, and showed slow increase of resistance with time.

Other Materials

The organic materials employed so far are either of the type which form strong absorbing bonds (acids, alcohols) or are of high enough molecular weight so that the absorbed layer is stable. However, some materials (e.g., acetone, benzene, chloroform) were found not to absorb on the Al–Al oxide surface in sufficient quantity to produce detectable spectra. These and others like them can probably be absorbed by first cooling the samples as low as liquid nitrogen temperatures (77°K). After the proper exposure to organic vapor the top metal film could be deposited. This has not been done in the present experiments.

Measurement Technique

The measurement which one wishes to make is essentially a careful study of the current–voltage relationship of a tunnel structure. It is well known that this can be done in a very sensitive way by using derivative techniques ([3]). We therefore use the method of harmonic generation to measure the second derivative of the *I–V* curve. The essence of this method is to detect changes in conductivity by detecting the generation of second harmonics where the *I–V* curve is nonlinear. The circuit we used yields an AC signal voltage whose amplitude is proportional to the second derivative of current with respect to voltage. The proportionality factor contains the junction impedance as a factor, but this did not vary greatly over the ranges studied. A schematic of the apparatus is shown in Fig. 4.

We chose a primary frequency of 50 kHz and then detected at 100 kHz. This frequency was used to minimize any excess noise from the sample and because the amplifiers used had optimum noise figure in this frequency range. We have not made any detailed study of the excess noise from our

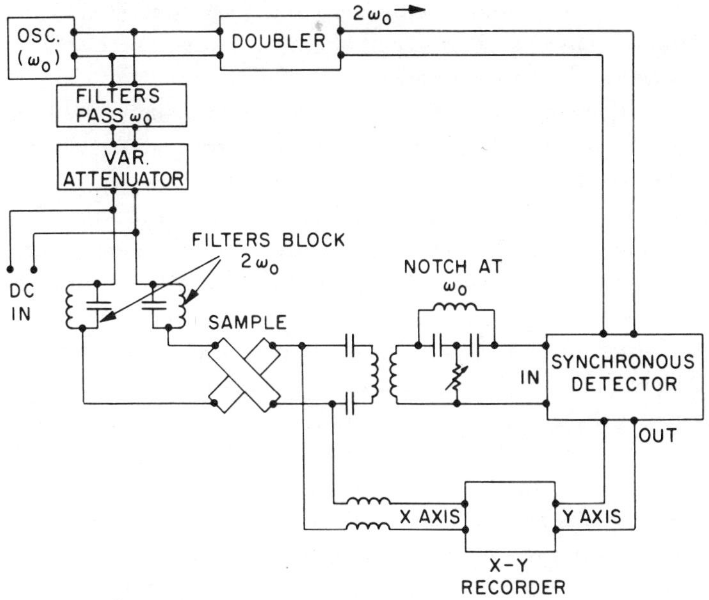

Fig. 4.

tunnel junctions, but it did appear that such noise dropped off with frequency. The large capacitance effects of the junction made it inconvenient for us to go to much higher frequencies to study this over a wide range.

The sample resistance plays an important rule in determining the overall sensitivity of such systems. The lower the sample resistance, the greater the available signal power. It is therefore desirable to have the sample impedance as low as possible consistent with required breakdown strength. In our case this impedance was about 100 ohms for a 1-mm^2 sample. One then uses a transformer to optimize the input impedance to the preamplifier.

In discussing the measurements we will refer to these second-derivative curves as spectra. One notes, in fact, that the curves very closely resemble conventional infrared absorption spectra. Thus the data are obtained in a very convenient form. The reason for this was seen in detail in the sections on line widths and mechanisms. It is enough to say here that the second-derivative curves can conveniently be thought of in terms of absorption lines. Thus, for example, if we wish to compare the relative overall strength of two inelastic processes, we compare the areas under the appropriate second-derivative curve.

The amount of 50-kHz modulation which can be used is determined by the line widths one is attempting to resolve. Thus for most of the spectra reported here modulations of between 1 and 5 mV were used. The smaller modulation results in smaller signal, of course, but was necessary to assure optimum resolution in some cases.

Conventional cryogenic techniques were used. Samples were mounted on a rod containing stainless steel outer conductor and copper inner conductor coaxial lead. This was immersed directly in the liquid refrigerant, liquid helium, liquid nitrogen, or dry ice and acetone, to take spectra at various temperatures. Temperatures as low as 0.9°K were attainable by pumping on the helium bath. The entire dewar was shielded with a μ-metal cylinder.

In constructing the experimental setup needed for measurement it is helpful to be able to test the quality of the apparatus. To test this, we would like to measure the overall noise figure F of the equipment. To do this, we can simply insert a resistor (wire-wound or thin-film) in place of the sample and carry out the following test. We examine the noise on the recorder with the resistor at room temperature and then cool the resistor to nitrogen temperature. If we had a perfect amplifier, the noise power should drop by 77/300 or about 1/4. More specifically, the noise figure is given by

$$F = \tfrac{3}{4}[\varrho/(\varrho - 1)],$$

where ϱ is the ratio of noise power at 300°K to that at 77°K. The noise power is proportional to the square of the r.m.s. deviation on the recorder. Noise figure is usually expressed in dB, which is $10 \log F$. Noise figures of 3 dB or better should be achieveable in such apparatus.

EXPERIMENTAL RESULTS

When we examine the curve d^2I/dV^2 versus V we see spectra characteristic of the molecular addition. They appear as shown in Figs. 5 and 6.

These spectra show the same structure as infrared spectra. If we deuterate, the voltage shifts down by $1/\sqrt{2}$, as we expect a harmonic oscillator to behave when the mass is doubled.

The spectra are quite sharp at helium temperature. As we raise the temperature, the lines broaden. This is due to the fact that the electron distribution in the metals becomes more diffuse at higher temperatures. At nitrogen temperature the linewidth is about 40 mV and at room temperature the linewidth is somewhat greater than 100 mV. The theoretical linewidth is about $6\,kT$ for normal metal electrodes. By using superconducting electrodes at low temperatures one should do better than this. We estimate

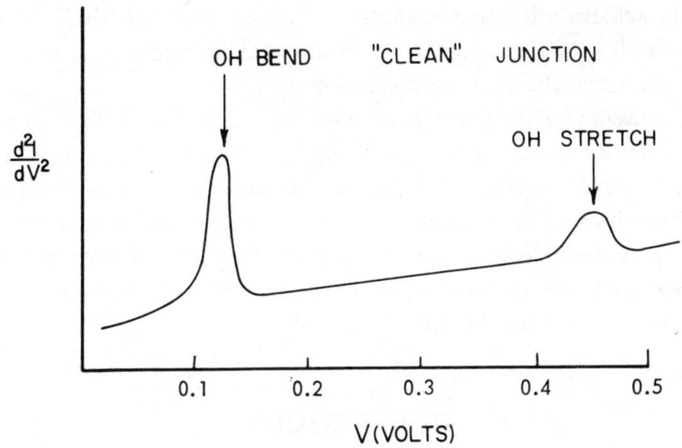

Fig. 5.

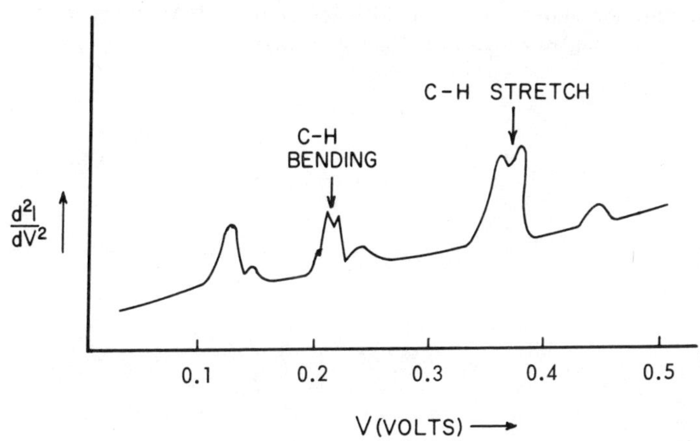

Fig. 6.

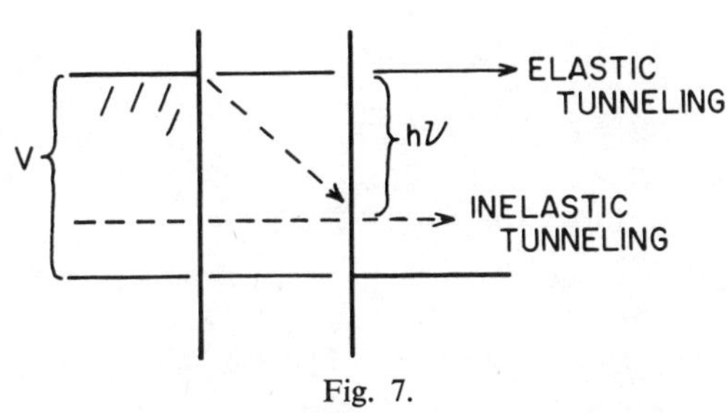

Fig. 7.

one could achieve a limiting resolution of about 30 μV at 0.3°K. Actually, most of the lines we examine have an inherent linewidth of a least 1 mV, so we cannot test the limiting resolution as yet.

We can summarize our experimental results in Fig. 7. We see that an electron loses an energy $h\nu$ to a molecule in the barrier. For this to happen eV must be equal to or greater than $h\nu$, so we have a series of thresholds which then show up as molecular spectra in the second derivative.

We have discussed losses to molecular vibrations. We should also be able to see electronic transitions, for example. This would generally require voltages in 1–2 V range, but this is feasible.

REFERENCES

1. R. C. Jaklevic and J. Lambe, Phys. Rev. Letters 17:1139 (1966); D. J. Scalapino and S. M. Marcus, Phys. Rev. Letters 18:459 (1967).
2. J. Miles and P. Smith, J. Electrochem. Soc. 110:12 (1963).
3. W. R. Patterson and J. Shewchun, Rev. Sci. Instr. 35:1704 (1964); D. E. Thomas and J. M. Rowell, Rev. Sci. Instr. 36:1301 (1965).

Chapter 18

Molecular Excitations in Barriers II

R. C. Jaklevic
and J. Lambe
Scientific Laboratory, Ford Motor Co.,
Dearborn, Michigan

INTRODUCTION

The process of inelastic tunneling between metallic electrodes can be visualized in terms of a rather simple physical picture ([1]). In the energy level diagram the two Fermi levels of the metals are separated by the energy eV, with V the applied voltage (Fig. 1).

According to energy conservation, the electrons in metal 1 are capable of making (elastic) horizontal transitions from one metal to another into empty states in the right-hand metal providing a voltage V is applied to the junction. To a first approximation, the current–voltage characteristic of this elastic tunneling is linear. However, if it is possible for a new transfer mechanism to exist, whereby an electron can tunnel from left to right and at the same time give energy to a local impurity state, a new channel for tunneling would open up. This new contribution can only occur if an empty state on the right is open for the tunneling. In the low temperature limit the threshold voltage $eV_0 = \hbar\omega_0$, where $\hbar\omega_0$ is the excitation energy of the impurity center. As V increases still further, current from this new process will continue to increase, since more and more final states for

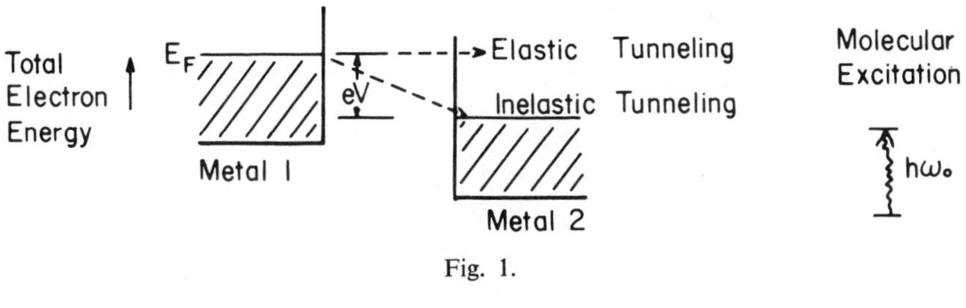

Fig. 1.

inelastic tunneling are brought into play. At the temperatures and excitation energies of interest we may assume that all of the impurities are in their ground state—therefore backward inelastic processes from right to left, or energy-increasing (anti-Stokes) processes, will not be significant. The onset of a new tunneling process causes an increase in the conductance at the voltage $eV \approx \hbar\omega_0$ which, plotted as d^2I/dV^2, would be a peak. The line shape will depend on the natural line shape of the excitation, the temperature broadening due to the thermal smearing of the Fermi surface of the metals, and, in the case of superconductors, density-of-states effects. These effects will be discussed below.

A simple theoretical calculation of the inelastic tunneling current has been given by Scalapino and Marcus (2), who also estimated the size of the effect for the case of excitation of IR-active vibrational modes. Using certain simplifying assumptions, this theory accounts for the main features of the observations. A simple presentation of this theory will be given. We will then discuss the additional possibility of exciting vibrational modes which show optical Raman activity and compare the size of the inelastic tunneling current caused by the two interaction mechanisms. Finally, we will discuss the effect of temperature broadening of the spectral lines and the possible improvement in resolution obtainable by the use of superconducting metal electrodes.

TUNNELING CURRENT

The simple model for tunneling proposed by Bardeen (3) was discussed by Kane in Chapter 1 of this volume. The tunneling rate between the electron states in the left and the right metals is given by the golden rule

$$w_{l \to r} = (2\pi/\hbar) \mid M_{rl} \mid^2 \delta(E_l - E_r). \quad (1)$$

Harrison (4), using a WKB approximation, showed that for a simple metal the matrix element for tunneling is

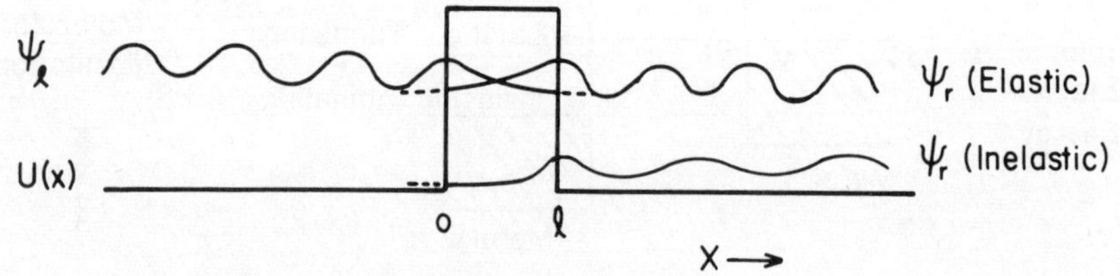

Fig. 2.

$$| M_{rl} |^2 = (h^2/16\pi) v_x^{(l)} v_x^{(r)} \exp\left(-2 \int_0^l K\,dx\right), \tag{2}$$

where $v_x^{(l)}$ and $v_x^{(r)}$ are the group velocities in the left and right metals, respectively, and K is the absolute value of the imaginary k-vector inside the barrier,

$$K = \{(2m/\hbar^2)[U(x) - (E - E_\perp)]\}^{1/2}, \tag{3}$$

where E is the total electron energy, $U(x)$ is the potential energy, and $E_\perp = \hbar^2 k_\perp{}^2/2m$ is the kinetic energy associated with motion parallel to the barrier (Fig. 2). The current for elastic tunneling is obtained by summing over all initial and final states (see Chapter 4):

$$I = \frac{4\pi me}{h^3} \int_{-\infty}^{\infty} dE[f(E) - f(E + eV)] \int_0^E dE_\perp \exp\left(-2 \int_0^l K\,dx\right), \tag{4}$$

$$f(E) = [1 + \exp(E/kT)]^{-1}.$$

For simple metals the integral over the transmission coefficient is appreciable only for electrons incident nearly perpendicular to the barrier and is determined primarily by the properties of the barrier [([4]), p. 87]. Therefore, to the approximation that we neglect the voltage dependence of the barrier penetration probability,

$$I = C \int_{-\infty}^{\infty} dE[f(E) - f(E + eV)] = CeV, \tag{5}$$

we get by direct integration

$$G_0 = dI/dV = \text{const},$$

i.e., normal elastic tunneling is ohmic in this approximation.

In the inelastic tunneling the electron is expected to tunnel from the state on the left to one on the right with lower energy, and at the same time the impurity will be excited from its ground state to an excited state, $\Psi_l(\text{el})\,\phi_0(\text{mol}) \to \Psi_r(\text{el})\,\phi_1(\text{mol})$.

The tunneling rate is again

$$w_{l\to r} = (2\pi/\hbar)\,| M_{rl} |^2\, \delta(E_l - E_r + \hbar\omega_0). \tag{6}$$

The simplifying assumption made ([2]) is that the electronic part of the tunneling matrix element can be approximated by the elastic tunneling expression (2), so that

$$M_{rl} \propto \langle \phi_1 \,|\, \exp - \int_0^l K\,dx \,|\, \phi_0 \rangle \tag{7}$$

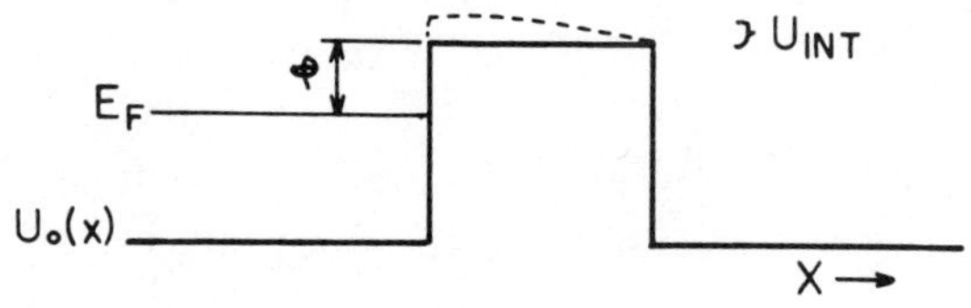

Fig. 3.

is left to describe excitation. This assumption neglects the very appreciable dependence of the electronic part on the final electron energy. This is because this barrier penetration probability for electrons is proportional to the overlap of right- and left-hand wave functions. The degree of overlap depends strongly on the energy of the electron. However, it is not expected that this assumption will affect the qualitative features of the results.

In this model excitations occur because of a dependence of the barrier height on the interaction of tunneling electrons with impurities, $U(x) = U_0(x) + U_{\text{int}}$, i.e., impurities are pictured as perturbing the height of the barrier (Fig. 3). If $U_0(x) - (E - E_\perp) \approx \varphi = \text{const}$, we may expand

$$K = K_0[1 + \tfrac{1}{2}(U_{\text{int}}/\varphi)], \qquad K_0 = (2m\varphi/\hbar^2)^{1/2}; \tag{8}$$

$$M_{lr} \propto \exp(-k_0 l) \langle \phi_0 \mid \phi_0 \rangle - \exp(-k_0 l) \left(\frac{K_0}{2\varphi} \right) \langle \phi_1 \mid \int_0^l U_{\text{int}}\, dx \mid \phi_0 \rangle . \tag{9}$$

The first term represents elastic tunneling, since $\langle \phi_0 \mid \phi_0 \rangle = 1$ for this case, and the second will produce inelastic tunneling.

The net current can be computed as before, and the inelastic part is (Fig. 4)

$$I_{\text{inel}} = G_0 \left(\frac{1}{e} \right) \frac{K_0^2}{4\varphi^2} \mid \langle \phi_1 \mid \int_0^l U_{\text{int}}\, dx \mid \phi_0 \rangle \mid^2$$
$$\times \int_{-\infty}^{\infty} f(E)[1 - f(E + eV - \hbar\omega_0)]\, dE. \tag{10}$$

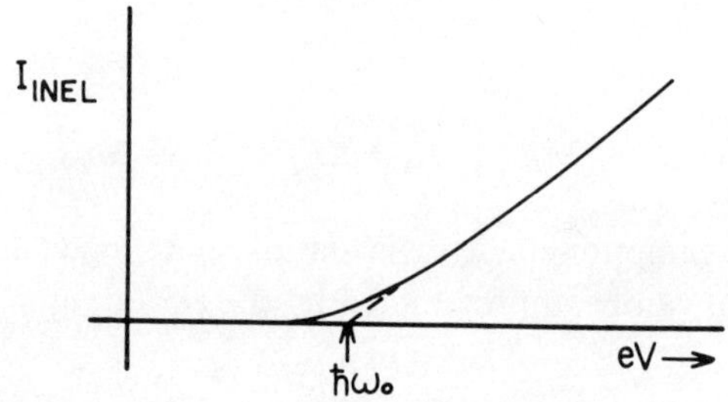

Fig. 4.

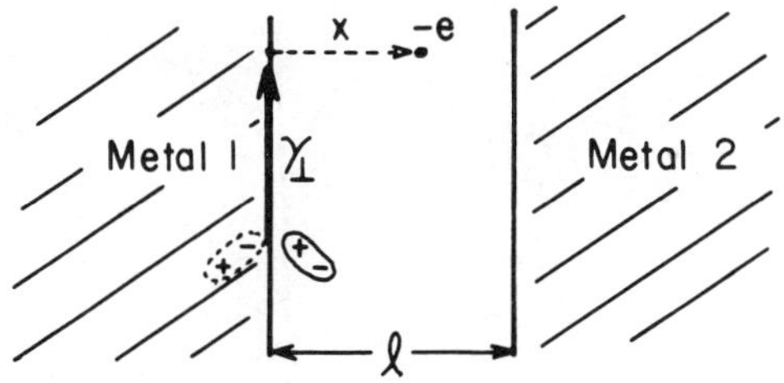

Fig. 5.

An important difference between this expression and the one for elastic current is the form of the Fermi-factor combination. The physical reason for the difference is that in elastic tunneling [Eq. (4)] both $l \to r$ and $r \to l$ tunneling are included, whereas for inelastic tunneling the excitation energies and temperatures of interest are such that all of the molecules will be in their ground state and therefore electron backflow (anti-Stokes) by inelastic processes will not occur. Hence, the combination of Fermi factors in (10) describes one-way current flow. By direct integration,

$$F(eV - \hbar\omega_0) \equiv \int_{-\infty}^{\infty} f(E)[1 - f(E + eV - \hbar\omega_0)]\, dE$$
$$= (eV - \hbar\omega_0) \frac{\exp[(eV - \hbar\omega_0)/kT]}{\exp[(eV - \hbar\omega_0)/kT] - 1}. \tag{11}$$

This is a "turning-on" function describing the onset of the inelastic tunneling, and contains the temperature dependence produced by smearing of the filled states around the Fermi energy. This current must be added to the elastic tunneling current.

INTERACTION MECHANISM

Electron-Dipole Interaction (ED)

In the interaction considered by Scalapino and Marcus there may be a dipole moment P associated with the molecular vibration located at one of the metals (Fig. 5). Taking into account the nearest image of the dipole, the interaction energy is, using the dipole approximation,

$$U_{\text{int}} = 2exP_x/(x^2 + r_\perp{}^2)^{3/2}. \tag{12}$$

Electron-Induced Dipole Interaction (EID)

The electric field of the electron is capable of inducing a dipole moment in the molecular species $P = \alpha E$ (Fig. 6). The polarizibility α in many

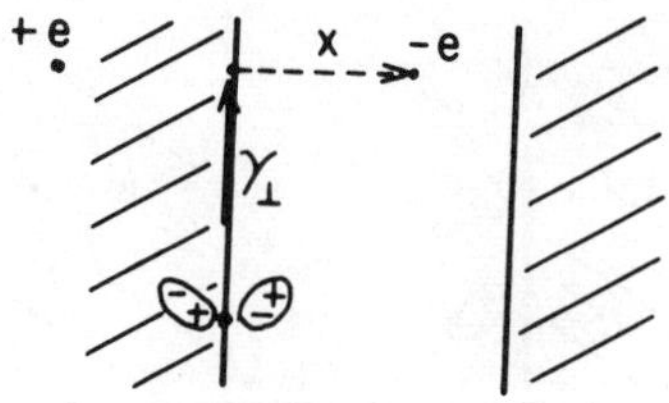

Fig. 6.

cases varies with the vibration of the molecule. Taking into account the nearest images, the interaction energy is

$$U_{\text{int}} = -\ 4e^2x^2\alpha/(x^2 + r_\perp{}^2)^3. \tag{13}$$

The inelastic current for these two cases is given by: for ED

$$I = NG_0\left[\frac{4\pi me}{\hbar^2\varphi}\ln\left|\frac{l}{r_0}\right|\right]\sum_m |\langle m \mid P_x \mid 0\rangle|^2 F(eV - \hbar\omega_m); \tag{14a}$$

for EID

$$I = NG_0\left[\int_0^\infty dr_\perp r_\perp{}^2 t^2(\varrho)\right]\frac{\pi me^3}{4\hbar^2\varphi l^6}\sum_m |\langle m \mid \alpha \mid 0\rangle|^2 F(eV - \hbar\omega_m), \tag{14b}$$

$$t^2(\varrho) = \frac{1}{\varrho^2}\left[\frac{1-\varrho^2}{1+\varrho^2} + \frac{1}{\varrho}\tan^{-1}\left(\frac{1}{\varrho}\right)\right], \qquad \varrho = \frac{r_\perp}{l};$$

where the factors in square brackets represent integration over the area of the junction to obtain the total inelastic contribution due to a single impurity and N is the total number of impurities. In ED the $\ln|\, l/r_0 \,|$ term results from noting that for $r_\perp \gtrsim l$ the effect is negligible, while r_0 is a cutoff near the molecule (~ 3 Å) where the dipole approximation will break down. In EID the expression is not so simple and can be estimated numerically. In addition, in order to conform to reality and take into account natural line shapes, a sum over a whole spectrum of final states is made.

It is easy to see the behavior at $T \approx 0$, where dI/dV is a step function: for ED

$$G = dI/dV = G_0NA\sum_m |\langle m \mid P_x \mid 0\rangle|^2 S_m(eV); \tag{15a}$$

for EID

$$G = G_0NB\sum_m |\langle m \mid \alpha \mid 0\rangle|^2 S_m(eV); \tag{15b}$$

where $S_m(eV)$ is a unit step at $eV = \hbar\omega_m$.

The intensity G_∞ of a spectra line (Fig. 7) is given by

ED: $$G_\infty = NG_0A\sum_m |\langle m \mid P_x \mid 0\rangle|^2, \tag{16a}$$

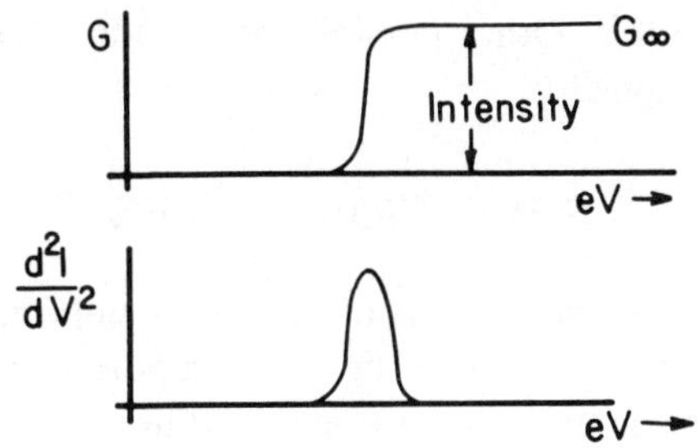

Fig. 7.

$$\text{EID:} \qquad G_\infty = NG_0B \sum_m |\langle m \mid \alpha \mid 0\rangle|^2. \tag{16b}$$

This represents the area under the second derivative curves, which are given by

$$\text{ED:} \qquad d^2I/dV^2 = NG_0A \sum_m |\langle m \mid P_x \mid 0\rangle|^2 \delta(eV - \hbar\omega_m), \tag{17a}$$

$$\text{EID:} \qquad d^2I/dV^2 = NG_0B \sum_m |\langle m \mid \alpha \mid 0\rangle|^2 \delta(eV - \hbar\omega_m). \tag{17b}$$

The quantities $\Sigma_m |\langle m \mid P_x \mid 0\rangle|^2$ and $\Sigma_m |\langle m \mid \alpha \mid 0\rangle|^2$ are directly proportional to the absorbtion coefficient of light (IR) and intensity of scattered light (Raman effect), respectively. The results show that the intensity of the inelastic tunneling is proportional to the analogous effects with light, and that the spectral shape of the lines are expected to be nearly the shape of the optical lines [Eqs. (17*a*) and (17*b*)]. This explains the very close resemblance of the inelastic tunneling spectra to optical data. To give an estimate of the relative size of the effects: $\Sigma_m |\langle m \mid P_x \mid 0\rangle|^2 \approx 10^{-38}$ (esu-cm)^2 (IR absorption data) and $\Sigma_m |\langle m \mid \alpha \mid 0\rangle|^2 \approx 10^{-50}$ cm^6 (from optical Raman data) for typical molecular excitations, and we have

$$\text{ED:} \qquad G_\infty/G_0 \approx 1\%,$$

$$\text{EID:} \qquad G_\infty/G_0 \approx (0.1\text{–}0.5)\%.$$

Thus, both mechanisms, when allowed, give intensities of sufficient size to be observed, and are nearly the same order of magnitude in size.

Other types of interaction should be possible, of course. By the same reasoning used for the electron–dipole interaction, it should be possible to excite electronic excitations in impurity atoms or molecules. The strength of these is typically 100 times greater than molecular dipole transitions, and so should be readily observable. The addition of such impurities to a junction would be desirable. Other possibilities include interaction with oxide vibrations and phonons.

SELECTION RULES

We have seen that the same matrix elements important for IR absorption and the Raman scattering of light are important for two interaction mechanisms in inelastic tunneling. In general, different selection rules govern these two processes ([5]). In fact, for molecules possessing a center of symmetry the two processes are mutually exclusive. On the other hand, for molecules possessing no symmetry they may both be allowed. The latter is probably the case for the experiments so far. Nevertheless, the possibility of observing this distinction exists.

To illustrate the optical IR and Raman effects, the simple classical arguments are valuable. Classically, for an oscillating electric dipole, resonance absorption of light occurs at a frequency ω_0. Generally, this dipole moment can be expanded in terms of the normal coordinate ξ of the vibrating molecule

$$P = P_0 + \left(\frac{dP}{d\xi}\right)_0 \xi + \ldots, \tag{18}$$

where $\xi = \xi_0 \cos \omega_0 t$ and P_0 is the permanent dipole moment which does not contribute to absorption. Thus, if $(dP/d\xi)_0$ is different from 0, the absorption can occur. Similiarly, in the Raman effect the field of the incident light $E = E_0 \cos \omega t$ induces a moment $P = \alpha E$, where α depends on the vibration coordinate:

$$\alpha = \alpha_0 + (d\alpha/d\xi)_0 \xi \ldots; \tag{19}$$

$$P = \alpha_0 E_0 \cos \omega t + (d\alpha/d\xi)_0 \xi_0 E_0 \tfrac{1}{2}[\cos(\omega - \omega_0)t + \cos(\omega + \omega_0)t]. \tag{20}$$

Therefore the scattered (reradiated) light will contain ω (Rayleigh scattering) and $\omega \pm \omega_0$ (Raman scattering). The factor $(d\alpha/d\xi)_0$ determines whether the Raman process is allowed or not. The simple example of the linear CO_2 molecule is illustrated in Fig. 8 for two of its modes of oscillation. The allowed and forbidden IR and Raman modes are indicated and may be inferred by simple symmetry considerations.

RESOLUTION

As with any spectrometer, it is important to consider resolution and possible limitations. In these experiments overall linewidth will be determ-

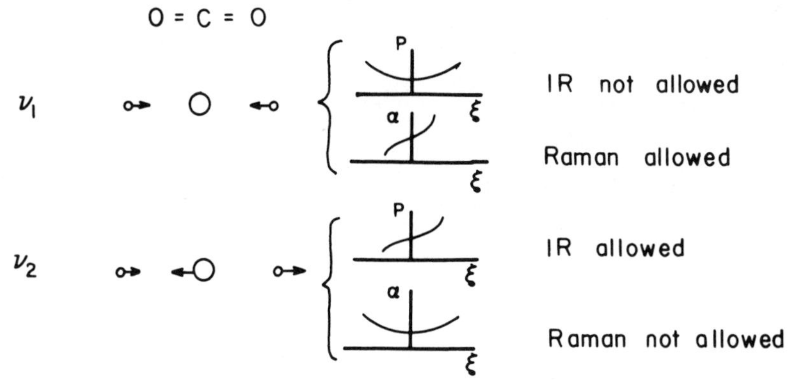

Fig. 8.

ined, by the natural width of the molecular excitation line, temperature broadening, and density-of-states effects due to the metal contacts (superconductor).

For normal metal electrodes the temperature broadening is important. Consider a single sharp molecular excitation. The current is $I=CF(eV-\hbar\omega_0)$. The second derivative is

$$\frac{d^2I}{dV^2} = C\left(\frac{e^2}{kT}\right)\left[e^v \frac{(v-2)e^v + (v+2)}{(e^v-1)^3}\right], \qquad v \equiv \frac{eV-\hbar\omega_0}{kT}, \tag{21}$$

which is a symmetrical bell-shaped peak of width 5.4 kT. The height of the peak is proportional to $(kT)^{-1}$, and the total area (intensity) is independent of temperature, as would be expected:

Room temperature	$5.4\,kT \approx 150$ meV
77 °K	$5.4\,kT \approx 38$ meV
4 °K	$5.4\,kT \approx 1.9$ meV
0.3 °K	$5.4\,kT \approx 0.15$ meV

Note that at room temperature, in agreement with experiment, the broadening is such that observation of the peaks is almost impossible, since the excitations occur over a range 100–500 meV. At 77°K the lines are easily observed, but not well-resolved, and at 4°K the natural width of most molecular lines is seen. Very sharp spectra would require still lower temperatures before temperature broadening could be avoided. Predictions are in close agreement with actual linewidth data.

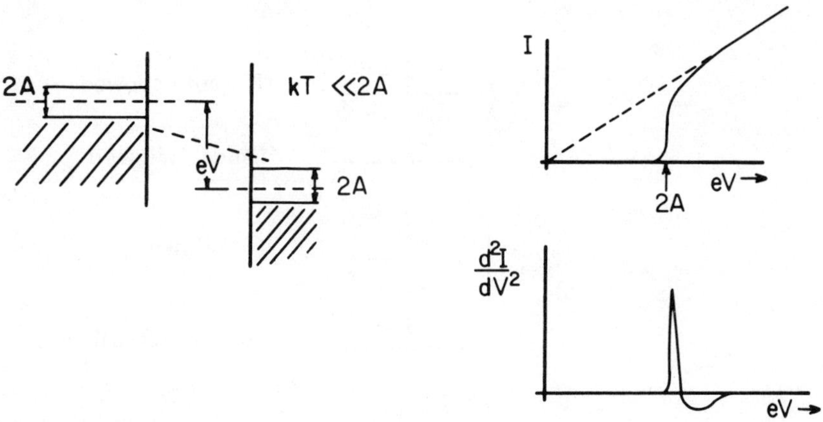

Fig. 9.

Another important point in this connection is the use of superconducting contacts. In this case the superconducting density of states must be included in the tunneling equations. These effects were discussed in previous chapters. Briefly, very sharp structure appears in the second derivative due to the very pronounced density-of-states peak at the edge of the superconducting energy gap. Figure 9 shows the energy spectrum of two superconducting electrodes making up the tunnel junction. Elastic tunneling curves showing how one observes the pronounced density-of-states effects are also shown. We note that the onset of current is very sharp, and in fact will determine the sharpness of the onset of inelastic tunneling processes as well. The expected effect on inelastic tunneling is shown in Fig. 10. Note, however, that inelastic tunneling can only occur if $eV \geq (\hbar\omega_0$

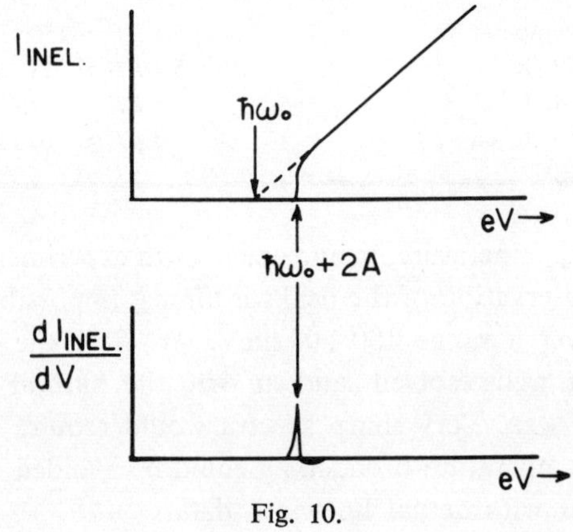

Fig. 10.

$+ 2\Delta$), so a small shift ($2\Delta \approx 2$ meV) in line position is expected. This has been observed. To estimate the resolution, there is the example of a Sn–Sn oxide–Sn junction measured by Giaever, where at 0.3°K the width of the onset was 0.030 meV. This is to be compared with 0.150 meV obtainable with normal metal contacts at the same temperature. This increased resolution has been observed in inelastic tunneling resolution when one or both metal electrodes are made superconducting.

REFERENCES

1. R. C. Jaklevic and J. Lambe, Phys. Rev. Letters 17:1139 (1966); Phys. Rev. 165:821 (1968).
2. D. J. Scalapino and S. M. Marcus, Phys. Rev. Letters 18:459 (1967).
3. J. Bardeen, Phys. Rev. Letters 6:57 (1961).
4. W. A. Harrison, Phys. Rev. 123:85 (1961).
5. G. Herzberg, Molecular Spectra and Molecular Structure, I and II, D. Van Nostrand Co., New York, 1945.
6. J. Giaever, H. R. Hart, Jr., and K. Megerle, Phys. Rev. 126:941 (1962).

Chapter 19

Tunneling Between Superconductors

I. Giaever

General Electric Research and Development Center
Schenectady, New York

INTRODUCTION

Over the last few years electron tunneling has become recognized as an important tool for studying the properties of superconductors, and I would like to start by reviewing some of the salient features of superconductivity. Many metals and alloys have the peculiar property that the resistance to electrical current becomes zero when the metal is cooled below a certain temperature, referred to as the superconducting transition temperature. Different metals have different transition temperatures; for example, for the metals considered here Pb becomes superconducting below 7.2°K, Sn below 3.7°K, In below 3.3°K, and Al below 1.2°K, and as far as we know Mg does not become superconducting below any temperature. (Note that experimentally we can never say that a metal is not a superconductor, because we cannot make the test at $T = 0$.)

It is not sufficient the temperature be low for a metal to enter into its superconducting state; the magnetic field must also be below a certain value, referred to as the critical field. Figure 1 shows the superconducting phase diagram for lead. Therefore when we perform a tunneling experiment on a superconductor both the temperature and the magnetic field will affect our results.

To understand what happens when superconductors are used in tunneling experiments, it is necessary to know something about the Bardeen–Cooper–Schrieffer (BCS) theory. The central feature of this theory is that a very significant change occurs in the electron density of states when a metal goes from its normal into its superconducting state. In an ordinary free-electron metal all the electrons will condense into a sphere in k space (wave number space), and the density of states, N, has the familiar square-

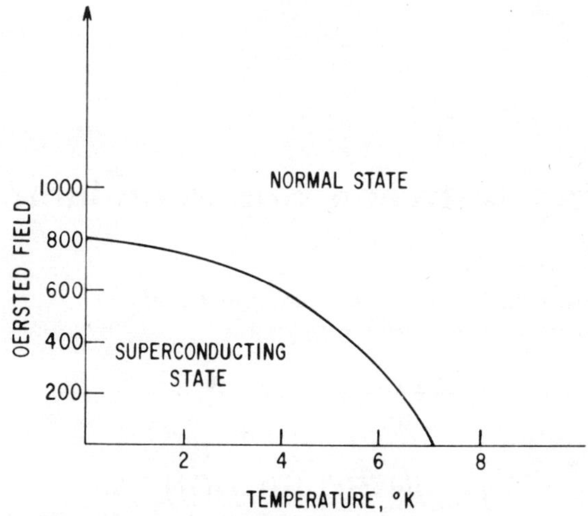

Fig. 1. A phase diagram for Pb showing the phase boundary in the magnetic field–temperature plane.

root dependence on the energy. At $T = 0$ the Fermi surface and wave number, k_F, in k space will be sharply defined, i.e., all states below k_F are occupied, while above k_F they are empty. In a superconductor, however, the BCS theory tells us that the electrons condense into so-called Cooper pairs. Because of this the Fermi level in k space is no longer sharply defined, i.e., there are states below k_F which are empty and states above k_F which are filled. This increases the kinetic energy of the electrons, but when the pairing energy, i.e., the binding energy of a Cooper pair, is taken into account the total energy of a superconductor is lower than that of a normal material. The binding energy produces a gap, 2Δ, in the excitation spectrum of the electrons, and the superconducting density of state, N_s, is simply related to the normal density of states by the following formula:

$$N_s = \frac{|E|}{(E^2 - \Delta^2)^{1/2}} N.$$

EXPERIMENTAL PROCEDURES

Scientists and engineers inexperienced in low-temperature work tend to believe that it is difficult to work with superconductors, and that the whole low-temperature field is pretty esoteric. That was indeed true in the past, and it is one reason for the slow development of our knowledge about superconductors. To obtain temperatures in the superconducting range, it is necessary to have access to liquid He, and it used to require great diligence and skill on the part of the researcher to liquify He; today all one needs is a little money, as liquid He is readily available from commercial sources.

To do a tunneling experiment at superconducting temperatures one needs first of all two Dewars, one inside the other (Fig. 2). A Dewar is nothing else than a large coffee jug. The outer Dewar is filled with liquid nitrogen or air. Liquid N_2 boils at 77°K and, at least in the United States, it is cheaper than milk. The inner Dewar is filled with liquid He. Liquid He boils at 4.2°K at atmospheric pressures and costs approximately the same as cheap cognac, and probably lasts about as long too! A couple of

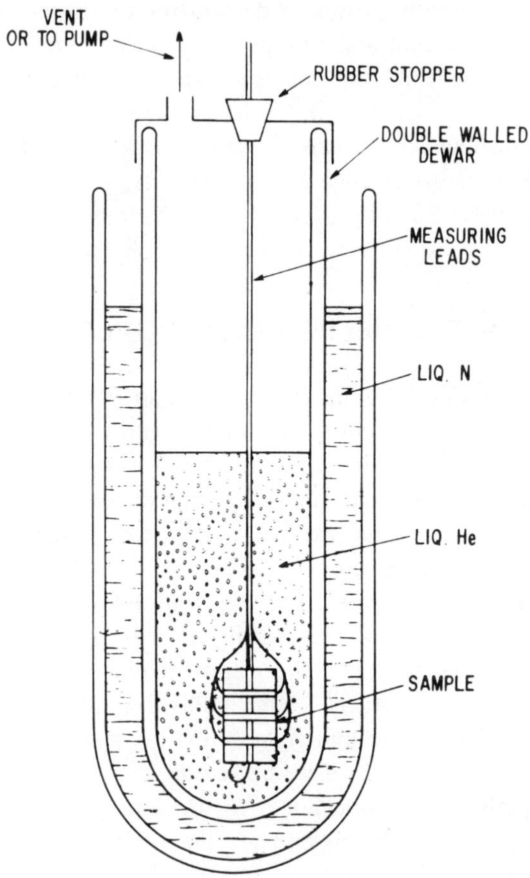

Fig. 2. A simple experimental setup for performing tunneling experiments at superconducting temperatures.

liters of He last approximately 4 hr in a reasonable Dewar system. If one wants to go below 4.2°K, one simply lets the He evaporate into a pressure lower than the atmospheric pressure. By pumping on the He with an ordinary pump one can easily go down to 1.5°K, and with a big pump down to maybe 0.9°K. Below that one must use different techniques, such as He^3 refrigerators. To measure the temperature of the He bath all one has to do is to measure the pressure above the bath, as the boiling point is a well-known function of the pressure. Therefore it is easy to measure the temperature to, say, three significant figures, and with care one can do much better.

The sample preparation is as I described in Chapter 3 of this volume. We prepare a capacitor-like arrangement by evaporating thin metal films, relying upon a naturally grown oxide film for the insulating layer. Next we solder on thin current and voltage leads to the sample and simply dip it, suspended by the leads, into the liquid He. We can now test the sample electrically, and if it is no good, we haul it out and try another one. In performing tunneling experiments the sample is the critical link. Therefore it is very important to arrange the experiments such that one can easily test many samples. Recently I tested over 50 samples in one day—and I might add they were all bad.

MODEL FOR TUNNELING INTO SUPERCONDUCTORS

The Bardeen–Cooper–Schrieffer theory is strictly a many-body theory, and we know that the reason for the energy gap is that the electrons are condensed into Cooper pairs. In spite of this I shall use a one-electron picture of the density of state in a superconductor, since intuitively this gives us most of the correct answers, and is very similar to the familiar model used for semiconductors. However, the model can be misused; Schrieffer gives a more correct but less intuitive model in Chapter 21. Figure 3 shows the density of states in an ordinary metal and how the density of states changes when the metal becomes superconducting. The energy gap, 2Δ, is of the order of a few millivolts, and the density of states goes to infinity right above and right below the gap. It is easy to understand the experimental results if we make one simple assumption—that the tunneling probability is proportional to the density of states in a superconductor. Bardeen was the first to use the "golden rule" in a tunneling problem, and he obtained

$$P_{LR} = (2\pi/\hbar)\,|\,M\,|^2 N_R(1 - f_R),$$

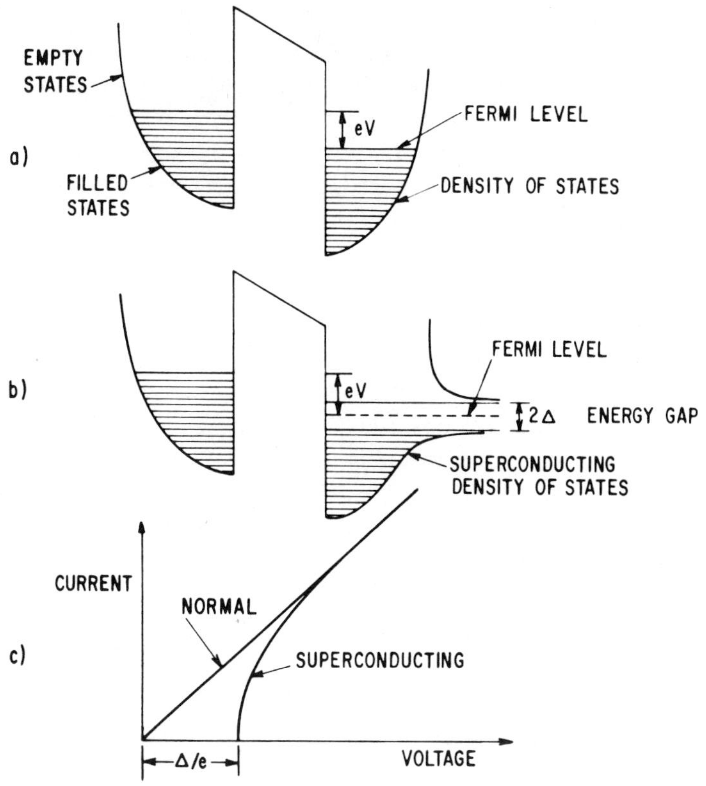

Fig. 3. Energy diagrams used in describing tunneling experiments. (*a*) Tunneling between two normal metals. Only electrons in the range eV can tunnel from left to right, because only these electrons face empty states. (*b*) Tunneling between a normal metal and a superconductor. A superconductor has a very different density of states and even an energy gap centered at the Fermi level. Only electrons in the range $eV - \Delta$ can tunnel from left to right, because only these electrons face empty states. The energy gap is a few millivolts wide, while the Fermi energy is several volts above the bottom of the band, and therefore the scale used here is very distorted. (*c*) Current–voltage characteristic of tunneling between two normal metals (the straight line) and tunneling between a normal metal and a superconductor. The temperature is assumed to be zero. At a finite temperature we must take excited electrons into account.

where P_{LR} is the tunneling probability for transition out of a state to the left of the barrier into a state on the right, $|M|^2$ is the matrix element for the transition, N_R is the density of states on the right side, and $1 - f_R$ is the probability that the state on the right side is empty. The matrix element $| M |^2$ contains a factor which is more or less inversely proportional to the density of states in a normal metal, and thus for normal metals the density of states cancels out. For superconductors one can show that it is proper to use the superconducting density of states in the above equation. If in addition we assume that the matrix element $| M |^2$ does not change when a metal becomes superconducting, the agreement between theory and experiment is excellent. It is not known why $| M |^2$ has the same value regardless of whether the metals are superconducting or normal. However, for our puposes here it is only important to remember that while the tunneling probability is very insensitive to changes in the density of states for normal metals, it is directly proportional to the density of states in superconductors.

Referring to Fig. 3 then, when both metals are normal we obtain a linear current–voltage characteristic. When one metal is superconducting, at absolute zero temperature no current can flow before we have applied sufficient voltage, $eV \geq \Delta$, such that electrons in the metal on the left find allowable and empty states to flow into in the superconductor on the right. The rapid increase in the current at $eV \approx \Delta$ is due to the large density of states close to the energy gap.

Of course, experiments cannot be carried out at absolute zero, but the following expression will give the current at finite temperatures as well:

$$I_{NS} = C_{NN} \int_{-\infty}^{\infty} \frac{| E |}{(E^2 - \Delta^2)^{1/2}} \langle f(E) - f(E + eV) \rangle \, dE, \tag{1}$$

where I_{NS} is the current between a normal metal and a superconductor, $f(E)$ the Fermi distribution, e the electron charge, and 2Δ the energy gap. Here C_{NN} is the conductance when both metals are normal, and since we are only interested in the difference between the current–voltage characteristics obtained for normal and superconducting metals, the exact form of C_{NN} is unimportant. This is one of the great advantages of tunneling into superconductors, and one reason it has become such a useful tool.

EXPERIMENTS

In Fig. 3 we saw that at absolute zero temperature no current could flow between a normal metal and a superconductor before $eV \geq \Delta$. For finite temperatures that is no longer true, and at finite temperatures Eq. (1) can be evaluated for $eV < \Delta$,

$$I_{NS} = 2C_{NN} \frac{\Delta}{e} \sum_{1}^{\infty} (-1)^{m+1} K_1\left(m \frac{\Delta}{kT}\right) \sinh\left(m \frac{eV}{kT}\right), \tag{2}$$

where K_1 is the modified Bessel function of the second kind, k the Boltzmann constant, and T the temperature. We shall first concentrate on the voltage range $eV < \Delta$. In particular, note that for $\Delta > eV \gg kT$ the above expression can be written as:

$$\ln I_{NS} = (e/kT)V + \alpha(\Delta, T), \tag{3}$$

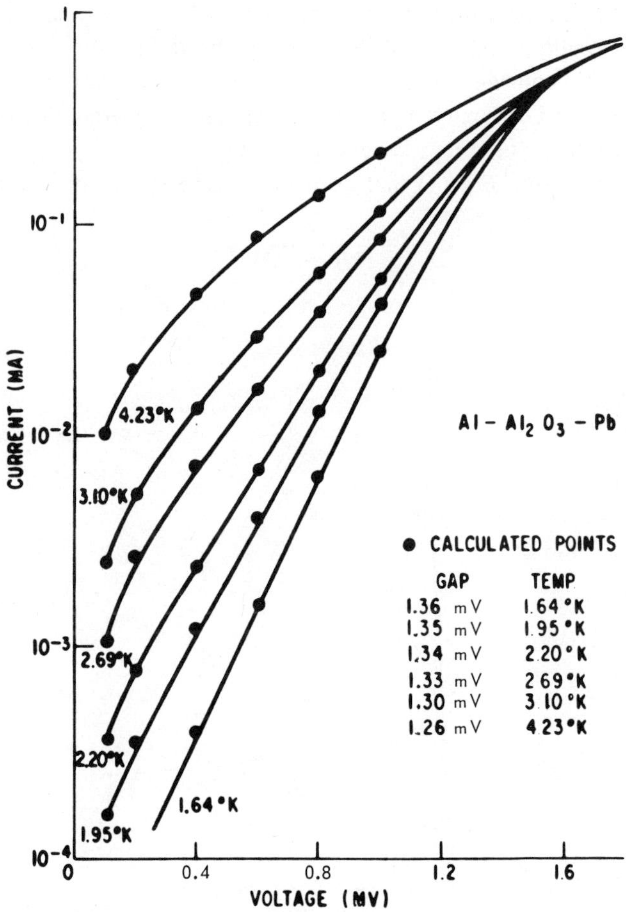

Fig. 4. This shows the current–voltages characteristic for applied voltages less than half the energy gap. At these temperatures Al is normal and Pb is superconducting. The slope is inversely proportional to the temperature, i.e., we are using an electron gas as a thermometer. If the temperature is measured independently, we can determine the ratio of the Boltzmann constant and the electron charge from the curves. From I. Giaever and K. Megerle, Phys. Rev. 122:1101 (1961).

i.e., if we plot $\ln I_{NS}$ against V the inverse of the slope gives us the correct temperature. Experiments to that point are displayed in Fig. 4. The energy gap has first been calculated using Eq. (2). For sufficiently low temperatures Eq. (3) becomes a good approximation, and $\ln I$ plotted against V gives a straight line. Since the slope is inversely proportional to temperature, we are actually using an electron gas as a thermometer! If in addition we measure the temperature, we can determine k/e this way as well. This is not an especially good way of determining the Boltzmann constant; however, Planck's constant divided by the electron charge was determined by a group at the University of Pennsylvania, using superconducting tunneling, and that measurement is very accurate.

It is seen from Eq. (1) that at absolute zero temperature we obtain

$$\frac{dI_{NS}}{dV} = C_{NN}\frac{eV}{[(eV)^2 - \Delta^2]^{1/2}}, \qquad eV > \Delta,$$
$$= 0, \qquad eV < \Delta. \tag{4}$$

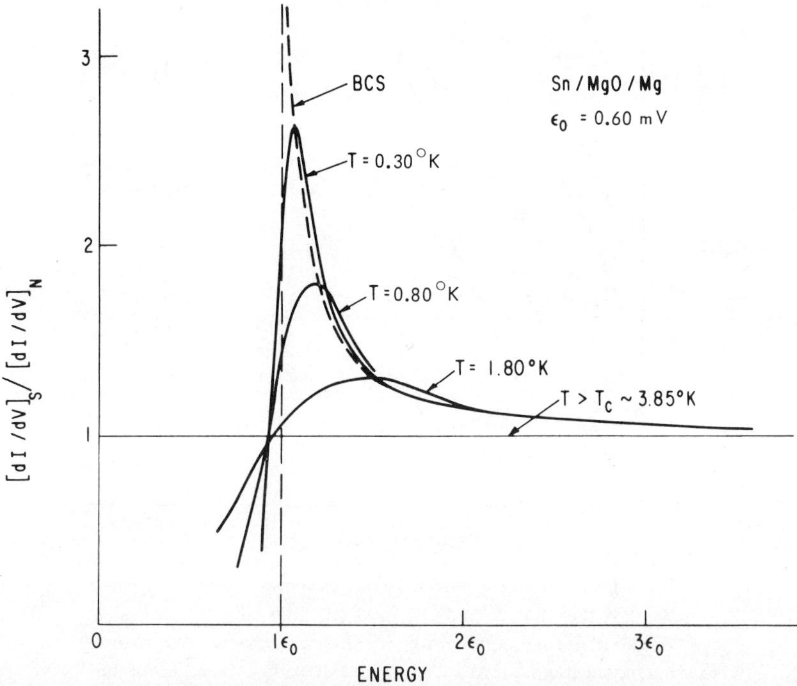

Fig. 5. The density of states in a superconductor obtained by taking the ratio of the derivatives of the current–voltage characteristics in the superconducting and normal states. As the smearing due to temperatures is reduced, the curves approach the BCS redictions. From I. Giaever, H. R. Hart, and K. Megerle, Phys. Rev. 126:941 (1962).

This is a direct consequence of the fact that the tunneling probability is directly proportional to the density of states in a superconductor. Experimentally, we can obtain the quantity dI_{NS}/dV directly by modulating I_{NS} by a small oscillating current and measuring the ΔV due to the modulation by using a lock-in amplifier. Thus we can determine the superconducting density of states very accurately. Figure 5 shows an experimental result compared to the BCS theory. As the temperature is lowered and therefore the smearing due to the temperature is diminished, the curves approach the BCS theory. Since in some sense one can say that the main purpose of an experimentalist is to find disagreements between experiments and

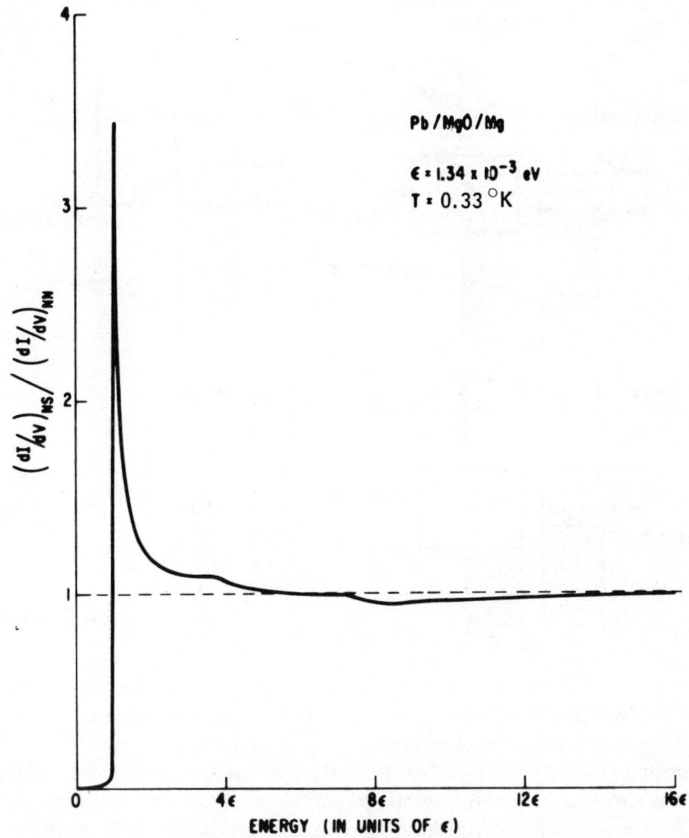

Fig. 6. The density of states for Pb, definitely different from the BCS theory. It turns out that the bumps in the curve reflect the phonon spectrum in Pb. Mg is a normal conductor. From I. Giaever, H. R. Hart, and K. Megerle, Phys. Rev. 126:941 (1962).

existing theories, from my point of view the good agreement was a disappointment.

Figure 6 shows a similar curve for Pb at low temperatures. It can clearly be seen that there are some small bumps in the curve, and therefore it does not agree with the BCS theory. I was very excited when we first found these bumps and made statements to the effect that the BCS theory must be wrong. As it turns out, a refinement of the BCS theory will give these bumps, at the same time it gives a good deal of information about the phonon spectrum in a superconductor (see Chapter 27).

We now consider the case where both metals are superconducting. The principle of the experiment is illustrated in Fig. 7 for two superconductors with different energy gaps. By simply keeping in mind that the

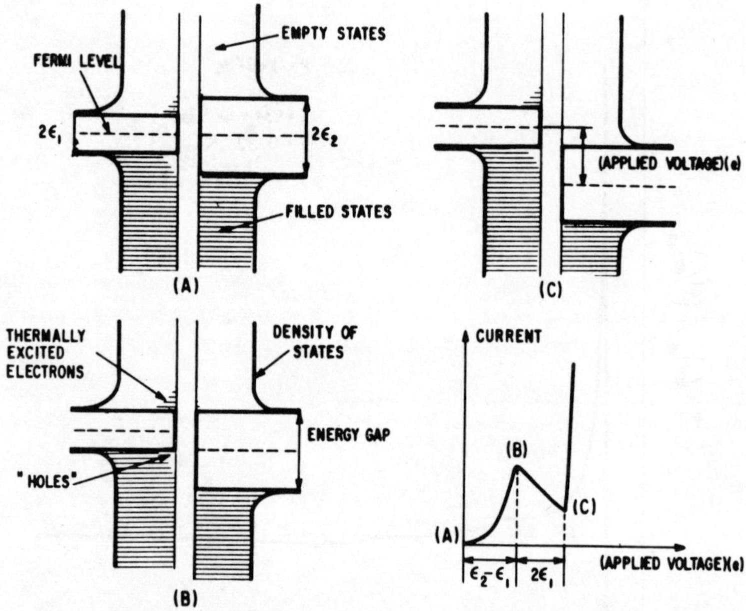

Fig. 7. Tunneling between two superconductors having different energy gaps at $T \neq 0°K$. (*A*) No voltage has been applied between the conductors. (*B*) As a voltage is applied, it becomes energetically possible for more and more of the thermally excited electrons to flow from the superconductor with the small gap into the superconductor with the large gap. Here all the electrons may tunnel. (*C*) As the voltage is increased, no more electrons come into play, and since the electrons face a lower density of states, the current actually drops as the voltage is increased, until electrons below the gap can tunnel. (*D*) The current–voltage characteristic. From I. Giaever, in: Proceedings of the International Conference on Solid State Circuits, Lewis Winner, New York, 1961.

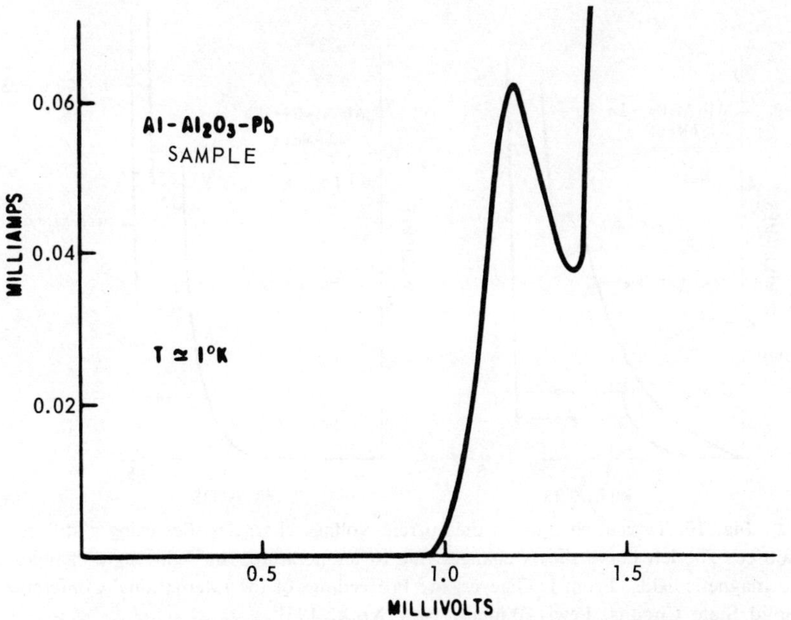

Fig. 8. Negative resistance characteristic obtained experimentally in tunneling between two different superconductors. From I. Giaever, in: Proceedings of the International Conference on Solid State Circuits, Lewis Winner, New York, 1961.

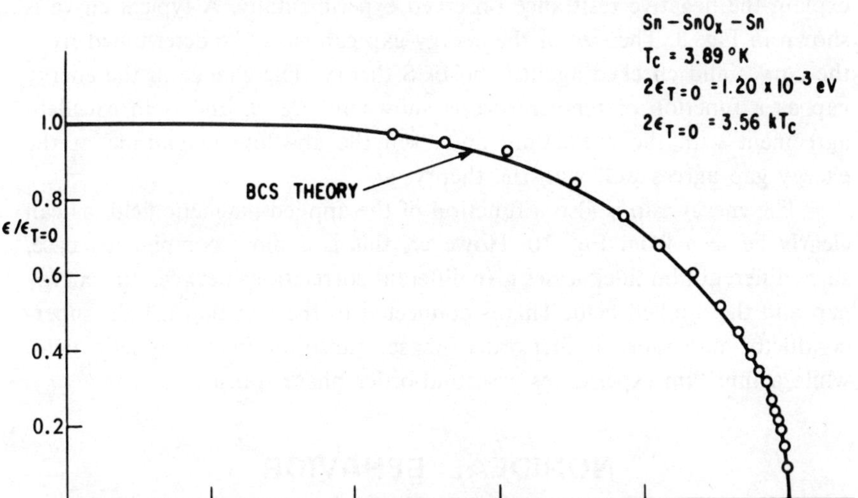

Fig. 9. Energy gap in Sn as a function of temperature. The circles are experimental points. From I. Giaever, in: Proceedings of the American Institute of Mining, Metallurgical, and Petroleum Engineers, February 18, 1962, Wiley, New York.

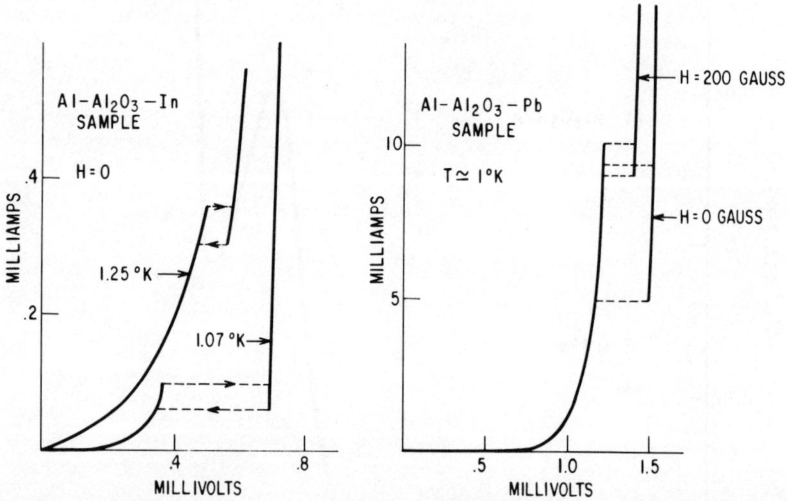

Fig. 10. Typical changes in the current–voltage characteristics using a stiff current source. The left curve shows changes due to temperature, the right curve changes due to magnetic fields. From I. Giaever, in: Proceedings of the International Conference on Solid State Circuits, Lewis Winner, New York, 1961.

tunneling probability is proportional to the density of states, we can easily explain the negative resistance observed experimentally. A typical curve is shown in Fig. 8. The size of the energy gap can easily be determined from the curves and checked against the BCS theory. The change in the energy gap as a function of temperature is shown in Fig. 9, and is in excellent agreement with the prediction, and even the absolute magnitude of the energy gap agrees well with the theory.

The energy gap is also a function of the applied magnetic field, as can clearly be seen from Fig. 10. However, this is a more complicated case, since different film thicknesses give different correlations between the energy gap and the applied field. This is connected to the fact that a bulk superconductor undergoes a first-order phase transition in a magnetic field, while a thin film experiences a second-order phase transition.

NONIDEAL BEHAVIOR

Rather than giving any sort of review of tunneling into superconductors, I shall point out a few places where we get nonideal behavior which is not very well understood. Figure 11 shows a current–voltage characteristic of a Sn–SnO–Sn sample. Theoretically the current should

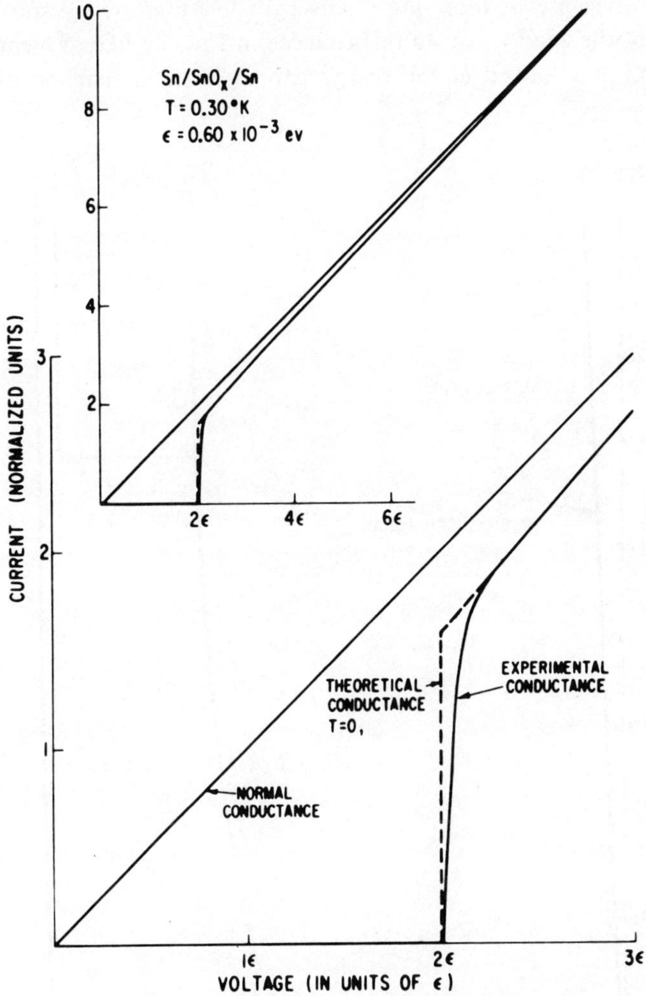

Fig. 11. Current–voltage characteristic of Sn–SnO–Sn sample. The current should jump discontinuously at $V = 2\Delta$ even when the temperature $T \neq 0$. From I. Giaever, H. R. Hart, and K. Megerle, Phys. Rev. 126:941 (1962).

make a discontinuous jump at $V = 2\Delta$ even at finite temperatures; experimentally, this never happens. Several explanations exist to explain this divergence from the theory. For example, because the excited electrons have a finite lifetime in the superconductor, the density of states becomes somewhat smeared out. Or, different crystal directions have different energy gaps, and because we look into a polycrystalline film, we see some sort of average of the energy gap. In this connection I would like to mention that Zavaritsky has looked at the energy gap in Sn as a function of crystal

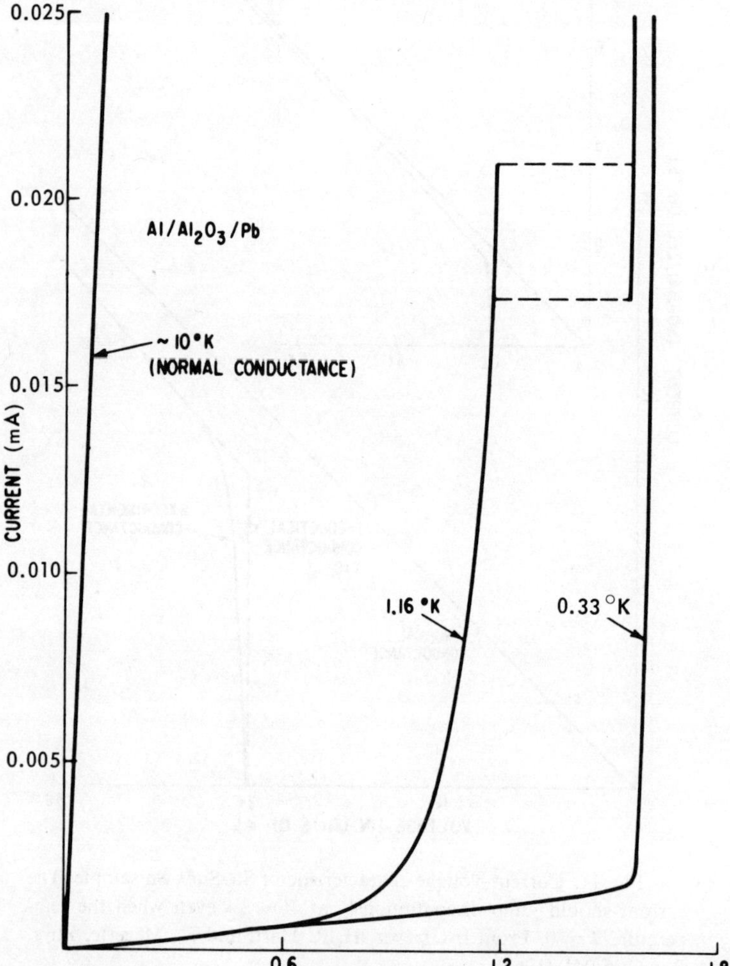

Fig. 12. Current–voltage characteristic, showing leakage current through the sample. The leakage current does not show up significantly until low temperatures; however, this sample is particularly good, and the leakage current is only a very small fraction of the total current at room temperature. Very often leakage current can account for 50–100% of the current at room temperature; 100% means that the sample is no good. From I. Giaever, H. R. Hart, and K. Megerle, Phys. Rev. 126:941 (1962).

directions and indeed found that the energy gap varies markedly with crystal direction. Very little work has been done in the area of tunneling into single crystals, obviously because it is difficult to do; however, there are many opportunities for extending Zavaritsky's work. Strain is known to change the transition temperature and therefore the energy gap, and thus the smearing may be due to a variable strain across the junction. Finally, of course, the simple BCS theory could be slightly incorrect, and perhaps a small modification will account for the phenomenon.

Figure 12 shows that not all current is tunneling. At 1.16°K we get a proper tunneling curve showing the negative resistance, while at 0.33°K the negative resistance has disappeared. This can be explained by assuming

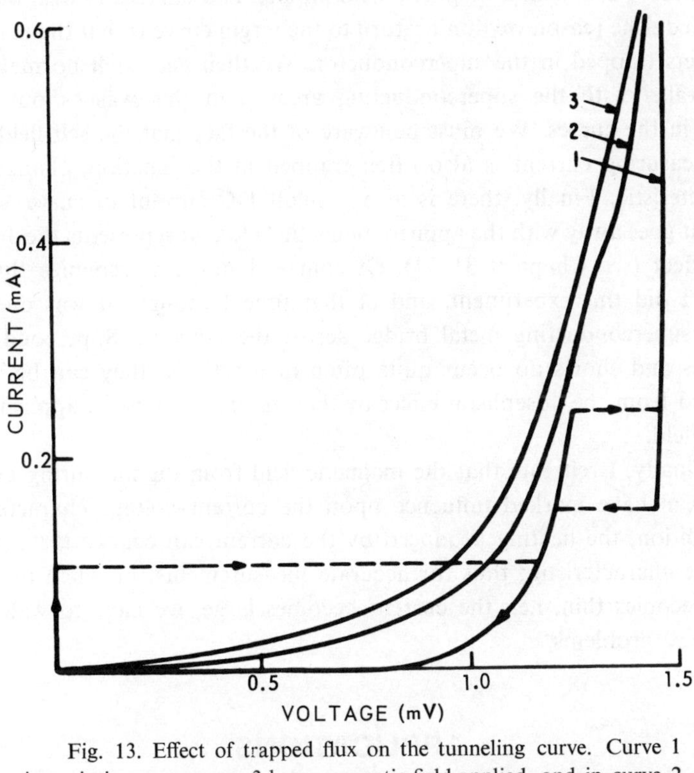

Fig. 13. Effect of trapped flux on the tunneling curve. Curve 1 is a virgin curve, curve 3 has a magnetic field applied, and in curve 2 the magnetic field has been removed. The negative resistance has disappeared because of the trapped flux in the films. In curve 1 we also have some resistanceless current which was once interpreted as a small metallic short; it is actually due to the Josephson effect. From I. Giaever and K. Megerle, Phys. Rev. 122:1101 (1961).

that a "leakage" current is flowing in parallel with the tunnel current. As seen from Eq. (4), the tunnel current depends exponentially on the temperature, and thus at sufficiently low temperature the leakage current becomes dominant. The origin of the leakage current is not entirely known—in all probability there are several causes—but at least in this sample the leakage current is relatively insensitive to temperature. This sample is particularly good, i.e., we have very little leakage current. However, it is not unusual to find samples where 80–90% of all the current flowing through the sample is due to mechanisms other than tunneling. Superconducting tunneling is one way of finding out which fraction of the current is due to tunneling and which fraction is not.

Figure 13 shows another sort of leakage current. Curve 1 is a virgin curve, curve 3 is with a magnetic field applied and curve 2 is with the field removed. The reason we don't return to the virgin curve is that the magnetic flux gets trapped in the superconductor. We then get small normal areas in parallel with the superconducting areas, and this washes out much detail in the curves. We must be aware of the fact that the self-field from the measuring current is also often trapped in the junction, changing its characteristic. Finally, there is also a small DC current in curve 1. This current goes away with the applied magnetic field and represents the Josephson effect (see Chapters 31–34). Of course, I did not recognize this fact when I did this experiment, and at that time I thought it was due to a small superconducting metal bridge across the junction. Superconducting bridges and shorts do occur quite often in junctions; they can be distinguished from the Josephson effect by their insensitivity to an applied magnetic field.

Finally, I reiterate that the magnetic field from the measuring current itself can have marked influence upon the current–voltage characteristic. In addition, the heating produced by the current can change the current–voltage characteristic; thus for accurate measurements, or when the junction becomes thin, i.e., the current becomes large, we have to watch out for these problems.

APPLICATIONS

The negative resistance characteristic obtained by performing a tunneling experiment between two superconductors represents an active device. Since one can change the characteristic by applying a magnetic field or by changing the temperature, one has essentially all the makings

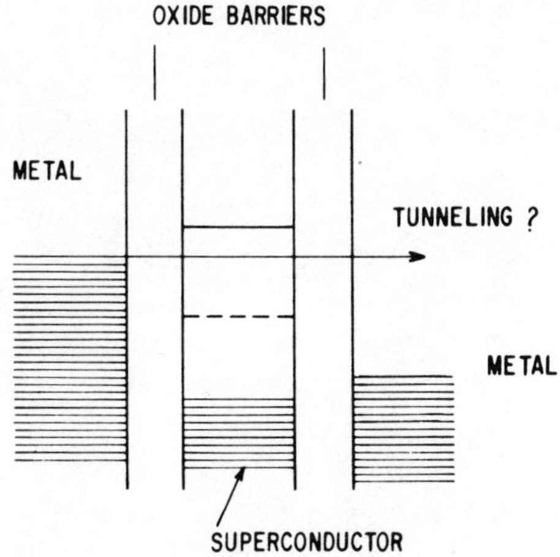

Fig. 14. Two metals separated by two oxide barriers and a superconductor. Experiment to see if it is possible to tunnel through the energy gap in a superconductor.

of a triode. A few oscillators and amplifiers have indeed been made using these films; however, they only operate at low temperatures. Had it not been for that fact, they would undoubtedly be in wide use today. At present they have a formidable competitor in the integrated semiconductor circuit business; hopefully they will eventually find special applications, particularly in environments where liquid He is used anyway.

Finally, I would like to propose a different tunneling experiment: an experiment to determine if it is possible to tunnel through a superconductor. Consider the idealized diagram in Fig. 14. Can an electron tunnel from the left, through the oxide layer, through the superconductor, through the second oxide layer, and into the second metal? If so, we have an ideal triode, because I can change the tunneling probability by changing the biasing potential of the superconductor. All I can say is that I have tried to do the experiment and have not been able to find such an effect.

Chapter 20

Tunneling Density of States—Experiment

J. M. Rowell
Bell Telephone Laboratories, Inc.
Murray Hill, New Jersey

In Chapter 19 of this volume Giaever showed how tunneling has been used to understand the properties of superconductors and measure the energy gap. In Chapter 21 Schrieffer presents a very elegant picture of what happens in the tunneling experiment when we take an electron from one metal and inject it into a superconductor where it can exist as a quasiparticle above or below the Fermi momentum k_F. In this case of metal-to-superconductor tunneling, at 0°K the derivative of the tunneling characteristic in the superconducting state $(dI/dV)_S$, divided by that in the normal state, $(dI/dV)_N$, is simply $N(E)/N(0)$, where $N(E)$ is the density of excited allowed states in the superconductor and $N(0)$ that in the normal metal. In this chapter I further discuss the development of the tunneling technique as a tool in the study of superconductivity, and I want to make the following three claims and try to substantiate them:

1. Tunneling is by far our most sensitive probe of the superconducting state.

2. Using tunneling we believe we have confirmed that the present theory of superconductivity is accurate to a few per cent, i.e., if we know a number of properties of the normal metal, then we can calculate the superconducting properties (T_c, H_c versus T, Δ_0, the tunneling characteristic, etc.) to an accuracy of a few per cent. Unfortunately, our knowledge of these required normal-metal properties is generally not as extensive as that of the superconducting properties.

3. Superconductivity, via an analysis of tunneling data, can tell us the normal-state properties mentioned in 2.

To return to the history of tunneling as applied to superconductivity, consider the situation in 1960 at the Low Temperature Conference in

Toronto. The BCS theory had been studied and compared with experiment for three years or so and was proving to be incredibly successful in explaining most of the properties of most of the soft superconductors. However, some were already uncomfortable about two superconductors, Pb and Hg, which they called the "bad actors," and which disagreed appreciably with some of the BCS predictions. For example, the critical field curve deviated most from the BCS parabola for these two materials (and tantalum, but this could be dismissed as a "hard" superconductor). The energy gap was also significantly larger than 3.5 kT_c for Pb and Hg. It was suggestive that of all the superconductors these two have the largest ratio of T_c to Debye temperature θ_D. This low θ_D indicates that a reasonable density of phonons of rather low energy can propagate in the metal, and in fact phonons with energies not much greater than the gap energy (for example, $\Delta_0 = 12°$ in Pb and θ_D, which is a measure of the maximum phonon energy, is 95°). There was the feeling that maybe the coupling between the electrons was too strong in these two cases to be described by the BCS model interaction. In fact, at the conference Anderson ([1]) stated that "in the 'bad' heavy metals there may be effects both of phonon scattering and of variation of energy gap with ε." It is interesting to note that at the same conference Giaever ([2]) reported some of his earliest tunneling measurements. As events have since shown, the problem was outlined and the experimental technique necessary for its solution discovered at practically the same time.

When Giaever extended his measurement of the density of states to 0.3°K (to reduce the "thermal smearing" of the characteristic which arises partly from excited quasiparticles in the superconductor but mostly from the smearing of the Fermi level in the normal metal) he found ([3]) that in Sn, In, and Al the density of states agreed well with the BCS dependence $|E|/(E^2 - \Delta^2)^{1/2}$. But it would be a dull world for the experimentalist if all results agreed with existing theories, and the measurement of a Mg–MgO–Pb junction at 0.33°K provided the clear deviations from the BCS plot shown in Fig. 1. Note that although the deviations are small compared to the large peak in density at Δ, they are in fact $\sim$10% changes in conductance over rather small energy ranges. The total energy range of the two structures is suggestively close to the Debye energy. As Schrieffer has shown, a density of states variation other than $|E|/(E^2 - \Delta^2)^{1/2}$ indicates that we must have an energy-dependent gap parameter $\Delta(E)$. Tunneling is unique in that we can measure the density of states, and hence $\Delta(E)$, in this high-energy range directly; in comparison, infrared absorption, which is useful in the region below and near the gap, is ineffective at high energies as the absorption is already too great at energies slightly greater than Δ.

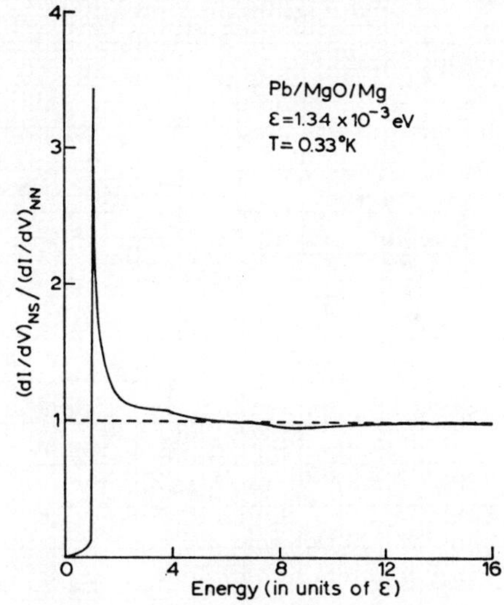

Fig. 1. The normalized conductance versus voltage for a Mg–*I*–Pb junction, measured at a temperature of 0.33°K. From I. Giaever *et al.* ([3]).

In order to produce an energy-dependent $\Delta(E)$, one must take a more realistic interaction than the BCS constant "V". Morel and Anderson ([4]) assumed an Einstein phonon spectrum at ω_c in the metal and pointed out that the phonon interaction was very short range in space and also retarded (i.e., occurred relatively slowly compared to the almost instantaneous Coulomb repulsion between the electrons). They obtained structure in $\Delta(\omega)$ at ω_c and its harmonics $2\omega_c, \ldots$. Swihart ([5]) and Culler *et al.* ([6]) used Einstein and Debye approximations to the phonon spectrum and obtained gap structure at θ_E or θ_D, but the resulting density of states bore little resemblance to the result of Giaever *et al.* ([3]).

About this time Chynoweth suggested that I study superconductor tunnel junctions using the derivative equipment he had been using in collaboration with Thomas and Logan for studies of semiconductor diodes. Our first results ([7]) showed the harmonic structure in density of states, and as we reduced the modulation signal of the equipment, the second derivative revealed differences in the structure near 4.5 mV and the structure at 8.5 mV. These energies are measured from the gap edge. This indicated that the 4.5-mV structure was due to the transverse peak in phonon density and the 8.5-mV structure to the longitudinal peak. Anderson insisted that improvement of the second-derivative data would reveal even more fine

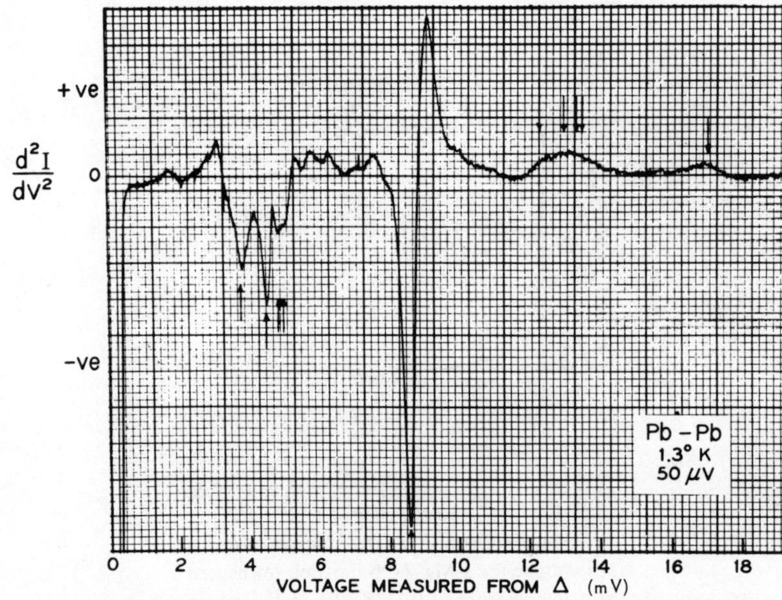

Fig. 2. Second derivative (d^2I/dV^2) of current with respect to voltage versus voltage for a Pb–I–Pb junction at 1.3°K. The arrows mark the voltages of sharp structure in the derivative near 4.5 and 8.5 mV and the sums and harmonics of these voltages near 13 and 17 mV.

structure and this was observed by first changing from Al–*I*–Pb to Pb–*I*–Pb junctions and then using the harmonic derivative circuit developed by Thomas ([8]) which allowed much smaller modulation signals to be used. The result is shown in Fig. 2, a trace taken with a modulating signal of 50 μV to avoid smearing the sharp structure.* When the neutron measurements by Brockhouse *et al.* ([9]) of the dispersion curves of lead were studied we realized that the transverse and longitudinal regions of high phonon density corresponded closely to the 4.5- and 8.5-mV structures. The sharp structures in this second derivative were also very close in energy to the critical points in the dispersion curves, the energies where E versus k has a maximum, a minimum, or a saddle point. These critical points were determined only along the symmetry directions.

It was obvious that in order to reproduce the density-of-states plot

* If one metal of the junction is superconducting, a modulation signal (peak-to-peak) less than kT is advisable (1°K $\approx$ 86 μV). When both films are superconducting the signal should be as small as possible. Recently for Pb–*I*–Pb junctions we used 40 μV peak-to-peak and otained plots with signal to noise comparable with Fig. 2.

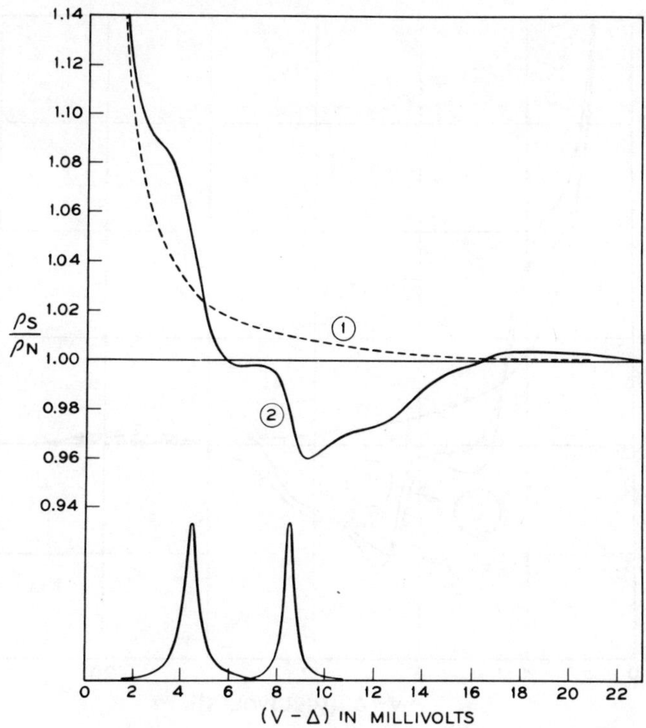

Fig. 3. The upper plots show the BCS density of states (dashed line) and the density of states in Pb determined by tunneling (solid line). The lower half of the figure shows the phonon density in Pb used in the calculation of Schrieffer *et al.* ([10,11]). From J. M. Rowell and L. Kopf, Phys. Rev. 137:A910 (1965).

in Pb a realistic phonon spectrum had to be estimated and introduced into the interaction. By studying the neutron measurements and the tunneling data the phonon spectrum of Fig. 3 was guessed at by Schrieffer, Scalapino and Wilkins (SSW) ([10,11]). They solved the gap equation using reasonable estimates of the electron–phonon coupling constant α^2 and the Coulomb repulsion U_c. The $\Delta(\omega)$ they obtained was a complex parameter, and they showed that the tunneling density of states is given by

$$N(0)\ \mathrm{Re}\{E/[E^2 - \Delta(E)^2]^{1/2}\}.$$

As can be seen from Fig. 4, the density of states they calculated was in remarkable agreement with the experimental result.

The sharp structure of the second-derivative plots was not, of course, contained in the model phonon spectrum of SSW, but Scalapino and An-

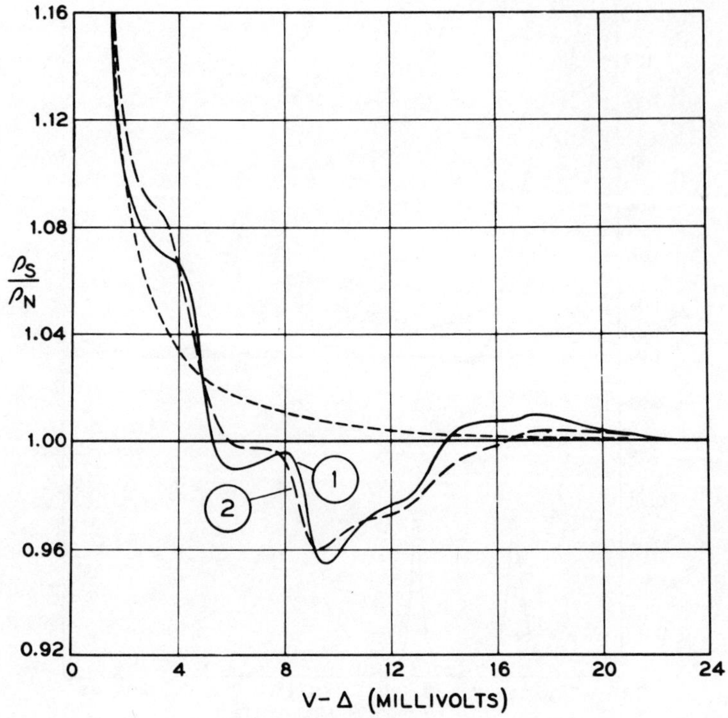

Fig. 4. Comparison of (1) calculated and (2) experimental densities of state in Pb. From J. M. Rowell and L. Kopf, Phys. Rev. 137:A910 (1965).

derson ([12]) showed how each type of critical point in $F(\omega)$ would give rise to singularities or jump discontinuities in d^2I/dV^2.

This was probably a good place to stop the experiments, as there was little doubt that the other properties of Pb and Hg could be explained with the use of the SSW interaction parameter and the Eliashberg gap equation, and in fact, the critical field deviations were soon reproduced by such a calculation ([13]). Structure in the density of states of lower T_c (weaker coupling) superconductors was also observed ([14]), showing that these phonon effects were not confined to the strong coupling Pb and Hg, but that they are just easiest to see in these two materials. It was, however, disturbing to realize that much of the information in the experimental data (particularly the second derivative) was being ignored when one assumed the Pb phonon spectrum to be simply the sum of two Lorentzian peaks. Obviously, if the phonon spectrum had been known in detail, including the critical points, the density of states and even the derivative $dN(\omega)/d\omega$ (corresponding to d^2I/dV^2) could have been calculated to compare with experiment. An

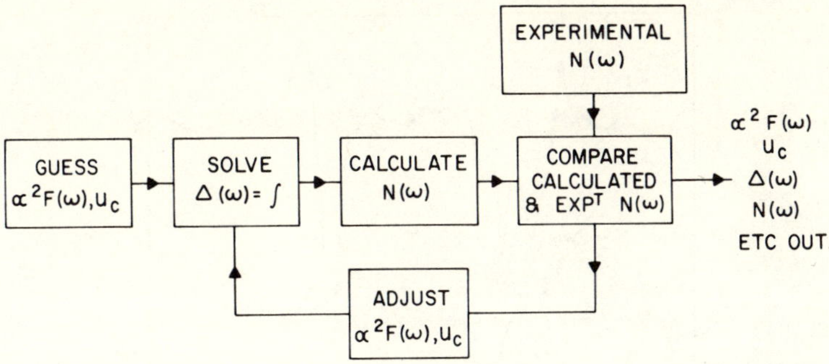

Fig. 5. Schematic of the method used to "invert the gap equation."

alternative approach was made by McMillan and Rowell ([15]), who succeeded in taking the experimental data and deducing from it the normal-state parameters $\alpha^2F(\omega)$ and U_c. The method used on the computer is summarized in Fig. 5. After 6–8 solutions of the gap equation they obtained an $\alpha^2F(\omega)$ which gave $N(\omega)_{\mathrm{calc}}$ in agreement with experiment within one part in 10^3. This result is shown in Fig. 6. Progress since 1965 includes some improvement in accuracy, calculation of other normal-state parameters to be mentioned later, and measurement of superconductors other than lead.

I would now like to stress some of the important features of the experimental method. As we now believe the theory of superconductivity to be accurate to a few per cent, the experiment is essentially a measurement of the *deviations* from the BCS density of states to an accuracy of a few per cent. For Pb this means a measurement of the *normalized* density of states to $\sim$1 in 10^3 and measurement of the gap Δ_0 to $\sim$1%. (To find the BCS density, Δ_0 must be measured.) For tin the required accuracy in the normalized derivative is $\sim$1 in 10^4, and for aluminum we estimate 1 in 10^6. These estimates for weak coupling materials can be made by noting that the strength of the deviations is $\sim(T_c/\theta_D)^2$.

My first point regarding the experiment, which is an extension of what Giaever noted in Chapter 19, is that one has to be somewhat selective about the junctions on which one chooses to spend a long time measuring and analyzing. Fortunately, there are a number of tests one can use to select a "good" junction before any derivatives are taken at low temperatures. After cooling the junctions to nitrogen temperatures the first of these tests is to measure the resistance for the different area junctions. If the resistance is not inversely proportional to area to $\pm$20%, you are probably in trouble. On cooling to helium temperatures the junction resistance should increase slightly—an appreciable decrease in any junction should

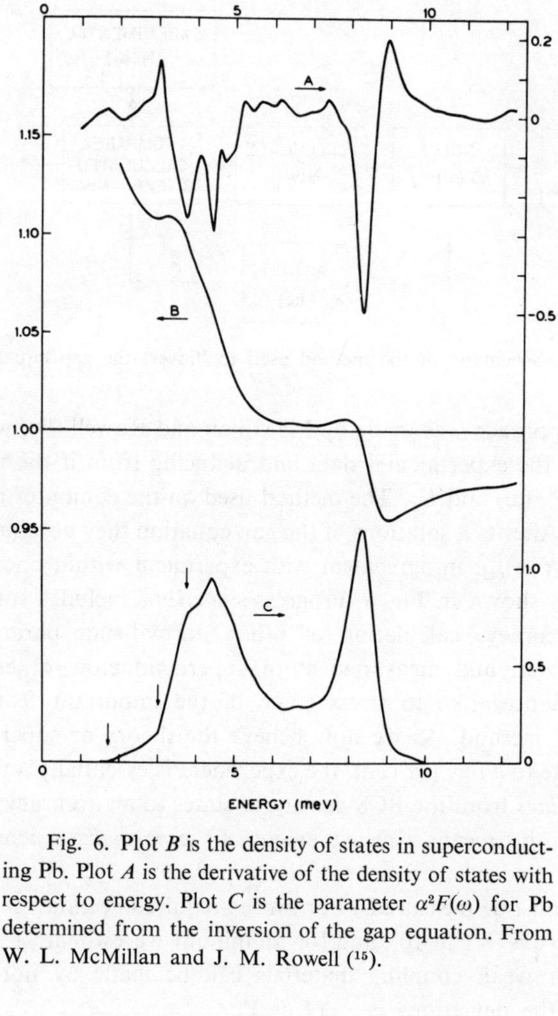

Fig. 6. Plot B is the density of states in superconducting Pb. Plot A is the derivative of the density of states with respect to energy. Plot C is the parameter $\alpha^2F(\omega)$ for Pb determined from the inversion of the gap equation. From W. L. McMillan and J. M. Rowell ([15]).

eliminate it from further consideration. Giaever mentions the rule, which hopefully will become more widely used, that the only test of tunneling is to put a superconductor on one side of the junction. To extend this observation of a gap we also require that at, say, 1°K for Al–I–M junctions (M = Pb, Sn, In, Tl; Al is superconducting) the conductance at zero bias is 10^{-3} of that when the junction is normal. This simply means that any nontunneling conductance (e.g., a small short) is 10^{-3} of the tunneling conductance and thus introduces negligible errors into the derivative measurement. If an experimenter cannot show a superconducting gap characteristic, we do not need to believe any other effects that his "tunnel junctions" exhibit; if the conductance at zero bias is not less than 10^{-1} of the normal state, I wouldn't believe even a gap measurement on such a junction.

If the junctions satisfy these tests, it is usually safe to proceed with further measurements. One last check for Al–*I*–S junctions is to examine the negative resistance region of width $2\Delta_{Al}$; the cusp at $\Delta_S - \Delta_{Al}$ should be well defined and the current onset at $\Delta_S + \Delta_{Al}$ should be sharp. Bad pickup in the measuring circuit can smear these features, and it is worth checking this if a poor *I*–*V* trace is obtained on a junction which otherwise looked usable.

The first measurement is that of T_c, which, while not absolutely necessary, is useful to know, especially in the case of films like Al and Sn which sometimes have surprisingly high transitions. The film T_c can be determined resistively, or by finding the temperature where the junction conductance at zero bias first begins to decrease. This method is capable of giving information about the uniformity of the film and its transition—for example, with Al we sometimes see a small decrease in conductance below 1.8°K and then a much bigger change below 1.2°K. Presumably the edges of the film have the 1.8° transition.

The second measurement is that of the gap, which, if one were to believe all the values given in the literature, is very hard to determine using tunneling. The main difficulty seems to be that no junction is ideal—for example, Al–*I*–S with an infinite cusp at $\Delta_S - \Delta_{Al}$ and a discontinuous jump in current at $\Delta_S + \Delta_{Al}$—so different workers pick different places on the smeared characteristic. Our method for Pb–*I*–Pb junctions is shown in Fig. 7; we find the voltage where the current is half its value extrapolated from voltages just greater than 2Δ ([16]). We find 2Δ is typically $2.795 \pm .005$ for Pb films ~1000 Å thick, which is 3% greater than the $2.705 \pm .005$ which one gets using the "minimum excitation energy" construction of Gasparovic *et al.* ([17]). The size of the "jump current" at 2Δ should be $\pi/4$ times the normal current at that voltage for weak coupling superconductors and $1.05 \times \pi/4$ for Pb ([16]). This is also a useful check of whether a good gap value has been chosen. For Al–*I*–S junctions it is actually quite hard to pick the gap accurately, particularly the $\Delta_S - \Delta_{Al}$ point. The best way seems to find the origin of the logarithmic singularity at $\Delta_S - \Delta_{Al}$ by trial and error. The $\Delta_{Al} + \Delta_S$ point is found in the same way as 2Δ is found for a two-superconductor junction.

The width of the rise in current at 2Δ in Pb junctions is considerably larger than that in Sn, In, or Al junctions, even if taken as a fraction of 2Δ. We found that the width decreased when we reduced the film thickness

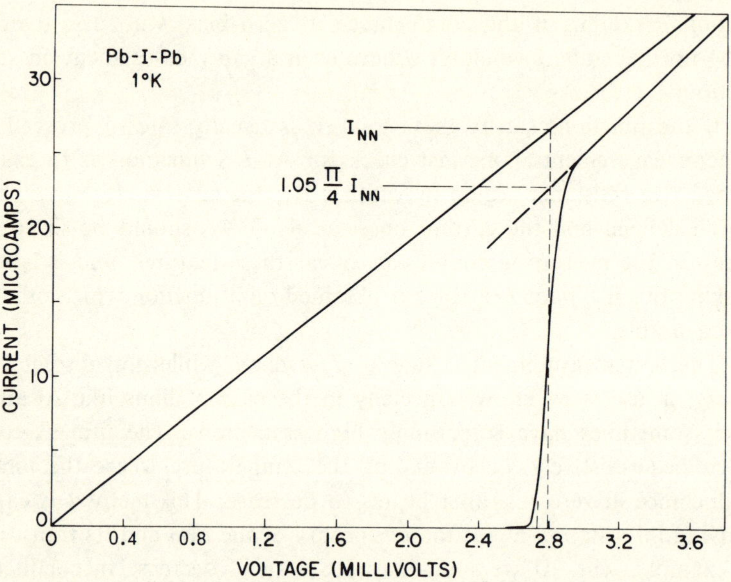

Fig. 7. Current–voltage characteristic for a Pb–I–Pb junction at 1°K. A magnetic field was applied to produce the normal-state characteristic.

from 2000 to 1000 Å, and the rise becomes very sharp indeed when we alloy In into the Pb. Similar results have been reported by Campbell *et al.* ([18]) for PbBi alloys. It would be easy to claim that these sharp gaps are due to increased isotropy, but apart from the "double gap structure" observed in thick Pb films [which in Al ([19]) and Cd ([20]) is due to film edges, so it's still an open question whether Pb is also an edge effect] there is no experimental evidence for anisotropic gaps in single-crystal lead, and in fact a point-contact tunneling experiment ([21]) has failed to find any variation in gap greater than ± 0.03 mV.

I am not going to describe in detail the equipment used to measure the derivatives, as this is a standard technique by now. With the first derivative, applying constant modulation δI and detecting the generated signal δV, one is much more likely to have problems with drift than with sensitivity, and the use of an AC bridge ([16,22]) is essential for materials other than Pb. The situation is the opposite for the second derivative—for example, one can detect 50 μV of first-derivative signal with good signal-to-noise ratio and without smearing the characteristic by the 50-μV modulation, but if one applies 50 μV of fundamental (500 cps in our case) and looks for the harmonic at 1000 cps to measure d^2I/dV^2, one finds the signals are $\sim 10^{-8}$ V or less ([8]) for most superconducting junctions. So one has to worry about signal to noise for the second derivative (if one increases the amplitude of the fundamental one loses some fine structure), and about drift in the first derivative, where one needs to know the normalized derivative to, say, 1 in 10^4.

The importance of this normalization was pointed out by Bermon and Ginsberg ([23]), but let me stress it again. One needs to know the ratio of the derivative measured in the superconducting state to that measured in the normal state at every voltage. The normal state can be produced by a magnetic field or by raising the temperature above T_c. Thus if one is going to read the superconducting derivative at 150 points across the voltage range, then the normal derivative must be known at the same 150 points with equal accuracy and without drift in the equipment between the times one reads the superconducting and the normal values, which may be an hour or more apart. Let me make this even more obvious by reference to Fig. 8, which shows the derivative traces for an Al–*I*–Sn junction. The first surprising thing is the marked asymmetry of Al junctions in the normal state, which I will discuss again in Chapter 27 on tunneling anomalies. We find that the Al negative bias is always the best for superconducting measurements, since the normal derivative is reasonably flat in that direction. From the expanded trace one can see that the normal conductance has a rather complicated voltage dependence. If one approximates the normal state by a straight line or smooth curve, one is obviously introducing serious errors

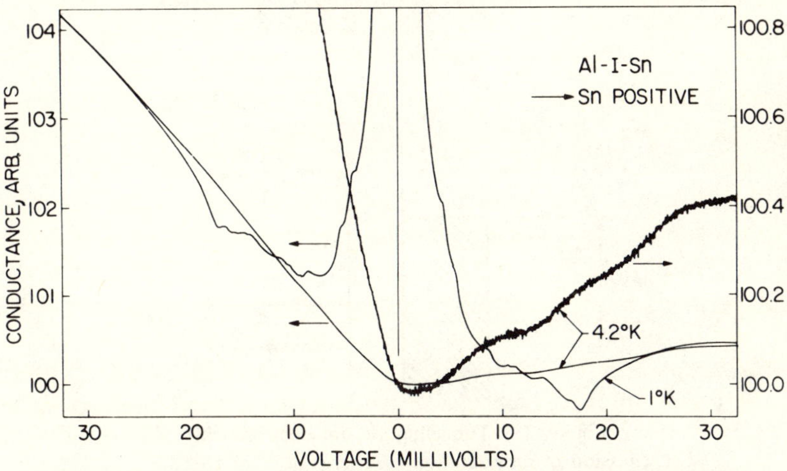

Fig. 8. Conductance versus voltage for an Al–*I*–Sn junction at 1°K (both Al and Sn superconducting) and at 4.2°K (both films normal). The trace showing some noise is an expansion of the normal-state conductance and the relevant scale is at the right.

into the value of the normalized derivative. Hence the conductance measurement is just as important a measurement in the normal state as in the superconducting state.

There is actually a small correction to the normalization arising from these structures in the normal-state derivative. We believe these occur when a phonon (of energy ω_p) in the metal is coupled into the tunneling process, giving rise to a small increase in conductance at ω_p. The nature of this coupling is not clear, but our studies of numerous junctions indicate clearly that it occurs. Now consider this phonon-emission process (Fig. 9) when one metal becomes superconducting. The emission now takes place at $eV = \Delta + \omega_p$ and there will be a much sharper conductance rise because of the peak in density of states just above Δ. One could say that even the "normal-state characteristic" changes when the metals become superconducting, although one doesn't see this change since the superconducting density of states dominates the conductance plot. This correction to the

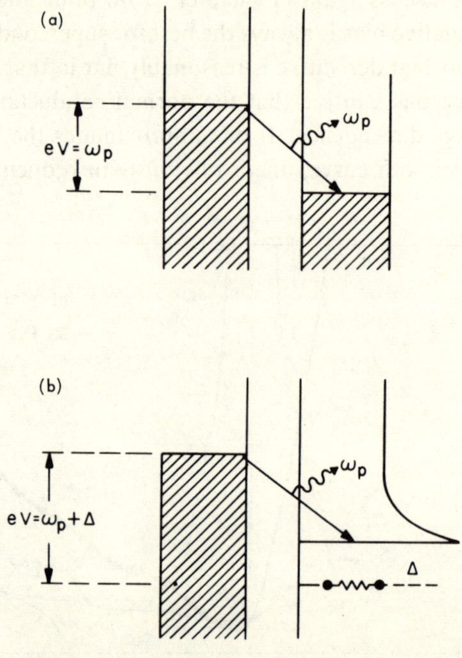

Fig. 9. (*a*) Tunneling of an electron with emission of an excitation of energy ω_p. (*b*) Demonstration that this process occurs at $eV = \omega_p + \Delta$ when one of the metals becomes superconducting.

normalization is negligible in Pb, where the superconducting effects are so large, but is not small for the more weakly coupled materials (Sn, In, Tl, etc.).

Finally, I would like to return to the experiment and briefly substantiate the claims I made at the beginning of this chapter. I do not believe anyone would dispute the claim that tunneling provides by far our most detailed knowledge of the superconducting state. I have shown one of the normal-state properties we can deduce from the data, namely α^2F. We also find the Coulomb pseudopotential, although its value is very sensitive to the accuracy of the experimental data. The self-energy for the normal state, which describes the interaction of the electron with the lattice, is also obtained directly, and we can calculate the enhancement of the electron mass due to this phonon interaction.

My last claim was that we could confirm the accuracy of the present theory of superconductivity. We do this by noting that we have used the experimental measurement of density of states only over the energy range 0–11 meV (from the gap) to find α^2F, $\Delta(\omega)$, etc. We can now calculate the expected density of states at higher energies, where structure in the density of states is due to multiphonon processes. In comparing this calculated density with experiment over the range 11–30 meV ([16]) we find agreement within $\pm.001$. This means, of course, that the Eliashberg gap equation correctly accounts for the single-phonon and multiphonon coupling processes to this accuracy.

REFERENCES

1. P. W. Anderson, in: Proceedings VII International Conference on Low Temperature Physics, Toronto, 1960, Univ. of Toronto Press, 1961, p. 298.
2. I. Giaever, *Ibid*, p. 327.
3. I. Giaever, H. R. Hart, and K. Megerle, Phys. Rev. 126:941 (1962).
4. P. Morel and P. W. Anderson, Phys. Rev. 125:1263 (1962).
5. J. C. Swihart, IBM J. Res. Develop. 6:14 (1962).
6. G. J. Culler, B. D. Fried, R. W. Huff, and J. R. Schrieffer, Phys. Rev. Letters 8:399 (1962).
7. J. M. Rowell, A. G. Chynoweth, and D. E. Thomas, Phys. Rev. Letters 9:59 (1962).
8. D. E. Thomas and J. M. Rowell, Rev. Sci. Instr. 36:1301 (1965).
9. B. N. Brockhouse, T. Arase, G. Caglioti, K. R. Rao, and A. D. B. Woods, Phys. Rev. 128:1099 (1962).
10. J. R. Schrieffer, D. J. Scalapino, and J. W. Wilkins, Phys. Rev. Letters 10:336 (1963).
11. D. J. Scalapino, J. R. Schrieffer, and J. W. Wilkins, Phys. Rev. 148:263 (1966).
12. D. J. Scalapino and P. W. Anderson, Phys. Rev. 133:A291 (1964).
13. J. C. Swihart, D. J. Scalapino, and Y. Wada, Phys. Rev. Letters 14:106 (1965).

14. J. M. Rowell and L. Kopf, Phys. Rev. 137:907 (1965); J. G. Adler, J. S. Rogers, and S. B. Woods, Can. J. Phys. 43:557 (1965).
15. W. L. McMillan and J. M. Rowell, Phys. Rev. Letters 14:108 (1965).
16. W. L. McMillan and J. M. Rowell, in: Treatise on Superconductivity (R. D. Parks, ed.), Marcel Dekker, New York, 1968.
17. R. F. Gasparovic, B. N. Taylor, and R. E. Eck, Solid State Commun. 4:59 (1966).
18. C. K. Campbell, R. C. Dynes, and D. G. Walmsley, Can. J. Phys. 44:2601 (1966); C. K. Cambell and D. G. Walmsley, Can. J. Phys. 45:159 (1967).
19. A. C. Thorsen, T. Wolfram, and L. E. Valby, Phys. Letters 25A:548 (1967).
20. P. Kumbhare, P. M. Tedrow, and D. M. Lee, Bull. Am. Phys. Soc. 12:77 (1967).
21. H. J. Levinstein, private communication.
22. J. G. Adler and J. E. Jackson, Rev. Sci. Inst. 37:1049 (1966).
23. S. Bermon and D. M. Ginsberg, Phys. Rev. 135:A306 (1964).

Chapter 21

Single-Particle Tunneling in Superconductors*

J. R. Schrieffer

Department of Physics
University of Pennsylvania
Philadelphia, Pennsylvania

INTRODUCTION

Tunneling experiments reveal a good deal about the nature of the superconducting state. Here we consider only single-particle tunneling. In this case one can summarize the results in terms of an effective tunneling density of states $N_T(E)$ which is analogous to the conventional density of states is normal metals and which can be compared rather directly with experiment. In the case of multiparticle tunneling, more complicated effects occur (e.g., the Josephson effect) which are treated elsewhere in this volume.

To treat single-particle tunneling we consider the idealized tunneling experiment shown in Fig. 1. Here metals A and/or B may be in the normal or superconducting state, depending on the temperature, applied magnetic field, etc. Ideally, one would like to predict the current I flowing through the junction as a function of the DC applied bias V. For convenience we

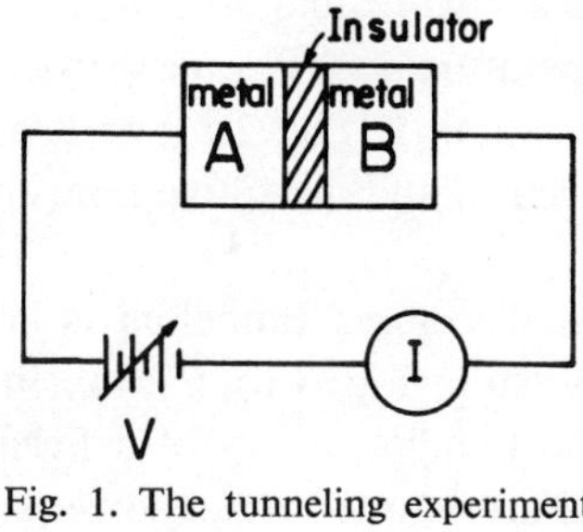

Fig. 1. The tunneling experiment.

* This work was supported in part by the Advanced Research Projects Agency.

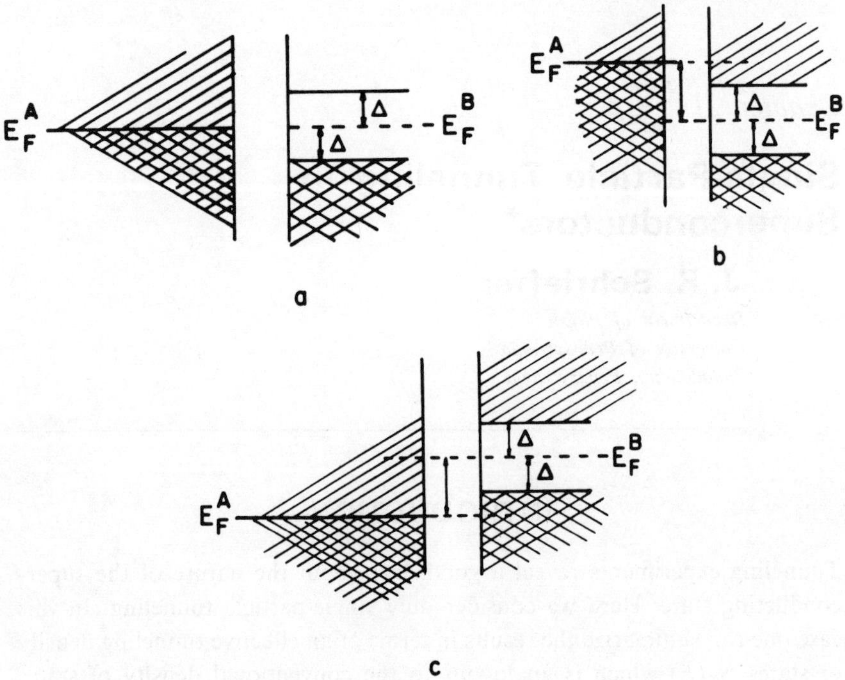

Fig. 2. (*a*) A normal metal–superconductor tunnel junction at zero bias described by the semiconductor model of the excitation spectrum of the superconductor. A normal–super junction with a bias V applied, leading to electron flow from A to B in (*b*) and electron flow from B to A (hole flow from A to B) in (*c*).

define V to be the electron charge $(-e)$ times the actual applied voltage. We first consider the case in which metal A is normal and metal B is superconducting. The semiconductor-type energy level diagram can be used to describe this case if one is careful to use the diagram correctly. For $V = 0$ the Fermi levels $E_F^{(A)}$ and $E_F^{(B)}$ are equal, as shown in Fig. 2(*a*). Considering for the moment the temperature $T = 0$, quasiparticles can flow between A and B only if V is greater than Δ [Fig. 2(*b*)] or less than $-\Delta$ [Fig. 2(*c*)], corresponding to quasielectron and quasihole flow into the superconductor, respectively.

An alternate way of discussing tunneling is in terms of the energy–momentum diagrams, shown in Fig. 3 for $T = 0$. In the normal metal one can speak in terms of the Landau theory of a Fermi liquid, in which the states of the Fermi liquid are in one-to-one correspondence with the states of the Fermi gas. At $T = 0$ all states with $|p| > p_F$ are occupied and those with $|p| < p_F$ are empty. In the normal phase an excited state can be described in terms of electrons in states with $|p| > p_F$ having positive excitation energies $\varepsilon_p^{(A)} \equiv E_p^{(A)} - E_F^{(A)} > 0$ and holes in states $|p| < p_F$ also having *positive* excitation energies, $-\varepsilon_p^{(A)} \equiv -(E_p^{(A)} - E_F^{(A)}) > 0$. Clearly, the total excitation energy is correctly given by the sum of these positive "elementary" excitation energies. This situation is illustrated in Fig. 3(*a*). Due to the sharpness of the Fermi surface at $T = 0$, as illustrated by the dotted curve of Fig. 3(*b*), one can only add electrons above the Fermi surface ($|p| > p_F$) and add holes below ($|p| < p_F$).

The situation is somewhat more complicated in a superconductor. As a consequence of the pairing interaction, the mean occupation numbers $\langle n_k \rangle$ of the Landau theory do not change abruptly near the Fermi surface even at $T = 0$, in contrast to a normal metal [see Fig. 3(*b*)]. As a result of the smearing of the Fermi surface in the superconducting phase, one cannot, as in the normal phase, think of a single-particle energy versus momentum curve in which all states below a Fermi surface in k space are occupied at $T = 0$ and all above it are empty. Instead, one should think of the excitation energy curves directly, i.e., the energy required to add an electron to the system (or remove an electron from the system) such that the total momentum of the system is increased (or decreased) by k. These curves (addition of a quasielectron or a quasihole) are shown in Fig. 3(*c*). Again the excitation energies are *positive* for both types of excitations, although we plot the quasielectron energy $E_k^{(\text{el})}$ positive upward and $E_k^{(\text{h})}$ positive downward to make as close contact as possible with the conventional E–k curve for a normal metal. In a superconductor the minimum energy to insert a quasi particle is $\Delta = E_{k_F}^{(\text{el})} = E_{k_F}^{(\text{h})}$. Now, as Δ goes to zero, the probability that one can inject an electron into a state with $|k| < k_F$, or a hole into a state with $|k| > k_F$, tends to zero, since in this limit the smearing of the Fermi surface goes to zero. From Fig. 3(*c*), one sees that the hole curve for $|k| < k_F$ and the electron curve for $|k| > k_F$ smoothly join as $\Delta \to 0$, giving just the excitation curve, Fig. 3(*a*), of the normal metal. The electron curve for $|k| < k_F$ and the hole curve for $|k| > k_F$ in Fig. 3(*c*) are irrelevant in this limit, since the probability for injection into these states vanishes as $\Delta \to 0$.

We note that for a given electron excitation energy there is a unique p (in our one-dimensional plot) in the normal metal, while there are *two* k states in the superconductor having the *same* energy. The same situation holds for the holes. This twofold degeneracy of states in the superconductor is of great importance in determining the tunneling characteristic.

Therefore the states above the energy gap in the semiconductor description (Fig. 2) correspond to the electron excitation curve of Fig. 3(*c*), and

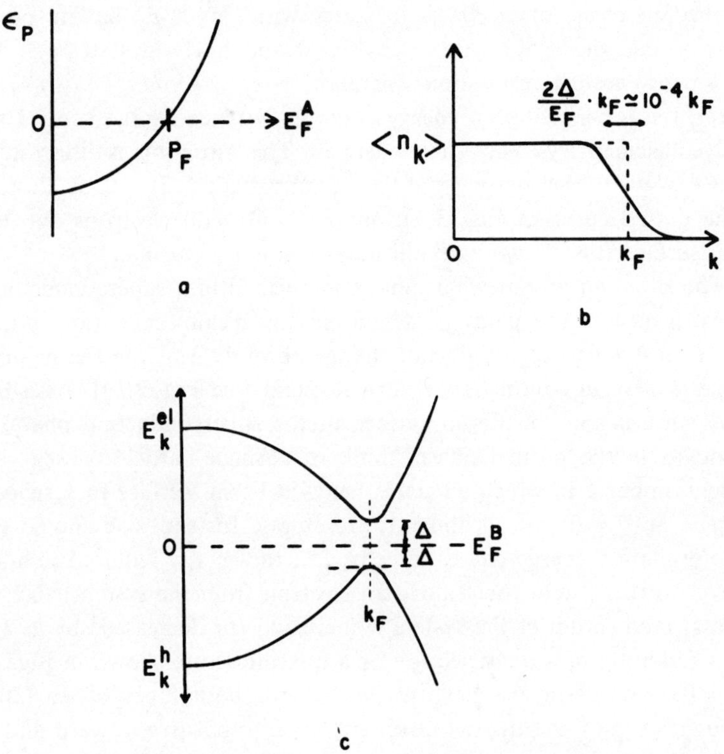

Fig. 3. (*a*) Energy versus momentum curve for electrons in a normal metal. (*b*) The average occupation numbers $\langle n_k \rangle$ at zero temperature plotted as a function of k for a normal metal (dashed) and for a superconductor (solid). The pairing interactions smear the Fermi surface of a superconductor even at $T = 0$. (*c*) The quasielectron and quasihole excitation energies plotted as functions of momentum for a superconductor. Quasihole energies are plotted positively downward. Due to the smearing of the Fermi surface, even at $T = 0$ a quasielectron or a quasihole can be added to any k state without violating the Pauli principle, as opposed to the situation in a normal metal.

one sees that *all* k states enter, both $|k| > k_F$ and $|k| < k_F$. The states below the energy gap in the semiconductor description correspond to the hole excitation curve of Fig. 3(*c*), and again *all* k states enter. This situation is in sharp distinction to real semiconductors, for which the states above and below the gap are truly of different character, arising from different single-particle energy bands. In applying the semiconductor diagrams to superconductors this feature must be kept in mind.

CALCULATION OF THE TUNNELING CURRENT

To calculate the current, we use the Hamiltonian formalism of tunneling as discussed first by Bardeen, and as formulated by Cohen, Falicov, and Phillips. The Hamiltonian has been justified to first order in the tunneling matrix element by Kadanoff using a Green's function treatment. Since we are interested only in terms of leading order in this matrix element, the Hamiltonian formalism suffices for our purposes. In this scheme one writes

$$H = H_A + H_B + H_T, \tag{1}$$

where H_A and H_B are the full many-body Hamiltonians for the isolated metals A and B and H_T represents a one-body coupling term which transfers single electrons between A and B. In terms of the creation and destruction operators for Bloch states in A (labeled ps) and in B (labeled ks) one has

$$H_\perp = \sum_{p,k,s} (T_{pk} c_{ks}^+ c_{ps} + T_{pk}^* c_{ps}^+ c_{ks}), \tag{2}$$

where T_{pk} is the tunneling matrix element, often approximated by the WKB expression. For a truly planar insulating layer the component of the momentum parallel to the surface is conserved.

The tunneling rate is given to order T^2 by the golden rule

$$w_{i\to f} = (2\pi/\hbar) \mid M_{if} \mid^2 \delta(\mathscr{E}_i - \mathscr{E}_f), \tag{3}$$

where i and f label the initial and final eigenstates of the zero-order Hamiltonian $H_A + H_B$ and M_{if} is the corresponding matrix element of H_T. If $V > 0$, so that electrons flow from A to B, let the states be

$$\begin{aligned} \mid i\rangle &= \mid 0\rangle_A \mid 0\rangle_B, \\ \mid f\rangle &= \mid \overline{ps}\rangle_A \mid ks\rangle_B, \end{aligned} \tag{4}$$

where $\mid 0\rangle_A$ is the ground state of A, $\mid \overline{ps}\rangle_A$ is the eigenstate of A with a hole in ps (holes will be represented by a bar), $\mid 0\rangle_B$ is the ground state of B, and $\mid ks\rangle_B$ is the eigenstate of B with a quasielectron in ks. This process is illustrated in Fig. 4, where k can be either k_1 or k_2. Energy conservation [arising from the delta function in Eq. (3)] requires that

$$\mathscr{E}_i - \mathscr{E}_f = \varepsilon_p + V - E_k = 0. \tag{5}$$

Since $\varepsilon_p < 0$ and $E_k > \Delta$, we see the process takes place only if $V > \Delta$.

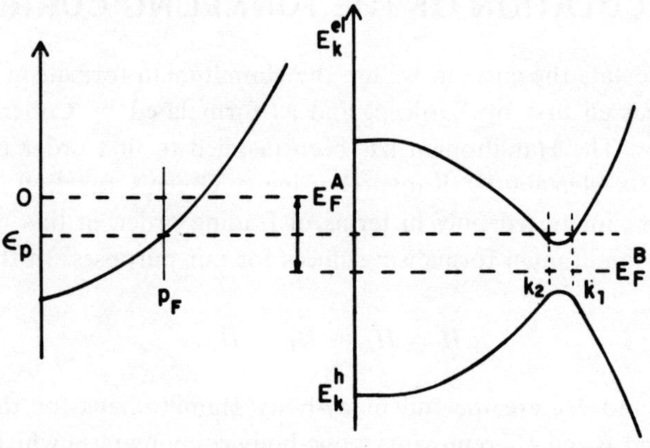

Fig. 4. A normal–super junction biased for electron flow from A to B, as described in the energy–momentum picture. Note the twofold degeneracy of the superconductor excitations. This degeneracy plays an important role in determining the shape of the tunneling characteristic.

The matrix element is

$$M_{if} = \langle f \mid H_T \mid i \rangle = T_{pk} \langle \overline{ps} \mid c_{ps} \mid 0 \rangle_A \langle ks \mid c_{ks}^+ \mid 0 \rangle_B . \tag{6}$$

Now, $\langle ps \mid c_{ps} \mid 0 \rangle_A = 1$, since we are describing the normal metal in terms of a single-particle Bloch scheme, while the pairing theory gives

$$\langle ks \mid c_{ks}^+ \mid 0 \rangle_B = u_k = \left[\frac{1}{2} \left(1 + \frac{\varepsilon_k}{E_k} \right) \right]^{1/2}, \tag{7}$$

with ε_k the Bloch state energy measured with respect to $E_F^{(B)}$. Therefore the tunneling rate from i to f is

$$w_{i \to f} = (2\pi/\hbar) | T_{pk} |^2 u_k^2 \, \delta(\varepsilon_p + V - E_k), \tag{8}$$

while the total tunneling rate for creation of a hole in p is

$$w_p = \sum_k (2\pi/\hbar) \mid T_{pk} \mid^2 u_k^2 \, \delta(\varepsilon_p + V - E_k). \tag{9}$$

From the pairing theory we know that

$$E_k^{(\mathrm{el})} = E_k^{(\mathrm{h})} = (\varepsilon_k^2 + \Delta^2)^{1/2}. \tag{10}$$

If k_1 satisfies the energy conservation condition,

$$\varepsilon_p + V = E_k, \tag{11}$$

then so does k_2, where $\varepsilon_{k_2} = -\varepsilon_{k_1}$, since from Eq. (10) $E_{k_2} = E_{k_1}$. Since $\varepsilon_{k_F} = 0$, we see that k_1 and k_2 are on opposite sides of the Fermi surface. Summing the contributions from these two degenerate states and assuming T_{pk} is essentially independent of k for the k's in question (this restriction is easily removed) one finds

$$\begin{aligned} w_p &= (2\pi/\hbar) \sum_k |T_{pk}|^2 u_k^2\, \delta(\varepsilon_p + V - E_k) \\ &= (2\pi/\hbar) |T_{pk}|^2 \sum_{k1\perp} \delta(\varepsilon_p + V - E_{k1}), \end{aligned} \tag{12}$$

where momentum conservation forces the components of p and k_1 parallel to the interface to be equal, and $k_{1\perp}$ is the component of k_1 perpendicular to the interface. In (12) the sum is over k_1 above the Fermi surface ($|k_1| > k_F$), since for each k_1 above the Fermi surface there is a term k_2 in the sum from below the Fermi surface such that $u_{k_1}^2 + u_{k_2}^2 = 1$. Thus it is the twofold degeneracy of the levels in the superconductor which leads to the of disapperrance the quasiparticle matrix element factor u_k^2 from the sum (12) if one writes the result in terms of a sum over states above the Fermi surface. It is conventional to rewrite (12) in terms of the density of tunneling states N_T by

$$\begin{aligned} \sum_{k1\perp} \delta(\varepsilon_p + V - E_k) &= \int \nu(k_\perp)\, \delta(\varepsilon_p + V - E_k)\, dk_\perp \\ &= \int \nu(k_\perp) \frac{m}{k_\perp} \frac{d\varepsilon_k}{dE_k}\, \delta(\varepsilon_p + V - E_k)\, dE_k \\ &= \left[\nu(k_\perp) \frac{m}{k_\perp} \frac{d\varepsilon_k}{dE_k}\right]_{Ek=\varepsilon p+V} \equiv \nu(k_\perp) \frac{m}{k_\perp} \frac{N_T(\varepsilon_p + V)}{N(0)}. \end{aligned} \tag{13}$$

Here N_T is defined in terms of the density of states at the Fermi surface in the normal metal, $N(0)$, by

$$N_T(E) = N(0)\, d\varepsilon/dE = N(0)[|E|/(E^2 - \Delta^2)^{1/2}]. \tag{14}$$

From Fig. 4 it is clear that dI/dV is proportional to the tunneling rate for electrons at the Fermi surface of A, i.e., $\varepsilon_p = 0$, so that

$$dI_{NS}/dV = \alpha N_T(V) = \alpha N(0)[|V|/(|V|^2 - \Delta^2)^{1/2}], \tag{15}$$

where α is a constant involving the average of $|T|^2$ and the $k_\perp$ factors of (13). If B is normal, Δ is zero and one has from (15)

$$dI_{NN}/dV = \alpha N(0). \tag{16}$$

Therefore one has the remarkably simple relation for the normalized conductance

$$g(V) \equiv \frac{dI_{NS}/dV}{dI_{NN}/dV} = \frac{|V|}{(V^2 - \Delta^2)^{1/2}}. \tag{17}$$

This is just the result one would derive if one naively used the semiconductor model with the conventional density of states for quasiparticles given by the pairing theory, $N(0)|E|/(E^2 - \Delta^2)^{1/2}$.

Enheartened by this simple result, one might assume that if Δ varies with k, the appropriate generalization of the expression (17) for g would be

$$g(V) = \frac{d\varepsilon}{dE}\bigg|_V = \frac{d[E^2 - \Delta^2(E)]^{1/2}}{dE}\bigg|_V = \frac{|E - \Delta(d\Delta/dE)|}{[E^2 - \Delta^2(E)]^{1/2}}\bigg|_{E=V}. \tag{18}$$

This is in fact incorrect, as we shall see below, and the correct expression is

$$g(V) = \mathrm{Re}\{|V|/[E^2 - \Delta^2(E)]^{1/2}\}, \tag{19}$$

where Δ can in general be complex.

For the reverse bias case holes flow from A to B (i.e., electrons from B to A) and the calculation proceeds just as above if one interchanges creation and destruction operators. Thus the matrix element entering from the superconductor is

$$\langle \overline{ks} \,|\, c_{ks} \,|\, 0\rangle = v_k = \left[\frac{1}{2}\left(1 - \frac{\varepsilon_k}{E_k}\right)\right]^{1/2}. \tag{20}$$

By squaring and summing on k one again has the twofold degenerate states k_1 and k_2, such that $v_{k_1}^2 + v_{k_2}^2 = 1$ and v_k^2 drops from the tunneling expression if k is summed only above the Fermi surface, as in the $V > 0$ case. One finds that (17) holds for either sign of V.

If one carries out the analysis for $T \neq 0$, other processes can occur—for example, thermally excited electrons in A can tunnel to states in B above the gap edge even for $|V| < \Delta$. Alternatively, certain tunneling processes allowed at $T = 0$ are forbidden by the Pauli principle, since some of the final states may be already occupied by thermally excited quasiparticles. When these effects are carefully worked out all the u^2 and v^2 factors disappear and one comes back to the simple semiconductor picture for the simple single-particle tunneling effect, as Giaever initially assumed in explaining his pioneering experimental results.

MANY-BODY FORMULATION OF TUNNELING THEORY

Early tunneling curves of Giaever and of Rowell on normal metal–superconducting lead junctions exhibited structure not accounted for by the above theory if Δ is taken to be constant. The structure occurred at voltages corresponding to the typical energies of the transverse and longitudinal phonons of lead plus Δ. Thus one had the possibility of a direct verification of the phonon mechanism of superconductivity in metals with such anomalies. To properly treat the energy dependence of Δ, one must take account of the retarded nature of the phonon interaction between electrons, since in typical metals the sound velocity is three orders of magnitude smaller than the Fermi velocity. Fortunately, the Green's function method of many-body theory handles this problem in a simple manner.

To see how this comes about, we see from the tunneling Hamiltonian and the golden rule that the exact expression for the transition rate for an electron to be withdrawn from state ps of A and inserted with total energy $E = \varepsilon_p + V$ into any state of B is proportional to

$$\sum_k |T_{pk}|^2 \sum_n |\langle n | c_{ks}^+ | 0\rangle_B|^2 \, \delta(E - \mathscr{E}_n^{(B)}), \tag{21}$$

where the $| n \rangle$ are the exact eigenstates of H_B

$$H_B | n\rangle = \mathscr{E}_n^{(B)} | n\rangle .$$

We set the initial ground-state energy of B equal to zero and measure single-particle energies in B relative to the Fermi energy. Therefore $\mathscr{E}_n^{(B)} \geq 0$. If the energy gap were perfectly sharp, one would have $\mathscr{E}_n^{(B)} > \Delta$. We suppress the index B for now.

While (21) allows one to calculate the tunnel current in principle, in practice it is impossible to determine the exact eigenstates $| n\rangle$ and the energies $\mathscr{E}_n$. Thus we desire a method to calculate the n sum in (21) directly.

Consider the *gedanken* experiment of adding an electron in Bloch state ks to B at time zero if B were initially in its ground state $| 0\rangle$. Then just after the electron has been added B is in the state

$$| ks, 0\rangle = c_{ks}^+ | 0\rangle . \tag{22}$$

Due to the many-body interactions, $| ks, 0\rangle$ is not an exact eigenstate of H. From elementary quantum mechanics one knows that after a time t has elapsed the state evolves to

$$| ks, t\rangle = e^{-iHt/\hbar} | ks, 0\rangle. \tag{23}$$

We now ask, what is the probability amplitude that the state at t is identical to that just after the particle was added? Let us call this probability amplitude

$$\begin{aligned} iG_s^{>}(k, t) &= \langle ks, 0 | ks, t\rangle, \qquad t > 0, \\ &= 0, \qquad\qquad\qquad\quad t < 0. \end{aligned} \tag{24}$$

Rewriting this expression, we have for $t > 0$

$$\begin{aligned} G_s^{>}(k, t) &= -i\langle 0 | c_{ks} e^{-iHt/\hbar} c_{ks}^{+} | 0\rangle \\ &= -i \sum_n \langle 0 | c_{ks} | n\rangle\langle n | c_{ks}^{+} | 0\rangle \exp(-i\mathscr{E}_n t/\hbar) \\ &= -i \sum_n | \langle n | c_{ks}^{+} | 0\rangle |^2 \exp(-i\mathscr{E}_n t/\hbar). \end{aligned} \tag{25}$$

By taking the time Fourier transform of (25) we get

$$G_s^{>}(k, E) = \int_{-\infty}^{\infty} G_s^{>}(k, t) e^{iEt/\hbar} = \sum_n \frac{| \langle n | c_{ks}^{+} | 0\rangle |^2}{E - \mathscr{E}_n + i\delta}, \tag{26}$$

where $\delta = 0^+$. By using the relation for a factor occurring under an integral,

$$\frac{1}{x \pm i\delta} = \frac{P}{x} \pm i\pi\delta(x), \tag{27}$$

where P signifies the principle part, one finds the desired sum

$$A_s^{>}(k, E) \equiv \sum_n | \langle n | c_{ks}^{+} | 0\rangle |^2 \, \delta(E - \mathscr{E}_n) = (1/\pi) \, | \operatorname{Im} G_s^{>}(k, E) |. \tag{28}$$

For hole injection (reverse-bias tunneling) one requires a slightly different sum

$$A_s^{<}(k, E) = \sum_n | \langle n | c_{ks} | 0\rangle |^2 \, \delta(E + \mathscr{E}_n). \tag{29}$$

This quantity can be calculated from the corresponding hole probability amplitude

$$G_s^<(k,t) \equiv 0, \qquad t>0$$
$$\equiv i\langle 0 \mid c_{ks}^{+} e^{iHt/\hbar} c_{ks} \mid 0\rangle \tag{30}$$
$$\equiv i\sum_n \mid \langle n \mid c_{ks} \mid 0\rangle \mid^2 \exp(i\mathscr{E}_n t/\hbar), \qquad t<0$$

by

$$A_s^<(k,E) = (1/\pi) \mid \operatorname{Im} G_s^<(k,E) \mid . \tag{31}$$

We note that $A^>$ and $A^<$ are nonzero only for $E > 0$ and $E < 0$, respectively, for our $T = 0$ case. From (24) and (30) we see that $G^> + G^<$ is the conventional time-ordered Green's function

$$G_s(k,t) = -i\langle 0 \mid T\{c_{ks}(t)c_{ks}^{+}(0)\} \mid 0\rangle, \tag{32}$$

where

$$c_{ks}(t) = e^{iHt/\hbar} c_{ks} e^{-iHt/\hbar} \tag{33}$$

and

$$T\{A(t)B(0)\} = A(t)B(0), \qquad t>0, \qquad = \pm B(0)A(t), \qquad t<0, \tag{34}$$

the $+$ and $-$ signs referring to Bose and Fermi statistics, respectively ($-$ in our case of electrons).

Thus the tunneling density of states is given by

$$N_T(E) = \sum_k (1/\pi) \mid \operatorname{Im} G(k,E) \mid , \tag{35}$$

where we suppress the spin index s. We can now use standard many-body techniques for calculating $G(k, E)$.

EXAMPLES OF GREEN'S FUNCTION FORMULATION

Free Particles

For this case it is well known that $N_T(E) = N(E)$, the density of single-particle states. However, for free particles

$$G(k,E) = 1/(E - \varepsilon_k + i\delta \operatorname{sgn} E), \qquad \operatorname{sgn} E = E/\mid E \mid . \tag{36}$$

Thus from (35) we find

$$N_T(E) = \sum_k (1/\pi) \mid -\pi \operatorname{sgn} E\, \delta(E-\varepsilon_k) \mid = \sum_k \delta(E-\varepsilon_k)$$
$$= \int N(\varepsilon)\, \delta(E-\varepsilon)\, d\varepsilon = N(E), \tag{37}$$

as required. For $E > 0$ this gives the electron injection density of states, while for $E < 0$ this gives the density of states for hole injection (since all states below the Fermi level are occupied at $T = 0$).

Decaying Particles

Let us make the usual Wigner–Weisskopf assumption of a single-lifetime-type decay of the initial state, i.e.,

$$\begin{aligned} iG^{>}(k,t) &= \langle k, 0 \mid k, t\rangle \sim \exp(-\Gamma t/\hbar)\exp(-i\varepsilon_k t/\hbar), \qquad && t > 0 \\ &= 0, && t < 0. \end{aligned} \tag{38}$$

Then

$$G^{>}(k,E) = 1/(E - \varepsilon_k + i\Gamma) \tag{39}$$

and the electron tunneling density of states is

$$N_T(E) = \sum_k \frac{1}{\pi}\frac{\Gamma}{(E-\varepsilon_k)^2 + \Gamma^2} = \int N(\varepsilon_k)\frac{\Gamma/\pi}{(E-\varepsilon_k)^2+\Gamma^2}\,d\varepsilon_k \approx N(E), \tag{40}$$

where we have assumed that $N(\varepsilon_k)$ varies slowly with ε_k on the scale of the level width Γ. This shows that a finite lifetime of the final state is not expected to appreciably affect the tunneling rate. The same result holds for decaying holes.

Complex Decay Curve

For the more complicated situation in which the initial state decays through many different channels one expects many different lifetimes to enter the problem. These are included by defining the self-energy $\Sigma(k, E)$ by

$$G(k,E) = 1/[E - \varepsilon_k - \Sigma(k,E)]. \tag{41}$$

In simple cases the real part Σ_1 gives the level shift and the imaginary part Σ_2 gives the decay rate associated with the interactions. In normal metals the strong electron–phonon interaction leads to large real and imaginary parts of $\Sigma(k, E)$, but Σ is slowly varying with k (i.e., $\partial\Sigma/\partial\varepsilon_k \sim \Sigma/E_F$), while Σ is rapidly varying with E (i.e., $\partial\Sigma/\partial E \sim \Sigma/\hbar\omega_D \gg \Sigma/E_F$, where ω_D is the Debye frequency). Therefore to order $\hbar\omega_D/E_F \sim 10^{-2}$ one can neglect the k dependence of Σ and obtain

$$N_T(E) = \sum_k \left| \frac{\Sigma_2(E)/\pi}{[E - \varepsilon_k - \Sigma_1(E)]^2 + \Sigma_2^{\,2}(E)} \right| \approx N(E), \tag{42}$$

where we again assume $N(E)$ varies slowly on the scale of energies of Σ, which for metals is of order $\hbar\omega_D \ll E_F$. Again, we retrieve the simple noninteracting result, despite the fact that the specific heat density of states (i.e., the quasiparticle density of states) is increased by the factor

$$\frac{m^*}{m} = \frac{1 - [\partial\Sigma(\varepsilon, E)/\partial E]}{1 + [\partial\Sigma(\varepsilon, E)/\partial\varepsilon]}\bigg|_{\varepsilon=E=0} \approx 1.5\text{–}2.0 \tag{43}$$

in most metals due to the electron–phonon interaction. Therefore if the tunneling matrix element and band density of states are slowly varying on the scale of $\hbar\omega_D$, phonon structure is expected in the tunneling characteristic only on the scale of $\hbar\omega_D/E_F$.

Weak Coupling Superconductors

We already saw that

$$\begin{aligned} \frac{N_T(E)}{N(0)} &= \frac{|E|}{(E^2 - \Delta^2)^{1/2}}, \qquad E > 0, \\ &= 0, \qquad E < 0, \end{aligned} \tag{44}$$

where $E > 0$ and < 0 correspond to electron and hole injection, respectively. One has from the pairing theory in the weak coupling limit

$$G(k, E) = \frac{u_k^2}{E - E_k + i\delta} + \frac{v_k^2}{E + E_k - i\delta}. \tag{45}$$

From this expression we find the tunneling density of states is

$$N_T(E) = \sum_k \{u_k^2\, \delta(E - E_k) + v_k^2\, \delta(E + E_k)\}. \tag{46}$$

As in the calculation of the tunneling current, we sum over the twofold degenerate states k_1 and k_2, having $\varepsilon_{k_1} = -\varepsilon_{k_2}$, in pairs so that the u_k^2 and v_k^2 factors each sum to unity. We then have

$$\begin{aligned} \frac{N_T(E)}{N(0)} &= \int_0^\infty d\varepsilon_k\, \delta(|E| - \varepsilon_k) = \left|\frac{d\varepsilon}{dE}\right| = \frac{|E|}{(E^2 - \Delta^2)^{1/2}}, \qquad E > 0, \\ &= 0, \qquad E < 0, \end{aligned} \tag{47}$$

as required. Notice that if $\Delta = \Delta(k, E)$, one must carefully distinguish the variation with respect to k and to E. As we will see, the strongly retarded nature of the phonon interaction between electrons ensures that in the normal metal Δ, like Σ, varies rapidly with E but slowly with k.

Strong Coupling Superconductors

There are two major shortcomings of the weak coupling theory of superconductivity discussed above. First, the interactions between electrons are assumed to be instantaneous, while we know that the phonons move with a phase velocity which is very small compared to the Fermi velocity, so that a phonon wave packet set up at one point in space by an electron takes a large time to propagate to another point in space where it influences the motion of another electron. Thus we expect retardation effects to play an important role. In addition, the finite lifetime of excited states is neglected in the weak coupling theory, and these damping effects can become important in determining the properties of the superconducting state. Since Hamiltonian dynamics cannot treat potentials which are retarded in time, the problem has been treated by the Green's function scheme. This problem was first studied by Nambu and by Eliashberg.

Here we merely summarize the results of the Green's function treatment. One finds that for a superconductor $G(k, E)$ is given by

$$G(k, E) = \frac{E + \varepsilon(k, E)}{Z(E)[E^2 - \varepsilon^2(k, E) - \Delta^2(E)]}, \tag{48}$$

where $Z(E)$ plays the role of a renormalization function through the relation

$$\varepsilon(k, E) = \varepsilon_k/Z(E). \tag{49}$$

The quantity $\Delta(E)$ is the analog of the energy gap parameter Δ_k of the weak coupling theory. In general, $Z(E)$ and $\Delta(E)$ are complex quantities, and at zero temperature they satisfy the coupled integral equations (the analog of the energy gap equation of the weak coupling theory),

$$\Delta(E) = \frac{1}{Z(E)} \int_0^{\hbar\omega_c} \mathrm{Re}\left\{\frac{\Delta(E')}{[E'^2 - \Delta^2(E')]^{1/2}}\right\}[K_+(E, E') - K_c]\, dE', \tag{50a}$$

$$[Z(E) - 1]E = \int_0^{\infty} \mathrm{Re}\left\{\frac{E'}{[E'^2 - \Delta^2(E')]^{1/2}}\right\} K_-(E, E')\, dE'. \tag{50b}$$

The interaction kernels are given by

$$K_\pm(E, E') = N(0) \int \alpha^2(\nu)g(\nu)\left\{\frac{1}{E' + E + \nu - i\delta} \pm \frac{1}{E' - E + \nu - i\delta}\right\} d\nu, \tag{51}$$

where $\alpha^2(\nu)g(\nu)$ is the product of the square of the electron–phonon matrix element and the phonon density of states averaged over all phonon modes of energy ν. The kernel K_c is the Coulomb pseudopotential. For a model in which the screened Coulomb matrix element times the density of electron states is taken to be a constant, $N(0)V_c$, in a band of width $2\hbar\omega_m$ centered at the Fermi surface and zero outside this band one has

$$K_c = \frac{N(0)V_c}{1 + N(0)V_c \log(\omega_m/\omega_c)}, \tag{52}$$

where ω_c is a cutoff which may be chosen for computational convenience, so long as $\omega_D \ll \omega_c < \omega_m$.

Numerical solutions of (50*a*), (50*b*), and (51) have been obtained for a variety of assumed forms for $\alpha^2(\nu)g(\nu)$. For Pb one knows from neutron inelastic scattering that $g(\nu)$ has peaks near 4 and 8.5 meV, corresponding to transverse and longitudinal phonons, respectively. Scalapino *et al.*, in an attempt to explain the tunneling results of Rowell and co-workers on superconducting Pb, assumed a simple two-Lorentzian form for $\alpha^2(\nu)g(\nu)$ and obtained the results shown in Figs. 5(*a*) and 5(*b*). By combining (35) with (48) one finds the tunneling density of states is given by the remarkably simple expression

$$N_T(E) = \text{Re}\{|E|/[E^2 - \Delta^2(E)]^{1/2}\}. \tag{53}$$

By combining the solutions shown in Figs. 5(*a*) and 5(*b*) with this relation one obtains the results shown in Fig. 5(*c*). The experimental results are shown for comparison, and one sees a remarkable agreement between theory and experiment even for this oversimplified model of $\alpha^2(\nu)g(\nu)$. More recently McMillan has inverted Eqs. (50*a*), (50*b*), (51), and (53) to determine $\alpha^2(\nu)g(\nu)$ and K_c directly from the tunneling characteristic. In general, the results are in reasonably good agreement with those of neutron inelastic scattering.

The above treatment can be readily extended to finite temperature, although the general features of the results, i.e., wiggles of dI/dV for V of order Δ plus the energy of a peak of $\alpha^2(\nu)g(\nu)$, persist. For Pb, a strong coupling superconductor, the phonon anomalies occur on a scale of a few per cent, while the effects are smaller for weaker coupling materials.

CONCLUSION

While the theory of single-particle tunneling in superconductors can be handled in a quantitative manner regardless of the electron–phonon coupling strength, there remains the problem of excitations on or near the

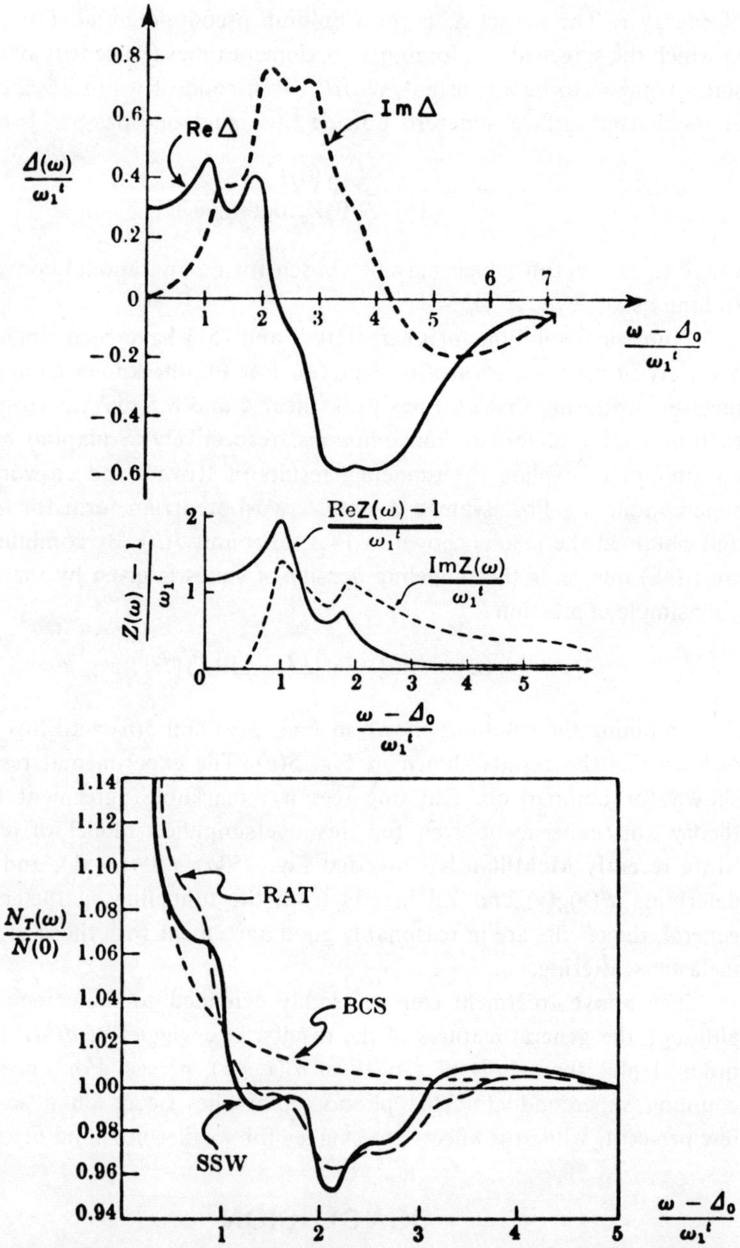

Fig. 5. (*a*) The real and imaginary parts of the gap function plotted as a function of energy for the case of a simple model of lead. (*b*) Comparison of the tunneling density of states predicted theoretically on the basis of the gap shown in (*a*) with the experimental results of Rowel *et al.* on lead.

interface leading to additional structure, as first discussed by Lambe and Jaklevic. There is also the question of whether $\alpha^2(\nu)g(\nu)$ for phonons in the vicinity of the interface, where the tunneling process takes place, is appreciably distorted from the bulk value of this function. The close agreement with the neutron results suggest the distortion is not large. Nevertheless, a finite value of $\alpha^2(\nu)g(\nu)$ is deduced from the tunneling data on many materials for ν greater than the upper cutoff of the bulk phonon spectrum. A detailed understanding of this surface–bulk interplay must await further work. It is clear, however, that single-particle tunneling in superconductors is a very rich phenomenon for gaining insight into the structure of the superconducting state as well as for providing a powerful tool for investigating parameters important in the normal-state properties of the material [e.g., $\alpha^2(\nu)g(\nu)$] and interface excitations.

REFERENCES

An extensive list of references related to the above discussion is contained in J. R. Schrieffer, Theory of Superconductivity, W. A. Benjamin, Inc., New York, 1964, pp. 78–87, 188–193. A more recent review of tunneling in superconductors is given by W. L. McMillan and J. M. Rowell in: Treatise on Superconductivity (R. D. Parks, ed.), Marcel Dekker, New York, 1968.

Chapter 22

Many-Body Theory of Tunneling: Polarons in Schottky Junctions

G. D. Mahan*

General Electric Research and Development Center
Schenectady, New York

The purpose of this chapter is twofold. The main effort is to continue the discussion of the many-body theory which was started by Schrieffer in Chapter 21 of this volume. Following this a particular example of a many-body effect will be given. This example occurs in metal–semiconductor (Schottky) tunneling, and arises from the electron–phonon interaction.

Many-body effects are usually not considered in semiconductor tunneling theory. The main shape and magnitude of the tunneling current or conductance is determined by one-electron processes. Most theoretical effort has rightfully gone into predicting the one-electron tunneling curves. The many-body effects which occur generally appear as merely small bumps and wiggles on the tunneling curves; yet this substructure does appear, and many-body theory must be employed in order to understand it.

Figure 1 shows a junction between two systems with different Fermi degeneracies E_{FL} and E_{FR}, where E_L and E_R are the electron energies from the respective band minima on either side of the junction. Since these band minima are not the same when tunneling between dissimilar solids, it is convenient to measure energy from the respective Fermi energies, $\varepsilon_L = E_L - E_{FL}$ and $\varepsilon_R = E_R - E_{FR}$. Energy conservation requires that $\varepsilon_L = \varepsilon_R + eV$. We also use p and k for the wave vectors of electrons on the two sides of the junction.

The tunneling current is evaluated by starting from an effective Hamiltonian ([1])

$$\mathscr{H} = \mathscr{H}_R + \mathscr{H}_L + \mathscr{H}_T,$$

* Present address: Physics Department, University of Oregon, Eugene, Oregon.

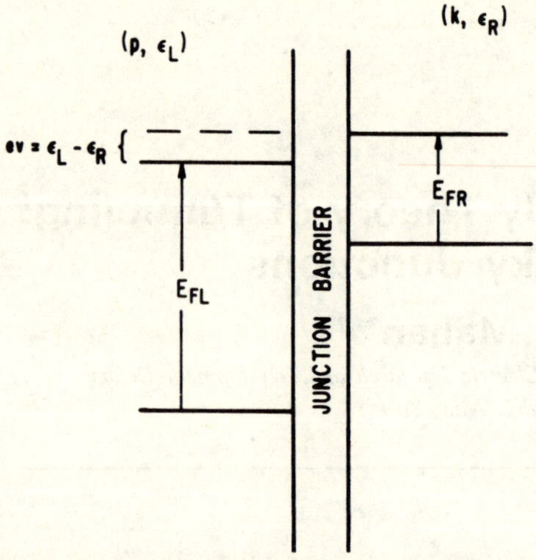

Fig. 1.

where $\mathscr{H}_R$ and $\mathscr{H}_L$ are the Hamiltonians for the two sides of the junction. They contain terms which describe the various interactions an electron can have. The tunneling term transfers electrons across the barrier

$$\mathscr{H}_T = \sum_{k,p} [T_{k,p} C_k{}^+ C_p + T^*_{k,p} C_p{}^+ C_k], \tag{1}$$

where $T_{k,p}$ is proportional to the matrix element for electron transfer. The $C_k{}^+$ and $C_p{}^+$ are creation operators for electrons on the two sides of the junction.

The tunneling current can be evaluated from (1) by using the golden rule,

$$I = 4\pi e \sum_{k,p} | T_{k,p} |^2 \, \delta(\varepsilon_L - \varepsilon_R - eV)[f(\varepsilon_L) - f(\varepsilon_R)], \tag{2}$$

where

$$f(\varepsilon) = 1/(e^{\beta\varepsilon} + 1).$$

This expression contains no many-body effects beyond those in the self-consistent Hartree approximation. Scattering processes which occur in the junction are not included either. We now consider how many-body effects alter this simple result.

In writing down (2) one has assumed a fixed relationship between the momentum and energy. In a free-electron gas this relationship is

$$E_R = \hbar^2 k^2 / 2m. \tag{3}$$

When interactions are taken into account such a relationship is no longer valid. Many-body effects are just those processes which cause an electron to interact and scatter. Scattering causes damping, which smears out the energy–momentum dispersion relation (3). This smearing can be described quantitatively by introducing the "spectral distribution function" $A(k, \varepsilon)$. This distribution function describes the probability that an electron has an energy ε for each value of k. When there are no damping mechanisms for the electron the spectral function is a Dirac delta function $A^{(0)}$,

$$A^{(0)}(k, \varepsilon) = 2\pi[\varepsilon - (\hbar^2 k^2/2m) + E_F].$$

In the presence of damping mechanisms the probability distribution may be rounded over as shown in Fig. 2.

Now, in calculating the properties of a system we must treat ε and k as separate variables and sum over them both. By inserting $A(k, \varepsilon)$ into these summations the properly weighted probability between k and ε is maintained. It is useful to keep in mind the two integrations

$$\begin{aligned} 1 &= \int \frac{d\varepsilon}{2\pi} A(k, \varepsilon), \\ \varrho(\varepsilon) &= \frac{2}{2\pi} \int \frac{d^3k}{(2\pi)^3} A(k, \varepsilon). \end{aligned} \tag{4}$$

The first of these is just a sum rule on the normalization of the distribution function. The second is the definition of the density of states $\varrho(\varepsilon)$. The density of states is just the probability that an electron has energy ε. The definition (4) is obvious when one realizes that one evaluates the probability

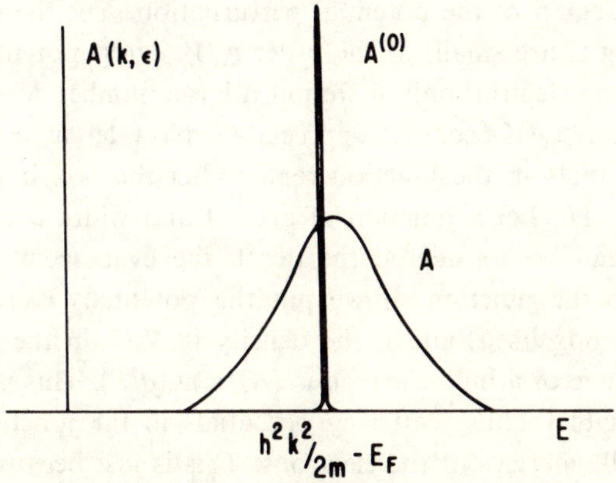

Fig. 2.

$A(k, \varepsilon)$ for each value of k and sums over all k values. The extra factor of two is for spin degeneracy.

The way in which the many-body effects are to be included in evaluating the tunneling current should now be apparent. Again one just sums over energy and wave vector separately and includes the spectral functions which relate k and ε,

$$I = 4\pi e \sum_{k,p} | T_{k,p} |^2 \int \frac{d\varepsilon_R}{2\pi} A(k, \varepsilon_R) \int \frac{d\varepsilon_L}{2\pi} A(p, \varepsilon_L) \times \delta(\varepsilon_L - \varepsilon_R - eV)[f(\varepsilon_L) - f(\varepsilon_R)]. \tag{5}$$

This result was first derived by Schrieffer *et al.* ([2]). It applies to tunneling in normal metals as well as in superconductors. Note that in the absence of interactions, when one uses $A^{(0)}$ for A, this immediately reduces to (2).

Now that we have obtained Eq. (5) we must discuss its limitations. It includes some, but unfortunately not all, many-body effects. The spectral density describes the interactions of an electron in the bulk of the solid on either side of the junction. Scattering processes in the *bulk* material, such as from electron–electron or electron–phonon interactions, can change $A(k, \varepsilon)$. But scattering processes which occur in the *junction* do not affect $A(k, \varepsilon)$. Thus the result (5) describes many-body processes which alter the energies of the electron in the bulk of the crystal, but it does not include scattering processes in the junction itself.

This last point deserves further elaboration. As an example, consider an electron in a large box of volume V. Assume the potential that the electron sees in the box is a constant except in a small region v_0 where the potential is different. The wave function of the electron can be altered greatly in the region of the potential perturbation. Yet the changes in the electron's energies are small, of the order v_0/V. These potentials will affect the energy of the electron only if we put a large number N_0 of them in the box, such that $N_0 v_0/V$ becomes appreciable. Now let us consider putting the same potentials in the junction region. Let the box have area A and length L, $V = AL$. Let a junction of area A and width d be on one end. The width d can be considered the depth the evanescent wave function penetrates into the junction. If we put the potentials v_0 in the junction but not in the box itself, and if the density is N_0/V in the junction, then the energy change of a bulk electron is $(N_0/V)v_0(d/L)$. This is of order d/L, which is negligible. Thus scattering potentials in the junction do not influence the bulk energies of the electrons. This is just because the junction region is such a small fraction of the bulk volume V.

However, these potentials in the junction can influence the tunneling. Since the electronic wave function is altered by the potentials, and since the wave function penetration into the junction determines the tunneling probability, the tunneling is affected. These changes in the wave function affect $T_{k,p}$. Thus one can divide many-body processes in tunneling into two categories: (1) interactions in the bulk material, which affect the energy of the electron—these alter $A(k, \varepsilon)$ and are included in Eq. (5); and (2) scattering processes in the junction, which change the form of the transmission coefficient $T_{k,p}$. These are not included in (5) unless we expand our definition of $T_{k,p}$. This can also be done by many-body theory, and this will be discussed in Chapter 28 by Duke.

For the remainder of this chapter, we will be concerned with applying (5) to Schottky junctions. The polar interaction between electrons and LO phonons in heavily doped semiconductors causes interesting structure in the electron dispersion relation. This causes structure in the spectral function $A(k, \varepsilon)$, which shows up in the Schottky tunneling characteristics. One can use Eq. (5) to describe this many-body effect.

Let us now consider how to evaluate $A(k, \varepsilon)$. First one must define another function, $\Sigma(k, \varepsilon)$ which is the electron self-energy. This has a real and an imaginary part

$$\Sigma(k, \varepsilon) = \Sigma_R(k, \varepsilon) + i\Sigma_I(k, \varepsilon).$$

The two parts Σ_R and Σ_I can be understood in the following way. Assume that the electron (actually, one should talk of quasi particles) is in a state (k, ε), where wave vector and energy are separate variables. Then $\Sigma_R(k, \varepsilon)$ is the energy an electron acquires from its various interactions: electron–electron, electron–phonon, etc., and Σ_I describes the scattering rate of an electron. If $\tau(k, \varepsilon)$ is a lifetime for an electron with (k, ε), then

$$\hbar/\tau(k, \varepsilon) = -2\Sigma_I(k, \varepsilon) > 0.$$

The quantity Σ_I is always negative. The functions Σ_R and Σ_I are not independent, but are related by a Kramers–Kronig dispersion relation. The retarded Green's function ([3]) is

$$G(k, \varepsilon) = \frac{1}{\varepsilon - (\hbar k^2/2m) + E_F - \Sigma(k, \varepsilon)},$$

and

$$A(k, \varepsilon) = -2 \operatorname{Im} G(k, \varepsilon) = \frac{-2\Sigma_I(k, \varepsilon)}{[\varepsilon - (\hbar k^2/2m) + E_F - \Sigma_R]^2 + (\Sigma_I)^2}. \tag{6}$$

Note that if Σ_R and Σ_I are both constants, $A(k, \varepsilon)$ just has a Lorentzian shape. However, the self-energies which lead to interesting structure are never constant.

In a degenerate electron gas, such as a metal or a heavily doped semiconductor, the interactions of electrons with optical phonons lead to an interesting self-energy,

$$\Sigma(k, \varepsilon) = g(k)\left[\ln\left|\frac{\varepsilon - \omega_0}{\varepsilon + \omega_0}\right| - i\pi\theta(|\varepsilon| - \omega_0)\right]. \tag{7}$$

The coefficient $g(k)$ is calculated from the screened electron–phonon interaction. In metals Englesberg and Schrieffer ([4]) found that g was essentially a constant because of the strong screening. As a consequence, when they used (6) and (7) for evaluating the density of states (4) they found that $\varrho(\varepsilon)$ was a constant. For normal metals the density of states only has structure when $\Sigma(k, \varepsilon)$ has k dependence.

In a heavily doped semiconductor the screening of the free carriers is not as strong. Mahan and Duke ([5]) found that $g(k)$ has a k dependence, which leads to corresponding structure in the density of states at $\varepsilon = \pm\omega_0$. The bumps at $\varepsilon = \pm\omega_0$ arise from the logarithmic singularities in the real part of $\Sigma(k, \varepsilon)$ in (7). The advantage of the many-body formalism is that it allows both the real and imaginary parts of the self-energy to be included correctly. This is an asset in tunneling calculations, since phonons not only can scatter electrons while tunneling, but the self-energy terms are also important.

Figure 3 shows Schottky tunneling measurements by Conley and Mahan ([6]) of the metal Au on n-type GaAs. Structure in the incremental resistance (dV/dI) measurement is observed at biases $V = \pm 0.036$ eV. The LO phonon energy in GaAs is $\omega_0 = 0.036$ eV, so the observed structure is a polaron effect, arising from the polaron self-energy in (7). This also appears quite different than the LO phonon structure which is observed in p–n junctions ([7]). In p–n junctions the dI/dV curves show a threshold at $V = +\omega_0$. Here phonons are being emitted in the junction region while the electron is scattering from one band minimum to another. Since these scatterings only occur in the junction region, they only alter the transmission coefficient $T_{k,p}$ and not the bulk energies of the electrons. The self-energy terms (7) are also present, but they lead to smaller effects which are masked by the stronger threshold processes. On the other hand, Schottky tunneling is primarily a one-band process and scattering between bands does not occur with any observable strength. Hence one can now observe the self-

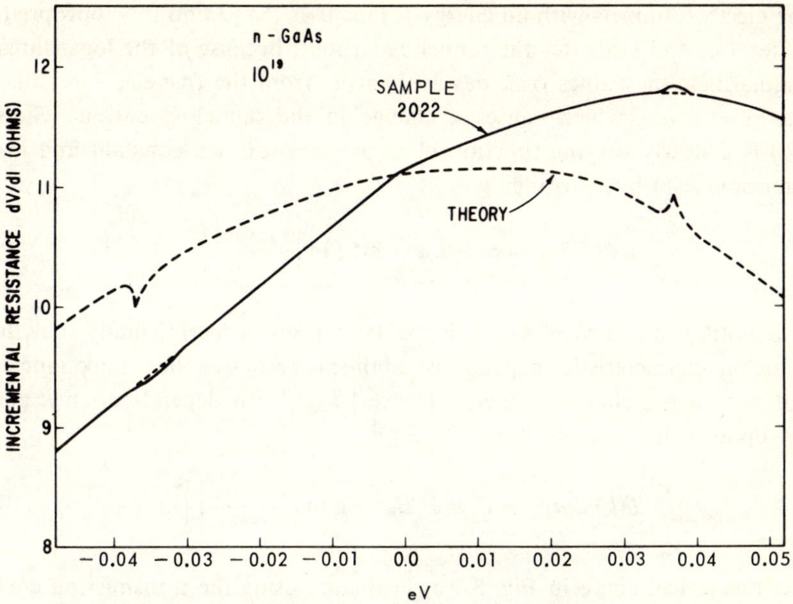

Fig. 3. Schottky tunneling measurements. From Conley and Mahan ([6]).

energy term (7) which is caused by electron scattering by LO phonons within a band minimum.

At first one might think that the observed structure in Fig. 3 is just caused by singularities in the density of states. This interpretation is not correct because it predicts the wrong sign for the logarithmic singularities. In addition, the observed structure is too large to be explained by the density of states.

The bumps in Fig. 3 arise because the transmission coefficient $|T_{p,k}|^2$ is wave-vector dependent. If we ignore damping,* the spectral function (6) becomes

$$A(k, \varepsilon) = 2\pi\left[\varepsilon - (\hbar k^2/2m) + E_F - g(k) \ln \left| \frac{\varepsilon - \omega_0}{\varepsilon + \omega_0} \right| \right].$$

This requires that the energy–wave-vector dispersion curve of the electron is given by

$$\varepsilon = (\hbar k^2/2m) - E_F + g(k) \ln \left| \frac{\varepsilon - \omega_0}{\varepsilon + \omega_0} \right|. \tag{8}$$

* That is, we ignore the Im Σ and assume that the structure is just caused by the logarithmic singularities in the Re Σ. Including Im Σ may change the theoretical curve for $|\varepsilon| > \omega_0$.

The electron tunnels with an energy ε. One uses (8) to find the appropriate value of k, and evaluates the tunneling current. Because of the logarithmic singularities, the values of k deviate sharply from the free-electron values near $\varepsilon = \pm\omega_0$, which causes a change in the tunneling current. Since $g(k)$ is a slowly varying function of k, one can call it a constant and just determine $k(\varepsilon)$ by rearranging

$$\hbar k^2/2m = \varepsilon + E_F - g(0) \ln \left| \frac{\varepsilon - \omega_0}{\varepsilon + \omega_0} \right|. \tag{9}$$

In Schottky junctions of GaAs it has been found experimentally that the tunneling characteristics depend just on the properties of the semiconductor and not on the choice of metal. Hence $| T_{k,p} |^2$ just depends upon k and not upon p; let us write it as $T(k^2/2m)^2$,

$$T(k^2/2m)^2 = T\left[\varepsilon + E_F - g \ln \left| \frac{\varepsilon - \omega_0}{\varepsilon + \omega_0} \right| \right]^2. \tag{10}$$

The theoretical curve in Fig. 3 was evaluated using the transmission coefficient (10). The k^2 dependence of T was evaluated by a WKBJ integral which employed the two-band model and the parabolic potential of Schottky barriers. The details of this calculation are given in ([6]). One sees that the theory predicts the anomalies with the correct sign and magnitude. There are no adjustable parameters in this calculation, since the effective mass, band gap, polaron constant, etc., of GaAs are all well-known numbers.

Thus many-body effects can be observed in tunneling in semiconductors. The logarithmic singularities in the polaron self-energy, originally predicted by Englesberg and Schrieffer, are observed in Schottky tunneling. They are observed, not through the density of states, but because the transmission coefficient is strongly wave-vector dependent. This contrasts sharply with tunneling between metals, where T^2 is nearly constant and structure has to arise through the density of states.

ACKNOWLEDGMENTS

I wish to thank Dr. J. W. Conley for his constant collaboration through all phases of the Schottky tunneling investigations. I also thank Dr. B. Stölan for assistance in preparing a preliminary draft of this chapter.

REFERENCES

1. M. H. Cohen, L. M. Falicov, and J. C. Phillips, Phys. Rev. Letters 8:316 (1962).
2. J. R. Schrieffer, D. J. Scalapino, and J. W. Wilkins, Phys. Rev. Letters 10:336 (1963).
3. A. A. Abrikosov, L. P. Gorgov, and I. E. Dzyaloshinski, Quantum Field Theory in Statistical Physics, Prentice Hall, Englewood Cliffs, New Jersey, 1963.
4. S. Englesberg and J. R. Schrieffer, Phys. Rev. 131:993 (1963).
5. G. D. Mahan and C. B. Duke, Phys. Rev. 149:705 (1966).
6. J. W. Conley and G. D. Mahan, Phys. Rev. (sched. August 15, 1967).
7. R. N. Hall and J. H. Racette, J. Appl. Phys. 32:2078 (1961); R. N. Hall, in: Proceedings of the International Conference on Semiconductor Physics, Prague, 1960, Academic Press, New York, 1961.

Chapter 23

Geometrical Resonances in the Tunneling Characteristics of Thick Superconducting Films*

W. J. Tomasch†

Atomics International
A Division of North American Aviation, Inc.
Canoga Park, California

BASIC EFFECT

In addition to the presence of an energy gap, the excitation spectrum of the BCS ground state differs from that of the corresponding normal state by virtue of a twofold energy–momentum degeneracy; *two* wave vectors satisfying $k_1(E) < k_F < k_2(E)$ are associated with a *single* excitation energy $E > \Delta$ (Fig. 1). One anticipates that additional elementary excitations (quasiparticles) can result from the coherent composition of these degenerate states. The propagation velocity of such composite excitations can be the same as that of single BCS excitations, both rapidly approaching v_F with increasing E. In this limited sense, composite excitations (away from the gap) resemble normal-state excitations. Because of interference effects, however, composite excitations may transport very little momentum compared to $\hbar k_F$ and therefore may have de Broglie wavelengths much longer than $\lambda_F = 2\pi/k_F$. In this sense there is a profound difference between such excitations and corresponding normal-state excitations, i.e., those which occur in the absence of a pairing interaction. In the superconductor, standing waves can be established across *very thick* films, effects having been observed with In films as thick as 30 μ. These resonances influence the density-of-states function, thereby giving rise to structure in the

* Based on work supported by Division of Research, Metallurgy and Materials Programs, U.S. Atomic Energy Commission, Contract No. AT(04-3)-701.

† Present address: Department of Physics, University of Notre Dame, Notre Dame, Indiana.

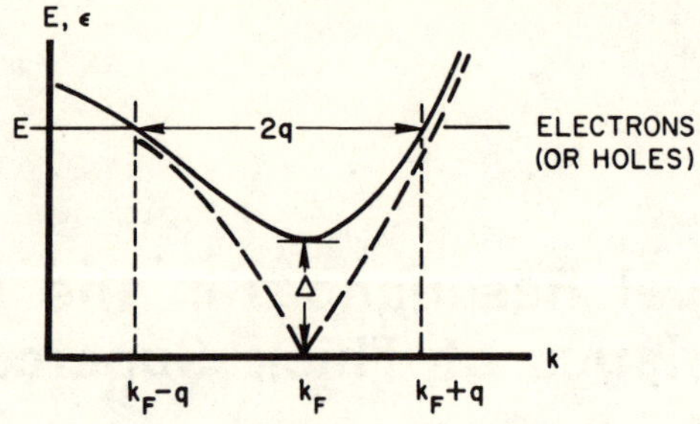

Fig. 1. Excitation spectrum of the BCS ground state. Energies are reckoned relative to the Fermi level E_F; ε (dashed curve) and $E = [\Delta^2 + \varepsilon^2]^{1/2}$ (solid curve) refer to the normal and superconductive states, respectively. A single excitation energy E is associated with two wave vectors $k_1 = k_F - q$ and $k_2 = k_F + q$.

Giaever tunneling current between a superconductive film S (Pb, In, or Sn) some microns thick and a much thinner superconductive Al film serving as analyzer. Systematic studies of geometrical tunneling resonances of this type yield experimental determinations of $E(\mathbf{k})$ and $\mathbf{v}_F$, as well as information involving various boundary phenomena, presumably related to the proximity effect.

EXPERIMENTAL CONSIDERATIONS

Electrical characteristics of cruciform Giaever diodes (Al–AlO_x–S) manufactured by the evaporation method are investigated in the four-terminal configuration, synchronous harmonic detection techniques being used to display dV/dI, dI/dV, and d^2V/dI^2 as functions of bias V. All films studied have a preferred crystallographic orientation; for Pb the (111) direction tends to be normal to the substrate, while for In and Sn the (101) and (100) directions, respectively, are involved. Resistivity measurements disclose that the normal-state mean free path l in the films S ($T = 1.1$°K, $H \approx 2H_c$) is comparable to, or exceeds, the film thickness d. Figure 2 schematically shows a typical junction in cross section. Only two factors distinguish the present diodes from the type of units originally studied by Giaever: (1) the large thickness of one component film, S, and (2) the possibility of other than natural oxide overlayers on the free surface of the upper film. Film thicknesses are determined either by single-beam or multiple-beam optical interferometry.

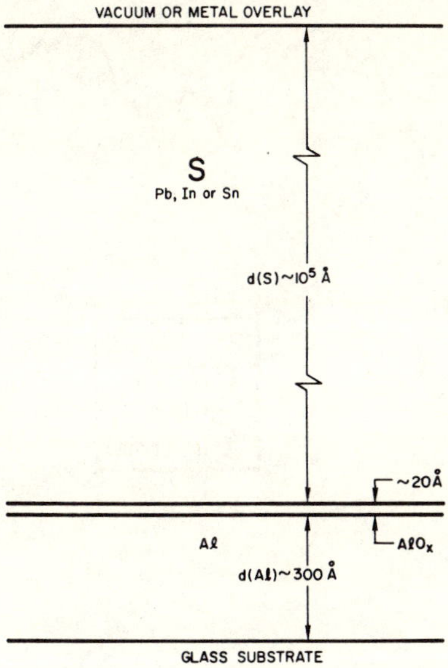

Fig. 2. Schematic cross section of typical diode junction.

HISTORICAL SKETCH

Geometrical resonances of the type under consideration were first observed in February 1965 ([1]) during the course of preliminary experiments designed to study the influence of alloying on the electron–phonon interaction in Pb. Measurements were being carried out with Pb films to certify the performance of a newly commissioned spectrometer. Micron-thick Pb films were employed as a precaution against observing "size" effects. In retrospect, this precaution virtually ensured the observation of the particular size effect in question. Figure 3 presents a plot of early data obtained with a 2.79-μ thick Pb film (natural oxide overlay). These data were found to be in agreement with the reported phonon structure of Pb with the exception that additional structural features (encircled) were observed near the gap edge. Rather than helping matters, refinements in technique emphasized the new structure (Fig. 4).

Several considerations suggested the presence of a genuine effect: (1) all diodes when studied carefully at 1.1°K displayed the structure to a greater or lesser degree, (2) reproducible structural features were observed employing sister diodes differing in junction area (A_J) and specific junction resistance (ϱ_J), and (3) high modulation levels caused the additional structure

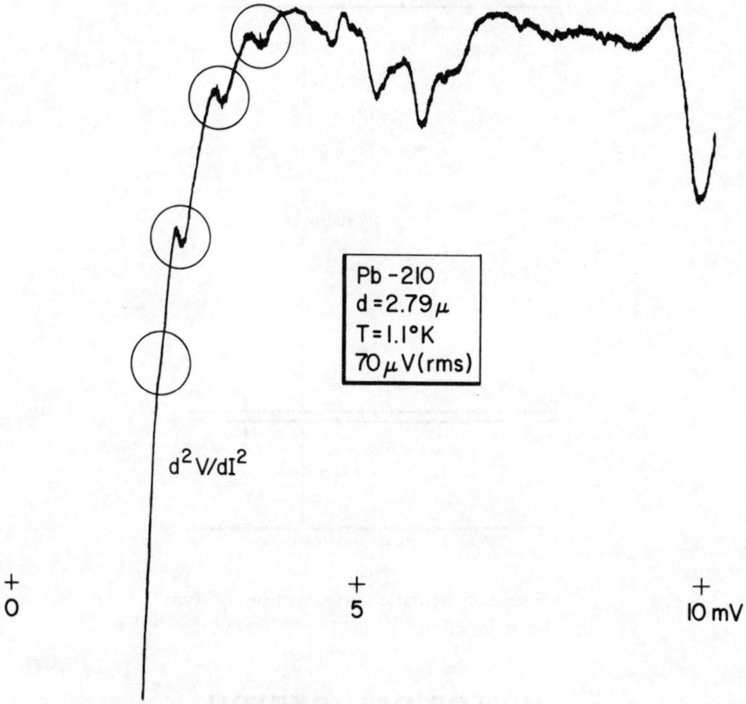

Fig. 3. Voltage dependence of second-derivative data for a 2.79-μ thick Pb film (natural oxide overlayer). Encircled features are geometrical resonances, remaining features are due to phonon effects.

to become washed-out. Further investigation disclosed the absence of additional structure in the characteristics of thin ($\sim$1000 Å) Pb films and thick Pb–5 at. % Tl alloy films, confirming the absence of instrumental difficulties and suggesting the relevance of having $d_{\rm Pb}$ and $l_{\rm Pb}$ in the micron range.

The data of Fig. 4 indicate that points of constant phase (V_η), say half-way down a resonance in d^2V/dI^2, have a uniform spacing ($h\nu$) in V for the particular diode employed. Repeated evaporation of similar masses of Pb repeatedly produced similar values of $h\nu$ independent of such parameters as A_J, ϱ_J, $d_{\rm Al}$, polarity of V, or heat treatment. Initial prejudice notwithstanding, one is virtually forced to examine the consequences of varying the amount of Pb evaporated, i.e., to examine the effect of $d_{\rm Pb}$ on $h\nu$. As Fig. 5 demonstrates, $h\nu$ is indeed a function of $d_{\rm Pb}$. The simple proportionality between $h\nu$ and $1/d_{\rm Pb}$ immediately suggests a standing wave of some description, in which case a characteristic propagation velocity can be obtained from the slope. Interpreted this way, the data of Fig. 5 (with no correction for renormalization effects) yield a velocity of 0.75×10^8

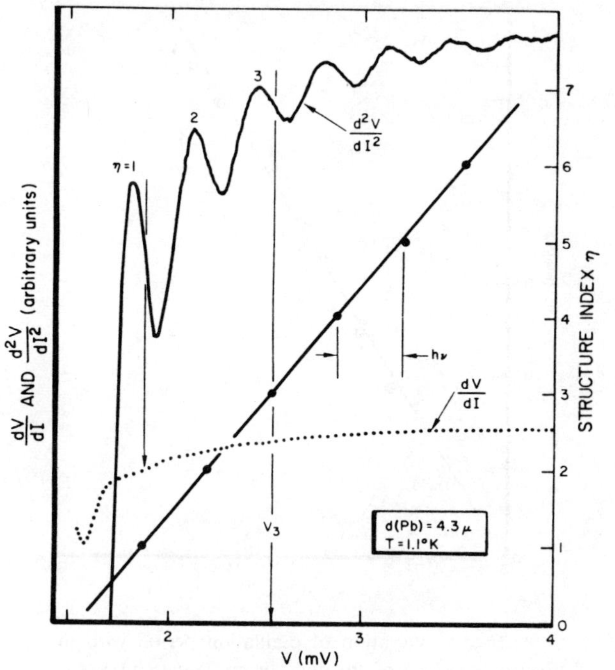

Fig. 4. First- and second-derivative data for 4.3-μ thick Pb film (natural oxide overlayer only). Modulation levels are 10 μV (r.m.s.) and 50 μV (r.m.s.) respectively. Integers η index resonances in d^2V/dI^2. Voltages V_η denote biases for which slope of d^2V/dI^2 has maximum negative value, corresponding to local maxima in dI/dV. From Tomasch (1).

cm/sec, and comparison with the anomalous skin effect (ASE) value of 0.50×10^8 cm/sec [Faber and Pippard (11)] certainly suggest that some type of *electronic standing wave* is involved. To date the effect has been observed with Pb, In (Fig. 6), and Sn (Fig. 7), and one now anticipates its occurrence to be widespread among superconductors capable of having l in the micron range. For In, the value of v_F obtained is in excellent agreement with recent cyclotron resonance work [Mina and Khaikin (10)], while for Sn there is reasonable agreement with ASE work [Faber and Pippard (11)]. In the case of Pb, however, v_F corrected for renormalization effects [McMillan and Anderson (3)] is very nearly just twice the ASE value. The origin of this discrepancy remains an open issue of some interest.

Two remaining ingredients essential to the basic experimental picture are indicated in Fig. 6. To begin with, the first resonance (encircled) does not index properly, i.e., the interval between the first two resonances is

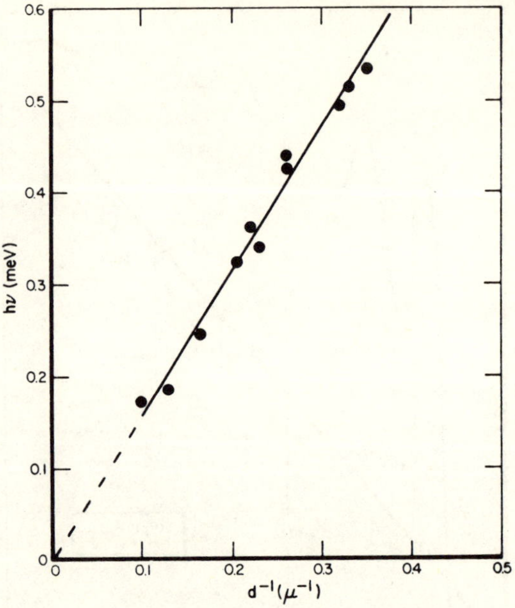

Fig. 5. Variation of oscillation period with inverse thickness of Pb film. From Tomasch ([1]).

considerably less than $h\nu$. Initially it appeared possible to view such behavior as a minor deviation from the rule. As more information became available, it became clear that a general tendency was involved, one which became more pronounced with increasing d_{In}. As we shall see, such deviations from uniform spacing were to provide a crucial clue to the identity of the electronic excitations involved. Finally, it was found that the strength of the structure could be enhanced substantially by evaporating a thin (1000–2000 Å) overlayer of Ag on the free surface of the thick film, even though the Ag was $d_S \approx 2$–$30\ \mu$ distant from the tunneling barrier (AlO_x). Again, such behavior is consistent with the notion of a standing wave, since, for example, the strength of an acoustic resonance in an organ pipe depends critically on the termination of the pipe, i.e., on the boundary condition imposed at the far boundary.

My initial speculations centered about quantum-assisted tunneling processes, with some family of long-wavelength collective excitations presumably serving as the quantum reservoir ($h\nu$, $2h\nu$, etc.), the general picture resembling the Dayem–Martin experiment. Such thinking preceded an appreciation of the fact that systematic deviations of the type mentioned invariably occur near the gap when thicker films ($d \geq 10\ \mu$) are employed.

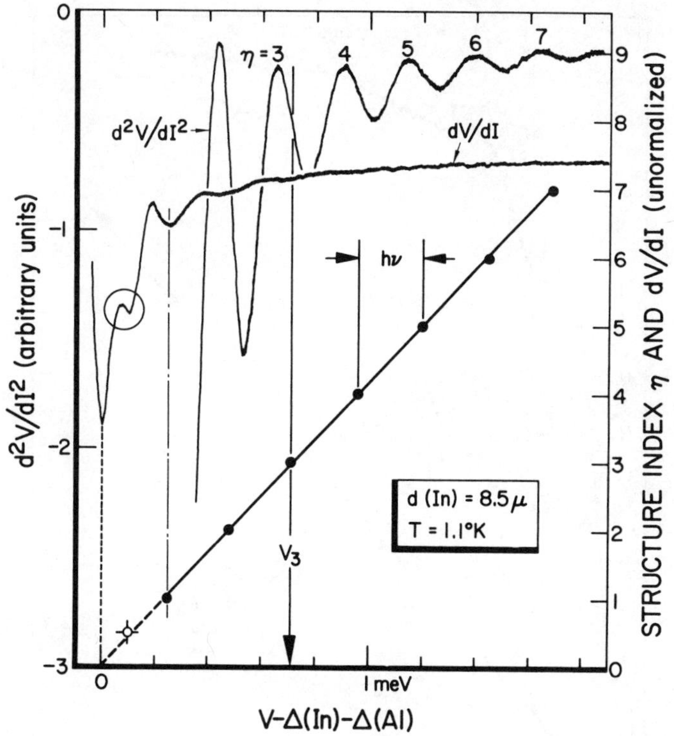

Fig. 6. First- and second-derivative data for a 8.5-μ thick Pb film with a 0.2-μ thick Ag overlay (Al–AlO_x–In/Ag). Modulation levels are 15 μV (r.m.s.) and 70 μV (r.m.s.) for first and second derivatives, respectively. Structural amplitude with Ag overlay is approximately five times larger than for sister diode without Ag. For the thin Ag overlays employed, $h\nu$ does not appear sensitive to the presence of Ag. Note that first resonance (encircled) does not index properly ($\eta \sim 1/2$) relative to remaining resonances ($1 \leq \eta \leq 7$). From Tomasch ([2]).

This appreciation was to be accelerated by a private communication from McMillan and Anderson describing their success in accounting for the In results (Fig. 6) by means of a "quasiparticle interference" picture based on the twofold degeneracy previously described. They had originally attempted an explanation of the earlier Pb data (Fig. 4), but with indifferent results. Upon becoming acquainted with the In data of Fig. 6, they were quickly able to account for the anomalous resonance (encircled) nearest the gap. Furthermore, the presence of a thin Ag overlayer on the In made the experimental situation correspond more nearly to their model, which required a pairing potential perturbation $\delta\Delta$ at the far surface. Encouraged,

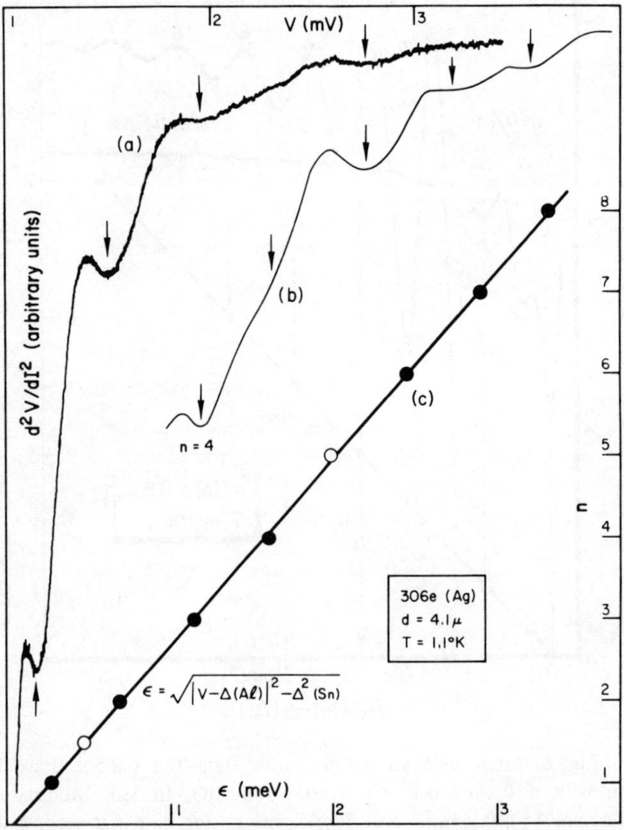

Fig. 7. Geometrical resonances in the tunneling characteristics of a 4.1-μ Sn film with 0.3-μ Ag overlay: (*a*) single trace of d^2V/dI^2 [maximum modulation $\sim 55\ \mu$V (r.m.s.)], (*b*) quieted sum of four independent traces, and (*c*) plot of $\varepsilon_n(n)$. Arrows identify points of constant phase (indexed by n) used for (*c*). Relatively weak structure is indicated by open circles. From Tomasch [7].

they returned to the Pb data (Fig. 4) and demonstrated that it too could be reconciled provided renormalization effects were included, although v_F was larger than the ASE value as previously mentioned.

THEORY AND EXPERIMENT

Choice of a regime which mixes degenerate excitations is of itself sufficient to fix the characteristic energy (voltage) distribution associated with a family of resonances. As a consequence, considerable heuristic

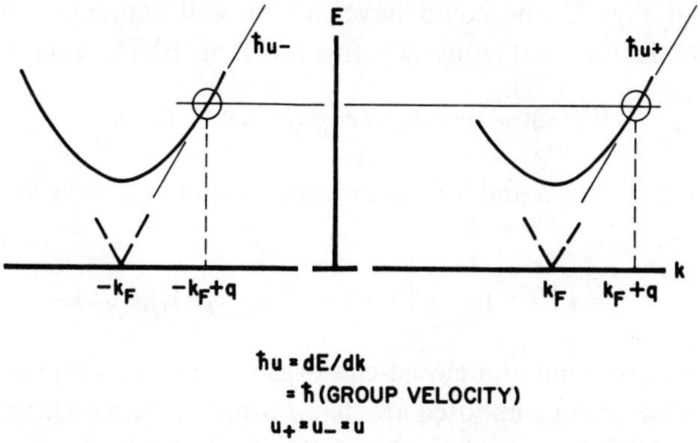

Fig. 8. Excitation spectrum for quasiparticles moving normal to the plane of the thick film. Positive k corresponds to motion away from the tunneling barrier (AlO_x).

insight can be gained from investigating even the most primitive one-dimensional model. To this end Fig. 8 depicts the excitation spectrum for quasiparticles moving normal to the plane of the thick film (S). Excitations $\psi(k)$ associated with $(k_F - q) = k_1$, $(k_F + q) = k_2$, $(-k_F + q) = -k_1$, and $(-k_F - q) = -k_2$ are energetically degenerate. The two associated with k_2 and $-k_1$ (encircled) have the same value of dE/dk, and hence the same group velocity $u > 0$. Presumably a linear combination,

$$\Psi(+) = \tfrac{1}{2}\psi(k_2) + \tfrac{1}{2}\psi(-k_1), \tag{1}$$

would also represent an admissible traveling wave having velocity u and whose general features remain unchanged with time. Predicating the simple form

$$\psi(E, k) = \exp i(kx - Et)/\hbar, \tag{2}$$

Eq. (1) becomes

$$\Psi(+) = (\cos k_F x) \exp i(qx - Et)/\hbar. \tag{3}$$

We see that $\Psi(+)$ is a Bloch wave of amplitude $\cos k_F x$. Although the $\cos k_F x$ modulation dominates the energy eigenvalue $(E_F + E)$, it does not contribute to the momentum expectation value $\langle p \rangle = \hbar q$. Since $k_F \gg q$, $\Psi(+)$ is associated with a wavelength Λ which is much larger than $\lambda(k_2) \doteq \lambda(-k_1) \doteq \lambda_F$:

$$(\Lambda/\lambda_F) = (k_F/q). \tag{4}$$

Returning to Fig. 8, one could have just as well combined $\psi(-k_2)$ and $\psi(k_1)$ to obtain the analogous negative-traveling Bloch wave $\Psi(-)$:

$$\Psi(-) = (\cos k_F x) \exp i(-qx - Et)/\hbar. \tag{5}$$

Superposition of $\Psi(+)$ and $\Psi(-)$ generates a standing wave Φ with nodes separated by $\Lambda/2$:

$$\Phi = \tfrac{1}{2}\Psi(+) - \tfrac{1}{2}\Psi(-) = i(\cos k_F x)(\sin qx). \tag{6}$$

In analogy to open-end and closed-end organ pipes, the simplest resonance conditions which can be imposed are that d either contain an integer number or an odd half-integer number of standing wave loops:

$$\begin{aligned} d &= j(\Lambda/2), \qquad (j = m, n), \\ m &= \tfrac{1}{2}, \tfrac{3}{2}, \tfrac{5}{2}, \ldots, \\ n &= 1, 2, 3, \ldots . \end{aligned} \tag{7}$$

Introducing these conditions into Eq. (6), the associated probability densities $\Phi_j{}^*\Phi_j$ can be written as

$$P_j = (\cos^2 k_F x)(\sin^2 j\pi x/d), \tag{8}$$

$$P_m(x = d) = 1, \tag{9}$$

$$P_n(x = d) = 0. \tag{10}$$

These results may be interpreted as follows: In the absence of a perturbation the present regime is capable of supplying long-wavelength excitations Φ_j which either allow quasiparticles to sample a macroscopic region ($\sim d/4j$) near the far boundary [$\Phi_m(d) = 1$], or to avoid such regions [$\Phi_n(d) = 0$]. If the pairing potential $\Delta(k)$ later becomes perturbed at the far boundary, say because of a metallic overlayer, this capability should facilitate an economical accomodation of the perturbation by the electronic system. One would therefore expect the density-of-states function to be enhanced at energy values corresponding to certain of these excitations. The characteristic distribution of such energies (voltages) is easily calculated within the free-electron approximation:

$$(E_F \pm \varepsilon) = \hbar^2(k_F + q)^2/2m^*, \tag{11}$$

$$\varepsilon \doteq \hbar v_F q = jhv_F/2d = \varepsilon_j, \tag{12}$$

$$E_j = [\Delta_S{}^2 + \varepsilon_j{}^2]^{1/2}. \tag{13}$$

Connection with experiment is made by means of the association

$$E_j = eV_j - \Delta_{\mathrm{Al}}. \tag{14}$$

[Inclusion of the term Δ_{Al} corrects for the uniform shift caused by the energy gap in Al, i.e., a bias of Δ_{Al}/e must be applied to bring the Al gap edge even with the Fermi level in S. Also, since j has not been given a specific significance, the V_j refer to voltages associated with features of constant phase, e.g., peaks (or dips) in dI/dV]. As indicated in Eq. (12), it is the associated ε_j of the normal state, and not the E_j (or V_j), which are uniformly spaced. The E_j (or V_j) approach this behavior only when $(\varepsilon_j/\Delta)^2 \gg 1$, i.e., for thinner films and/or higher j values. Rather than plotting $V_j(j)$, this suggests that ε_j be plotted instead:

$$\varepsilon_j = \{[eV_j - \Delta_{\mathrm{Al}}]^2 - \Delta_{\mathrm{S}}^2\}^{1/2} \tag{15}$$

$$= (hv_F/2d)j. \tag{16}$$

For very thick films, Eqs. (13) and (16) assert that $\varepsilon_j(j)$ must remain linear even though E_j (or V_j) may become nonlinear. Furthermore, the slope of $\varepsilon_j(j)$ should be directly related to v_F by

$$v_F = (2d/h)(d\varepsilon_j/dj). \tag{17}$$

To test these assertions, Fig. 9 presents first- and second-derivative data for a 30.0-μ thick In film, while Fig. 10 displays corresponding plots of $E_j(j)$ and $\varepsilon_j(j)$. Contrary to the minor deviation from linearity exhibited by $V_\eta(\eta)$ in Fig. 6 ($d = 8.5\ \mu$), the $E_j(j)$ results of Fig. 10 ($d = 30.0\ \mu$) amount to an absence of linearity, except perhaps as a limiting tendency. [By definition, $V_\eta(\eta)$ of Fig. 6 differs from $E_j(j)$ by only an additive constant.] On the other hand, $\varepsilon_j(j)$ is indeed observed to be linear, the value of v_F obtained via Eq. (17) being in close agreement with that obtained from recent cyclotron resonance measurements [Mina and Khaikin ([10])]. Results such as these provide the experimental justification for accepting a quasi-particle interference explanation of the tunnelling structure.

The expression

$$E_n = [\Delta_{\mathrm{S}}^2 + (nhv_F/2d)^2]^{1/2} \tag{13$'$}$$

was first derived by McMillan and Anderson ([3]) on the basis of a scattering model calculation involving coherent interference between degenerate excitations. Specifically, they found that the density of states $N(E)$ became

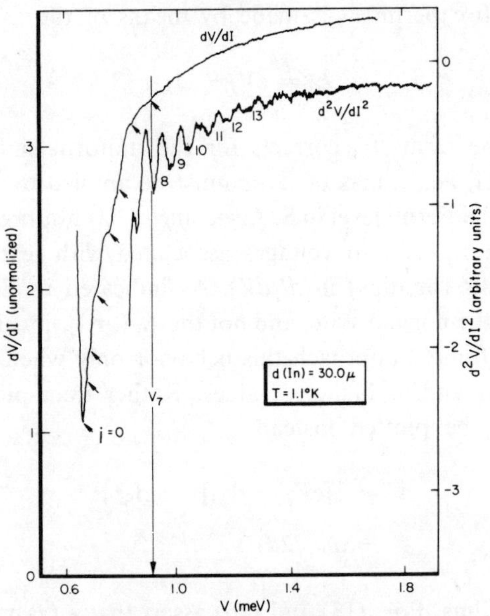

Fig. 9. First- and second-derivative data for a 30.0-μ thick In film. Modulation levels are 5–10 μV (r.m.s.) and 10–20 μV (r.m.s.), respectively. From Tomasch and Wolfram ([4]).

modulated as a result of introducing a discontinuous perturbation $\delta\Delta$ at $x = d$:

$$\delta N \propto (E\Delta/\varepsilon^2)[(\pi/2) - \mathrm{Si}(2d\varepsilon/hv_F)]\delta\Delta\,, \tag{18}$$

where $\mathrm{Si}(r)$ is the tabulated function

$$\mathrm{Si}(r) = \int_0^r (1/t)\sin t\,dt\,. \tag{19}$$

For $\delta\Delta > 0$, δN has maxima according to

$$\varepsilon = (hv_F/2d)n\,, \tag{20}$$

while for $\delta\Delta < 0$, δN has minima according to Eq. (20), i.e., has maxima according to

$$\varepsilon = (hv_F/2d)(n - \tfrac{1}{2})\,. \tag{21}$$

The conditions embodied in Eq. (20) and Eq. (21) are familiar, corresponding to Eq. (16) with j first set equal to n and then to m. Interpreting these results in terms of our heuristic discussion, increasing $\Delta(x = d)$

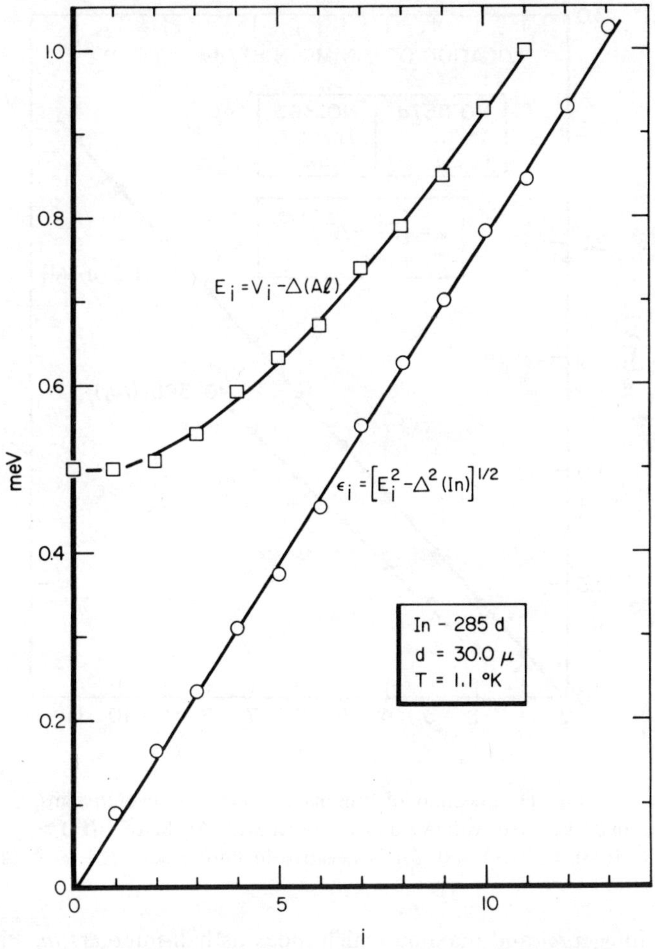

Fig. 10. Plots of $E_\eta(\eta)$ and $\varepsilon_\eta(\eta)$ obtained from data of Fig. 9. From Tomasch and Wolfram (4).

reduces the density of quasiparticles near $x = d$, while decreasing $\Delta(x = d)$ has the opposite effect. Both of these conclusions are pleasing in that they coincide with results based on physical reasoning.

Besides predicting the energy distribution of the resonances, Eq. (18) has several other interesting features which are subject to experimental verification. The first involves the relationship between the sign of $\delta\Delta$ and the way minima (or maxima) in $N(E)$ index. On the assumption that minima (or maxima) in dI/dV correspond to minima (or maxima) in $N(E)$, negative perturbations $\delta\Delta < 0$ should produce minima in dI/dV which

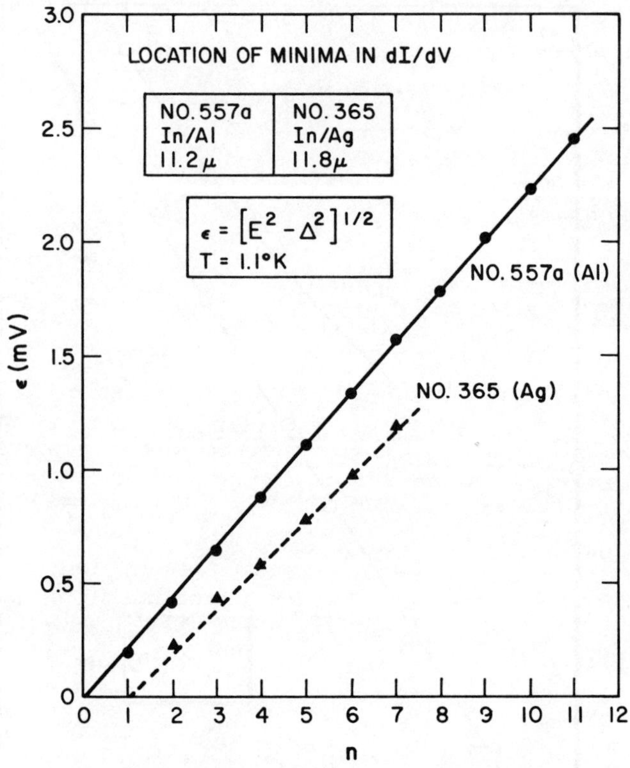

Fig. 11. Location of minima in dI/dV for two In films, one overlayed with Ag and the other with Al. Since $\Delta(\text{In}) > \Delta(\text{Al}) > \Delta(\text{Ag}) = 0$, $\delta\Delta$ is negative in both cases.

index as integers n and maxima which index as half-intergers m. Figure 11 indicates that In films ($d \sim 11.5\ \mu$) overlayed with either Ag or Al do behave this way. Figure 12 contains data for two In films ($d \sim 4.1\ \mu$), one with a Ag overlay and the other with no overlay (natural oxide only). With Ag the minima again index as n, or, equivalently, maxima index as m. With no overlay the situation becomes more complex in that two series are observed, the stronger of which has maxima which index *not* as m, but rather as n. Maxima of the weaker series index as m. One might be tempted to associate the strong series with patches on the far boundary for which $\delta\Delta > 0$, perhaps indicating that proximity of the natural oxide enhances superconductivity, as Ginzburg ([12]) has suggested. Because of the complexity of the no-overlay results, however, little can be said with certainty beyond that they do produce proper values of v_F.

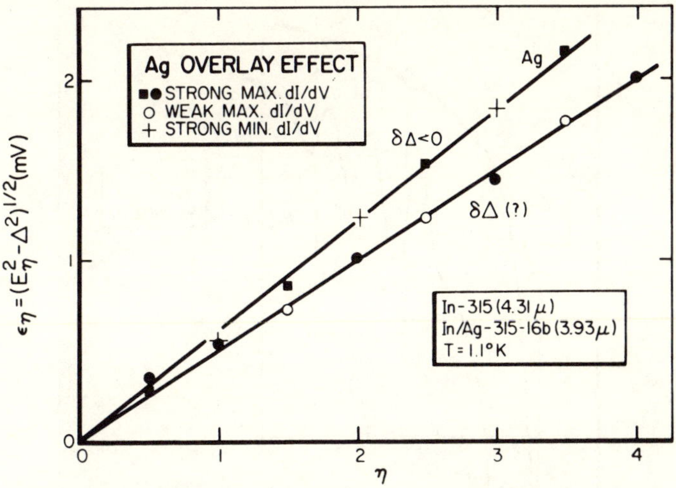

Fig. 12. Location of maxima and minima in dI/dV for two In films, one overlayed with Ag and the other with no overlay (natural oxide only).

Although attempts to observe the effects of a positive perturbation $\delta\Delta > 0$ by overlaying Pb onto In have not met with complete success, they have served to emphasize significant differences between model pictures involving extremely localized gap perturbations and the real experimental situation. Rather than having a nonzero perturbation $|\delta\Delta| > 0$ confined to a sharp boundary $x = d$, the actual situation more nearly involves having $|\delta\Delta| > 0$ over a boundary layer $d \leq x \leq d + d_0$, where d_0 is essentially the overlayer thickness. One now considers the behavior of a quasiparticle incident upon this layer. Transmission into the overlay is possible for all energies $E > \Delta_{\text{In}}$ provided that $\delta\Delta < 0$. For $\delta\Delta > 0$, however, transmission is energetically forbidden over a range $\Delta_{\text{In}} \leq E < \Delta_{\text{Pb}}$, since there are no allowed states $E < \Delta_{\text{Pb}}$ in the Pb overlay. Hence there is a threshold bias $V_t = \Delta_{\text{Al}} + \Delta_{\text{Pb}}$ below which quasiparticles suffer total internal reflection prior to significant overlayer penetration. One naturally wonders if the structure observed below V_t differs from that above V_t. Figure 13 presents data for an 8.4-μ thick In film with a Pb overlay. Although somewhat complex, geometrical resonance structure is observed below V_t. In all of the cases studied to date the structure terminates more-or-less abruptly near V_t even though other features of the data may vary. Such behavior could be due to the presence of a Pb–In diffusion layer, which is very effective in incoherently scattering quasiparticles which penetrate the layer ($V > V_t$). Quasiparticles which undergo total internal reflection in clean In ($V < V_t$)

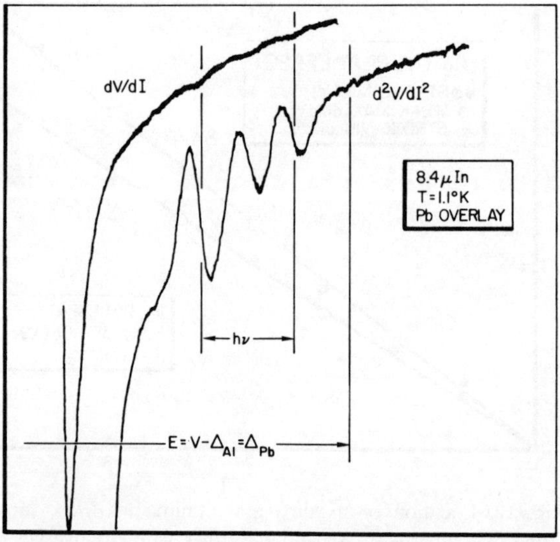

Fig. 13. First- and second-derivative data for an 8.4-μ thick In film with a thin ($\sim$ 2000 Å) Pb overlay. Geometrical resonance structure is missing above the threshold bias $V_t = \Delta_{\text{Al}} + \Delta_{\text{Pb}}$ for quasiparticle transmission into the Pb overlay region. Taken from: W. J. Tomasch, "Tunneling from In Films with Superconductive and Ferromagnetic Overlays," Phys. Letters 26A:379 (1968).

avoid such scattering. Two observations which tend to support this tentative explanation can be made: First, details of the observed structure change as a diode is vacuum aged at room temperature, suggesting that Pb–In interdiffusion is indeed involved. No analogous behavior is observed when Ag and Al overlays are employed. Finally, it is known that $\Delta_{\text{alloy}} \approx \Delta_{\text{Pb}}$ for Pb–In alloys over a rather wide concentration of In, which could account for V_t remaining constant at the proper value despite other differences and/or changes in the structure.

With regard to the enhancement capability of various overlay specie, the experimental facts are as follows: Although subject to variation from diode to diode, In films without overlay (natural oxide only) have exhibited dI/dV structure as strong as a few per cent. Effects as large as 10% have been observed with Ag overlays, the weakest results being as strong as the strongest without overlay. As Fig. 14 indicates, effects as large as 35% can be obtained with Al overlays. Aluminum generally produces structure which is three to five times stronger than that obtained with Ag. According to Eq. (18), the strength of the dI/dV structure should vary as $|\,\delta\Delta\,| = |\,\Delta_{\text{In}}$

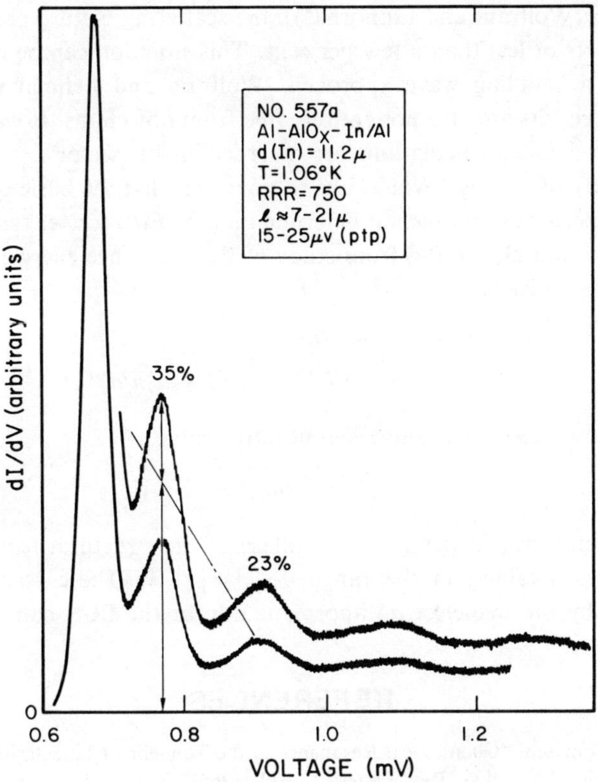

Fig. 14. First-derivative data for a 11.2-μ thick In film with a 2000-Å Al overlay. Largest peak (lower curve) occurs at $V = \Delta_{\text{In}} + \Delta_{\text{Al}}$ and is not a resonance. Slanting dashed line bisects leading edge of successive oscillations. Resonance strengths (in per cent) are given by the ratio of the indicated dimensions. Taken from W. J. Tomasch, "Tunneling from In Films with Superconductive and Ferromagnetic Overlays," Phys. Letters 26A:379 (1968).

$- \Delta_{\text{Ag,Al}} |$, and hence Al ($| \delta\Delta | \approx 0.40$ meV) should be somewhat *less* effective than Ag ($| \delta\Delta | \approx 0.52$ meV), all other things being equal. In view of possible alloying and transmission effects, it may well be that a direct comparison between the Ag and Al results is not justified and that the discrepancy between theory and experiment stems from practical considerations. Nevertheless, it would be reassuring to find clear cut experimental situations in which a proper harmony exists between theory and experiment. There is still another discrepancy, which involves the overall strength of the observed structure. For parameters within the validity of the model [Wolfram and Einhorn ([8])], the scattering picture predicts maximum effects of less than a few per cent. This situation can be remedied by adopting a standing wave approach [Wolfram and Lehman ([9])], but to date such results are also not entirely free from objections. It would appear that more detailed calculations are in order in this sector.

By way of closing I would like to point out that the basic quantization $q = j\pi/d$ permits experimental determination of $E(k)$ for real superconductors [Maki and Griffin ([5])] from study of the resonance energies as a function of film thickness:

$$\begin{aligned} eV_j - \Delta_{\mathrm{Al}} &= E(k_j) \\ &= E\{k_F + [(j + \alpha)\pi/d]\}, \end{aligned} \tag{22}$$

where j now takes on positive *and* negative values,

$$j = \pm m, \pm n, \tag{23}$$

and the constant α depends on the phase of the structural feature investigated, always falling in the range $0 \leq |\alpha| \leq \frac{1}{2}$. These results are not restricted by the free-electron approximation or the BCS pairing approximation.

REFERENCES

1. W. J. Tomasch, "Geometrical Resonance in the Tunneling Characteristics of Superconducting Pb," Phys. Rev. Letters 15:672 (1965).
2. W. J. Tomasch, "Geometrical Resonance and Boundary Effects in Tunneling from Superconducting In," Phys. Rev. Letters 16:16 (1966).
3. W. L. McMillan and P. W. Anderson, "Theory of Geometrical Resonances in the Tunneling Characteristics of Thick Films of Superconductors," Phys. Rev. Letters 16:85 (1966).
4. W. J. Tomasch and T. Wolfram, "Energy Spacing of Geometrical Resonance Structure in Very Thick Films of Superconducting In," Phys. Rev. Letters 16:352 (1966).
5. K. Maki and A. Griffin, "Tomasch Effect as a Probe of the Dispersion Relation of Electrons in Gapless Superconductors," Phys. Rev. 150:356 (1966).
6. T. Wolfram and G. W. Lehman, "Macroscopic Quantization in Superconducting Films," Phys. Letters 21:631 (1966).
7. W. J. Tomasch, "Geometrical Resonance Effect in Superconducting Sn," Phys. Letters 23:204 (1966).
8. T. Wolfram and M. B. Einhorn, "Surface States and Additional Structure in the McMillan–Anderson Model for the Tomasch Effect," Phys. Rev. Letters 17:966 (1966)
9. T. Wolfram and G. W. Lehman, "Theory of the Tomasch Effect," Phys. Letters 24:101 (1966).
10. R. T. Mina and M. S. Khaikin, Zh. Eksperim. i Teor. Fiz. 51:62–86 (1966).
11. T. E. Faber and A. B. Pippard, Proc. Roy. Soc. A231:336 (1955).
12. V. L. Ginzburg, Phys. Letters 13:101 (1964).

Chapter 24

Multiparticle Tunneling*

J. W. Wilkins†

Laboratory of Atomic and Solid State Physics
Cornell University
Ithaca, New York

INTRODUCTION

Genuine multiparticle tunneling (actually, only two-particle) has probably been observed only once. Accordingly, we will emphasize the pedagogic aspects of the problem, returning reluctantly at the end to hard practicalities.

First we will review the properties of single-particle excitations in a superconductor with a view to providing a qualitative description of the one- and two-particle tunneling processes. One-particle processes, including the Josephson effect, are of first order in the *square* of the tunneling matrix element, while two-particle processes are of second order.

Details of the calculation (which have not previously been published) of the two-particle tunneling current are presented below (p. 338) and in the Appendix. Comparison with experiment reveals the inadequacy of the theory. In the last section we return to the first-order process, in particular to the AC Josephson effect, which may prove to be the source of the multiparticle-like structure in the tunneling data.

SINGLE-PARTICLE EXCITATIONS IN A SUPERCONDUCTOR

Let us first review the single-particle excitations of a superconductor at zero temperature. In a normal metal a single particle can be added to the ground state of the system without creating any excitations if the energy

* Work supported in part by the Office of Naval Research under contract NONR-401(38), Technical Report No. 24.

† Alfred P. Sloan Foundation Fellow.

of the added particle is equal to the chemical potential μ. The single-particle energy ε_p of an added particle with momentum p in a normal system is conveniently defined with respect to the chemical potential. [For non-interacting fermions $\varepsilon_p = p^2/2m - \mu$.] In a superconductor single-particle excitations have a definite momentum, obey Fermi statistics, and possess an energy E_p measured relative to the chemical potential (or Fermi energy at zero temperature) given by

$$E_p = (\varepsilon_p{}^2 + \Delta^2)^{1/2} \tag{1}$$

in the BCS theory. The spectrum of the single-particle excitations in a superconductor, which we shall call quasiparticles, plotted schematically in Fig. 1(a) as a function of momentum, has a gap Δ at the Fermi surface. The resulting density of states relative to the normal metal density of states at the Fermi surface $N_T(E_p)/N(0)$ $(= d\varepsilon_p/dE_p)$, plotted along the abscissa of Fig. 1(b), has a singularity at $E_p = \Delta$, $[N_T(E)/N(0) = E/(E^2 - \Delta^2)^{1/2}]$. These two properties of the quasiparticles—the *gap* in the excitation spectrum and the *singularity* in the density of states—are sufficient to explain the qualitative features of the one- and two-particle tunneling processes at zero temperature using the excitation plot shown in Fig. 1(a). An additional feature of this plot is that the pair states $(k\uparrow, -k\downarrow)$ may be placed along the abscissa—i.e., at the chemical potential. This is because a pair state of energy $2\,\mu$ can be added to the ground state of a superconductor without creating any excitations. We refer to these pair states at the chemical potential as the *superfluid*.

The excitation spectrum plot can be used as a guide to quantitative

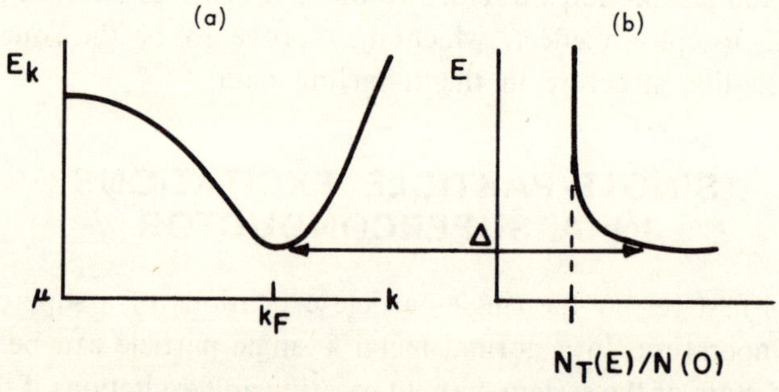

Fig. 1. (a) Single-particle excitation energy relative to chemical potential as a function of momentum. (b) Density of single-particle states in a superconductor relative to free-electron density of states (plotted along abscissa); if a particle is added at energy gap Δ, it "sees" a singularity in the density of states.

calculations if we multiply the relevant matrix element for the equivalent normal metal process by a factor u_p for each quasiparticle of momentum p which is "added" to the system and v_p for each one "removed," where

$$u_p{}^2 = \tfrac{1}{2}[1 + (\varepsilon_p/E_p)] = 1 - v_p{}^2. \tag{2}$$

Observe that for a normal metal (where $E_p = |\,\varepsilon_p\,|$) the factor u_p is 0 for momenta below the Fermi surface and 1 for those above, which is consistent with the fact that all single-particle states (for a noninteracting normal fermion system at zero temperature) are filled below the Fermi surface. For the superconductor the Fermi surface is slightly smeared, as indicated by the nonzero value of u_p below the surface or of v_p above. It should be noted that Eq. (2) defines only the magnitude of u_p. Within the BCS theory the phase of u_p (and v_p) for a *single* superconductor is quite arbitrary. (In the strong-coupling extension of the BCS theory the phase is a function of momentum, or, more properly, energy.) But the *relative phase* between two superconductors which are weakly coupled, by, for example, a thin oxide barrier, is *not* arbitrary and may be influenced by applied electric and magnetic fields. (The equations governing and involving this relative phase are presented in Chapter 32 of this volume, in Scalapino's discussion of the Josephson effect.)

Finally, the effects of finite temperatures on tunneling can be discussed. Roughly speaking, the effects are twofold. First, since the gap is a function of temperature, the voltage dependence of the tunneling current scales with the gap. Second, the quasiparticle states become occupied as the temperature increases, allowing new tunneling processes to occur.

QUALITATIVE DESCRIPTION OF TUNNELING PROCESSES

Let us first review the conventional first-order (or one-particle) tunneling process between superconductors, which was discussed by Schrieffer in Chapter 21. A superconductor S_l is biased by an applied potential (times the electronic charge) V with respect to a superconductor S_r at zero temperature. The two superconductors are separated by a thin oxide film. The conventional single-particle tunneling process present at zero temperature is illustrated in Fig. 2 by the extraction of a single electron from the superfluid (pairs) in S_l, leaving a quasiparticle in state **k**. The electron tunnels through the oxide to become a quasiparticle in state **p** in S_r. Energy conservation requires that the energy of the initial state measured relative

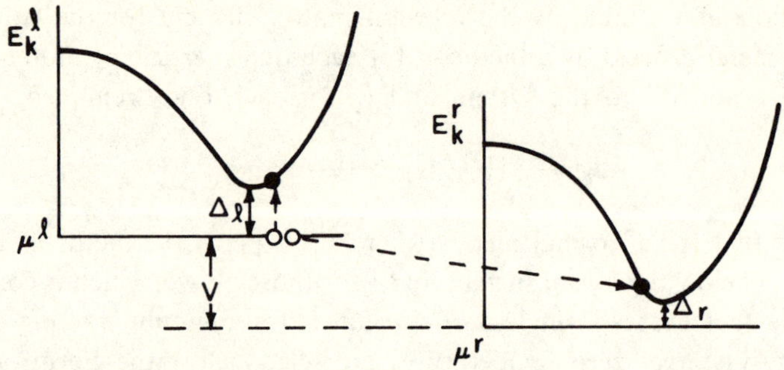

Fig. 2. Schematic quasiparticle spectra for two superconductors illustrating conventional single-particle tunneling.

to the chemical potential of say, S_r—i.e., the energy $2V$ of the pair-state broken up in the tunneling process—is equal to the energy of the two quasiparticles ($E_k^l + V$ and E_p^r) in the final state. Consequently,

$$V = E_k^l + E_p^r > \Delta_l + \Delta_r,$$

and the tunneling process does not occur below the threshold voltage ($\Delta_l + \Delta_r$). But at this threshold each of the quasiparticles is being added at the bottom of the excitation spectrum, where the density of states has a singularity. The product of these two singularities gives rise to a *discontinuous* onset in the current at $V = \Delta_l + \Delta_r$ which is observed experimentally. As the voltage increases beyond threshold, the density of final states increases, so that the current I increases, until for voltages large compared to Δ the current is linear in the voltage, as would (theoretically) be the case for tunneling between normal metals.

At finite temperatures (but below the transition temperatures) the onset voltage decreases, since the gaps decrease with increasing temperature. In addition, the sharpness of the onset softens. More interestingly, if the gap energies are unequal ($\Delta_r > \Delta_l$, say), then direct quasiparticle tunneling occurs when $V = \Delta_r - \Delta_l$, as shown in Fig. 3. There is a discrepancy between the observed and calculated tunneling current due to this process ([1]), the reasons for which are poorly understood.

There is one more tunneling process between two superconductors which is first-order—i.e., the current is proportional to the square of the tunneling matrix element—and that is the Josephson effect. Under no applied voltage a DC current composed of pairs flows from one superconductor to the other. The direction of the current flow across the oxide is set

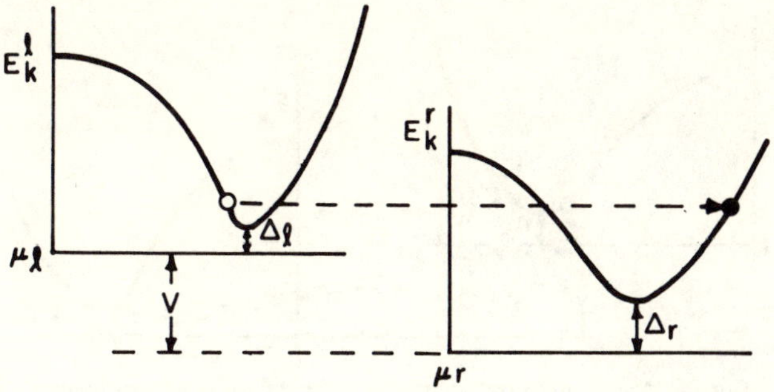

Fig. 3. A direct quasiparticle tunneling process which only occurs at finite temperature (or if quasiparticles are injected). Threshold voltage (times electronic charge) is $\Delta_r - \Delta_l$.

either by some external current source or by some small voltage applied sufficiently long to adjust the relative phase (mentioned previously) between the two superconductors. If a DC bias voltage is applied, then an AC Josephson current flows with frequency equal to $2V/h$. That the current is proportional to only the square of the tunneling matrix element we will argue elsewhere in this chapter (see also Chapter 32 by Scalapino).

Now let us consider two-particle or second-order tunneling processes, i.e., those which give rise to currents proportional to the fourth power of the tunneling matrix element. A typical process is shown in Fig. 4. Two electrons are extracted from the superfluid of the left-hand superconductor S_l and tunnel through the barrier to two quasiparticle states (**k** and **k**$'$) in the right-hand superconductor S_r. In the process the number of superfluid pairs in S_l is decreased by one, and no quasiparticle states are excited

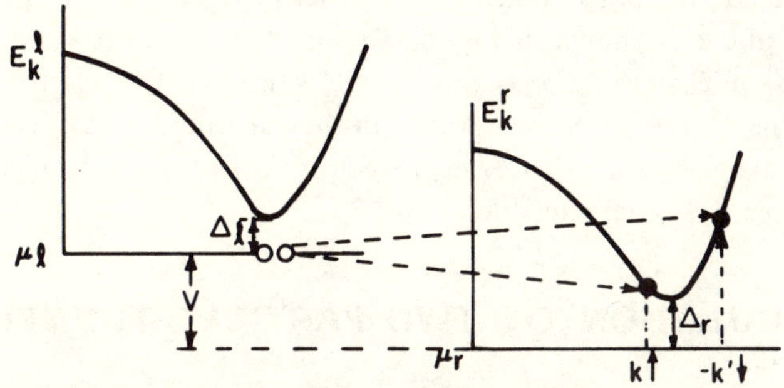

Fig. 4. A process contributing to the two-particle tunneling current between two superconductors. The threshold voltage (times electronic charge) is Δ_r.

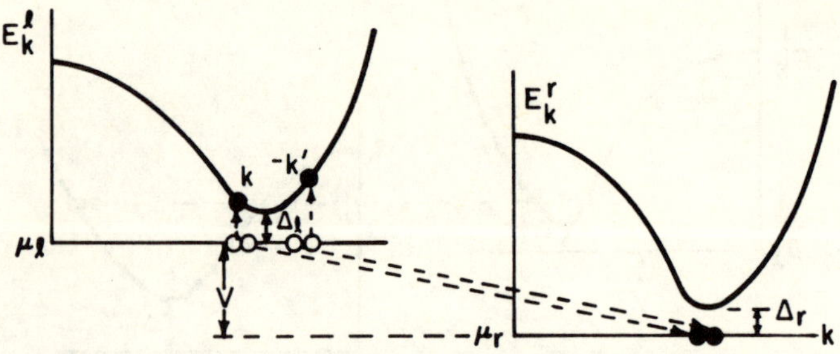

Fig. 5. A process contributing to the two-particle tunneling current between two superconductors. The threshold voltage (times electronic charge) is Δ_l.

in S_l. From conservation of energy we have $2V = E_k^r + E_{k'}^r > 2\Delta_r$; thus the threshold voltage is Δ_r. At this onset each tunneling particle "sees" a singularity in the density of states. The product of these two singularities, as in the conventional single-particle tunneling process, gives rise to a *discontinuous* onset in the current.

We can now qualitatively understand more of the structure of the I–V characteristic. As the voltage increases, the phase space for the two-particle final state increases as $(V - \Delta_r)$ to some power (initially just to first power). Thus the current should increase rapidly as V increases beyond Δ_r. Finally, when $V = \Delta_r + \Delta_l$ there is, as we shall see, the possibility of conservation of energy in the intermediate state. This is reflected in a strong singularity in the current at $V = \Delta_l + \Delta_r$ which, in practice, is masked by the much larger discontinuous onset of the one-particle tunneling process. We shall defer comparison with experiment until we have discussed the calculation in more detail.

There are, of course, other second-order processes. One similar to that discussed above is shown in Fig. 5. Two electrons are extracted from the superfluid of S_l leaving two quasiparticles in S_l. The electrons tunnel through the barrier to become a pair in the superfluid of S_r. This process clearly has an onset at $V = \Delta_l$. Let us turn now to more detailed calculations of the I–V characteristics.

CALCULATION OF TWO-PARTICLE TUNNELING

The second-order tunneling current is proportional to the transition probability per unit time for two electrons to tunnel from one conductor (S_l, say) to the other (S_r). In the time-independent perturbation theory this probability can be written as

$$\omega_{r\leftarrow l} = \frac{2\pi}{\hbar} \sum_F \left| \sum_M \frac{\langle F \mid H_T \mid M\rangle\langle M \mid H_T \mid I\rangle}{E_M - E_I} \right|^2 \delta(E_F - E_I), \tag{3}$$

where F, M, and I refer to the final, intermediate, and initial states, respectively, and H_T is the tunneling Hamiltonian given (for simplicity) by

$$H_T = \sum_{rl} T_{rl}(C_{k_r}^+ C_{k_l} + C_{k_l}^+ C_{k_r}). \tag{4}$$

Before considering the superconducting case we shall find it instructive to consider the case of second-order tunneling in normal metals using the independent-particle model. In particular, the rate for such tunneling turns out to be identically zero precisely because the particles are independent. If there are no interactions between the particles, then each particle must conserve energy separately in tunneling. But this process is already included in the first-order processes, so the second-order process is zero in the independent-particle model. (Fortunately, quantum-mechanical perturbation theory has this physics built-in, so that the correct answer arises from the calculation willy-nilly.) In the superconductor the quasiparticles are not independent. For example, if one member of a pair tunnels, the other remains as a quasiparticle with energy greater than Δ. Then energy is not conserved in the intermediate state, and the second-order tunneling current is not zero between two superconductors. The following more detailed calculations will substantiate these statements.

A characteristic two-particle tunneling process between normal metals is depicted in Fig. 6. The double lines of the excitation spectrum for mo-

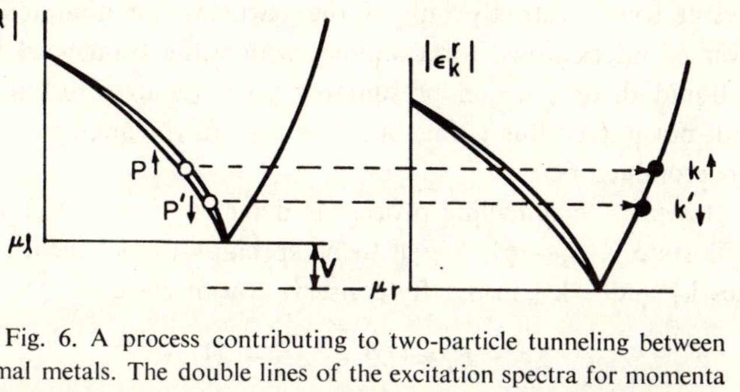

Fig. 6. A process contributing to two-particle tunneling between normal metals. The double lines of the excitation spectra for momenta below k_F indicate that all such states are filled. In a superconductor the probability of a quasi particle state being filled is $\nu_p{}^2 = [1 - (\varepsilon_p/E_p)]/2$.

mentum below the Fermi surface indicate that all those states are filled (at zero temperature). Metal M_l is biased by an applied voltage V with respect to metal M_r. Two particles $\mathbf{p}\uparrow$ and $\mathbf{p}'\downarrow$ with energies $|\varepsilon_p|$ and $|\varepsilon_{p'}|$, respectively, measured relative to chemical potential μ_l, tunnel from M_l to states $\mathbf{k}\uparrow$ and $\mathbf{k}'\downarrow$ with energies ε_k and $\varepsilon_{k'}$, respectively, in M_r. (Note: the spins of the particles have been chosen to make it clear that $\mathbf{k}$ tunnels to $\mathbf{p}$ and $\mathbf{k}'$ to $\mathbf{p}'$.) Conservation of energy, implied by the delta function in (3), requires that

$$E_F - E_I = \varepsilon_k + \varepsilon_{k'} - (2V + |\varepsilon_p| + |\varepsilon_{p'}|) = 0. \tag{5}$$

There are *two* intermediate states which connect the given initial and final states:

1. $\mathbf{p}\uparrow$ tunnels to $\mathbf{k}\uparrow$, the energy denominator for which is given by $E_{M_1} - E_I = \varepsilon_k - |\varepsilon_p| - V$, and then $\mathbf{p}'\downarrow$ tunnels to $\mathbf{k}'\downarrow$.

2. $\mathbf{p}'\downarrow$ tunnels to $\mathbf{k}'\downarrow$, for which $E_{M_2} - E_I = \varepsilon_{k'} - |\varepsilon_{p'}| - V$. Then $\mathbf{p}\uparrow$ tunnels to $\mathbf{k}\uparrow$.

The sum must be done over all the appropriate intermediate states before the effective matrix element is squared. Accordingly,

$$\sum_M \frac{\langle F | H_T | M\rangle\langle M | H_T | I\rangle}{E_M - E_I} = T_{rl}^2\left\{\frac{1}{\varepsilon_k - |\varepsilon_p| - V} + \frac{1}{\varepsilon_{k'} - |\varepsilon_{p'}| - V}\right\}$$

$$= T_{rl}^2\left\{\frac{\varepsilon_k + \varepsilon_{k'} - (2V + |\varepsilon_p| + |\varepsilon_{p'}|)}{(\varepsilon_k - |\varepsilon_p| - V)(\varepsilon_{k'} - |\varepsilon_{p'}| - V)}\right\}.$$

But the numerator of this expression is identically zero because of overall conservation of energy [Eq. (5)], and hence the tunneling current is zero. This result follows strictly only if the particles are noninteracting. The inclusion of interactions, for example, within the framework of Landau Fermi liquid theory, would presumably give rise to some finite current. We shall not pursue this point, but now turn to the analogous calculation for superconductors.

A two-particle tunneling process is depicted in Fig. 4. Two electrons in a pair state in S_l—($\mathbf{p}\uparrow$, $-\mathbf{p}\downarrow$), to be specific—tunnel through the oxide to states $\mathbf{k}\uparrow$ and $-\mathbf{k}'\downarrow$ in S_r. By conservation of energy

$$E_F - E_I = E_k^r + E_{k'}^r - 2V = 0. \tag{6}$$

Again as in the normal metal case there are two intermediate states connecting the given initial and final states:

1. One member of the pair state $\mathbf{p}\uparrow$ in S_l tunnels to the quasiparticle state $\mathbf{k}\uparrow$ in S_r, leaving behind the other member of the pair $-\mathbf{p}\downarrow$ in a quasiparticle state in S_l. Subsequently this quasiparticle tunnels to the state $-\mathbf{k}'\downarrow$ in S_r. Accordingly, the energy denominator in (3) is

$$E_{M_1} - E_I = E_p{}^l + E_k{}^r - V.$$

Furthermore, the matrix element has a factor u (v) for each quasiparticle added (removed). Since the quasiparticle $-\mathbf{p}\downarrow$ is added and then removed, the appropriate set of factors is $u_k{}^r$, $u_{k'}^r$, $u_p{}^l$, and $v_p{}^l$.

2. The member $-\mathbf{p}\downarrow$ of the pair in S_l tunnels to $-\mathbf{k}'\downarrow$ in S_r, leaving behind a quasiparticle $p\uparrow$ in S_l, which subsequently tunnels to $\mathbf{k}\uparrow$ in S_r. In this case

$$E_{M_2} - E_I = E_{k'}^r + E_p{}^l - V,$$

and the factors multiplying the matrix element are the same as they were in the first intermediate state above.

The effective tunneling matrix can thus be written

$$\begin{aligned}
&\sum_M \frac{\langle F \mid H_T \mid M\rangle\langle M \mid H_T \mid I\rangle}{E_M - E_I} \\
&\quad = T_{rl}^2 u_k{}^r u_{k'}^r \sum_p u_p{}^l v_p{}^l \left\{\frac{1}{E_k{}^r - V + E_p{}^l} + \frac{1}{E_{k'}^r - V + E_p{}^l}\right\} \\
&\quad = T_{rl}^2 u_k{}^r u_{k'}^r \sum_p \frac{\Delta_l}{-(V - E_k{}^r)^2 + (E_p{}^l)^2}. \qquad (7)
\end{aligned}$$

In the second equality the overall conservation of energy, Eq. (6), has been used together with the definitions of u_p and v_p, Eq. (2) [i.e., $u_p v_p = \Delta/2E_p$]. The integral over momentum $\mathbf{p}$ can be done. Eventually, the tunneling current can be shown (see the Appendix) to be proportional to

$$| T_{rl} |^4 \int_{\Delta_r}^{2V-\Delta_r} dE \, \frac{N_r(E) N_r(2V - E)}{\Delta_l^2 - (E - V)^2}. \qquad (8)$$

This integral can also be done analytically, as shown in the Appendix, but its general properties can be deduced by inspection. A threshold voltage of Δ_r is consistent with the limits of integration. As in the case of conventional one-particle tunneling, both densities of states are singular (while the denominator is finite) at this threshold voltage, giving rise to a *discontinuous* onset, as pointed out in our original discussion. As the voltage increases, the current increases, initially linearly in $(V - \Delta_r)$ due to the in-

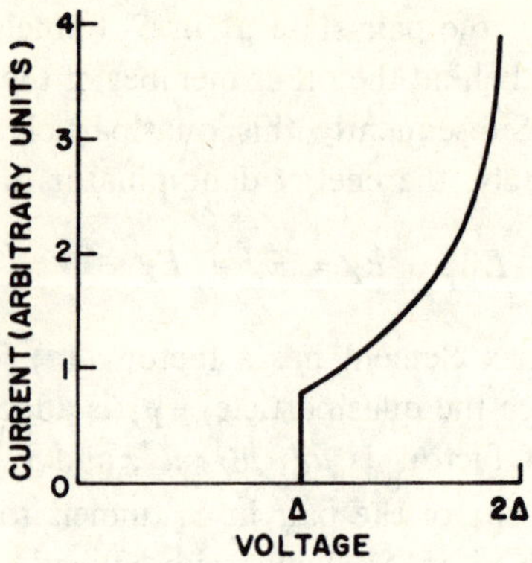

Fig. 7. Current–voltage characteristic for the double-particle tunneling processes between identical superconductors.

creasing two-particle final-state phase space and then more rapidly as the denominator becomes singular. The tunneling current is plotted in Fig. 7 as a function of voltage.

As we mentioned previously, there is another process (shown in Fig. 5) contributing to the two-particle current, which onsets at $V = \Delta_l$. Thus we can roughly anticipate the I–V characteristics for two different superconductors separated by a thin oxide film at zero temperature. Suppose $\Delta_l < \Delta_r$. When $V = \Delta_l$ there will be a discontinuous rise in the current. As V increases, I increases faster than linearly (in $V - \Delta_l$) until $V = \Delta_r$, where there is another discontinuous rise in I. An even faster rise in I continues with increasing V, until finally at $V = \Delta_r + \Delta_l$ there is a very rapid increase which is masked by the onset of the single-particle tunneling process. Note that the current is independent of the sign of the bias voltage. In addition, we would expect little temperature dependence except for the voltage scale following the temperature dependence of the gap.

COMPARISON WITH EXPERIMENT

There does seem to be one case, at least, where the two-particle tunneling process described in the preceding section was seen experimentally. Tunneling experiments by Taylor and Burstein ([2])* on lead sandwiches with an aluminum oxide barrier showed a polarity-independent temperature-insensitive excess current with onset at a voltage equal to that of Δ_{Pb}.

* The theory for the observed effect was presented by Schrieffer and Wilkins ([3]).

To obtain agreement between the observed magnitudes of the one- and two-particle currents, it was necessary to assume that the oxide films were patchy, having small regions of thinner oxide through which the bulk of the two-particle current passed. Otherwise the calculated ratio of the two-particle current to the one-particle at their respective threshold voltages [this ratio being proportional to $|T|^2$, i.e., to exp(-constant/oxide thickness)], would be far smaller than the observed ratio. The values of the

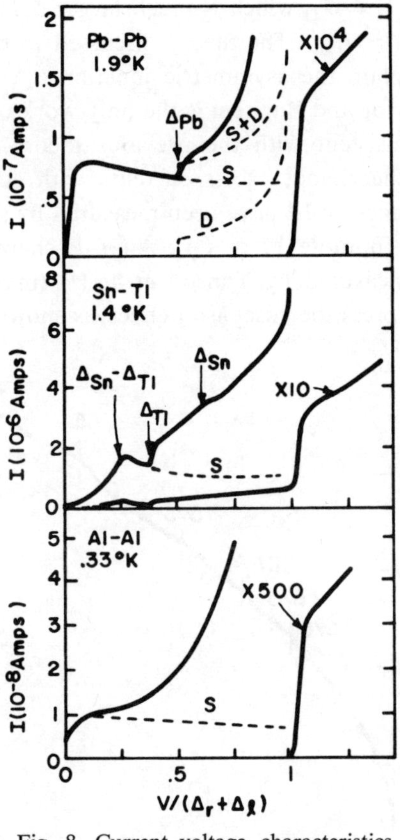

Fig. 8. Current–voltage characteristics for Pb–Pb, Sn–Tl, and Al–Al junctions as measured by Taylor and Burstein (²). The solid lines represent the experimental data and the dashed lines marked S represent the theoretical single-particle tunneling curves. For the Pb junction the theoretical double-particle tunneling current, marked D, and the sum of the single- and double-particle currents, marked S + D, are also shown.

oxide thicknesses and of the ratio of the thin and thick areas required to obtain agreement with the observed current were not unreasonable in the case of the Pb sandwich.

The *I–V* characteristics (of Taylor and Burstein) for sandwiches of Pb–AlO–Pb, Sn–SnO–Tl, and, Al–AlO–Al are shown in Fig. 8. In the aluminum sandwiches no excess current at $V = \Delta_{\text{Al}}$ is observed, possibly because the oxide formed in this case is quite uniform and hence the two-particle current is unobservably small. In the Sn–Tl sandwich there is a fairly clear onset at $V = \Delta_{\text{Tl}}$ which is much larger than any structure that may be present at $V = \Delta_{\text{Sn}}$. The theory discussed in the previous section is inadequate to explain the asymmetric junction.

The data of Taylor and Burštein is the only work of which this author is aware that is consistent with the previous theory—which, in fact, it inspired. On the other hand, excess currents with structure at voltages which are subharmonics of 2Δ have been measured by several investigators [([4–6]); also ([7]) see footnote 17 of ([9])]. An *I–V* characteristic for a Sn–SnO–Sn sandwich measured by Yanson *et al.* ([4]) (in a magnetic field of 33 G in order to surpress the Josephson effect) is shown in Fig. 9, and has

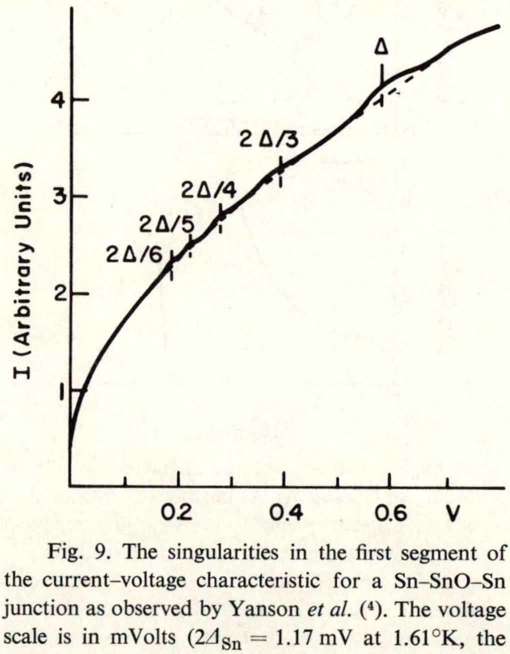

Fig. 9. The singularities in the first segment of the current–voltage characteristic for a Sn–SnO–Sn junction as observed by Yanson *et al.* ([4]). The voltage scale is in mVolts ($2\Delta_{\text{Sn}} = 1.17$ mV at 1.61°K, the temperature of the measurement). A magnetic field of 33 G was applied to suppress the DC Josephson effect.

singularities at Δ, $2\Delta/3$, $2\Delta/4$, $2\Delta/5$, and $2\Delta/6$ superimposed on a large excess current. These current peaks might be associated with multiple-particle tunneling. However, aside from the fact that the relative amplitude of the peaks (for example, the peak at $V = 2\Delta/4$ compared to that at Δ) is much larger than theory could explain, the shape of the peak is quite different from that predicted. In particular, the supposed two-particle current (i.e., that in excess of the background current indicated by a dotted line on Fig. 9) has an onset before $V = \Delta$, where it has its maximum. In summary [see, e.g., ([11])] almost all the structure in the I–V characteristics for voltages at subharmonics of 2Δ *cannot* be explained by multiparticle tunneling, which predicts both the wrong size and the wrong shape for the singularities observed. In addition, the subharmonic structure is superimposed on a large temperature-independent excess current (about 1/10 to 1/2 of the normal-state current), which suggests the presence of metallic shorts in the tunneling junction. To search for alternate explanations we turn to a more detailed discussion of one-particle tunneling, especially the Josephson effect.

FIRST-ORDER PROCESSES AND JOSEPHSON EFFECT

In this section we will present an argument (based on time-independent perturbation theory) which, while not completely rigorous, although it can be made so, demonstrates that the Josephson effect is proportional to $|T_{rl}|^2$, as is the conventional one-particle tunneling current. The transition probability per unit time out of some initial state I due to a perturbation H_T, which of course we take to be the tunneling Hamiltonian, Eq. (4), is given by

$$\omega_i = 2\pi \sum_F \langle I \mid H_T \mid F \rangle \langle F \mid H_T \mid I \rangle \, \delta(E_I - E_F),$$

where F refers to the appropriate final states and $\hbar = 1$. The delta function has the integral representation

$$\delta(E_I - E_F) = \int_{-\infty}^{\infty} \frac{dt}{2\pi} \exp i(E_I - E_F)t.$$

In the Heisenberg representation for the operators, where, for example,

$$H_T(t) = e^{iHt} H_T(0) e^{-iHt}$$

(H is the Hamiltonian for the system excluding H_T but including any applied bias voltage), the transition rate can be written

$$\omega_i = \int_{-\infty}^{\infty} dt \langle I \mid H_T(t) H_T(0) \mid I \rangle ,$$

where the completeness of the final states has been invoked. More explicitly,

$$\omega_i = \int_{-\infty}^{\infty} dt \langle I \mid \sum_{lr} T_{rl} [C_r^+(t) C_l(t) + C_l^+(t) C_r(t)] \\ \times \sum_{l'r'} T_{r'l'} [C_{-r'}^{\pm}(0) C_{-l'}(0) + C_{-l'}^{\pm}(0) C_{-r'}(0)] \mid I \rangle , \quad (9)$$

where the subscripts r, l, etc. stand for the momenta in the right- and left-hand conductors. In Eq. (9) if we take $-l' = l$ and $-r' = r$, the term proportional to $C_r^+(t) C_l(t) C_l^+(0) C_r(0)$ gives rise to a current flowing from S_r to S_l, while the term $C_l^+(t) C_r(t) C_r^+(0) C_l(0)$ to one flowing from S_l to S_r, as shown in Chapter 21 by Schrieffer.

On the other hand, the two terms remaining in (9), if we take $l' = l$ and $r' = r$, are proportional to

$$\langle I \mid C_r^+(t) C_l(t) C_{-r}^{\pm}(0) C_{-l}(0) \mid I \rangle \quad (10a)$$

and

$$\langle I \mid C_l^+(t) C_r(t) C_{-l}^{\pm}(0) C_{-r}(0) \mid I \rangle ; \quad (10b)$$

these might appear to be identically zero if the number of particles on each side of the barrier were fixed in the initial state. But we observe that in the case of *zero* bias the system of two superconductors with n_l pairs in S_l and n_r pairs in S_r is degenerate, with $(n_l + m)$ pairs in S_l and $(n_r - m)$ pairs in S_r, where m is an integer. The electrostatic energy (proportional to m^2), which tends to split the degeneracy, is very small and will be neglected in this analysis.* Accordingly, the initial state constructed out of a coherent superposition of states with fixed $n_r + n_l$ but varying $m = \frac{1}{2}(n_l - n_r)$ is of the form

$$\mid I_\alpha \rangle = \sum_m e^{im\alpha} \mid I_m \rangle ,$$

where α is the phase of the coherent state I_α. The effect of electric and magnetic fields on α is discussed in Chapter 21 by Scalapino. If we define P_r (P_r^+) as an operator which adds (removes) a pair to (from) S_r without

* This point is discussed by Scalapino in Chapter 32 together with the effect of thermal fluctuations on the stability of the Josephson current.

creating or destroying any quasiparticle excitations, then two obvious properties of these operators are

$$P_l^+P_r \mid I_m\rangle = \mid I_{m+1}\rangle$$

and

$$e^{-i\alpha}P_l^+P_r \mid I_\alpha\rangle = \mid I_\alpha\rangle .$$

Using these two relations, we can write (10a) successively as

$$\langle I_\alpha \mid C_r^+(t)C_l(t)C_{-r}^+(0)C_{-l}(0)P_l^+P_re^{-i\alpha} \mid I_\alpha\rangle$$

and then*

$$e^{-i\alpha}\langle I_m \mid C_r^+(t)C_{-r}^+(0)P_rC_{-l}(0)C_l(t)P_l^+ \mid I_m\rangle . \tag{11}$$

In the second step we have neglected the fact that operators at different times cannot be trivially exchanged in position. However, this technical difficulty is easily surmounted by a more sophisticated formalism ([8,9]) which need not be invoked in our qualitative argument. In fact, the evaluation of (10a) need not be continued, since the integral of (11) over all time is clearly finite. Specifically, the current arising from (10a) turns out to be given by $+ie^{-i\alpha}J_s/2$, while that from (10b) by $-ie^{i\alpha}J_s/2$, so that the total current is given by $J_s \sin \alpha$. Here J_s is proportional to $\mid T_{rl} \mid^2$, as noted before, and is a function of Δ, going to zero as Δ does.

Most interesting is the case when a finite bias voltage is applied. Then the initial states $\mid I_m\rangle$ are no longer degenerate, but are separated in energy by $2V$; hence any pair current between S_l and S_r will be an AC current with angular frequency $2V/\hbar = \omega_J$. There are several aspects of the AC Josephson effect which should be noted.

1. The AC current is generally complex, the real and imaginary parts being related by Kramers–Kronig relations. The imaginary part of current is zero until $\hbar\omega = \Delta_l + \Delta_r$, where there is a discontinuous onset in the imaginary part with a corresponding singularity in the real part of the current, which was pointed out by Riedel ([10]). A corresponding singularity in the real part of the AC current associated with conventional one-particle tunneling processes has been noted by Werthamer ([9]). In this case a direct transition corresponding to a conventional one-particle tunneling process

* The calculation now proceeds by the insertion of a complete set of states between the operators P_r and $C_{-l}(0)$. However, since only the state I_m of this set has a finite matrix element, Eq. (11) can be reduced to properties of the ground state of a superconductor. For details see ([8]).

assisted by a photon gives rise to a discontinuous onset in the imaginary part of the current.

2. An AC current produces an AC electric field in the sandwich—primarily in the oxide, but also "leaking" into the superconductors. The current is then driven by the total potential—i.e., by the applied DC voltage V plus the generated AC voltage $V_1 \sin \omega_J t$, where V_1 must be determined self-consistently. Since $\exp(iA \sin \omega_J t) = \Sigma_n J_n(A) \exp(in\omega_J t)$, the radiation in the sandwich occurs at multiples of the Josephson frequency.

The modulation of the AC current by the induced radiation gives rise to a DC component of the current which, as Werthamer has suggested, may have peaks when the voltage is such that the singular part of the AC current is strongly coupled to the radiation.

A useful diagram for predicting the values of the voltage at which these peaks in the DC current may appear is shown in Fig. 10(a). If there were no induced radiation, only an AC current would flow for $V < \Delta_l + \Delta_r$. However, a DC current could flow if the radiation supplied the additional energy required to conserve energy in the process—specifically, if

$$2V + n(2V) = \Delta_r + V + \Delta_l, \qquad \text{or} \qquad V = (\Delta_l + \Delta_r)/(2n + 1).$$

This mechanism produces peaks at only *odd* subharmonics of $\Delta_l + \Delta_r$. Another mechanism is the absorption of induced radiation, which "leaks" into one of the superconductors to promote a tunneling pair into quasi-particle states. This mechanism, shown in Fig. 10(b), produces a peak in the DC current when

$$2V + n(2V) = 2\Delta_r, \qquad \text{or} \qquad V = 2\Delta_r/2(n + 1).$$

Similarly, there is another even series at $\Delta_l/(n + 1)$. It should be noted that all these mechanisms are speculative, since no detailed calculations have been performed.* It is not even clear that the peaks will have the right shape or size.† [We note that there are multiple-particle tunneling processes which could, in principle, produce subharmonic structure. For example, a second-order process ($\sim |T|^4$), involving a simultaneous Josephson tunneling and a conventional one-particle process, would onset at $V = (\Delta_l + \Delta_r)/3$. Such a process has presumably never been calculated.]

Rowell has recently done a more systematic investigation of the sub-

* Interesting effects occur if the AC radiation is applied externally. For this and other effects see footnotes 14–16 in (9).

† See the discussion on the last page of (9).

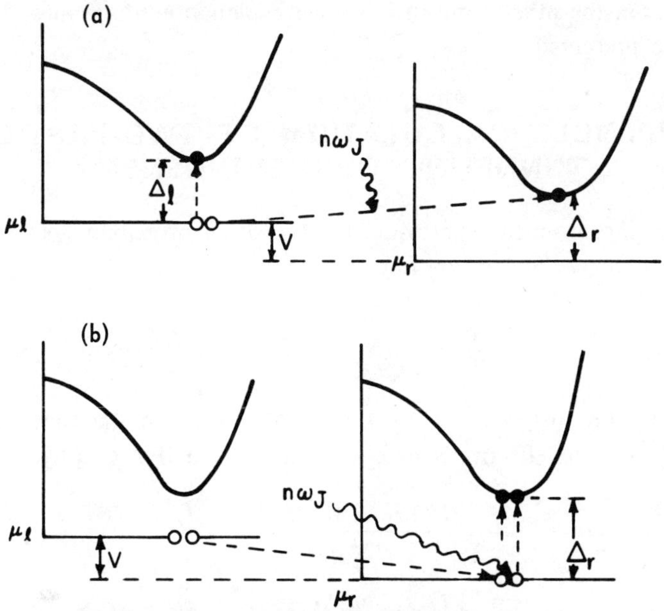

Fig. 10. Diagrams for predicting subharmonic structure in *I*–*V* characteristics. Photons at multiples of Josephson frequency $\omega_J = 2V/\hbar$ couple with AC currents to give a DC current. Peaks in the DC current may be related via the singularity in the real part of the AC current to the discontinuous onset of the imaginary part at $\hbar\omega = \Delta_r + \Delta_l$. These diagrams bypass such analysis by focusing directly on a possible photon-assisted DC process. The presumption is that the peak occurs at the onset of the process. (*a*) A Josephson photon in the barrier assists a conventional one-particle process. Energy conservations gives a peak at $(\Delta_r + \Delta_l)/(2n + 1)$. (*b*) A Josephson photon "leaks" into superconductor S_r to promote a tunneling AC Josephson pair to quasi particle states. Peak is at $\Delta_r/(n + 1)$.

harmonic structure of superconducting tunnel junctions ([11]). In junctions with large subharmonic excess currents he finds evidence of metallic shorts which could serve as good radiators at multiples of the Josephson frequency. Symmetric junctions have structure at $2\Delta/n$, as reported by other investigators. However, in asymmetric junctions he finds structure at $\Delta_l + \Delta_r$, Δ_l, Δ_r, and Δ_l/n, where Δ_l is the smaller gap, but *not* at $(\Delta_l + \Delta_r)/(2n + 1)$ or Δ_r/n as might be predicted.

In conclusion, it seems likely that most of the "multiparticle," or, preferably, subharmonic, tunneling structure which has been observed is connected with the interaction of AC currents with induced Josephson radiation. On the other hand, no detailed explanation along these lines has yet been presented.

APPENDIX. CALCULATION OF TWO-PARTICLE TUNNELING CURRENT DENSITY

We start from the expression for transition probability per unit time for the process depicted in Fig. 5,

$$\frac{2\pi}{\hbar} \sum_F |\langle F \mid H_{\text{eff}} \mid I\rangle|^2 \, \delta(E_F - E_I), \tag{A.1}$$

where, by Eq. (6), $E_F - E_I = E_k^r + E_{k'}^r - 2V$, and the sum is over $\mathbf{k}$ and $\mathbf{k}'$. Here the effective matrix element is given [Eq. (7)] by

$$\begin{aligned} \langle F \mid H_{\text{eff}} \mid I\rangle &= \sum_M \frac{\langle F \mid H_T \mid M\rangle\langle M \mid H_T \mid I\rangle}{E_M - E_I} \\ &= \sum_p T_{kp} T_{pk'} u_k^r u_{k'}^r \frac{\Delta_l}{(E_p{}^l)^2 - (V - E_k{}^r)^2}. \end{aligned} \tag{A.2}$$

Assuming a barrier appropriate to the WKB approximation and specular transmission, Harrison ([12]) showed that

$$|T_{kp}|^2 = \frac{1}{4\pi^2} \frac{\delta_{k_\parallel, p_\parallel}}{\varrho_\perp{}^r \varrho_\perp{}^l} \exp\left\{-2 \int_{x_r}^{x_l} \varkappa_\perp(x)\, dx\right\},$$

where $\varrho_\perp^{(r,l)}$ are the one-dimensional densities of states in metals (r, l) in a direction perpendicular to the barrier. The delta function in the components of $\mathbf{k}$ parallel to the barrier reflects the assumption of specular transmission. In the exponential x_r and x_l refer to the classical turning points, for a given energy of the tunneling particle, of the barrier potential $u(x)$ in metals l and r, respectively. Here $\varkappa_\perp(x)$ is given by $[2mu(x) - p_\parallel{}^2]^{1/2}/\hbar$.

Accordingly, the sum over $\mathbf{p}$ collapses into a sum over $p_\perp > 0$ and can be written as

$$\sum_{p_\perp > 0} \frac{1}{\varrho_\perp{}^l} \frac{\Delta_l}{(E_p{}^l)^2 - (V - E_k{}^r)^2} = \frac{1}{2} \int_{-\infty}^{\infty} d\varepsilon_p{}^l \frac{\Delta_l}{(\varepsilon_p{}^l)^2 + [\Delta_l{}^2 - (V - E_k{}^r)^2]}.$$

This integral can be evaluated simply within the BCS model, where the energy gap parameter is taken as a constant for all energies (relative to the chemical potential) which are less than some cutoff ω_c and as zero elsewhere. Then for $V - E_k^r < \Delta_l$, which is the interesting region, the integral is given by

$$\frac{\pi}{2} \frac{\Delta_l}{\Delta_l^2 - (V - E_k^r)^2} + O\left(\frac{\Delta_l}{\omega_c}\right). \tag{A.3}$$

The second term is neglected. We can now substitute (A.3) into (A.2) and square it to complete the calculation of (A.1). This would give the tunneling rate into the sum of all final quasiparticle states $\mathbf{k}\uparrow$ and $\mathbf{k}'\downarrow$ in S_r. There is an equivalent process into the final states $\mathbf{k}\downarrow$ and $\mathbf{k}'\uparrow$ in S_r. Thus the expression for the total transition probability is (A.1) multiplied by two. Thus

$$\omega_{r\leftarrow l} = \frac{2\pi}{\hbar} \frac{2\Delta_l^2}{(8\pi)^2} \sum_{k_\parallel} \exp\left\{-4 \int_{x_r}^{x_l} \varkappa_\perp(x)\, dx\right\} I \tag{A.4}$$

where*

$$I = \sum_{k_\perp > 0} \frac{(u_k^r)^2}{\varrho_\perp^r(k)} \sum_{k_\perp' > 0} \frac{(u_{k'}^r)^2}{\varrho_\perp^r(k')} \frac{\delta(2V - E_k^r - E_{k'}^r)}{\Delta_l^2 - (V - E_k^r)^2}. \tag{A.5}$$

Using those arguments we can write

$$\sum_{k_\parallel} \exp\left\{-4 \int_{x_l}^{x_r} \varkappa_\perp(x)\, dx\right\} = \frac{A_\parallel}{4\pi} \beta t^2,$$

where $t = \exp(-2d(2m\varphi)^{1/2}/\hbar)$ and $\beta = [(2m\varphi)^{1/2}/\hbar + 1/(4d)]/2d$. Here $A_\parallel$ is the area of the barrier, φ the effective metal–barrier work function, and d the effective thickness of the barrier.

Since except for $(u_k)^2$ and $(u_{k'})^2$ all factors in the integrand of I are even in ε_k^r and $\varepsilon_{k'}^r$, the u^2 each contribute a factor of $\frac{1}{2}$ to I. Thus (A.5) becomes

$$\begin{aligned} I &= \left(\frac{1}{2}\right)^2 \int_{-0}^{\infty} d\varepsilon_k \int_{-0}^{\infty} d\varepsilon_{k'} \frac{\delta(2V - E_k^r - E_{k'}^r)}{\Delta_l^2 - (V - E_k^r)^2} \\ &= \frac{1}{4} \int_{\Delta_r}^{2V - \Delta_r} dE \frac{N_r(E) N_r(2V - E)}{\Delta_l^2 - (E - V)^2}, \end{aligned} \tag{A.6}$$

where $N_r(E) = E/(E^2 - \Delta_r^2)^{1/2}$ is the BCS density of states. The integral (A.6) can be done analytically. The transformation $E = V + (V - \Delta_r)X$ converts (A.6) into a sum of elliptic integrals.

We write down these analytic forms for the case $\Delta_l = \Delta_r = \Delta$. Then the process depicted in Fig. 6 contributes equally to the current density due to the two-particle tunneling process $J^{(2)}$ which is given by

* The sum over $k_\parallel$ has been separated out by arguments given in Appendix B of ([13]).

$$J^{(2)}(V) = \frac{e\Delta}{\pi\hbar}\beta\left(\frac{t}{16}\right)^2 A_2\left(\frac{V-\Delta}{\Delta}\right). \tag{A.7}$$

Here $A_2(X)$ is normalized to unity at $X = 0^+$ and is given by

$$\frac{\pi}{4}A_2(X) = \frac{1}{2+X}K\left(\frac{X}{2+X}\right) + X\Pi\left(-X^2, \frac{X}{2+X}\right),$$

where $K(k)$ and $\Pi(\varrho, k^2)$ are elliptic integrals ([14]). In Fig. 8 $\pi A_2[(V/\Delta) - 1]/4$ is plotted as a function of V. For comparison the current density for the single-particle tunneling process at zero temperature is given by

$$J^{(1)}(V) = \frac{e\Delta}{2\pi\hbar}\alpha\left(\frac{t}{16}\right)A_1\left(\frac{V-2\Delta}{2\Delta}\right), \tag{A.8}$$

where $\alpha = [(2m\varphi)^{1/2}/\hbar + 1/2d]/d \approx 2\alpha$. Here $A_1(X)$ is normalized to unity at $X = 0^+$ and is given explicitly by

$$\frac{\pi}{4}A_1(X) = (2+X)\,E\left(\frac{X}{2+X}\right) - \left(\frac{3+2X}{2+X}\right)K\left(\frac{X}{2+X}\right),$$

where $K(k)$ and $E(k)$ are elliptic integrals. Since $\beta \simeq \alpha/2$, the ratio of the two currents (A.7) and (A.8) at their respective thresholds is

$$J^{(2)}(\Delta)/J^{(1)}(2\Delta) = t/16.$$

REFERENCES

1. S. Shapiro *et al.*, IBM J. Res. Develop. 6:34 (1962).
2. B. N. Taylor and E. Burstein, Phys. Rev. Letters 10:14–17 (1963).
3. J. R. Schrieffer and J. W. Wilkins, Phys. Rev. Letters 10:17–20 (1963).
4. I. K. Yanson, V. M. Svistunov, and I. M. Dmitrenko, Soviet Phys.–JETP 20:1404 (1965).
5. S. M. Marcus, Phys. Letters 19:623 (1966); 20:236 (1966).
6. G. I. Rochlin and D. H. Douglass, Jr., Phys. Rev. Letters 16:359 (1966); A. Zawadowski, Phys. Letters 23:225 (1966).
7. Unpublished work of I. Giaever *et al.*, J. M. Rowell, and others.
8. V. Ambegaokar and A. Baratoff, Phys. Rev. Letters 10:486 (1963).
9. N. R. Werthamer, Phys. Rev. 147:255 (1966).
10. E. Riedel, Z. Naturforsch. 19A:1634 (1964).
11. J. M. Rowell, this volume.
12. W. A. Harrison, Phys. Rev. 123:85 (1961).
13. D. J. Scalapino, J. R. Schrieffer, and J. W. Wilkins, Phys. Rev. 148:263 (1966).
14. W. Gröbner and N. Hofreiter, Integraltafel, Springer-Verlag, Vienna, 1961.

Chapter 25

Photon-Assisted Single–Particle Tunneling Between Superconductors

Glen E. Everett

University of California
Riverside, California
and U. S. Naval Ordnance Laboratory
Corona, California

INTRODUCTION AND SUMMARY OF EXPERIMENTAL RESULTS

Before attempting to describe what effect a microwave photon might be expected to have on the tunneling current I would like to review briefly the gross features of the usual single-particle tunneling between dissimilar superconductors. If we assume that the energy gaps are $\varepsilon_{g1} = 2\Delta_1$ and $\varepsilon_{g2} = 2\Delta_2$ and for convenience take $\Delta_1 > \Delta_2$, then for $T = 0°$K there will be no tunneling current for an applied voltage $eV_0 < \Delta_1 + \Delta_2$. At $eV_0 = \Delta_1 + \Delta_2$ there will be an abrupt rise, with an approximately linear increase in the current for all values of $eV_0 \geq \Delta_1 + \Delta_2$. The effect of finite temperatures on this abrupt rise, or tunneling edge, will be to broaden the transition and give it a definite width. In addition, due to the thermal excitation of electrons across the gap there will be electrons able to tunnel at voltages less than $eV_0 = \Delta_1 + \Delta_2$, giving rise to a cusp in the current with the peak occurring at $eV_0 = \Delta_1 - \Delta_2$. The magnitude of the current below the tunneling edge, being related to the number of thermally excited electrons, can be made arbitrarily small by going to progressively lower temperatures.

If a microwave field is applied to the superconductors, and if we assume that the photon energy $\hbar\omega$ is much less than the energy gap for either superconductor in order to avoid the added complication of direct excitation across the gap, then at zero applied voltage one-photon absorption processes would not be allowed. If the applied voltage were $eV_0 = \Delta_1 + \Delta_2 - \hbar\omega$, as illustrated in Fig. 1(a), the electron could be excited to a virtual state in

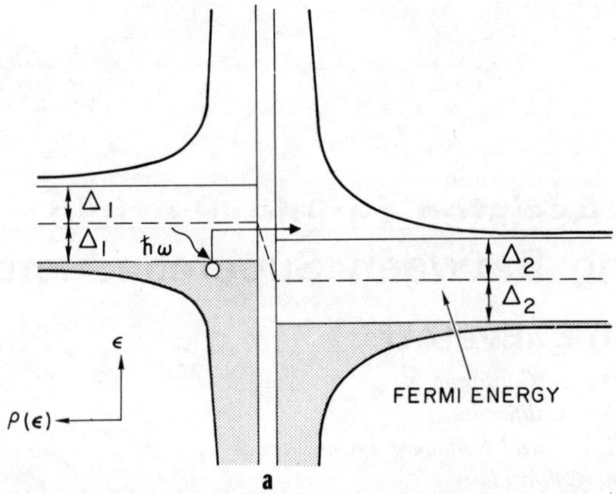

Fig. 1 (*a*). Schematic illustration of photon-assisted tunneling at an applied bias $eV_0 = \Delta_1 + \Delta_2 - \hbar\omega$. The electron, by absorbing a photon, is excited into a virtual state in the gap and subsequently tunnels into a region of high density of final states.

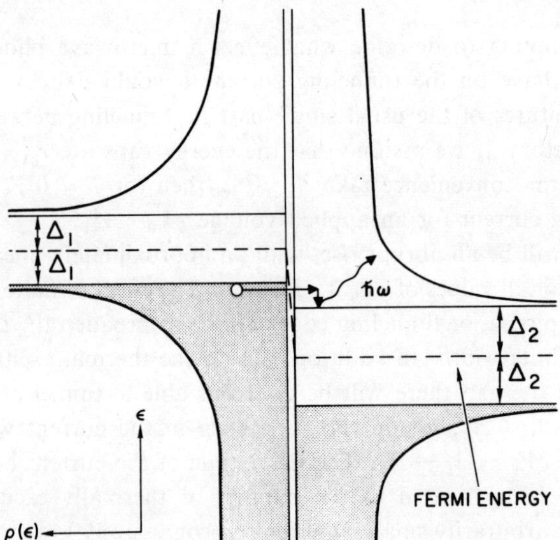

Fig. 1 (*b*). Schematic illustration of photon-assisted tunneling at an applied bias $eV_0 = \Delta_1 + \Delta_2 + \hbar\omega$. The electron tunnels into the second superconductor, and then by stimulated emission it loses the energy $\hbar\omega$. Since the density of final states is higher, an enhanced tunneling probability results.

the gap by absorbing one photon and then make an allowed tunneling transition into the second superconductor, and the process would be allowed. We would expect to see the onset of tunneling at $eV_0 = \Delta_1 + \Delta_2 - \hbar\omega$. Similarly, we would expect to see an effect at $eV_0 = \Delta_1 + \Delta_2 + \hbar\omega$, since by first tunneling and then emitting a photon by stimulated emission the electron would see a greater density of final states, as illustrated in Fig. 1(*b*). For the case of $T \neq 0°$ it should also be possible to observe such processes in the vicinity of the cusp in the current $eV_0 = \Delta_1 - \Delta_2 \pm \hbar\omega$.

This effect has been observed, and was first reported by Dayem and Martin ([1]). The effect of a microwave field on the single-particle tunneling was subsequently observed and used by Giaever ([2]) as a detector for the radiation emitted by a Josephson junction. In the work of Dayem and Martin the features described above were observed, but in addition there were steps at $eV_0 = \Delta_1 + \Delta_2 \pm n\hbar\omega$, where at the microwave power levels used n was as large as 5 or 6. The interpretation of these steps with $n \neq 1$ would then be that of multiple-photon absorption and emission processes.

The nature of the photon-assisted single particle tunneling is shown in Fig. 2 for the case of a tin–lead junction at 1.4°K. The current I has been plotted versus the voltage V with and without microwave power. For no power the usual tunneling edge at about 1.9 mV is observed. With micro-

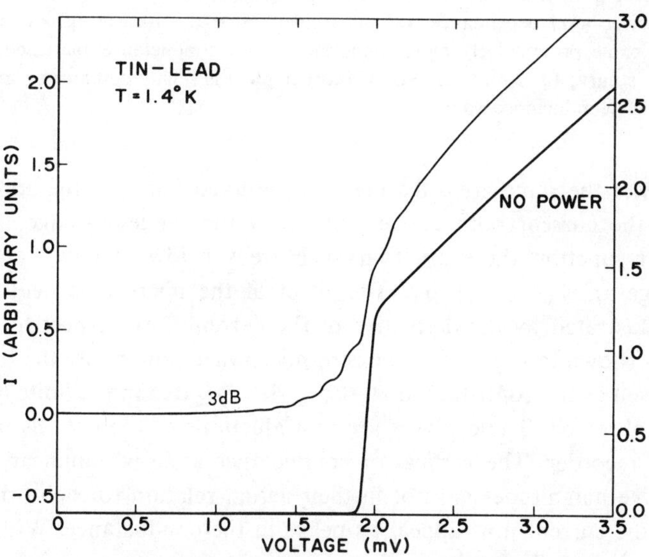

Fig. 2. The current versus voltage for a Sn–Pb junction with and without microwave power; the point of zero current has been displaced for clarity.

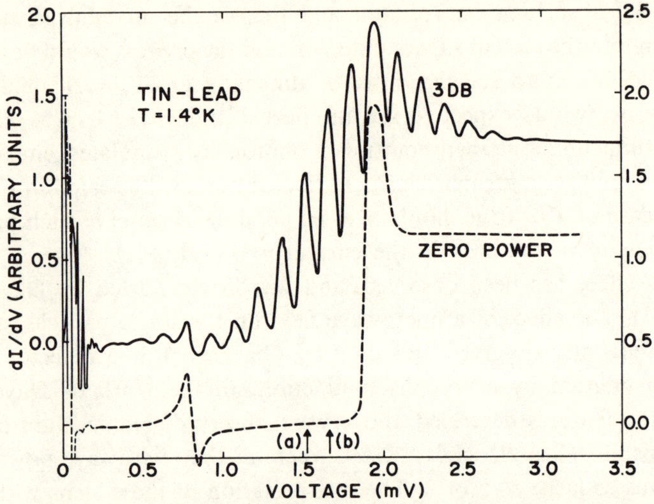

Fig. 3. The conductance dI/dV versus voltage with and without microwave power for the same Sn–Pb junction as shown in Fig. 2. The peak in the zero power conductance at about 1.9 mV is the tunneling edge. The effect of the microwave power has been to induce additional peaks in the conductance, the spacing being the photon energy $\hbar\omega/e$. The arrows (*b*) and (*a*) indicate the peaks at two and three photon energies below the tunneling edge, for which the detailed power level dependance was determined. Sine the measuring circuit became progressively more nonlinear as the conductance increased, this curve (a reproduction of a recorder plot) is highly compressed at high conductance values.

wave power the same gross features are observed but with the addition of steps in the current both above and below the tunneling edge. For this particular junction there are four such steps below and two above the main edge at $eV_0 = \Delta_1 + \Delta_2$. The effect of the microwave field is more clearly illustrated by the derivative of the current with respect to voltage, dI/dV, as shown in Fig. 3. For the zero microwave power case the tunneling edge as well as the contribution at $eV_0 = \Delta_1 - \Delta_2$ due to the finite temperature are observed. These curves are reproductions of the data as taken on the X–Y recorder. The vertical or conductance scale is nonlinear, so that the relative magnitudes are not in their actual relation to each other. The steps in the current now appear as peaks in the conductance. With microwave power, the photon-induced peaks are clearly observable out to $n = 6$ above the tunneling edge, and while they are clearly observable below the edge, there is obviously some sort of interference effect around $eV_0 = \Delta_1$

$-\Delta_2$. Above the edge the position of the induced peaks is given exactly by $eV_0 = \Delta_1 + \Delta_2 + n\hbar\omega$, there being no phase shift or constant displacement of the peaks with respect to the tunneling edge. Also shown in this figure near zero bias is the microwave interaction with the Josephson currents ([3]). Their width and positions clearly indicate that this is an entirely different process, about which nothing more will be said beyond the indication of their presence. Finally, the arrows labeled (*a*) and (*b*) denote the peaks at $N = 3$ and $N = 2$ photon energies below the tunneling edge, for which careful and detailed power level dependences were taken. This choice was convenient because there is no zero microwave conductance to consider.

The interference effect noted in Fig. 3 was observed by Dayem and Martin ([1]) as a phase shift or inexact fitting of the peaks to the expression $eV_0 = \Delta_1 + \Delta_2 - n\hbar\omega$ below the tunneling edge. Their data was for indium–aluminum films in which the energy gap E_g(Al) for the aluminum is almost exactly equal to twice the photon energy used. Figure 4 shows the effect of carefully adjusting the microwave power level such that while the photon-induced processes at $\Delta_1 + \Delta_2$ and $\Delta_1 - \Delta_2$ are clearly observable, the power is low enough so that the two sets of peaks do not overlap. This shows that there are photon-assisted processes associated with the cusp at

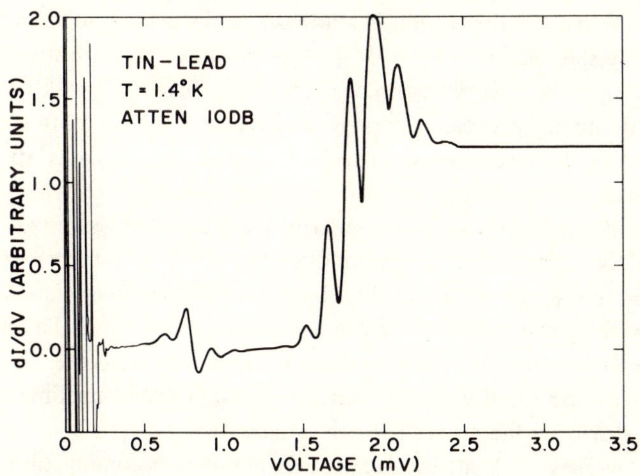

Fig. 4. The conductance of a Sn–Pb junction versus voltage for which the microwave power level has been adjusted to demonstrate the distinct effects of the photon interaction associated with the tunneling edge at 1.9 mV and the current cusp at 0.7 mV. The additional structure at lower bias voltages is the microwave interaction with the Josephson currents.

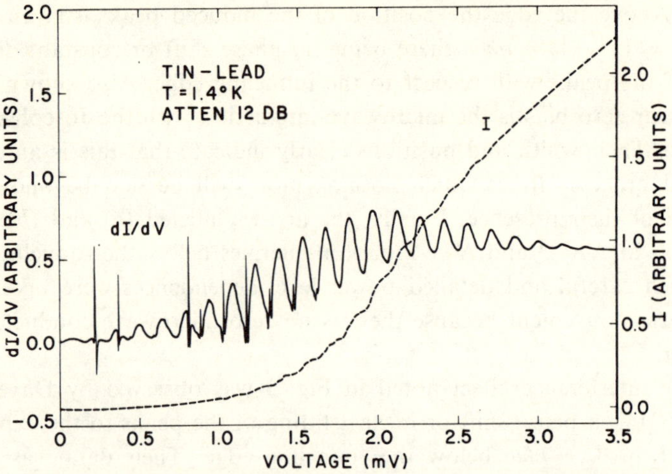

Fig. 5. The tunneling current and conductance versus voltage at a high microwave power level. This illustrates the general decrease in the magnitude of the peak conductance and also the monotonic decrease in amplitude of successively higher-order peaks. The superimposed spikes in the conductance are the result of the microwave interaction with the Josephson current.

$\Delta_1 - \Delta_2$. When this separation is made the peaks below the main tunneling edge satisfy the relation $eV_0 = \Delta_1 + \Delta_2 - n\hbar\omega$ exactly with no phase shift required. Thus we have shown that there is no phase shift with the photon-assisted tunneling process, although an apparent phase shift can appear when the processes associated with $\Delta_1 + \Delta_2$ and $\Delta_1 - \Delta_2$ overlap at higher power levels.

The limiting behavior of the microwave interaction is shown in Fig. 5 for both the current I and the conductance dI/dV. At higher microwave power levels the number of peaks observable in the conductance becomes progressively greater; this curve does not represent a limit. The amplitude of the main conductance peak at the tunneling edge decreases, while those induced increase rapidly with power, go through broad maxima, and then decrease slowly as the power is further increased. The peaks corresponding to higher values of N all behave in this manner, becoming observable at successively higher power levels but never becoming equal in amplitude to a peak of lower index N. It is to be noted in anticipation of the next section that we have selected particular peaks and measured their power level dependence carefully to determine if there is an interval in which their amplitude might approach zero. This behavior has not been observed. One additional point is worth noting, namely, the very sharp lines occurring on this curve. These are the result of the photon interaction with the Josephson current. I shall not discuss them further except to note their very narrow width and different character from that of the photon-assisted single-particle tunneling effect.

EXPERIMENTAL TECHNIQUES

These experiments were performed on films prepared in the usual manner, namely, the first strip was evaporated, and allowed to oxidize and then the second strip was deposited in such a way as to form a cross. The substrate for these films was a very thin sheet of Mylar stretched tightly over an aluminum plate. This combination of junction and backing plate then formed the end wall of a cylindrical resonator operating in a TE_{111} mode, as shown in Fig. 6. The operating frequency was 36.0 GHz. The choice of Mylar as a substrate is particularly convenient because of its low dielectric loss and the fact that thin sheets are very easy to work with. This made possible the mounting of the junction without a significant

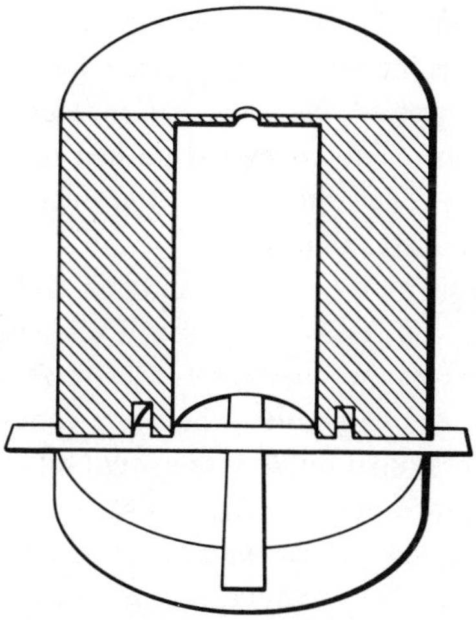

Fig. 6. A sectional view of the microwave cavity showing the tunneling junction in position against the end wall. The superconducting strips are deposited on a thin sheet of Mylar and further insulated from the cavity as described in the text, the details having been omitted for clarity.

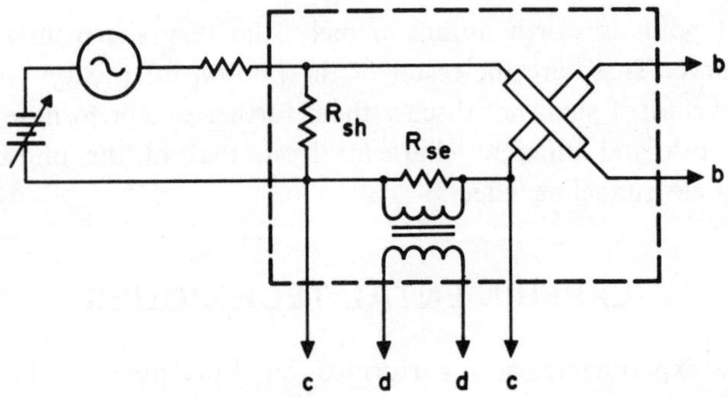

Fig. 7. Simplified diagram of the measuring circuit. The components inside the dashed lines are located in the Dewar and as close to the junction as possible. The conductance is computed from the voltage measured across the terminals (d, d).

reduction of the cavity Q. With this configuration the junctions were located in a region of maximum rf magnetic field and minimum electric field. While with suitable modifications in design the junction could be located in a region of maximum electric field, this has not been done. The entire assembly is then mounted on a section of low thermal conductivity waveguide and suspended in a Dewar, the liquid helium being allowed to fill the cavity and come in intimate contact with the junction.

The essential features of the electrical measuring circuit are shown in Fig. 7, those portions enclosed by the dashed line being located in the Dewar near the junction. The voltage across the terminals (b, b), the junction voltage, is suitably amplified and used to drive the X axis of an X–Y recorder. The current I or (by using a transformer and an additional small-amplitude AC excitation in connection with a phase-sensitive detector) the conductance dI/dV was measured across the small series resistor R_{se}. The tendency of the circuit to oscillate in regions of bias where the conductance was negative was eliminated by approximating with the shunting resistor R_{sh} a constant voltage source at the junction. With R_{sh} and R_{se} sufficiently small such that their reciprocals were always larger than the magnitude of dI/dV the circuit is always stable and data could be obtained for all values of the bias voltage. Finally, I note that the magnitude of the AC voltage across the junction depends on the value of conductance of the junction relative to that of the series resistor R_{se}. This quantity depended on the bias voltage for the values we used, necessitating a calibration to convert apparent into absolute conductance. The data shown are as taken directly, the correction having been applied in the power level dependence of the conductance.

TIEN-GORDON MODEL

Following the work of Dayem and Martin (1) a theoretical treatment of this effect was given by Tien and Gordon (4). They considered two possible formulations of the problem, one based on the vector potential **A** and the other assuming that the total effect of the microwave field was to induce a time-dependent potential difference between the two metals. Since the two treatments lead to essentially the same qualitative results, we will summarize the results for just the latter case. The approach is to assume that one side of the barrier is held at a fixed potential, the other side being modulated with respect to it by an amount

$$eV_m(t) = eV_m \cos \omega t, \tag{1}$$

as shown schematically in Fig. 8. This additional potential energy is added to the original Hamiltonian and the new wave functions are calculated. From these wave functions the new density of states function is determined to be

$$\varrho'(E) = \sum_{n=-\infty}^{\infty} \varrho(E + n\hbar\omega) J_n^2(\alpha), \tag{2}$$

where $\varrho(E)$ is the original density of states in the absence of a microwave field, $J_n(\alpha)$ is the usual Bessel function of order n, and α is given by $\alpha = eV_m/\hbar\omega$. Inserting this into the expression for the tunneling current, they get

$$I = C \sum_{n=-\infty}^{\infty} J_n^2(\alpha) \int_{-\infty}^{\infty} [f(E-eV_0) - f(E+n\hbar\omega)]\varrho_1(E-eV_0)\varrho_2(E+n\hbar\omega)\, dE, \tag{3}$$

$$I = C \sum_{n=-\infty}^{\infty} J_n^2(\alpha) I_0(eV_0 + n\hbar\omega), \tag{3'}$$

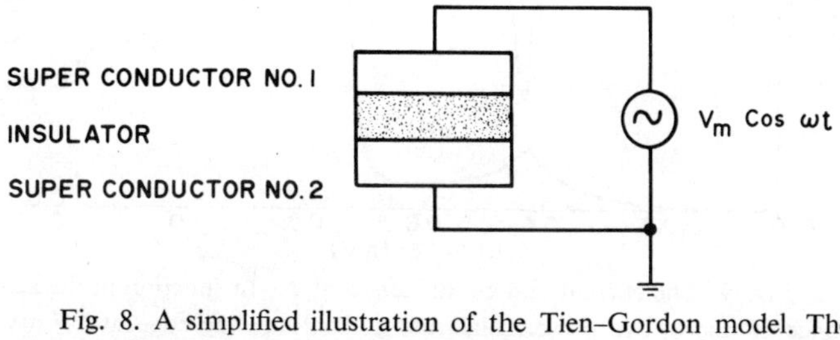

Fig. 8. A simplified illustration of the Tien–Gordon model. The second superconductor is regarded as being held at a fixed potential and the first modulated with respect to it.

where $I_0(eV_0)$ is just the usual tunneling current in the absence of a microwave voltage. Qualitatively, this predicts a tunneling current that is the superposition of a series of current steps occurring at $eV_0 = \Delta_1 + \Delta_2 \pm n\hbar\omega$, each step having an amplitude proportional to $J_n^2(\alpha)$. Alternatively, if we calculate the conductance by taking the derivative, we get

$$\frac{dI}{dV_0} = \sum_{n=-\infty}^{\infty} J_n^2(\alpha) \frac{d}{dV_0} [I_0(eV_0 + n\hbar\omega)]. \tag{4}$$

The conductance should consist of a series of peaks of amplitude $J_n^2(\alpha)$ displaced from the main conductance peak at $eV_0 = \Delta_1 + \Delta_2$ by an amount $\pm n\hbar\omega$, each induced peak having exactly the same shape as the original peak in the conductance.

We have measured the tunneling in aluminum–indium junctions; an example of the current and conductance in the absence of a microwave field is shown in Fig. 9. The usual features are the main tunneling edge at about 0.7 mV and the cusp in the current due to the finite temperature at about 0.4 mV. In particular, we would like to call attention to the shape of the conductance, its sharp peak followed by a broad but lower amplitude peak, without digressing to consider the origin. In the presence of a micro-

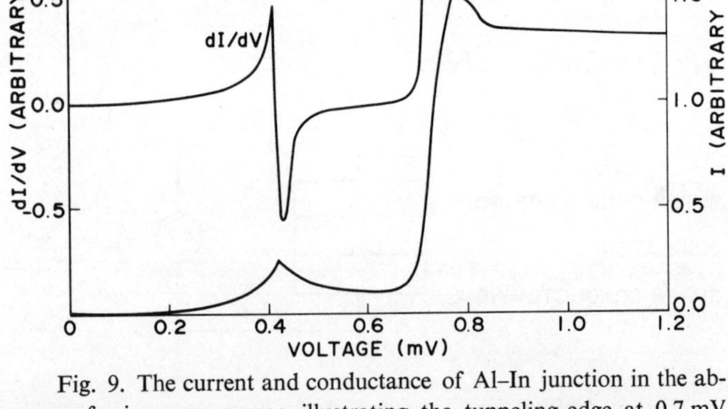

Fig. 9. The current and conductance of Al–In junction in the absence of microwave power, illustrating the tunneling edge at 0.7 mV and the current cusp at about 0.4 mV. The zeros of I and dI/dV have been displaced for clarity.

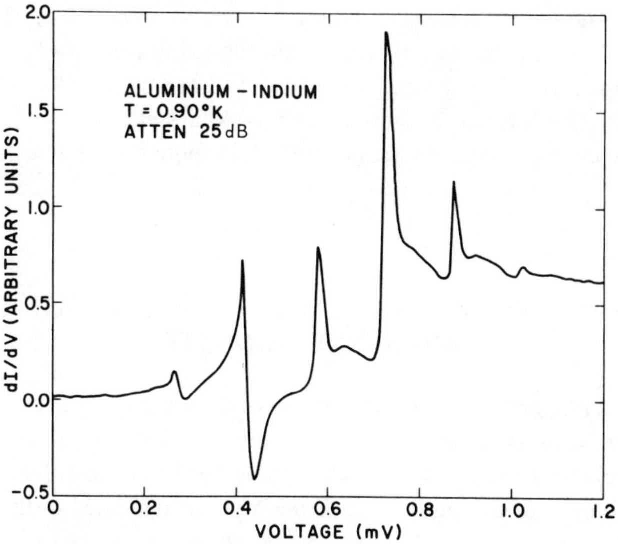

Fig. 10. The conductance of the same Al–In junction as shown in Fig. 9 in a microwave field. The photon-induced peaks above and below the tunneling edge are more clearly resolved than for Sn–Pb junctions and display the characteristic shape of the conductance at the tunneling edge.

wave field the conductance has the same structure of induced peaks spaced just a photon energy apart, as shown in Fig. 10, but in this system the peaks are better resolved. The peaks at one photon energy above and below the tunneling edge can be seen to have the same shape as the main conductance peak, with even the additional broad peak being reproduced in the induced structure. In addition, the shape of the peak at about 0.25 mV, one photon energy below the current cusp at $\Delta_1 - \Delta_2$, reproduces the shape of this structure. We can conclude, therefore, that the predictions of the Tien–Gordon model as regards the positions and shapes of the photon-induced peaks are in very good agreement with experiment. The dependence of the magnitude of the conductance peaks, as shown in Fig. 5, is poor. Equation (4) predicts that the amplitude of the N-photon process should be proportional to $J_N^2(\alpha)$. For large α and a particular value of N this should allow this peak to be of lesser magnitude than the one immediately above or below it. Tien and Gordon ([4]) have adjusted α to give the approximate magnitude of the steps for particular microwave power levels and find that the values of the microwave fields required would be much larger than those actually used in the experiment. In particular, a reasonable fit to most of one set of experimental data is obtained with $\alpha = 2$. This results in $J_0^2(\alpha = 2) \approx 0$, predicting that the main tunneling edge at $eV_0 = \Delta_1 + \Delta_2$ should go to zero, a result that is not observed experimentally. The results of Tien and Gordon can be summarized as accurately predicting the positions of the current steps and qualitatively describing the relative amplitudes but proving inadequate in describing either the detailed qualitative or quantitative results.

MODIFIED THEORETICAL MODEL AND COMPARISON WITH EXPERIMENT

The Tien–Gordon model for the microwave interaction with the tunneling current assumes that one of the superconductors can be regarded as being held at a fixed potential while the other has a time-dependent potential difference measured relative to the first. As indicated in the previous section, this leads to a modulation in the density of states for this superconductor and is given by Eq. (2), with the current given in Eq. (3). This modulation applies to both the occupied and unoccupied states, but in the Tien–Gordon theory it is assumed that only one superconductor, without specifying which one, is so affected. Since this reference of zero potential is somewhat arbitrary, we have elected to assume that both superconductors feel an oscillating potential, the reference being chosen to make this amplitude equal in each metal, as shown in Fig. 11. As a result, the density of states in both superconductors is modulated and has the form of Eq. (3). The resulting form of the density of states for dissimilar superconductors is illustrated schematically in Fig. 12 by the solid lines, the dashed lines representing the situation at zero applied microwave power. This represents the behavior at zero bias and at $T = 0°K$. (We have elected

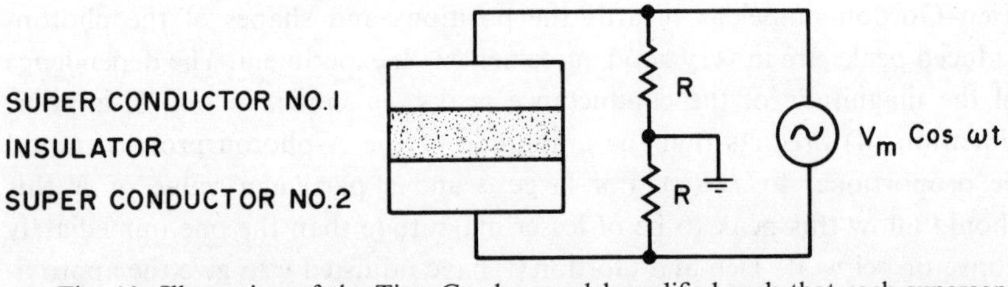

Fig. 11. Illustration of the Tien–Gordon model modified such that each superconductor is voltage-modulated with respect to a fixed reference potential. For this case the reference has been chosen so that each superconductor is modulated by the same amount.

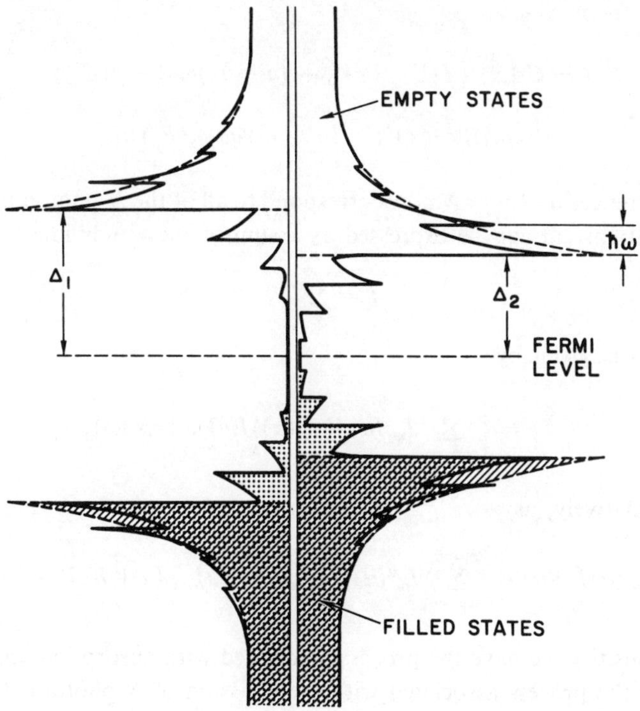

Fig. 12. The density of states of the superconductors resulting from the microwave modulation. The dashed lines represent the original density of states.

to illustrate the peaks in the density of states as being sharp with well-defined edges but finite in their magnitudes for convenience in illustration and to reflect the finite conductance observed experimentally.) Inserting this density of states into the usual integral for the tunneling current, we obtain

$$I' = C \sum_{m,n=-\infty}^{m,n=\infty} J_m^2(\alpha) J_n^2(\alpha) \int_{-\infty}^{\infty} [f(E + m\hbar\omega - eV_0) - f(E + n\hbar\omega)] \times \varrho_1(E + m\hbar\omega - eV_0)\varrho_2(E + n\hbar\omega)\, dE. \tag{5}$$

This result is more convenient when transformed and rewritten in terms of one integral representing the current as

$$I' = \sum_{n,m=-\infty}^{n,m=\infty} J_n^2(\alpha) I[eV_0 + (n - m)\hbar\omega], \tag{6}$$

where

$$I = C\int_{-\infty}^{\infty} \{f[E' - eV_0 + (n-m)\hbar\omega] - f(E')\}$$
$$\times\, \varrho_1[E' - eV_0 - (n-m)\hbar\omega]\varrho_2(E')\, dE'. \tag{7}$$

The differences $n - m = N$ now correspond to all of the N photon processes. The total current can be expressed as a sum of these individual processes

$$I = \sum_N I_N, \tag{8}$$

where I_N is given by

$$I_N = \left\{\sum_{m=-\infty}^{m=\infty} J_m{}^2(\alpha)J_{N+m}^2(\alpha)\right\} I(eV_0 + N\hbar\omega), \tag{9}$$

or, alternatively, as

$$I_N = \left\{J_0{}^2(\alpha)J_N{}^2(\alpha) + \sum_{m=-\infty}^{m=\infty} J_m{}^2(\alpha)[J_{N+m}^2(\alpha) + J_{N-m}^2(\alpha)]\right\} I(eV_0 + N\hbar\omega). \tag{10}$$

For N positive we have the process associated with absorption and with N negative the process associated with the emission of N photons. For small values of α the behavior is dominated by the first term in Eq. (10), giving for the current

$$I_N \approx J_N{}^2(\alpha)I(eV + N\hbar\omega), \tag{11}$$

the result of Tien and Gordon ([4]). If we call the tunneling at $eV_0 = \Delta_1 + \Delta_2$ the $N = 0$ edge, this result indicates that when the modulation of the density of states is weak, i.e., low microwave powers, the N-photon process is dominated by the first-order process, namely, transitions from $n = 0$ in superconductor 1 to $n = N$ in superconductor 2 and $n = N$ in superconductor 1 to $n = 0$ in superconductor 2. The other contributions, being proportional to $J_m{}^2J_{N\pm m}^2$, do not contribute measurably. As α increases, the zero-order Bessel function does decrease toward zero, but at the same time the other contributions increase and keep the total current associated with the N-photon process from ever going to zero. While this is the behavior obtained by considering all of the possible contributions to the N-photon process, the total current is given by the sum of the various multiple processes as given by Eq. (8). We have computed coefficients of the various functions I_N as they appear in Eqs. (9) and (10). A further computation of the general function $I(eV_0 + N\hbar\omega)$ and its derivative dI/dV_0 can be obtained from Werthamer ([5]), with the usual result that at 0°K the conductance at $eV_0 = \Delta_1 + \Delta_2$ is discontinuous. For the contribution due to the other terms we have made use of the measured dependence of dI/dV such as is shown in Fig. 3 for zero microwave power. In particular, at one photon energy above the edge the conductance is essentially a constant. The total conductance was then computed as follows: the term under consideration, I_N of Eq. (10), was calculated taking $I(eV_0 + N\hbar\omega)$ as unity and to this was added the terms appropriate to $N + 1$, $N + 2$, etc., but multiplied by the ratio of the conductance at one photon energy above the edge to that at the edge as determined experimentally. This ratio, being essentially constant for all of the terms, is not determined with great precision owing to the extremely nonlinear response of our measuring system for very large values of conductance. It does lie between 0.05 and 0.20. Using this ratio as an adjustable parameter, the best fit to our data was obtained with a value of 0.15. This function has been indicated for $N = 0$, 1, 2, 3, and 4 in Fig. 13. Also plotted for purposes of comparison are $J_0^2(\alpha)$ and $J_2^2(\alpha)$, the expected

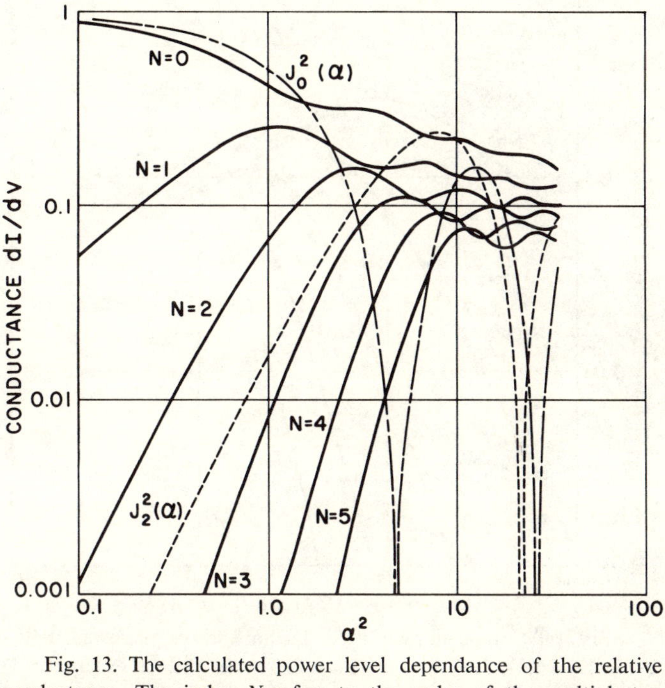

Fig. 13. The calculated power level dependance of the relative conductance. The index N refers to the order of the multiphoton process. The Bessel functions J_0^2 and J_2^2, the Tien–Gordon predicted dependance, are shown by broken lines for reference purposes.

Tien–Gordon results. For the specific case of $N = 2$ the essential feature predicted by Eq. (10) is that this function rises to a maximum value largely as a result of J_2^2, goes through a maximum for a 40% lower value of α than the Tien–Gordon prediction, and then decreases slowly, with some evidence of very shallow maxima and minima as α continues to increase. For $N = 1$ and higher values of N the behavior is very similar, the main maximum in the conductance occurring at a lower value of α for $N = 1$ and at progressively higher values for increasing values of N. The total response of the system is for any particular peak associated with an N-photon process to rise rapidly like $J_N^2(\alpha)$ for small values of α, to go through a broad maximum, and then to decrease slowly as the power is further increased. The detailed quantitative comparison shown in Fig. 14 indicates that the agreement with theory is good except at the very highest power levels used, where the conductance shows a tendency to saturate, but then continues to rise slowly. This rise is a consequence of finite temperature and heating effects.

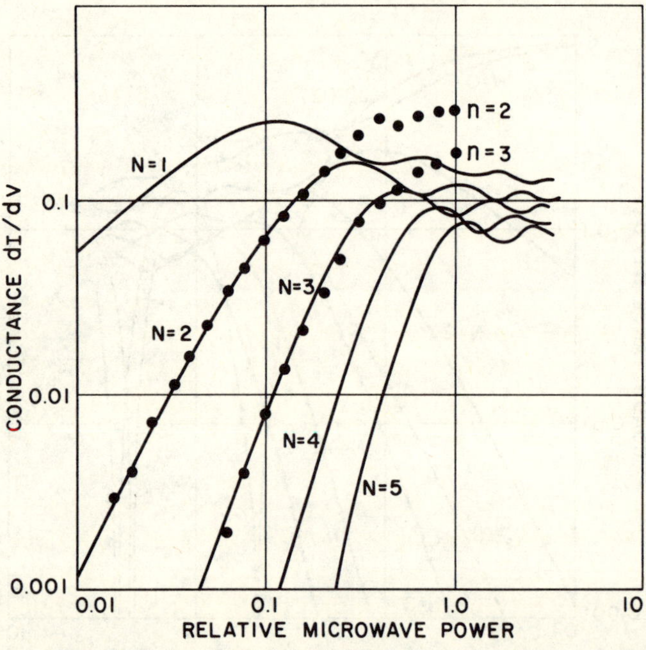

Fig. 14. Computed relative conductance versus microwave power for $N = 1$, 2, 3, 4, and 5. The solid circles are the experimental points for the peaks at two and three photon energies below the tunneling edge, i.e., at the positions, indicated by the arrows (*b*) and (*a*) in Fig. 3.

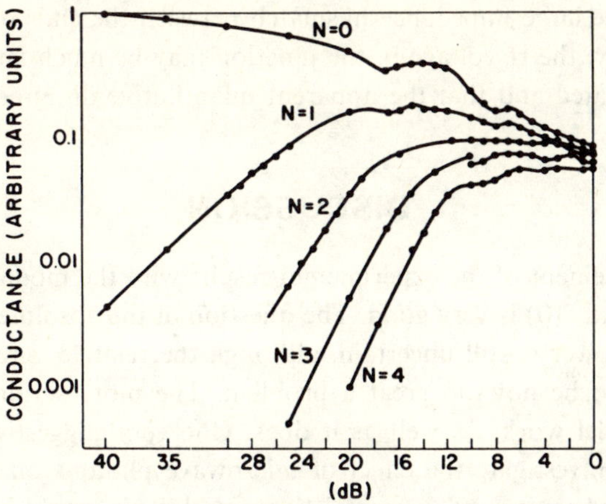

Fig. 15. The conductance versus power for peaks at $N = 0$, 1, 2, 3, and 4. In this junction there were no microwave heating effects. The solid lines drawn through the points are for clarity and do not represent the theoretical curves.

With increasing microwave power we have observed that some films can be driven normal due to microwave heating. The extra rise in the conductance is most likely due to the extra conductance below the tunneling edge that results when the temperature of the junction is caused to rise toward the transition temperature.

Subsequent to this work we have prepared additional junctions that have not exhibited the heating effect. Measurements of the power level dependence of the conductance for the peaks $N = 0, 1, 2, 3$, and 4 were made and the results are shown in Fig. 15. A solid line has been drawn through the experimental points for clarity, but a detailed fit with the theory has not been made. A comparison with the theoretical results shown in Fig. 13 indicates that the agreement is satisfactory. The remaining problem associated with this theory is related to the actual magnitudes of α, and hence to the electric fields required to produce these interactions. In their paper Tien and Gordon ([4]) found that the values of α required for an approximate fit to the magnitude of the tunneling steps were approximately two orders of magnitude larger than would be possible from considerations of the magnitude of the microwave power, the cavity Q, and the location of the sample in the cavity. Werthamer ([5]) on the other hand, has pointed out that the relevant voltage is the actual voltage in the tunneling junctions, and not that in an equivalent section of waveguide. He conjectures that because of the large impedance mismatch between the ends of the junction and the cavity, the rf voltage in the junction may be much larger than previously estimated and that the apparent quantitative disagreement may be spurious.

DISCUSSION

The agreement of the experimental results with the model as expressed in Eqs. (8) and (10) is very good. The question of the absolute values of the microwave power is still uncertain, although the relative agreement makes this appear to be not too great a problem. The more serious question is why our model works as well as it does. One very suggestive experiment would be to investigate the effect of microwave photons on the tunneling for the case where one of the metals is normal. This would have the effect of yielding just the original Tien–Gordon results. We have attempted some preliminary experiments in this direction but have been limited by fact that the width of the tunneling edge was too large at our lowest temperatures relative to the microwave photon energy. The effect is a lack of resolution that we are attempting to overcome by going to higher frequencies, although working at lower temperatures should also help to solve this problem.

A final comment on our model is in order. Its essential feature is that each side of the junction is behaving in the same way. This is somewhat difficult to justify on the basis of just a simple potential difference between the two superconductors. It would be much easier to understand if the action of the electric field **E** or the vector potential **A** inside the individual superconductors were the actual controlling mechanism for the interaction. This will require a complete reformulation of this problem and has not yet been done.

ACKNOWLEDGMENT

This work has been done in collaboration with Mr. Charles F. Cook of the U. S. Naval Ordnance Laboratory, Corona, California.

REFERENCES

1. A. Dayem and R. J. Martin, Phys. Rev. Letters 8:246 (1962).
2. I. Giaever, Phys. Rev. Letters 14:904 (1965).
3. S. Shapiro, A. Janus, and S. Holly, Rev. Mod. Phys. 36:223 (1964).
4. P. K. Tien and J. Gordon, Phys. Rev. 129:647 (1963).
5. N. Werthamer, Phys. Rev. 147:255 (1966).

Chapter 26

Phonon Generation and Detection by Single–Particle Tunneling in Superconductors*

W. Eisenmenger

III. Physikalisches Institut
der Universität Göttingen
Göttingen, West Germany

INTRODUCTION

In single-particle tunneling between superconductors energy from the voltage source is supplied to the tunneling junction in the form of quasiparticle excitations within the superconductors adjoining the tunneling barrier ([1,2]).

Simply speaking, the decay of these excitations is expected to produce heat. Since heat corresponds to incoherent phonon radiation with Planck's energy distribution (temperature well below Debye temperature), it is required that primary phonons directly emitted via electron–phonon interactions in quasiparticle decay undergo further phonon–electron and phonon–phonon interactions until the phonon energy distribution arrives at thermal equilibrium.

By proper experimental conditions it might be possible to prevent thermalization of primary decay phonons and to observe properties of their energy distribution.

EXPECTED ENERGY DISTRIBUTION OF DECAY PHONONS

According to Tewordt ([3,4]) deexcitation occurs in two steps predominantly via incoherent phonon emission. This is illustrated by the simple semiconductor model for tunneling between two identical superconductors

* Work underlying this chapter has been performed at Bell Telephone Laboratories, Murray Hill, N.J. in 1966 and 1967, on leave from the University of Göttingen.

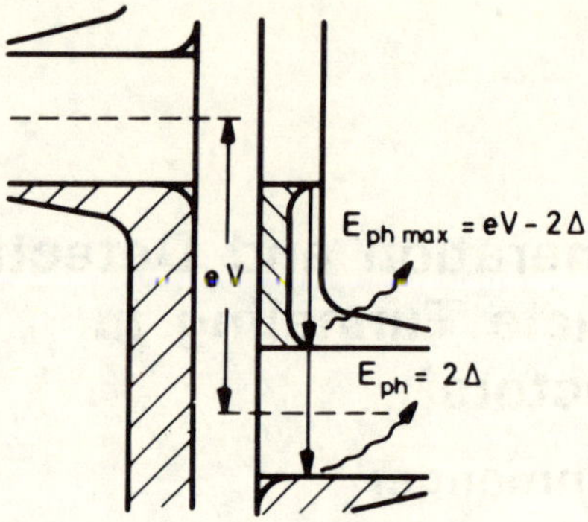

Fig. 1. Phonon generation, $eV \geq 2\Delta$.

(Fig. 1). If a battery voltage $V > 2\Delta/e$ (2Δ is the energy gap, e the electron charge) is applied to the junction of two identical superconductors, the energies of injected quasiparticles range from Δ to $eV - \Delta$ with respect to the Fermi level. Relaxation occurs as a decay of quasiparticles to the upper gap edge via single-phonon emission, the continuous energy spectrum of relaxation phonons extending from 0 to $eV - 2\Delta$. The succeeding recombination of quasiparticles to Cooper pairs results in recombination phonons of energy 2Δ. Recombination phonon frequencies range from 10^{11} to 10^{12} cps depending on the superconducting energy gap. For tin as used in our phonon experiments, the frequency amounts to 2.8×10^{11} cps, corresponding to an energy gap of 1.15 meV at 1.2°K. This frequency is about one order of magnitude higher than the main thermal phonon frequency at 1.2°K.

Relaxation and recombination lifetimes are of the order of 10^{-9} sec ([5]) and 10^{-7} sec ([5,6]), respectively, indicating that the phonon spectra should be fairly well defined. Contributions of photon emission in relaxation ([3]) and recombination ([7]) appear negligibly small compared to phonon emission.

For experiments concerning relaxation and recombination phonons it is necessary that these phonons are directly transmitted into a substrate with little further phonon interactions and finally received by a frequency-selective detector. For tunneling junctions in form of thin (1000 Å) stripes evaporated on a sapphire crystal this condition is at least partly fulfilled for phonons propagating perpendicular to the metal–sapphire interface. Estimates of the phonon mean free path in normal metals with respect to the predominant electron interaction amount to 10^{-5} to 10^{-4} cm at low temperatures. This estimate is based on ultrasonic measurements in tin ([8]), which have been extrapolated according to theory ([9]) to frequencies of 3×10^{11} cps. Information on the phonon mean free path in sapphire can be obtained by heat conductivity measurements ([10]); a value of the order of 1 cm is indicated. In addition, heat pulse experiments at 4°K ([11,12]) proved that phonons of 10^{11} cps are propagating in sapphire with little attenuation. Finally, the expected incoherent phonon radiation need to be probed by a detector with frequency- or energy-selective properties. These requirements are, in principle met by a phonon quantum detector.

PHONON DETECTION BY INTERACTION WITH COOPER PAIRS

According to a suggestion of Burstein *et al.* ([7]), superconductor tunnel junctions may be used for photon detection via breaking of Cooper pairs. Obviously, the corresponding process also applies for phonons of energy $E \geq 2\Delta$.

A tunneling junction with two identical superconductors is biased in the voltage range of thermal quasiparticle tunneling $0 < eV < 2\Delta$ (see Fig. 2). Incident phonon radiation of energy $E \geq 2\Delta$ will be strongly absorbed by breaking Cooper pairs, thereby increasing the number of excited quasiparticles. This results in a corresponding change of the tunneling current. Since only phonons of $E \geq 2\Delta$ can contribute to pair breaking, detection has frequency-selective properties. Phonon absorption taking account of pair breaking has been calculated by Tewordt ([3]) and Bobetic ([13]). At $T = 0$ phonon absorption disappears for phonon energies below 2Δ as predicted by BCS theory as well. At 2Δ the absorption increases discontinuously to a value of 1.57 times the normal-state absorption, and then with increasing energy monotonically approaches the normal-state value. At finite temperatures, e.g., $T/T_c = 0.5$, this behaviour is not changed appreciably, the difference being a small absorption also for energies below

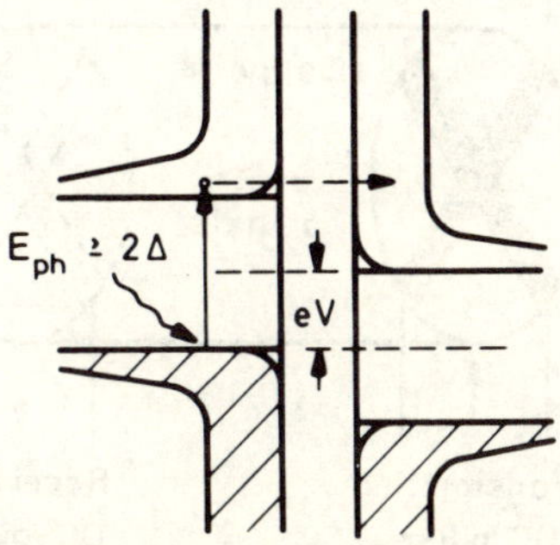

Fig. 2. Phonon detection, $0 < eV < 2\Delta$.

2Δ amounting to about 0.05 the normal-state value and a small shift in the position of the discontinuous absorption increase corresponding to the temperature dependence of the energy gap. Absorption at energies below 2Δ results from phonon interactions with thermally excited quasiparticles, and of course does not contribute to phonon detection. The same order of magnitude of phonon absorption via pair breaking and of normal-state absorption indicates that even with thin superconducting films used in a tunneling junction phonon detection with high sensitivity should be possible (compare the preceding estimate on the phonon mean free path in tin).

EXPERIMENTAL PROCEDURE

In the experiments reported in ([14]) phonon generation and detection was performed using two identical Sn *I* Sn junctions (strip width 1 mm; thickness: bottom 1000 Å, top 3000 Å) evaporated on the optically polished end surfaces of a cylindrical sapphire rod of 1-cm length and 1-cm diameter. This arrangement (Fig. 3) allows for pulse measurements in order to check the phonon nature of observed signals by time-of-flight determination. A single crystal of synthetic sapphire with low dislocation density, 10^2–10^3 cm^{-2}, has been used because of the small phonon scattering by lattice imperfections it affords. It is further of importance that phonons generated

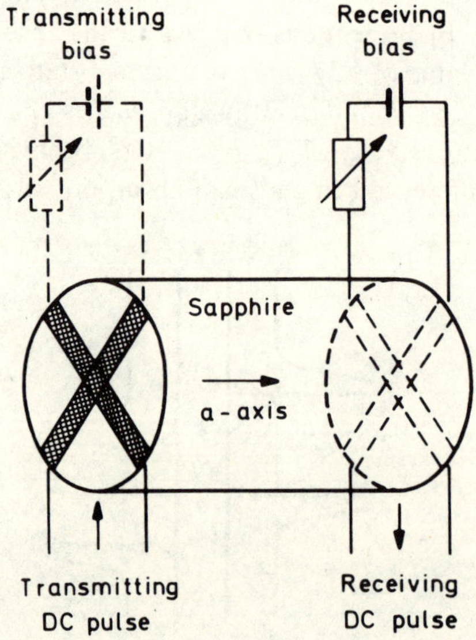

Fig. 3. Sapphire crystal with tunneling junctions.

in the junction are transmitted to the sapphire with little reflection. An estimate of the acoustical mismatch at the tin–sapphire interface results in an energy transfer of about 85% for longitudinal and transverse phonons of normal incidence. The sample in Fig. 3 is in direct contact to liquid helium at temperatures of 1.2°K.

Measurements of the DC *I–V* characteristics of the junctions are performed in the familiar four-terminal technique. In a corresponding way it is also possible to obtain the characteristics under pulsed conditions.

For phonon propagation measurements DC current pulses with duration from 0.1 to 0.5 μsec and a repetition rate of 10 kHz are applied to the generating junction. In addition a variable DC bias can be superimposed. The detecting junction is operated with a current bias (high ohmic source) at a junction voltage of about 0.8 mV. Under these conditions detection results in voltage pulses at the free terminals of the junction. After low-noise wideband amplification the signals are fed to a sampling scope and finally plotted with the help of an *X–Y* recorder (Fig. 4). Further noise reduction is accomplished by a low-pass filter at the *X–Y* recorder input.

Since junctions with low ohmic tunneling resistance, typically 1–10 m-ohm outside the gap region, are preferable as generator and detector, matching transformers are used with both junctions. A magnetic field parallel to the diode plane ranging from 10 to 15 G serves to suppress Josephson DC currents and corresponding instabilities.

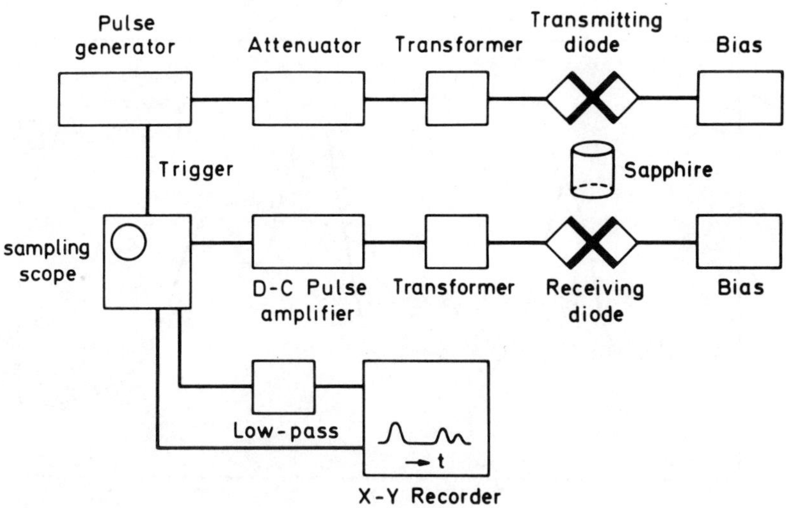

Fig. 4. Electronic setup.

EXPERIMENTAL RESULTS

Time-of-Flight Measurement

Figure 5 shows phonon signals as function of time propagating parallel to the *a* axis in sapphire. The first signal from the left corresponds to magnetic feedthrough and may serve as indication for zero delay. Subsequent signals correspond to the weak longitudinal mode and the strong fast transverse mode. The time delay of the last signal is too large to be attributed to a slow transverse mode; instead, diffuse scattering of the fast transverse mode at the fine-ground cylindrical sidewall of the sapphire crystal does account well for the observed time delay. Better time resolution is obtained at an elevated temperature of 1.9°K with an improved pulse system [15] (see Fig. 6). In this measurement Pb–*I*–Pb was used for generation and Sn–*I*–Sn for detection, the sapphire length in the *a* direction amounting to 1.2 cm. Time-of-flight evaluation for this measurement does agree with ultrasonic velocity data [16] in sapphire within $\pm 3\%$ for all three modes in the *a* direction. The slow transverse signal (weak) is now well separated from the strong fast transverse pulse and the sidewall scattering contribution. The shape and time delay of the sidewall scattering signal are in agreement with calculations using ultrasonic velocity data for the fast transverse mode assuming that scattering occurs only once for each phonon ray. An estimate of the corresponding signal amplitude in relation to the fast transverse signal amplitude indicates an omnidirectional or Lambert cosinus

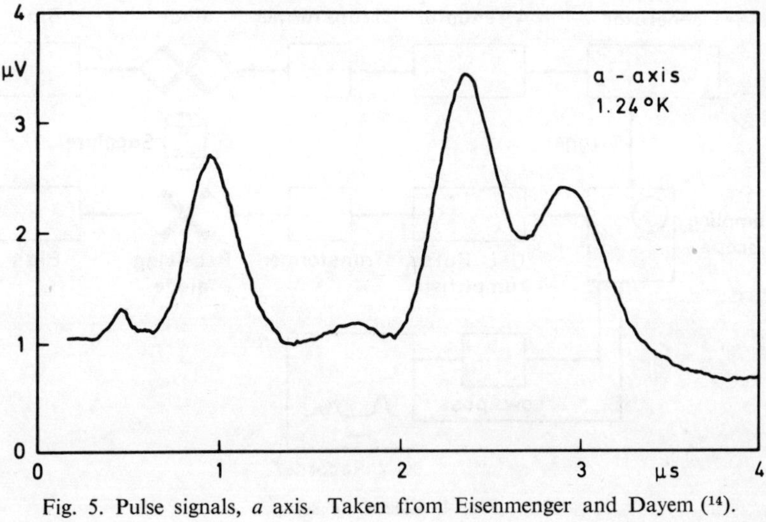

Fig. 5. Pulse signals, *a* axis. Taken from Eisenmenger and Dayem [14].

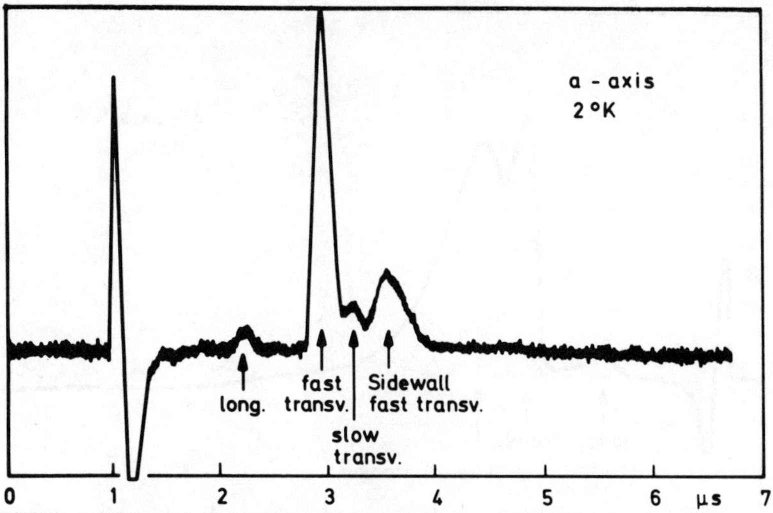

Fig. 6. Pulse signals, improved time resolution, *a* axis.

law radiation distribution at the transmitter. This gives evidence for incoherent phonon emission from the generator. In corresponding pulse measurements for phonon propagation in *c*-axis sapphire (Fig. 7) only one transverse signal is observed, in agreement with the transverse wave degeneracy for this crystal orientation. Again, the time delay of the last

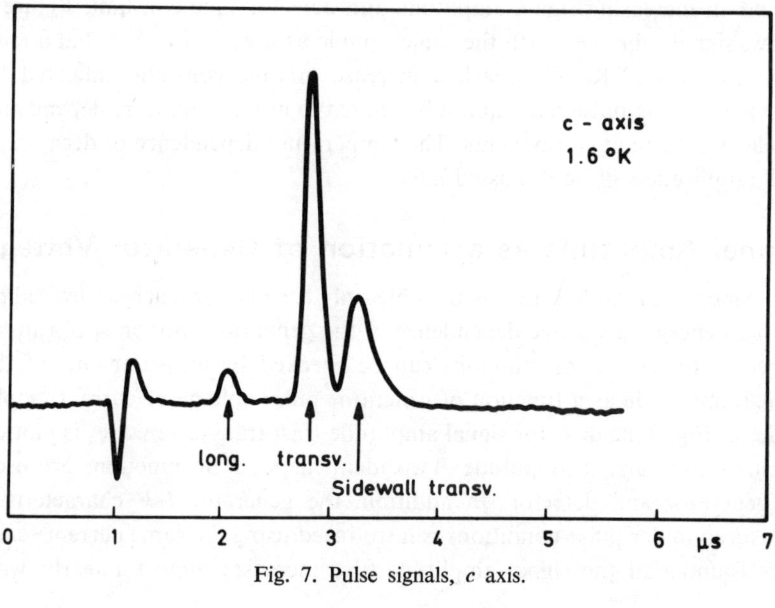

Fig. 7. Pulse signals, *c* axis.

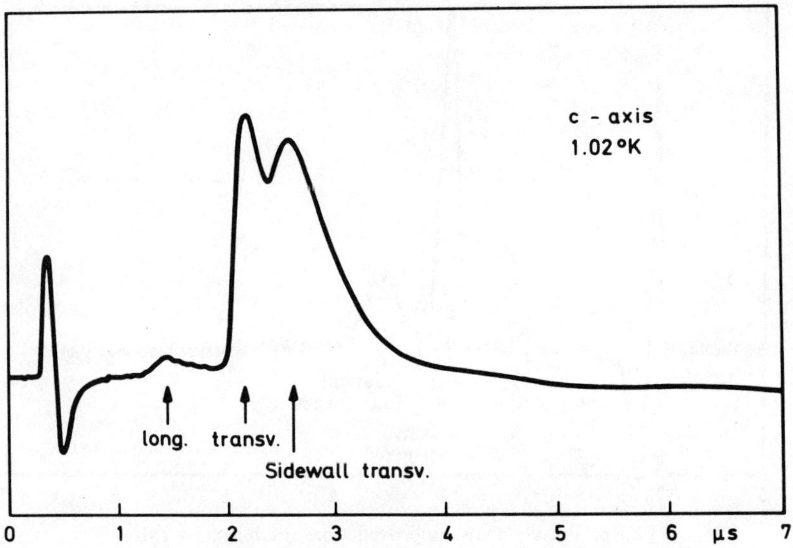

Fig. 8. Pulse signals, *c* axis, low temperature.

signal resulting from sidewall scattering can be calculated to be in agreement with the scattering of the transverse wave.

In all pulse measurements there is an appreciable influence of temperature on pulse amplitude and time resolution. Lowering temperature results in increasing signal amplitude and detector time constant. Figure 8 shows signals obtained with the same sample as used in Fig. 7 but at a temperature of 1.02°K. The marked increase in time constant indicated by slowly decaying pulse tails cannot be related to any temperature dependence of electrical circuit components. The temperature dependence of decay time and amplitude will be discussed later.

Signal Amplitude as a Function of Generator Voltage

Since phonon detection is possible only for phonon energies exceeding the gap energy, a voltage dependence of the generator phonon spectrum as expected for relaxation phonons can be checked by measurements of the signal amplitude as a function of generator pulse voltage or current amplitude. In Fig. 9 the detector signal amplitude (fast transverse wave) is plotted versus pulse current amplitude. Two identical Sn–I–Sn junctions are used as generator and detector. In addition, the generator I–V characteristic measured under pulse conditions is introduced using the same current scale. It is found that the signal amplitude first increases almost linearly with

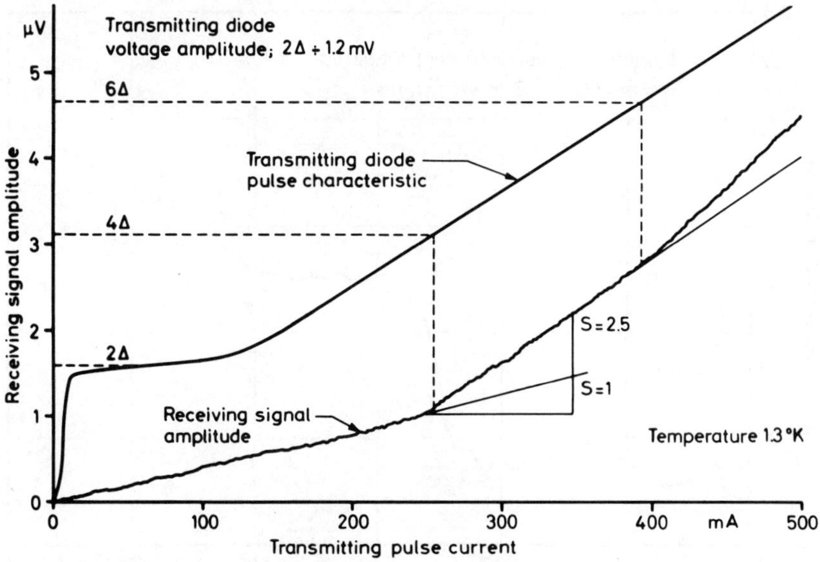

Fig. 9. Detector signal amplitude versus generator pulse current (voltage). Taken from Eisenmenger and Dayem ([14]).

transmitter current. At a generator voltage corresponding to 4Δ in tin a sharp change in slope is succeeded again by a linear increase up to a generator voltage corresponding to 6Δ. From this point on an almost linear increase in slope is observed. This is more clearly brought out in the derivative with respect to generator current as shown in Fig. 10. To take the derivative a constant current pulse of 25 mA in addition to a variable DC bias current is applied to the generating junction. The signal in Fig. 10 is plotted versus bias current indicating junction voltages corresponding to 4Δ and 6Δ. Since only pulse signals from the detector are amplified and recorded, the DC background of phonon radiation does not show up. In terms of the semiconductor model (Fig. 1) this experimental condition roughly corresponds to the injection of a constant number of quasiparticles in the range of maximum energy, the latter being determined by the bias-dependent DC voltage drop across the junction.

For an interpretation of the findings shown in Fig. 9 and Fig. 10 we have now to remember the voltage dependence of the upper energy limit for relaxation phonons, $E_{\max} = eV - 2\Delta$, together with the condition that the detector is only sensitive to phonon energies larger than 2Δ. This implies that relaxation phonons do only contribute to the receiving signal if the generator voltage is larger than $4\Delta/e$. In contrast, the contribution of re-

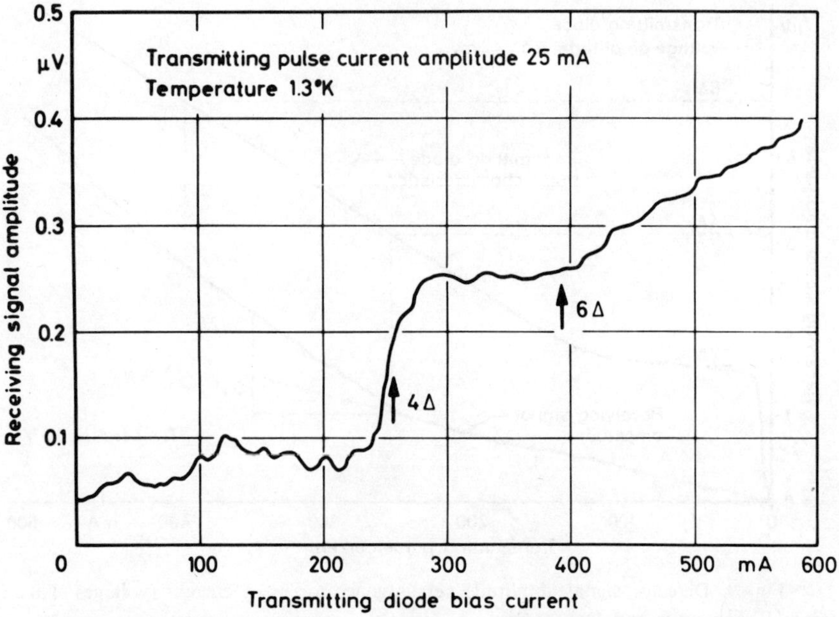

Fig. 10. Derivative of the detector signal of Fig. 9 plotted versus generator bias current (voltage) at constant generator pulse current. Taken from Eisenmenger and Dayem ([14]).

combination phonons with energies centered at 2Δ does not depend on the battery voltage. As long as V is larger than $2\Delta/e$,e. g., exceeding the range of thermally excited quasiparticle tunneling, the rate of recombination phonon emission is directly proportional to the single-particle tunneling current.

The changes in slope and in the step height at 4Δ in Fig. 9 and Fig. 10, respectively, are determined by the ratio of the number of relaxation to recombination phonons emitted. According to the semiconductor model (Fig. 1) we would expect this ratio to be 1. More reliable conclusions might be drawn from the correct excitation diagram for tunneling between superconductors ([1,2]) under the condition of zero temperature and equal gap energies. For one electron to tunnel a Cooper pair must be broken up leaving one excitation on each side of the barrier with a total energy of eV referred to the ground state. The total energy with respect to the upper gap edges amounts to $eV - 2\Delta$. Correspondingly, in relaxation two phonons of total energy $eV - 2\Delta$ are emitted. The energies of the individual phonons can range from 0 to $eV - 2\Delta$. In order to allow for recombination, a second Cooper pair must be broken up altogether, resulting in two excitations on each side. The decay of these excitations produces four relaxation phonons and two recombination phonons, resulting in a ratio of two to one.

The same conclusion can be arrived at by simply considering energy conservation. If relaxation phonons have a continuous energy distribution between 0 and $eV - 2\Delta$, they must be emitted in pairs of total energy $eV - 2\Delta$. This plus the recombination phonon energy 2Δ add up to the energy eV supplied in the tunneling of one electron.

Comparing the relaxation–recombination phonon ratio of two to one with the relative step height at 4Δ in Fig. 10, reasonable agreement is found.

In trying to explain the structure at 6Δ we find that this may be more involved than the simple interpretation applicable at 4Δ. Basically, the 6Δ structure indicates the possibility of two phonons with energy 2Δ being emitted instead of one relaxation phonon with energy 4Δ.

Temperature Dependence of Signal Amplitude and Time Constant

In discussing the effect of temperature variation we have to distinguish among properties of the phonon generator, the sapphire crystal, and the detecting junction.

The I–V characteristic of superconductor tunneling junctions well below T_c and with Δ approaching Δ_0 essentially becomes temperature independent. An exception is to be made with respect to the voltage range $0 < V < 2\Delta/e$, where the tunneling current is mainly determined by the number of thermally excited quasiparticles. Since this current is small compared to the tunneling current at voltages $V > 2\Delta/e$ (comprising in good junctions about 1%), it can be neglected in phonon generation. Temperature changes in a range well below T_c will therefore have little or no influence on the primary spectrum emitted in the relaxation and recombination of quasiparticles.

Phonon propagation inside the sapphire crystal may in principle be affected by temperature-dependent scattering of thermal lattice waves. Extrapolating ultrasonic absorption data ([17]) according to theory ([18]) to a frequency of 3×10^{11} cps and a temperature of 2°K results in a phonon mean free path of the fast transverse mode exceeding 10^3 cm. From this we expect that temperature-dependent phonon scattering in sapphire will be negligibly small, for instance, compared to temperature-independent scattering by lattice imperfections [compare also ([11])].

Since experiment ([15]) indicates a marked increase in signal amplitude with decreasing temperature (Fig. 11), this temperature dependence must be a property of the detection junction.

With incident phonons of energy $E \geq 2\Delta$ being absorbed by breaking

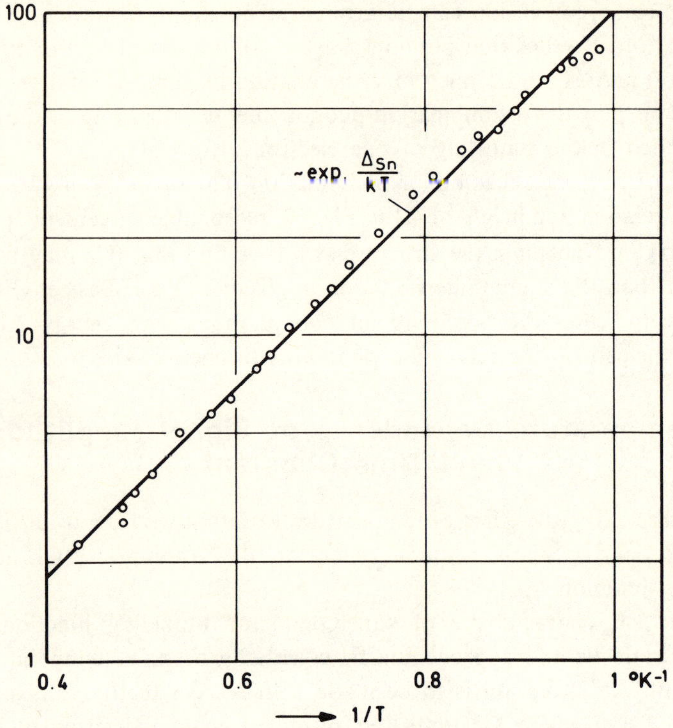

Fig. 11. Signal amplitude versus temperature, $\tau_{\text{pulse}} > \tau_{\text{eff}}$.

Cooper pairs in the detector, the steady-state increase from thermal equilibrium of the number of excited quasiparticles δn amounts to

$$\delta n = 2n_{\text{ph}}\tau_{\text{eff}}, \tag{1}$$

where n_{ph} corresponds to the number of absorbed phonons per unit time, each phonon creating two quasiparticles. According to ([13]), this absorption also shows little temperature dependence for $T/T_c < 0.5$. Here τ_{eff} is the effective recombination lifetime of quasiparticles. The term "effective" indicates the possibility of reabsorption of phonons emitted by recombination in the detector [see also ([19])]. The steady-state expression is also valid for pulse measurements as long as τ_{eff} is small compared to the pulse duration. The receiver current i_R (dynamical short-circuit current) is determined by the number of additional quasiparticles δn per unit volume v:

$$i_R = (\delta n/v)C, \tag{2}$$

where C is a temperature-independent tunneling constant. Together with

Eq. (1) this results in

$$i_R = \tau_{\text{eff}}(2n_{\text{ph}}/v)C, \tag{3}$$

with τ_{eff} the only temperature-dependent factor.

Since τ_{eff} exceeds the recombination lifetime by only a constant factor, its temperature dependence is predicted ([6,19]) to be roughly proportional to $(1/\sqrt{T})e^{\Delta/RT}$ at low temperatures, or inversely proportional to the number of thermally excited quasiparticles. As shown in Fig. 11, for measurements with Sn–I–Sn as detector, i_R exhibits the expected temperature dependence.

In experiments with generator pulse duration comparable to τ_{eff} a corresponding time constant of the detector should show up directly in the pulse decay time. This has been observed ([15]) at low temperature (Fig. 8). Evaluations at different temperatures with respect to the absolute value of τ_{eff}, taking account of the electronic system contribution to the pulse shape, are presented in Fig. 12. Again agreement with the expected temperature dependence of τ_{eff} is found. This result, together with the same temperature dependence of i_R, gives support to the assumptions made that

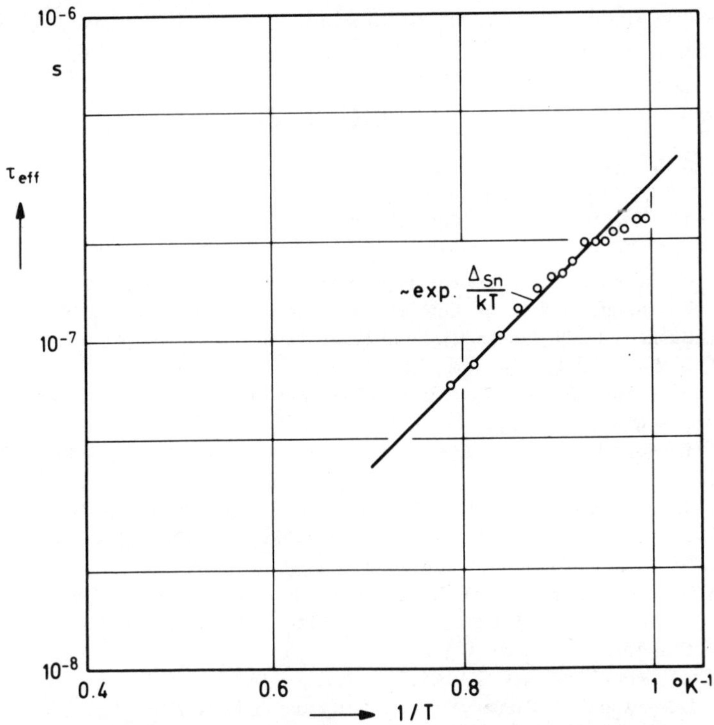

Fig. 12. Signal decay time (time constant) versus temperature.

there are comparatively small temperature dependences of the generator phonon radiation, the phonon absorption inside the sapphire, and the phonon absorption in the detector.

CONCLUSION

In reviewing the experiments we should summarize the arguments indicating that the observed phonon signals are due to nonthermal phonon radiation emitted in the decay and recombination of injected quasiparticles.

It appears very difficult to explain the observed structure in the dependence of the signal amplitude as a function of the generator voltage in terms of heat pulses, e.g., phonons with Planck's distribution. An estimate of the possible heat effect arrives at much smaller signal amplitudes than observed in ([14]). In addition, only little temperature dependence of the signal amplitude is to be expected. In contrast to this, quantum processes provide a straightforward understanding of the experimental findings.

Finally, it might be worthwhile to mention that phonon–electron interactions, especially the phonon detection by pair breaking, indicate the close analogy to corresponding photon interactions as, e.g., in the photoelectric effect.

REFERENCES

1. J. R. Schrieffer, this volume, Chapter 21.
2. J. Wilkins, this volume, Chapter 24.
3. L. Tewordt, Phys. Rev. 127:371 (1962).
4. L. Tewordt, Phys. Rev. 128:12 (1962).
5. B. I. Miller and A. H. Dayem, Phys. Rev. Letters 18:1000 (1967).
6. J. R. Schrieffer and D. M. Ginsberg, Phys. Rev. Letters 8:207 (1962).
7. E. Burstein, D. N. Langenberg, and B. N. Taylor, Phys. Rev. Letters 6:92 (1961).
8. W. P. Mason and H. Bömmel, Acoust. Soc. Am. 28:930 (1956).
9. W. P. Mason, Physical Acoustics and the Properties of Solids, D. Van Nostrand Inc., New York, 1958, p. 323ff.
10. R. Berman, E. L. Foster, and J. M. Ziman, Proc. Roy. Soc. (London) A231:130 (1955).
11. R. J. v. Gutfeld and A. H. Nethercot, Phys. Rev. Letters 12:641 (1964).
12. R. J. v. Gutfeld and A. H. Nethercot, Phys. Rev. Letters 17:868 (1966).
13. V. M. Bobetic, Phys. Rev. 136:A1535 (1964).
14. W. Eisenmenger and A. H. Dayem, Phys. Rev. Letters 18:125 (1967).
15. W. Eisenmenger, A. H. Dayem, and B. I. Miller, unpublished work.
16. B. T. Bernstein, J. Appl. Phys. 34:169 (1963).
17. J. de Klerk, Phys. Rev. 139:A1635 (1965).
18. L. Landau and G. Rumer, Phys. Z. Sowjetunion 11:18 (1937).
19. A. Rothwarf and B. N. Taylor, Phys. Rev. Letters 19:27 (1967).

Chapter 27

Tunneling Anomalies—Experiment

J. M. Rowell

Bell Telephone Laboratories, Inc.
Murray Hill, New Jersey

In this Chapter I first briefly summarize all the deviations from ideal tunneling behavior, starting at high biases, and then consider the anomaly at zero bias as probably the most interesting of the excitation processes that that can occur in the barrier.

As an anomaly is obviously defined as a departure from some ideal behavior, which we claim we understand, we should first discuss what this ideal behavior is. In earlier chapters it was shown that if we take two metals separated by an insulating barrier, then the tunnel current at low voltages is ohmic (directly proportional to the applied voltage) and at high voltages depends exponentially on voltage. In fact, when one carefully measures conductance versus voltage the ohmic region is nonexistent and the exponential region only sets in for biases >2 V. As most of the interesting tunneling measurements are over the range 0–1 V, we have to look a little more carefully at these intermediate voltages.

Consider the tunnel junctions of Fig. 1. For simplicity we consider first the junction which is symmetrical when $V = 0$. The net tunnel current from metal 1 to 2 can be described as the difference of two terms ([1]). We have a tunnel current from 1 to 2 given by

$$J_{1,2} = J_0 \bar{\varphi} \exp(-A\bar{\varphi}^{1/2}), \tag{1}$$

where ([1]) should be consulted for A, and $\bar{\varphi}$ is the average barrier height as shown in the figure. There is also a tunnel current from metal 2 to 1, which one might consider as opposing the bias,

$$J_{2,1} = J_0(\bar{\varphi} + eV) \exp[-A(\bar{\varphi} + eV)^{1/2}], \tag{2}$$

and the net current is

$$J = J_{1,2} - J_{2,1}. \tag{3}$$

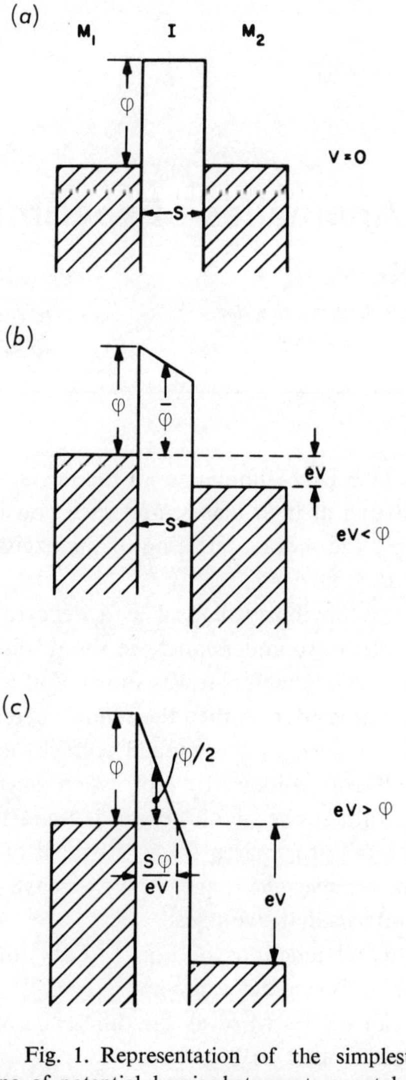

Fig. 1. Representation of the simplest type of potential barrier between two metals. (*a*) The barrier is symmetrical for $V = 0$. Shown are the average height and thickness of the barrier for (*b*) intermediate ($eV < \varphi$) and (*c*) high ($eV > \varphi$) voltage ranges.

At very high voltages ($eV > \bar{\varphi}$) such that electrons are injected into the conduction band of the insulator (sometimes called field emission or Fowler–Nordheim tunneling) the average barrier height is $\varphi/2$ (independent of bias), but the effective width of the barrier changes with bias and is $S\varphi/eV$. In this region the current depends approximately exponentially on voltage.

In the intermediate voltage range ($eV < \varphi$) the width S of the barrier is unchanged with voltage, but the height varies, $\bar{\varphi} = \varphi - (eV/2)$, as is clear from Fig. 1(*b*). Substituting this into (3) we have

$$J = J_0\left\{\left(\varphi - \frac{eV}{2}\right)\exp\left[-A\left(\varphi - \frac{eV}{2}\right)^{1/2}\right] - \left(\varphi + \frac{eV}{2}\right) \times \exp\left[-A\left(\varphi + \frac{eV}{2}\right)^{1/2}\right]\right\}. \tag{4}$$

An interesting limit of this expression for the fairly low voltages of interest here has been considered by Simmons (2). On expanding the exponentials and dropping terms of V^4 or higher order, he obtains

$$I = \alpha(V + \gamma V^3), \tag{5}$$

where α and γ are constants for each junction. The first term $I = \alpha V$ is, of course, the familiar ohmic region; the terms in V^2 have dropped out, and the next term is the V^3 current term as found by Knauss and Breslow (3). A more sensitive way to study this voltage dependence is to measure conductance as a function of voltage,

$$dI/dV = G(V) = \alpha(1 + 3\gamma V^2), \tag{6}$$

where α is now the constant conductance expected at low bias and $3\alpha\gamma V^2$ is a parabolic dependence symmetric about zero bias, as illustrated in Fig. 2. This gives the minimum conductance for the junction exactly at zero bias.

We find that our junctions of the type Al–*I*–*M* (where M = Pb, Sn, In) can be accurately described by a constant plus parabolic conductance up to ~0.5 V except that it is our experience that the parabola is *never centered at zero bias*, being shifted to the *M* positive bias in all cases. As an example, Fig. 3 shows an Al–*I*–Sn junction measured to ±1.2 V; the plot is symmetric about a bias of 125 mV in the Sn positive direction. This offset of the minimum G is a little puzzling, as it is not mentioned as a result of any theoretical tunneling calculation however complex the barrier shape. However, as it seems to be a basic asymmetry of the Al–*I*–M system, I think it must arise from the asymmetry of the barrier, the 125 mV being proportional to the difference in barrier height on the Al and Sn sides of the insulator. Thus the barrier is not of the type shown in Fig. 1(*a*) but similar to that shown in Fig. 4. The only other inherent asymmetry I can think of is the density

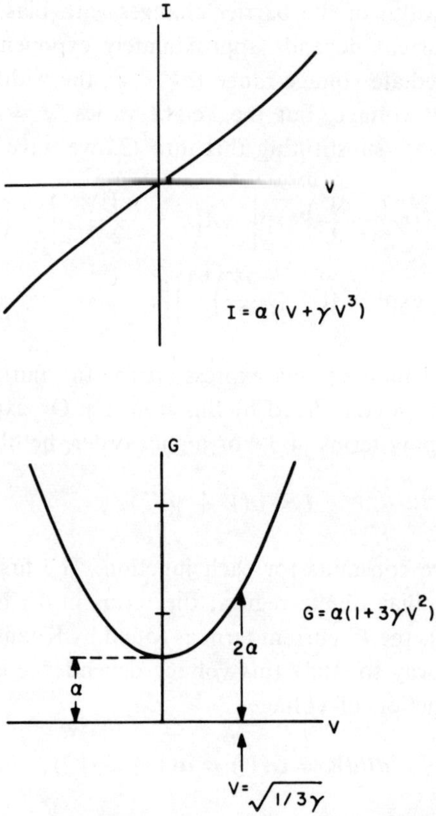

Fig. 2. Current and conductance versus voltage for an ideal symmetrical barrier of the type shown in Fig. 1 (*a*). The conductance is a constant plus a parabolic term which is symmetric about $V = 0$.

of states in the metals, but I wouldn't expect such a slowly varying G to be a density of states effect.*

* In the discussion following my talk at the Advanced Studies Institute in Risø a number of people objected to these statements, pointing out that no calculation of I versus V for asymmetric barriers gave this offset of the minumum conductance from $V = 0$. W. F. Brinkman has recently indicated to me that the asymmetry mostly arises from the tunneling transmission probability

$$D(E_x) = \exp\left\{\frac{-4\pi}{h}\int_{S_1}^{S_2}[2m(V(x)-E_x)]^{\frac{1}{2}}\,dx\right\},$$

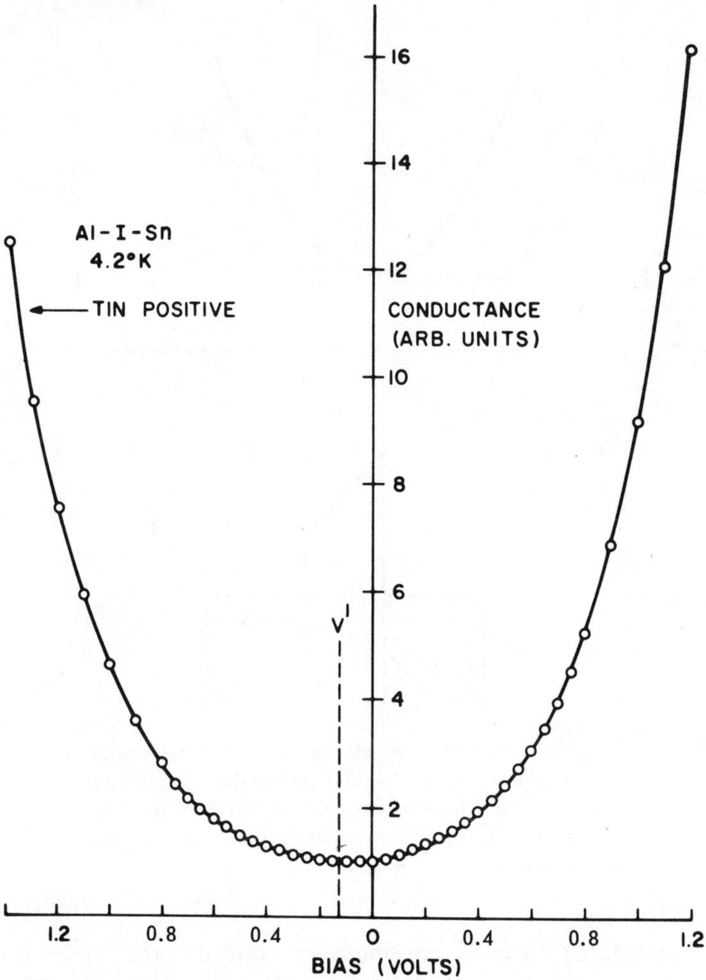

Fig. 3. Conductance versus voltage for an Al–I–Sn junction at 4.2°K. For voltages > 400 mV the plot is symmetric about $V = -125$ mV.

which is not the same when biases $+V$ and $-V$ are applied to an asymmetric junction. In the work of Simmons (1) this integral is simplified by assuming an average barrier height $\bar{\varphi}$. Once this is done the asymmetry of $D(E_x)$ is ignored. The approach of Stratton (4) is to expand $\ln D(E_x)$ for energies close to the Fermi energy. This approach leads to expressions for conductance of the form

$$G = a + bV + cV^2.$$

The linear term bV is only zero for symmetric barriers, and a finite value is of course an alternative way of saying that the parabola cV^2 is offset from $V = 0$. For a discussion of the asymmetry of tunneling I–V characteristics see (5).

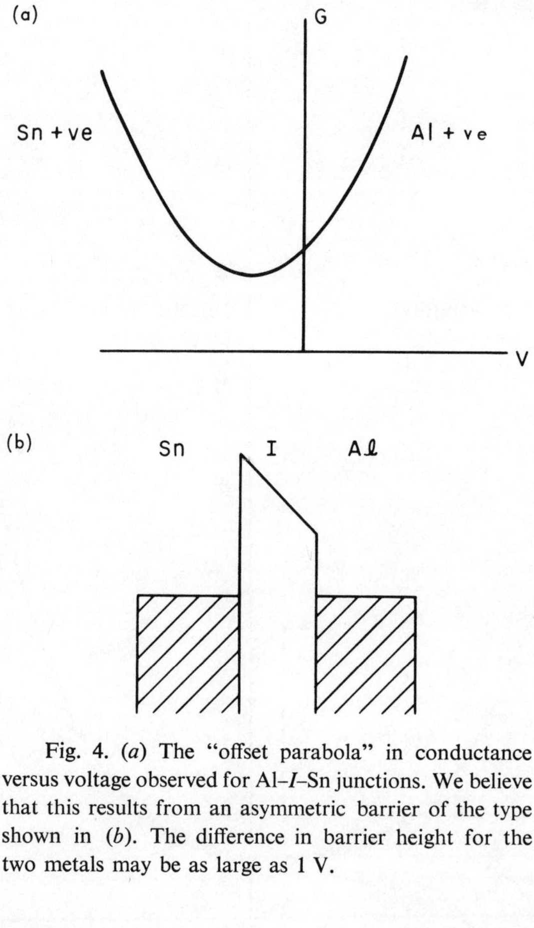

Fig. 4. (*a*) The "offset parabola" in conductance versus voltage observed for Al–*I*–Sn junctions. We believe that this results from an asymmetric barrier of the type shown in (*b*). The difference in barrier height for the two metals may be as large as 1 V.

Assuming that the offset can simply be ascribed to the barrier, it seems reasonable to write

$$G(V) = \alpha[1 + 3\gamma(V - V_0)^2]. \tag{7}$$

to describe the behavior of Fig. 3. It is convenient to introduce the voltage $V' = V - V_0$ measured relative to the voltage at minimum G. A plot of G versus V' is shown in Fig. 5. The parabolic dependence is evident but is affected slightly for $|V| < 400$ mV by the barrier excitations to be discussed later. To check the voltage range of the V'^2 term we have plotted $G(V) - \alpha = G'$ versus V'^2 in Fig. 6 for both biases. The low voltage points are again affected by excitations (as we will see, these excitations give rise to small increases in conductance symmetric about $V = 0$, whereas the parabola is

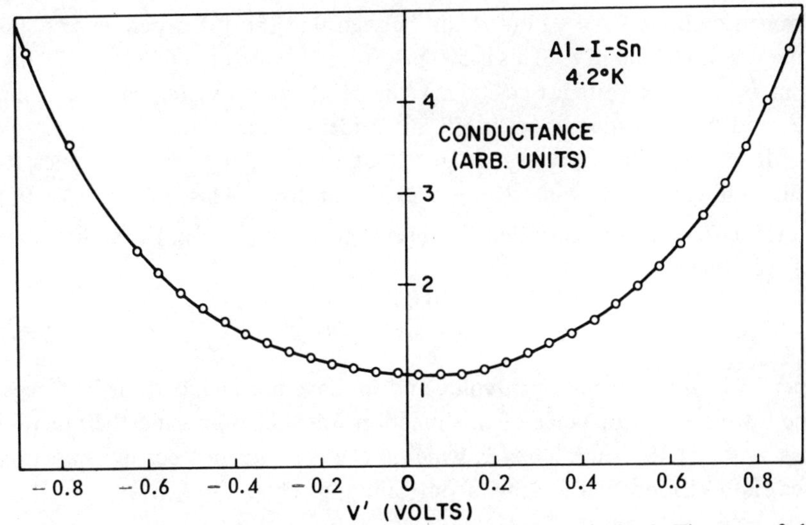

Fig. 5. Conductance versus voltage V' for the junction of Fig. 3. The zero of this V' voltage scale is at $V = -125$ mV in Fig. 3.

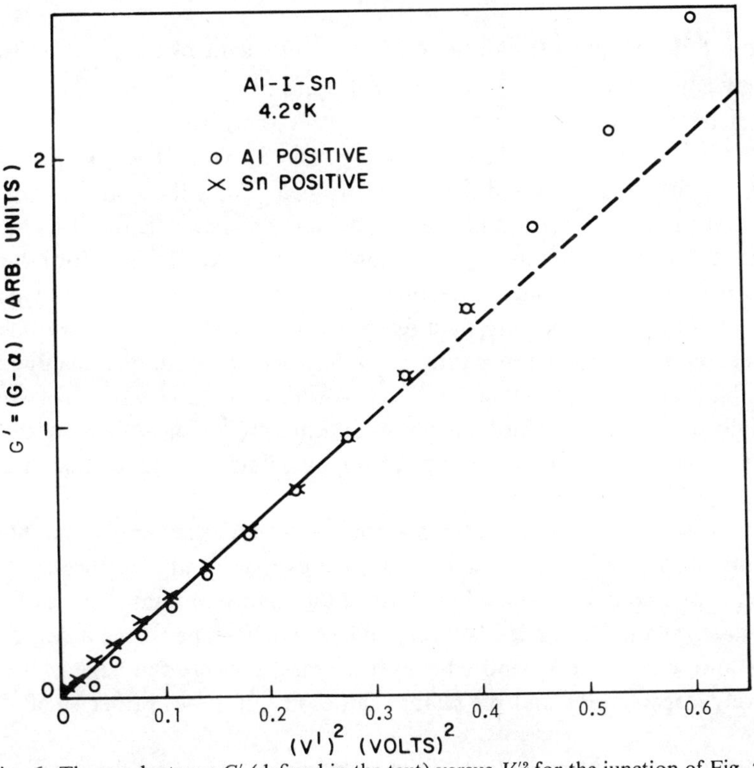

Fig. 6. The conductance G' (defined in the text) versus V'^2 for the junction of Fig. 5.

symmetric about $V' = 0$) but it can be seen that the V'^2 dependence holds quite well up to 500 mV. (This is 500 mV from $V' = 0$, or -625 to $+375$ mV from $V = 0$). At voltages > 500 mV the conductance increases faster than V'^2, and higher order terms in V' are evidently important.

It can also be seen from Fig. 6 that the conductance G doubles its minimum value (α) when $V'^2 = 0.29\ V^2$, or for a bias of 540 mV. But from Eq. (6) it can be seen that this bias is given by $\sqrt{1/3\gamma}$. From Simmons ([2]) we have

$$\gamma = \left(\frac{0.0115}{\bar{\varphi}} S^2 - \frac{0.0315}{\bar{\varphi}^{3/2}} S\right) \approx \frac{0.0115}{\bar{\varphi}} S^2, \tag{8}$$

where S is in angstroms, φ in volts, and we have not included image forces. The low-voltage resistance of this junction was 95.8 ohms and the junction area was 1.51 10^{-3} cm^2. Thus α, which is the conductance per unit junction area, is 6.91 mhos/cm^2. Again from Simmons ([2])

$$J = \left[3.16 \times 10^{10} \frac{\bar{\varphi}^{1/2}}{S} \exp(-1.025\, S\bar{\varphi}^{1/2})\right] V = \alpha V. \tag{9}$$

From Eqs. (8) and (9) and these two measurements of the junction (minimum conductance α and voltage $\sqrt{1/3\gamma}$, where the conductance is 2α) we determine that $S = 14$ Å and $\varphi = 1.95$ V, both very reasonable values.

Let me summarize this tunneling behavior, which I will call ideal, so we can then discuss some deviations from it. First, the conductance is a constant plus a V^2 term, but the minimum conductance is offset from zero bias. Thus there is *not* an ohmic region near zero bias. In fact, for a bias of 100 mV in the Al positive direction G increases by $\sim$8% over its value at $V = 0$ and would decrease by a few per cent at 100 mV in the Sn positive direction if no barrier excitations took place. Second, the conductance doubles for a bias of 540 mV, consistent with reasonable values for barrier height and thickness. Third, higher order terms are important above 500 mV. This general behavior seems typical for our aluminum junctions with Pb, Sn, and In counterelectrodes.

To come now to deviations from this ideal behavior, we might first think of band structure effects. It is obvious that band structure is clearly seen in semiconductor tunneling,* where the change in total electron density of states at the band edge is large, and we might hope to see a small conductance change at a band edge even in metals where the edge is at a relatively large voltage and the change in density of states rather small. This

* How does one explain the Esaki diode characteristic if this is not the case?

should give an increase in conductance for one direction of bias only, i.e., an asymmetric effect. There is to date no observation of band edges in metals using tunneling, but there is some work ([6])* on semimetals which looks encouraging. The assumption that tunneling into a band edge gives a peak in conductance seems questionable, however. The strongest structure one might expect would be a step increase in dG/dV for a tunneling onset into a band or a step decrease in dG/dV when a filled band uncrosses the Fermi level. I should stress that I believe these band edge effects will be small structure superimposed on an approximately V^2 background, as shown by Hauser ([6]).

Jaklevic and Lambe described in Chapters 17 and 18 of this volume how tunneling electrons can excite vibrations (frequency ν) of molecules trapped in the oxide layer. This gives rise to conductance increases at $eV = h\nu$, and this occurs symmetrically about zero bias. In fact, all barrier excitation processes give rise to conductance increases which are symmetric. As far as I know, only band edge effects are asymmetric. Figure 7 shows a conductance plot for an Al–I–Sn junction made in an oil-diffusion pump system and oxidized by exposure to laboratory air. At first sight, this looks impossibly different from the simple parabola of Fig. 3, but we have taken the Sn positive direction and measured the conductance in much greater detail. The detail is hard to see in the Al positive direction since the conductance rises faster (compare the conductance scales of Fig. 3 and 7), but the excitation structure is symmetric about $V = 0$. Somewhere lower than this plot one can imagine the constant G plus V^2 term and regard the structures of Fig. 7 as increases above this parabola. The main increases are 0–35 mV, 110–125 mV 170–200 mV, and 350–370 mV. The increases above 100 mV can be identified as OH and CH vibrations by comparing with the d^2I/dV^2 plots of Lambe and Jaklevic; the region 0–35 mV will be discussed later. It can be seen that the maximum conductance increase is that at 360 mV, being $\sim 3\%$, so these are not observed on the overall conductance plot of Fig. 3, although they do affect slightly the symmetry of the V'^2 plot about the minimum conductance voltage. As stated above, this is because the parabola is symmetric about -125 mV; the molecular excitations are symmetric about $V = 0$.

* Similar results for junctions made on bulk bismuth were shown by Giaever ([7]) at the tunneling conference. The earlier result of Esaki and Stiles ([8]) is very different from those of Giaever and Hauser. As there is no proof of tunneling in the case of Esaki and Stiles, their results must remain suspect until reproduced with a junction showing an energy gap in the counterelectrode.

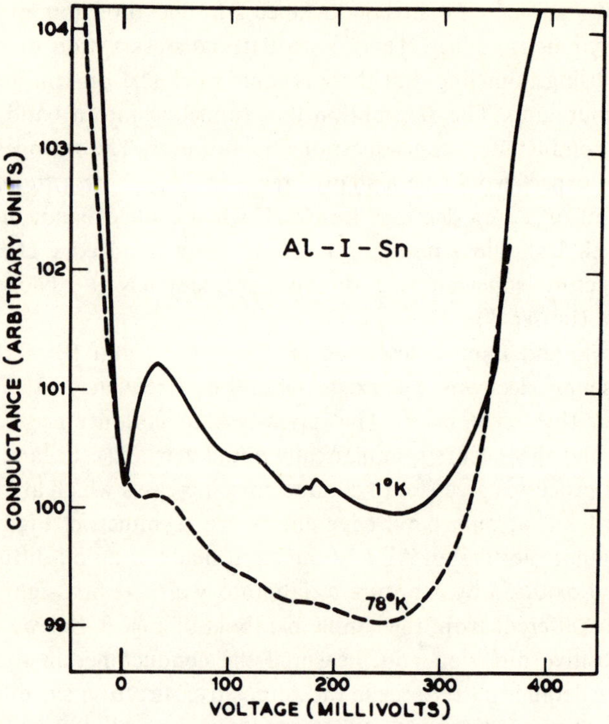

Fig. 7. Conductance versus voltage for an Al–I–Sn junction at 78°K and at 1°K (a magnetic field was applied to destroy superconductivity in the films). The bias shown in tin positive. Note that in this figure the conductance increases only 4% from its value at $V = 0$, while in Fig. 3 it increases 13 ×.

At lower voltages we observe the excitation of vibrations of the oxide itself ([9]). These can be regarded as oxide phonons or vibrations of the oxide molecules. This phenomenon is particularly striking in lead oxide junctions, and Fig. 8 shows first and second derivatives for a Pb–I–Pb junction. It is believed that the second-derivative peaks between 25 and 60 mV are vibrations of the lead oxide, probably the oxygen atom frequencies; the peak at 17 mV is possibly a Pb atom frequency. Peaks can also be seen near 5 and 9 meV, which are characteristic phonon energies for bulk lead. We have also measured Al–I–Pb junctions and find that the only common peaks are those less than 10 mV, supporting the suggestion that somehow these vibrations of the surface of the Pb film are coupled into the tunneling process. In Al–I–Pb junctions there is also a weak peak between 30 and 40 mV, which corresponds to phonon energies in Al. We substituted

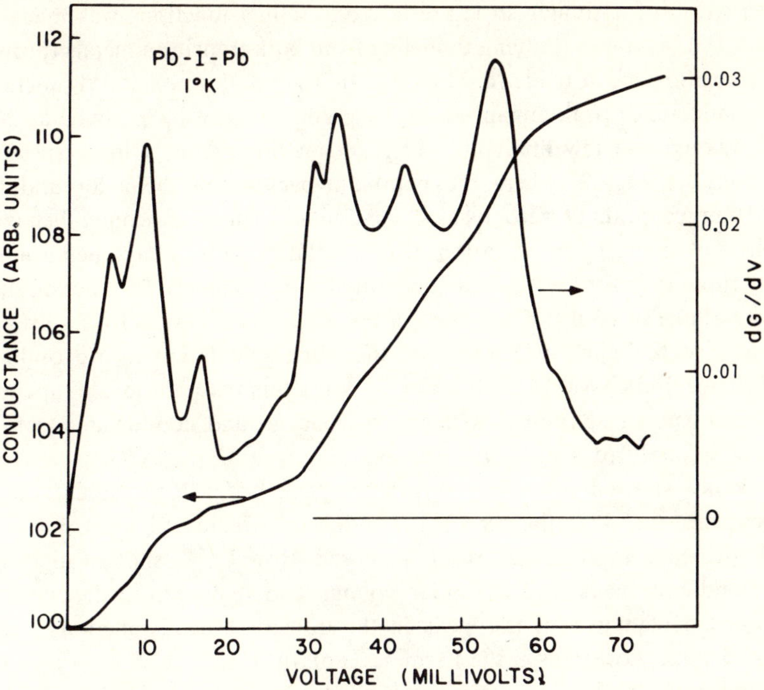

Fig. 8. Conductance and derivative of conductance (d^2I/dV^2) for a Pb–*I*–Pb junction at 1°K. A magnetic field was used to quench superconductivity.

Sn and In for Pb in Al–*I*–M junctions and found structures at the phonon energies of metal M in both cases. In the Al–*I*–Sn junction of Fig. 7 the rise in conductance from 0 to 35 mV is due to these metal phonon excitations.

To summarize these "finite-bias anomalies": we might hope to observe band edges, and have observed the excitation of molecules in the oxide, the oxide itself, and the metal films adjacent to the oxide. All these excitation processes give rise to conductance increases ~1%. We now turn to the zero-bias anomaly, which is a symmetric structure exactly at zero bias, only a few millivolts wide, and may locally increase the conductance by ~10%. Compared to the excitation process described above it is a strong effect.

The history of this effect goes back to 1960, when Hall and co-workers ([10]) noticed a dip in conductance at zero bias in the tunneling characteristic of III–V diodes. Chynoweth *et al.* ([11]) observed a similar dip in Si diodes which were relatively lightly doped, but in heavily doped diodes the structure in conductance was a peak instead of a dip. The peak increased the conductance locally by at most 5% and was only a few millivolts wide. The first observation of a similar structure in oxide tunnel junctions was made by Wyatt ([12]), who was studying tunneling from bulk transition metals through the transition metal oxide to Al or Ag. In Ta–tantalum oxide–Al junctions the conductance peak enhanced the local conductance by as much as 20% and was again a few millivolts wide. To put this anomaly in perspective, we consider Fig. 9, where the parabolic background behavior and the conductance peak at zero bias are presented. The background behavior itself is of interest, the variation of conductance with voltage being much more rapid than for the Al–*I*–Sn junction of Fig. 3. The conductance doubles for a voltage of 65 mV. The junction resistance was 6 ohms for an area of about 2×10^{-3} cm² and, following the same calculation as we outlined earlier, we find $\bar{\varphi} \approx 0.13$ V and $S \approx 30$ Å for this tantalum oxide junction. It is not known at present whether this background conductance can be ascribed simply to such a low barrier.

Wyatt showed that the conductance peak had a logarithmic voltage dependence and that the zero bias conductance increased logarithmically with decrease in temperature. Logan and Rowell ([13]) showed that the semiconductor peak had a similar voltage and temperature dependence. The first explanation of this peak in the transition-metal junctions was a peak in density of states at the Fermi level of Ta. However, it was suggested by Anderson ([14]) that magnetic scattering at the metal–metal oxide interface might be important, and independently Kim ([15]) calculated one such mag-

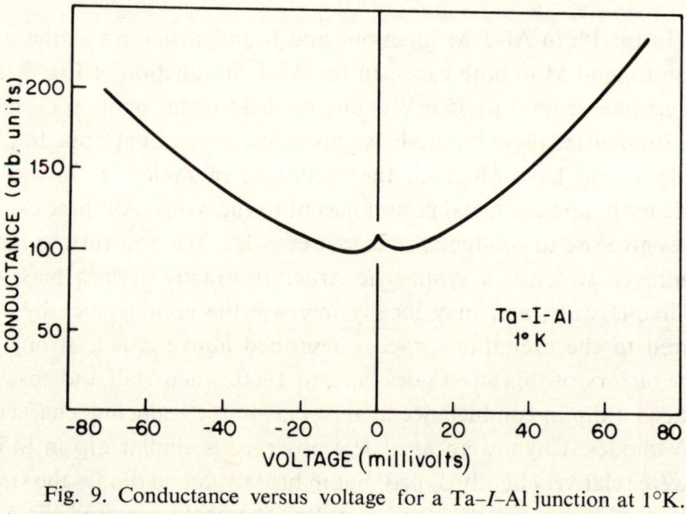

Fig. 9. Conductance versus voltage for a Ta–*I*–Al junction at 1°K. A magnetic field sufficient to destroy superconductivity was applied. From Shen and Rowell ([22]).

netic scattering term which had logarithmic T and V dependence but appeared to fourth order in the tunneling matrix element (M^4). Further experiments by Rowell and Shen ([14]) indicated that the magnitude of the anomaly varied with the surface preparation of the Ta and with the oxidation procedure (clearly implicating the oxide as the origin of the anomaly) but that it was independent of junction resistance, hence eliminating any terms to higher order in M than M^2.

The scattering of electrons from localized magnetic moments in dilute alloys was shown by Kondo ([17]) to explain the resistivity minimum at low temperatures and logarithmic increase in resistivity for temperatures lower

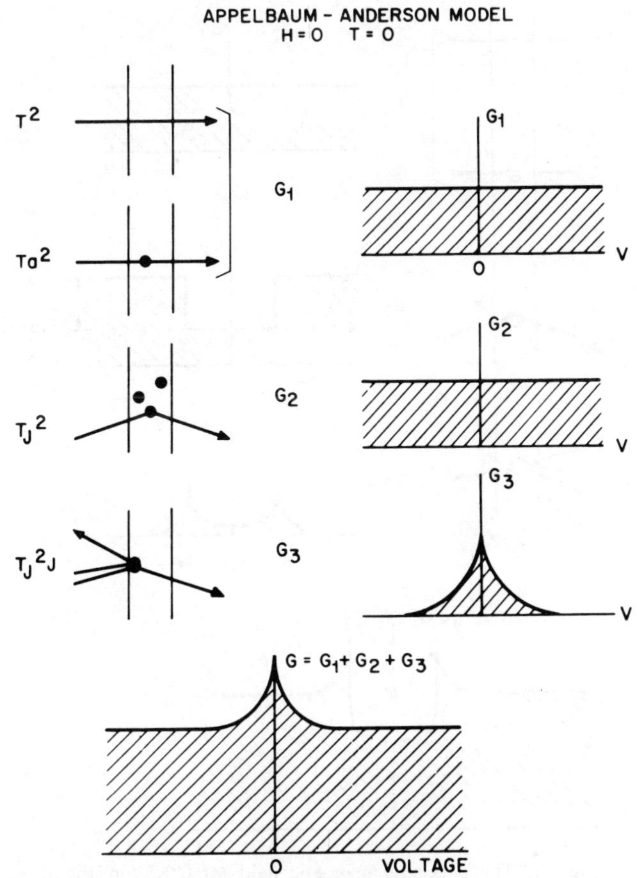

Fig. 10. Very schematic representation of the tunneling and scattering processes that occur in a junction containing localized magnetic impurities. From Shen and Rowell ([22]).

than the minimum. This treatment of the bulk scattering problem was applied to tunneling by Appelbaum ([18]), who assumed that a number of noninteracting localized magnetic impurities existed at the metal–oxide interface of the transition-metal junctions.

The tunneling conductance for such a junction can be regarded as the sum of three parts, as shown schematically in Fig. 10. The first, G_1, arises from electrons which either go straight through the oxide without seeing the impurities or are simply scattered without any spin interaction. This

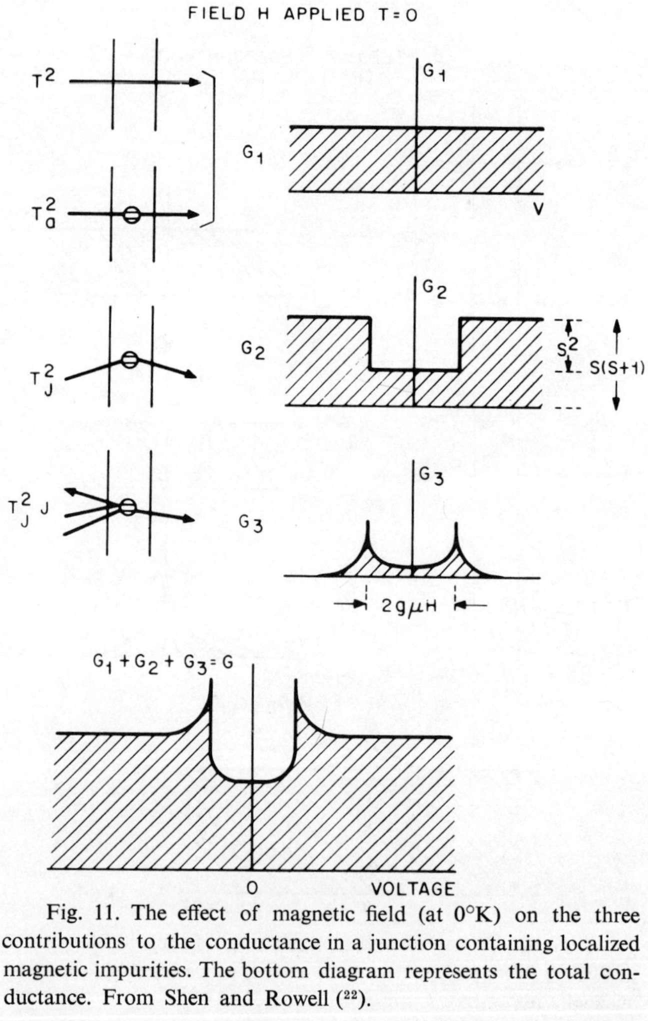

Fig. 11. The effect of magnetic field (at 0°K) on the three contributions to the conductance in a junction containing localized magnetic impurities. The bottom diagram represents the total conductance. From Shen and Rowell ([22]).

is the term we expect in any tunnel junction. The unusual parts are the second, G_2, which is the current due to electrons which scatter from Ta to Al and in doing so flip the spin of the impurity, and the third, G_3, which is an interference between these transmitted electrons and the electrons from Ta which are spin-flip scattered back into the Ta. It is this third interference term which Appelbaum showed gave a logarithmic conductance peak which Anderson ([19]) demonstrated could be the right order of magnitude. This of itself would not be convincing proof of the correctness of the calculation, but Appelbaum ([20]) also predicted the magnetic field dependence of these three conductance terms. The first, G_1, is independent of magnetic field, but the "transmission term" G_2 is strongly affected by field, as shown in Fig. 11. This follows because an applied field, H, will split the Zeeman levels of the impurity by $g\mu H$, and a tunneling electron which flips this spin will also have to lose energy $g\mu H$, in exact analogy with the other emission processes discussed above. Thus a hole of width $2g\mu H$ appears in the G_2 term. The G_3 term is expected to split into two peaks, also $2g\mu H$ apart. Figure 11 shows the shape of the total conductance for $T = 0$ and applied field H. The resulting conductance at $T = 1.4°K$ was calculated by Appelbaum and is shown in Fig. 12. Note that the split "peaks" produced by the field are *not separated by* $2g\mu H$. In fact, the dominant effect of the field is to reduce the G_2 term and decrease the conductance. Thermal smearing then gives a depression of conductance which extends beyond $2g\mu H$. Indeed, Fig. 12 shows that $2g\mu H$ is appreciably smaller than the width of the calculated "hole" in conductance and approximately one-half the separation of the "peaks" produced by this depression. Thus, although Appelbaum does not point this out, it is the G_2 term which dominates his magnetic field dependence. The only field effect attributable to the G_3 term is the *increase* in conductance with field at higher voltages.

Experimental measurements ([21]) of the field dependence for a number of junctions are shown in Fig. 13; the similarity to Appelbaum's calculation is readily apparent. An analysis of the data has been carried out ([22]) assuming that only the G_2 field dependence is important, and reasonable values are obtained for the "g factor" of the impurity. It seems to me that there is little doubt that localized magnetic impurities are present in these junctions and that the scattering model of Appelbaum explains the resulting conductance peak in the tunneling characteristic.

Now I would like to briefly discuss the "giant anomaly" I observed in chromium oxide junctions. I chose this system because the oxide is known to be magnetic, but hardly expected the strong resistance peak in the tunneling characteristic shown in Fig. 14. Actually, it's less dramatic if one

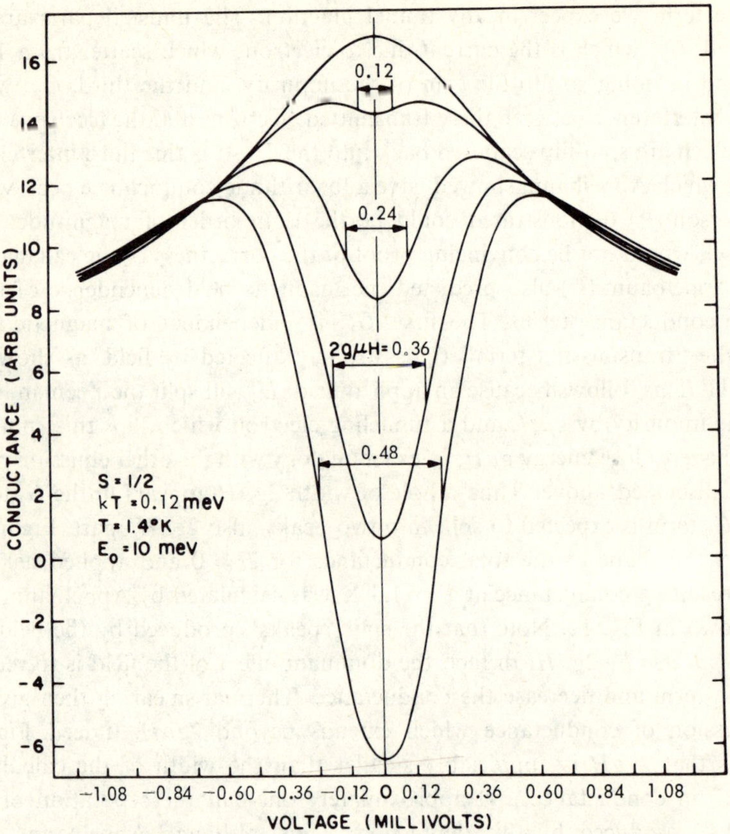

Fig. 12. The effect of magnetic field on the zero-bias tunneling anomaly (at $T = 1.4°K$) as calculated by Appelbaum. The four values of $2g\mu H$ are shown on the figure.

plots conductance against voltage, as that appears as roughly a straight line ([22]), but it's a very rapid conductance change compared to any other type of junction. Attempts have been made to explain this "giant anomaly" in terms of strong coupling between the tunneling electron and the impurities ([23,24]), but to date no one has dared to plot the experimental data and his theory on the same page or deduce any significant number from such a comparison, so personally I still consider the data to be unexplained, although there can be little doubt that it is the result of magnetic scattering.* It is a little surprising that after all the discussion in this volume of the characteristics of tunneling—for example, ohmic current–voltage depend-

* See, for example, the very interesting results on "doped" aluminum oxide junctions ([25]).

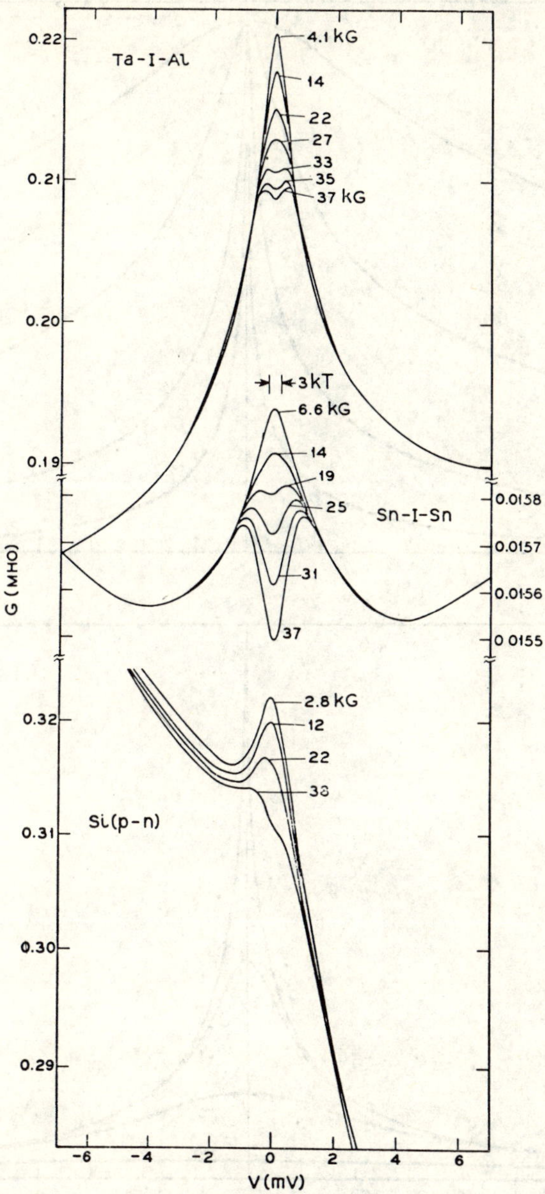

Fig. 13. Experimental results for the dependence of the zero-bias anomaly on magnetic field. The junctions are Ta–I–Al, Sn–I–Sn, and a silicon p–n diode. The temperature was 1.4°K and the voltage corresponding to 3 kT is shown on the figure. From Shen and Rowell ([21]).

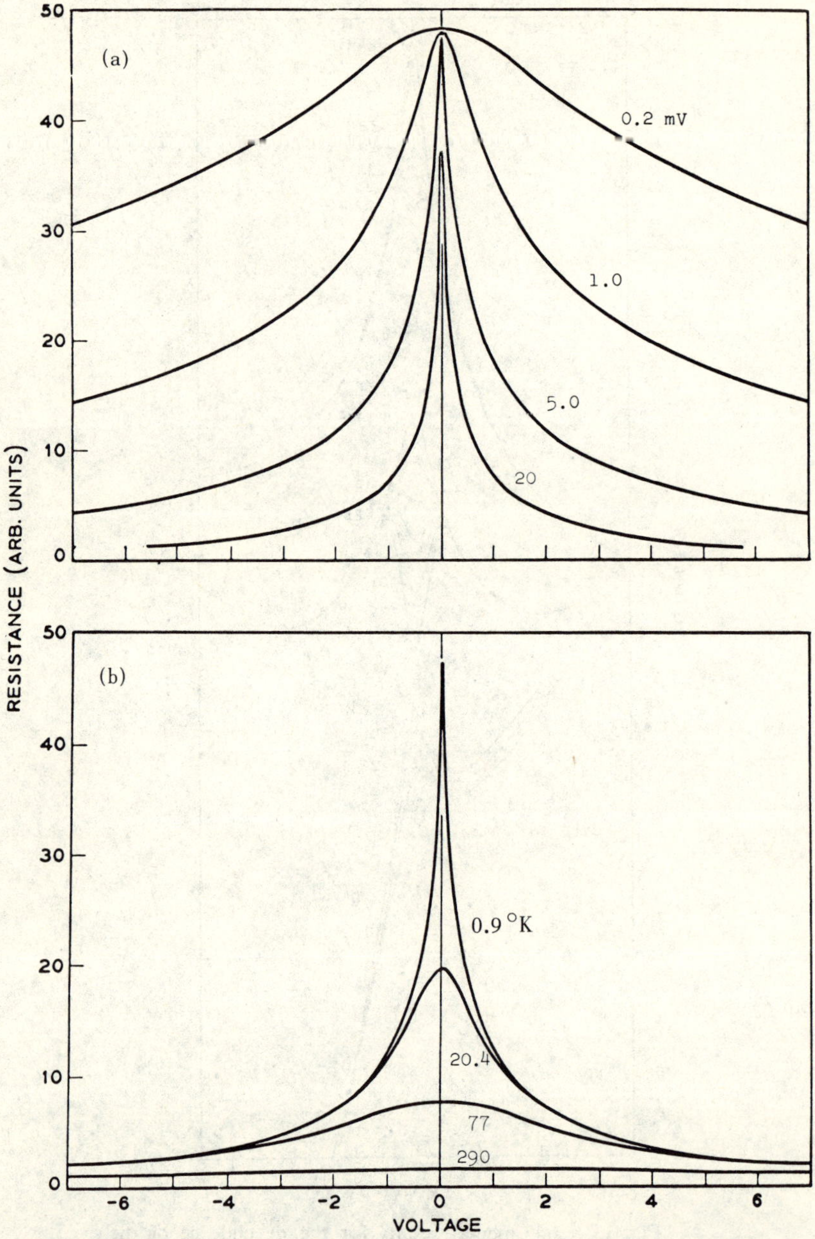

Fig. 14. Dynamic resistance versus voltage for a Cr–*I*–Ag junction. (*a*) For a temperature of 0.9°K; the numbers by each curve give the voltage in millivolts per division of the lower scale. (*b*) The voltage dependence at the four temperatures indicated, the voltage scale being 10 mV/division. From Rowell and Shen [16].

ence and temperature independence— I can now show a junction with such a strong voltage and temperature dependence and say that this is tunneling. One might doubt that this is tunneling at all. We thought of this possibility —it would have been the easiest explanation of the characteristic—and made a Cr–*I*–Pb junction which showed the Pb superconducting gap very clearly ([22]) in addition to a characteristic at higher voltages similar to that of Fig. 14. But of course, that only proves one has tunneling up to 3 mV, and as far as I can see, one only infers that one has a tunneling mechanism for all higher voltages. In fact, our measurements in the region of the phonon energies in Pb ([22]) showed smaller effects in conductance than would be found in an Al–*I*–Pb junction, and the simplest interpretation of this is to say that not all the current flow is via tunneling. I don't think I really believe that, but it's worth thinking about.

For example, one could argue that if in Al–*I*–Sn junctions with no magnetic impurities we have a barrier height ~2 V—and in Ta–*I*–Al junctions with a conductance peak anomaly the barrier is reduced to only 0.13 V—then perhaps in the chromium junctions, where the magnetic oxide effects seem so much stronger, we have a very low barrier indeed, say 10 mV.* In this case we would not expect the density of states in Pb to be reflected in the tunneling characteristic in the usual way, as the applied bias is now comparable to the barrier height.

I would like to make a comment about another "resistance peak" anomaly, the conductance dip in III–V diodes ([10]) which started this field of tunneling anomalies. This has been ascribed by Duke *et al.* ([27]) (also see Chapter 28) to phonon emission in the diode depletion region, in the same way as we say the conductance rises from 0 to 30 mV in Fig. 7 and from 0 to 15 mV in Fig. 8 are due to the emission of phonons of the metal films adjacent to the oxide. There are three things which seem odd about this explanation. First, the conductance dip in semiconductor diodes is much larger than any of the emission processes in oxide junctions. Second, the oxide junctions rather clearly reflect the oxide phonon density as steps in conductance or peaks in d^2I/dV^2. The semiconductor phonon density also has a number of well-defined peaks, but except for structure at the $k = 0$ longitudinal optical phonon energy the conductance dip has no sudden changes in slope. Third, as most phonon spectra can be approximated by a Debye spectrum for low energies, I would have expected the conductance dip, if it is due to phonon emission, to be rather flat at low voltages.

* This suggestion has been made recently by Christopher *et al.* ([26]) to explain their results for Fe–*I*–Fe junctions.

Experimentally, it often looks too singular for $V < 1$ mV. In view of this I would like to see more experimental work before I believe that the zero-bias conductance dip anomaly in III–V diodes is satisfactorily explained by phonon emission in the barrier.

In conclusion, it appears that we understand the tunneling characteristic for Al–*I*–M junctions for biases up to ∼1 V and realize that emission processes of various types in the barrier give rise to small conductance increases above the background parabola. The conductance peak zero-bias anomaly is satisfactorily explained by the magnetic scattering theory of Appelbaum, but the resistance peaks remain, for me, unexplained at present.

REFERENCES

1. J. G. Simmons, J. Appl. Phys. 34:1793 (1963), 34:2581 (1963).
2. J. G. Simmons, J. Appl. Phys. 34:238 (1963).
3. H. P. Knauss and R. A. Breslow, Proc. IRE 50:1834 (1962).
4. R. Stratton, J. Phys. Chem. Solids 23:1177 (1962).
5. T. E. Hartman, J. Appl. Phys. 35:3283 (1964).
6. J. J. Hauser and L. R. Testardi, Phys. Rev. Letters 20:12 (1968).
7. I. Giaever, Solid State Commun. Vol. 5 (unpublished).
8. L. Esaki and P. J. Stiles, Phys. Rev. Letters 14:902 (1965).
9. J. M. Rowell and W. L. McMillan, Bull. Am. Phys. Soc. 12:77 (1967).
10. R. N. Hall, J. H. Racette, and H. Ehrenreich, Phys. Rev. Letters 4:456 (1960); R. N. Hall, in: Proc. International Conference on Semiconductor Physics, Prague, 1960, Academic Press, New York, 1961, p. 193.
11. A. G. Chynoweth, R. A. Logan, and D. E. Thomas, Phys. Rev. 125:877 (1962).
12. A. F. G. Wyatt, Phys. Rev. Letters 13:401 (1964).
13. R. A. Logan and J. M. Rowell, Phys. Rev. Letters 13:404 (1964).
14. P. W. Anderson, unpublished.
15. D. J. Kim, Phys. Letters 18:215 (1965).
16. J. M. Rowell and L. Y. L. Shen, Phys. Rev. Letters 17:15 (1966).
17. J. Kondo, Prog. Theor. Phys. (Kyoto) 32:37 (1964).
18. J. Appelbaum, Phys. Rev. Letters 17:91 (1966).
19. P. W. Anderson, Phys. Rev. Letters 17:95 (1966).
20. J. Appelbaum, Phys. Rev. 154:633 (1967).
21. L. Y. L. Shen and J. M. Rowell, Solid State Commun. 5:189 (1967).
22. L. Y. L. Shen and J. M. Rowell, Phys. Rev. 165:566 (1968).
23. J. Appelbaum, J. C. Phillips, and G. Tzouras, Phys. Rev. 160:554 (1967).
24. J. Solyom and A. Zawadowski, Physik kondensierten Materie 7:325, 342 (1968).
25. F. Mezei, Phys. Letters 25A:534 (1967); A. F. G. Wyatt and D. J. Lythall, *ibid.* 25A:541 (1967).
26. J. E. Christopher *et al.*, Phys. Rev. 172:485 (1968).
27. C. B. Duke, S. D. Silverstein, and A. J. Bennet, Phys. Rev. Letters 19:315 (1967).

Chapter 28

A Unified Theory of Zero-Bias Anomalies and Energy-Loss Mechanisms in the Barrier

C. B. Duke

Research and Development Center, General Electric
Schenectady, New York

During the past six years a variety of experiments have been performed which reveal a rapidly varying tunnel conductance on an energy scale of from 5 to 100 meV. All of these variations occur at too small a bias ($V \lesssim 100$ mV) or are too sharp to be attributed to variations in the junction potential with bias. In this chapter I shall outline a general theory in which these rapid variations in the differential tunnel conductance are attributed either to the excitation of internal degrees of freedom of the barrier or to self-energy effects in the electrodes.

Interest in such rapid conductance changes, collectively referred to by the generic title of "anomalies," originally was stimulated by the observation of narrow conductance minima in III–V tunnel diodes by Hall *et al.* ([1]). These particular anomalies have a stormy theoretical as well as experimental history. Proposed explanations attributing them to polaron formation or LO phonon density-of-states variations enjoyed popularity, but were demonstrated to be incorrect by Mahan and Duke ([2]). Furthermore, some of the conductance minima near zero bias are due to the contacts, as pointed out by Vul *et al.* ([3]). In that work, however, the interpretation of the *p–n* junction minima in terms of density-of-states variations is incorrect.

In the years from 1964 to the present both the variety of "anomalies" and the systems in which they were observed multiplied rapidly. Further work on III–V tunnel diodes is discussed in ([2,3]). Logan and Rowell ([4]) considered Si and Ge *p–n* diodes. Metal–semiconductor (GaAs, Ge) contact measurements illustrating the zero-bias conductance minima have been reported by Conley and Tiemann ([5]) and by Conley and Mahan ([6]). From these measurements emerged three systematics of the observed minima: they are stronger in *p*- than in *n*-type material, they disappear for acceptor doping $n_A \gtrsim 5 \times 10^{19}\ \text{cm}^{-3}$, and their width, $\Delta V \approx 8$ meV, is not changed appreciably upon doping. The temperature dependence of the line shapes was investigated by Vul *et al.* ([3]) and by Shewchun and Williams ([7]).

In Chapter 27 of this volume Rowell discussed the experiments on metal–insulator–metal (M–*I*–M) junctions in detail. We recall that in Ta–*I*–Al and Sn–*I*–Sn a conductance maximum is observed at zero bias and exhibits a width $\Delta V \lesssim 3$ meV ([8]). Much larger conductance minima have been reported by Rowell in Cr–*I*–Ag and Cr–*I*–Pb junctions and by Coleman in Al–*I*–Fe junctions ([9]). A host of fine structures in other M–*I*–M junctions below 500 meV includes the identification of steps in the conductance due to the excitation of vibrational states of water molecules ([10]) and tentative identification of optical phonon excitation in the oxide and acoustic phonon excitation in the metal ([11]).

The theoretical interpretations of most of these results have been given piecemeal. Appelbaum ([12]) and Anderson ([13]) attributed the zero-bias conductance maxima to the resonant excitation of spin-fiip transitions of magnetic impurities in the oxide. Jaklevic and Lambe interpreted their measurements of step discontinuities in the conductance as due to the excitation of vibronic states of adsorbed water molecules in the junction. Recently Duke *et al.* ([14]) have attributed the conductance minima in III–V tunnel diodes and metal–semiconductor (M–S) contacts to phonon excitation in the barrier. All of these interpretations exhibit a common feature: they invoke a mechanism in which the tunneling electron excites some internal degree of freedom of the barrier. For tunneling through paramagnetic impurities the excitation of the spin-flip transition involves an energy loss only in the presence of a magnetic field. For the excitation of excited states of individual impurities or of collective excitations of the barrier as a whole, the electron transfers energy to the internal excitation of the barrier. Therefore we see that the tunneling electron may act as a spectroscopic probe of the internal degrees of freedom of the *barrier* in normal-metal junctions, rather than as a probe of the bulk metals as in the usual applications of superconducting junctions. The important conservation law in inelastic tunneling spectroscopy is that of energy conservation. For the (incoherent) interaction with individual impurities in the junction there is no concomitant momentum conservation law. For the coherent excitation of collective modes of the junction by the tunneling electron we find a conservation law on the component of momentum parallel to the junction plane. Therefore inelastic tunneling spectroscopy of collective modes differs from, e.g., slow neutron spectroscopy of these same modes, not only by use of a different coupling mechanism, but also by the absence or modification of the role of momentum conservation in the excitation process. The tunneling processes in which the $\mathbf{k}_{\parallel}$ conservation rule are most important are those in which the initial and final states of the electron suffer a large shift in $\mathbf{k}_{\parallel}$. In Chapter 13 Kleinman discussed this aspect of tunneling in indirect band-gap semiconductor tunnel diodes for the coherent excitation of phonons. The incoherent phonon excitation processes are affected by the large shift in $\mathbf{k}_{\parallel}$ between the initial and final states via both a reduction in the strength of these processes and a narrowing of the associated conductance minima near zero bias. Our subsequent discussion is confined to cases in which the electron (hole) energies in the initial and final states of the tunneling process are centered around the same point in momentum space.

Prior to a detailed discussion of our theoretical model let us examine the underlying features of any model of tunneling via energy-loss mechanisms. For our qualitative considerations we view tunneling as the energy-loss process shown in Fig. 1.

For a continuous spectrum of excitation energies of density $N(\varepsilon)$ we see from Fig. 1 that all excitations with energy $\varepsilon = \hbar\omega_p < |eV|$ can be created. Every tunneling electron creates excitations, so we expect the inelastic conductance to be roughly proportional to the number of such excitations.

$$\Delta G(eV) \propto \int_0^{|eV|} N(\varepsilon)\, d\varepsilon. \tag{1a}$$

More precisely, the proportionality of ΔG to the integrated density of states up to energy eV occurs because as the bias is increased from eV to $eV + d(eV)$, each tunneling electron on the left-hand side of Fig. 1 has more inelastic tunneling channels available to it. The electron of energy ε above

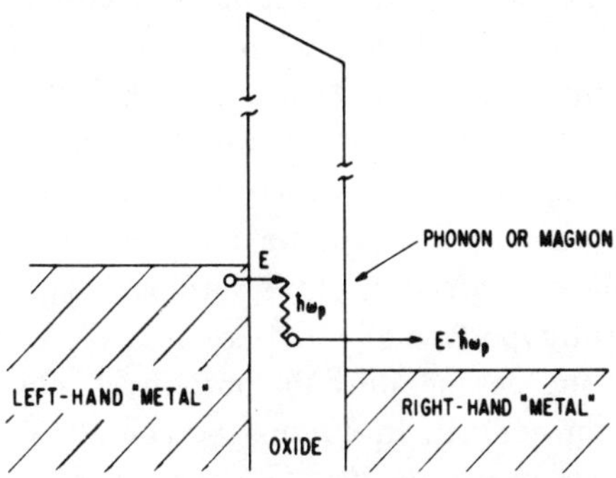

Fig. 1. Schematic illustration of tunneling via an energy-loss mechanism in which an excitation of energy $\hbar\omega_p$ is created.

the Fermi energy of the right-hand side has $N(\varepsilon)$ additional new channels open to it. Thus the integral in Eq. (1a) is interpreted as summing up these added channels for all the electrons which can tunnel. It follows immediately that the derivative of the conductance is directly proportional to the density of states of the continuum excitations at the bias eV:

$$d(\Delta G)/d(eV) \propto N(eV). \tag{1b}$$

These observations are evidently quite independent of the dynamic origin of the excitations to which the electrons transfer energy. They are the consequences of any second-order perturbation theory in which the electron is coupled to excitations of spectral density $N(\varepsilon)$ in the junction. LO phonons ([14]) and the hydroxyl excitations ([10]) exhibit $N(\varepsilon) \approx N_0\, \delta(\varepsilon - \varepsilon_0)$, so that Eq. (1$a$) gives step-like increases in the conductivity. Acoustical excitations starting at zero energy give conductance minima at zero bias with $\Delta G \approx (eV)^n$ for small values of eV, where n is determined by the dynamics of the electron–excitation coupling.

The major deficiency of our qualitative picture is the complete neglect of the dynamics of either the excitation spectrum itself or of the coupling of the electron to this spectrum. It is precisely the effects of these dynamics which distinguish between the various types of rapid changes in the conductance and make the observation of these changes a useful spectroscopic tool. The explanation of the Jaklevic–Lambe experiment ([10]) needs no further essential refinement. The various descriptions of zero-bias conductance minima differ depending on the character of the excitation spectrum as reflected in $N(\varepsilon)$ (e.g., magnon versus phonon excitation) and, of more importance, depending on the details of the electron–excitation coupling. The Appelbaum–Anderson ([12,13]) description of a conductance maximum at zero bias is based on a resonance coupling of the tunneling electron to zero-energy spin-flip excitations. Hence this conductance "anomaly" is interpreted using a very special model for the dynamics of the electron–excitation coupling, and is a consequence of the peculiarities of these dynamics. Therefore in order to proceed, we must develop a model within the context of which the consequences of the various dynamics can be explored.

Consider a prototype "metal"–barrier–"metal" system in which either "metal" could be the degenerate bulk portion of a p–n tunnel diode or metal–semiconductor contact. In writing the Hamiltonian to describe tunneling in this system we must give a description of the dynamics in the barrier region which encompasses the possibility of electron excitation of either excited states of individual impurities or collective modes of the "barrier" as a whole. We use the Heisenberg Hamiltonian:

$$\mathscr{H} = \mathscr{H}_R + \mathscr{H}_L + \mathscr{H}_B, \tag{2a}$$

$$\mathscr{H}_R = \sum_i [(-\hbar^2\nabla_i^2/2m) + V_{pR}(\mathbf{r}_i)], \tag{2b}$$

$$\mathscr{H}_L = \sum_i [(-\hbar^2\nabla_i^2/2m) + V_{pL}(\mathbf{r}_i)], \tag{2c}$$

$$\mathscr{H}_B = \sum_i [-\hbar^2\nabla_i^2/2m + V_B(\mathbf{r}_i)] + \sum_n g_n\mu_B S_n^z(t)H + \sum_{i,n} \{V_n[\mathbf{r}_i - \mathbf{R}_n(t);t] + \boldsymbol{\sigma}_i \cdot \mathbf{S}_n(t)J_n[\mathbf{r}_i - \mathbf{R}_n(t)]\}. \tag{2d}$$

The effect of the magnetic field $\mathbf{H}$ on the electrons has been neglected for convenience. No explicit dynamics of the collective phonon (magnon) modes is incorporated in Eqs. (2), because the selection of these dynamics depends on the system under consideration. For example, in a single-crystal *p–n* diode the phonons excited by the tunneling electron are characteristic of the diode as a whole. In a M–S contact they are characteristic of the semiconductor, and in an M–*I*–M junction acoustical phonons are probably characteristic of the layer structure as a whole, whereas optical phonons, which damp in the bulk metals, are likely characteristic only of the barrier "oxide." The subscript "*p*" indicates a periodic potential and $V_B(\mathbf{r})$ is the electrostatic potential in the barrier. The V_n and J_n denote the potentials seen by a tunneling electron due to the atoms in the barrier.

The possibility of internal excitation of barrier atoms is included via the explicit t dependence of $V_n[\mathbf{r} - \mathbf{R}_n(t); t]$. The possibility of phonon excitation is incorporated by the t dependence of the atomic positions,

$$\mathbf{R}_n(t) = \mathbf{R}_n^{(0)} + u_n(t). \tag{3a}$$

Magnon excitation in magnetically ordered systems and resonant-spin-flip transitions in paramagnetic systems arise from the t dependence of $\mathbf{S}_n(t)$ in the *s–d* exchange term.

Our first task is the extraction of the periodic part of the potential in the barrier and the subsequent derivation of the "effective" tunneling Hamiltonian due to this term. The periodic potential is extracted by writing

$$\sum_n V_n[\mathbf{r} - \mathbf{R}_n(t);t] \equiv \sum_n \{V_{fn}[\mathbf{r} - \mathbf{R}_n(t);t] + \bar{V}[\mathbf{r} - \mathbf{R}_n(t)]\}, \tag{4}$$

in which $\bar{V}(\mathbf{r} - \mathbf{R}_n)$ is the potential at the nth site in the absence of impurities. We then expand

$$\bar{V}[\mathbf{r} - \mathbf{R}_n(t)] = \bar{V}[\mathbf{r} - \mathbf{R}_n^{(0)}] + \vartheta(u_n). \tag{3b}$$

The neglected terms in Eq. (3*b*) represent the coherent interaction of the electrons with phonons in the oxide. The periodic potential in the oxide is given by

$$V_{pB}(\mathbf{r}) = \sum_n V[\mathbf{r} - \mathbf{R}_n^{(0)}]. \tag{5}$$

The treatment of the periodic potential alone is discussed in detail in Chapters 1, 2, 4 and 7. The one-electron eigenfunctions are determined from the equation:

$$[(-\hbar^2\nabla^2/2m) + V_{pi}(\mathbf{r})]\psi_{i;\mathbf{k}n}(\mathbf{r}) = \varepsilon_{i,n}(\mathbf{k})\psi_{i;\mathbf{k}n}(\mathbf{r}). \tag{6}$$

For the remainder of this chapter I shall use the effective-mass-approximation limit of the effective-tunneling-Hamiltonian method of Bardeen ([15]). The eigenfunctions in the presence of the junction are written in terms of the Bloch functions (6) as

$$\varphi_E(\mathbf{r}) = \sum_{\mathbf{k}} f_E(\mathbf{k})\psi_{\mathbf{k}n}(\mathbf{r}), \tag{7a}$$

with the $f_E(\mathbf{k})$ being evaluated from the effective-mass equations for its Fourier transform (we neglect field penetration into the "metals"),

$$\begin{aligned} &\varepsilon_{L,n_L}(-i\nabla)f_E(\mathbf{r}) = Ef_E(\mathbf{r}), && z < z_L \\ &[\varepsilon_B(-i\nabla) + V_B(\mathbf{r})]f_E(\mathbf{r}) = Ef_E(\mathbf{r}), && z_L < z < z_R, \\ &\varepsilon_{R,n_R}(-i\nabla)f_E(\mathbf{r}) = Ef_E(\mathbf{r}), && z > z_R, \end{aligned} \tag{7b}$$

assuming

$$\varepsilon_i(-i\nabla) \simeq -(\hbar^2\nabla^2/2m_i) + V_{0i} \tag{8}$$

and continuity of f and $(m_i^{-1}\, df/dz)$ at the boundaries $z = z_R, z_L$. We consider the envelope functions for the "effective" left- and right-hand wave functions to be given by

$$f_{E,i}(\mathbf{r}) = \frac{\exp(i\mathbf{k}_{\parallel,i}\cdot\boldsymbol{\rho})}{\sqrt{A}}\,\chi_{k_L}(z), \tag{9a}$$

$$E = (\hbar^2 k_{\parallel,i}^2/2m_i) + E_\perp(k_i) + V_{0i}. \tag{9b}$$

For both $i = R$ and L the boundary condition on the χ is that it be purely decaying for increased penetration into the junction, as shown in Fig. 2. In the Bardeen model the potential-barrier tunneling is described by

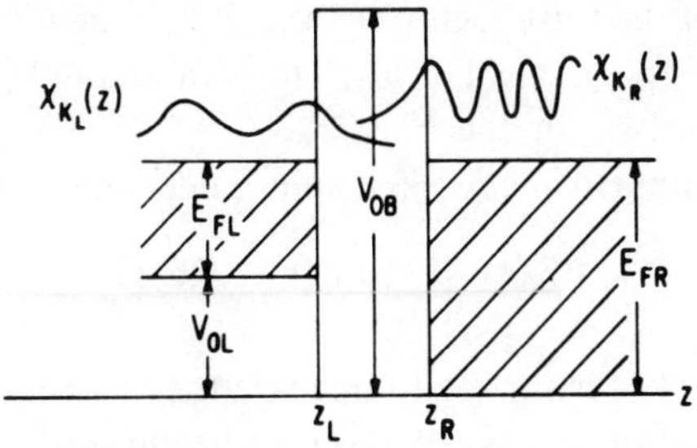

Fig. 2. Schematic illustration of the junction potential and effective tunneling wave functions in a metal–insulator–metal junction.

$$i(eV) = \frac{4\pi e}{\hbar} \sum_{\mathbf{k}_L, \mathbf{k}_R} |\langle \mathbf{k}_L | M | \mathbf{k}_R \rangle|^2 [f(\mathbf{k}_R) - f(\mathbf{k}_L)]\, \delta(E_L - E_R - eV), \tag{10a}$$

$$\begin{aligned}\langle \mathbf{k}_L | M | \mathbf{k}_R \rangle &= \langle \mathbf{k}_L | i\hbar J | \mathbf{k}_R \rangle \\ &= \frac{\hbar^2}{2m_b}\, \delta_{\mathbf{k}_{\|,L},\, \mathbf{k}_{\|,R}} \left[\chi_{k_R} \frac{d\bar{\chi}_{k_L}}{dz} - \bar{\chi}_{k_L} \frac{d\chi_{k_R}}{dz} \right]_{z_L < z < z_R}, \end{aligned} \tag{10b}$$

$$\int dz \chi_{k_i} \bar{\chi}_{q_i} = \langle k_i | q_i \rangle = \delta_{k_i, q_i}. \tag{10c}$$

$$f(\mathbf{k}_i) \equiv \{1 + \exp[(E_i - E_{Fi})/\varkappa T]\}^{-1} \tag{10d}$$

In our model we describe the inelastic tunneling processes by taking the matrix elements of the remainder of the barrier Hamiltonian

$$V_B \equiv \mathscr{H}_B - \sum_i \left[\frac{-\hbar^2 \nabla_i^2}{2m} + V_B(\mathbf{r}_i) + V_{pB}(\mathbf{r}_i) \right] \tag{11}$$

between the states χ_{k_i} localized on the two sides of the barrier. The current is calculated in second-order perturbation theory from expressions analogous to Eq. (10*a*) with the matrix element computed in an appropriate approximation. We shall treat summations over impurity positions in a superficial manner. A complete analysis is outlined in ([14]).

First consider features of the tunnel conductance caused by the excitation of internal degrees of freedom of individual impurities in the barrier. This topic includes both the Jaklevic–Lambe vibronic excitations and the Appelbaum–Anderson interpretation of the conductance maxima at zero bias. For this discussion we ignore the deviations of $\mathbf{R}_n(t)$ from $\mathbf{R}_n^{(0)}$ and consider that any paramagnetic impurities in the barrier do not interact with each other. In this limit the one-electron potential in the barrier is given by

$$V_B(\mathbf{r}) \approx \sum_n (V_{fn}[\mathbf{r} - \mathbf{R}_n^{(0)}; t] + \mathbf{S}_n(t) \cdot \{g_n \mu_B \mathbf{H} + \boldsymbol{\sigma} J_n[\mathbf{r} - \mathbf{R}_n^{(0)}]\}). \tag{12}$$

The dependence of the form factors V_f and J on n, denoted by the subscript, indicates that screening, local fields, etc. can require different potentials for different impurities in the barrier.

In the static approximation for spin-independent impurities,

$$V_B(\mathbf{r}) = \sum_n V_{fn}[\mathbf{r} - \mathbf{R}_n^{(0)}], \tag{13}$$

the effect of impurities is merely to increase the current through the barrier by virtue of the additional contribution to the current (10a) using [we shall try to follow a minor variant of the notation of Appelbaum ([16]) in order to facilitate comparison with that paper]

$$\begin{aligned} \langle \mathbf{k}_L \mid M \mid \mathbf{k}_R \rangle &\equiv \langle \mathbf{k}_L \mid T_a(R_\perp) \mid \mathbf{k}_R \rangle, \qquad z_L < R_\perp < z_R \\ &= \int d^3r \exp[-i(\mathbf{k}_{\parallel,L} - \mathbf{k}_{\parallel,R}) \cdot \boldsymbol{\rho}] \\ &\quad \times \bar{\chi}_{k_L}(z)\chi_{k_R}(z)V_{fn}(\boldsymbol{\rho}, z - R_\perp) \end{aligned} \tag{14a}$$

as the transition amplitude and taking

$$\left\langle | \langle \mathbf{k}_L \mid M \mid \mathbf{k}_R \rangle |^2 \right\rangle_{AV} = A \int_{z_L}^{z_R} dR_\perp C_\parallel(R_\perp) \mid \langle \mathbf{k}_L \mid T_a(R_\perp) \mid \mathbf{k}_R \rangle |^2, \tag{14b}$$

in which $C_\parallel(R_\perp)$ is the concentration per unit area of impurities, all with form factor $V_{fn}(\boldsymbol{\rho}, z)$, a distance from the plane of the junction denoted by $z_L < R_\perp < z_R$. The overlap matrix element, T_a, gives this term a slowly varying bias dependence similar but not identical to that of the potential-barrier conductance. It is worth noting that in p–n diodes and M–S contacts this contribution to the current can be large not because of the resonance-tunneling mechanism of Anderson, but because charge fluctuations in the barrier are not screened and hence the impurity potential is much stronger than in the bulk. This increased strength compensates for the low-impurity density. Thick Ta films are known to have a p–i–n structure. Therefore a similar enhancement mechanism may also be operative in Ta–I–M junctions. In the absence of a resonance effect due to bound states of the impurity potential the contribution of static impurity potentials to the conductance is slowly varying with bias, as shown in Fig. 3.

When the electron excites an internal degree of freedom of the impurity the potential is dynamic rather than static. To illustrate this case we follow Scalapino and Marcus ([17]) in noting that for a dipole excitation the dynamic potential is approximately given by

$$V_{fn}(\mathbf{r}, t) = \frac{2e\langle f \mid p_z \mid i \rangle z \exp(i\omega_0 t)}{[z^2 + \varrho^2]^{3/2}}, \tag{15}$$

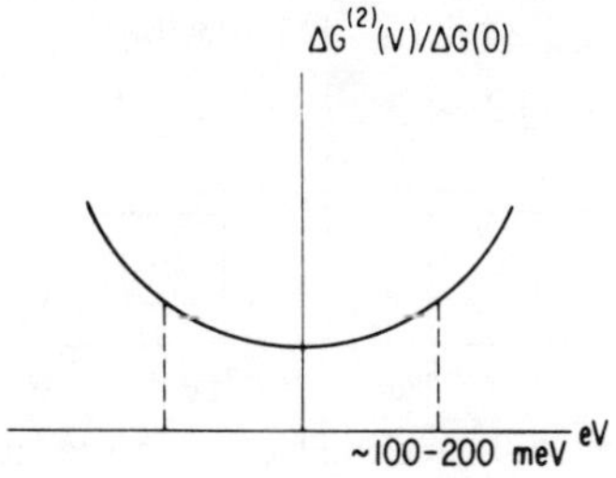

Fig. 3. Contribution to the conductance of a symmetric metal–insulator–metal junction due to static nonresonant impurity potentials in the barrier.

in which $\langle f \mid p_z \mid i \rangle$ is the intra-impurity dipole matrix element associated with the dipole-active mode of energy $\hbar\omega_0$. The second-order contribution to the current is evaluated from the "golden rule" expressions:

$$j_{RL}(eV; \varkappa T \ll \hbar\omega_0) = \frac{4\pi e N_i}{\hbar} \sum_{\mathbf{k}_L, \mathbf{k}_R} \mid \langle \mathbf{k}_L \mid T_a(\omega_0) \mid \mathbf{k}_R \rangle \mid^2 \times [1 - f(E_L)] f(E_R)\, \delta(E_L - E_R - eV + \hbar\omega_0) \quad (16a)$$

$$\langle \mathbf{k}_L \mid T_a(\omega_0) \mid \mathbf{k}_R \rangle \approx \int d^3r \exp[-i(\mathbf{k}_{\parallel,L} - \mathbf{k}_{\parallel,R}) \cdot \boldsymbol{\rho}] \bar{\chi}_{k_L}(z) \chi_{k_R}(z) \times \frac{2e\langle f \mid p_z \mid i \rangle (z - \bar{R}_L)}{[(z - R_\perp)^2 + \varrho^2]^{3/2}}. \quad (16b)$$

All of the hydroxyl impurities were taken to have the same potential, leading to a factor of N_i, the number of impurities in the junction. The Scalapino–Marcus analysis differs from ours in that they used the potential (15) in a WKBJ exponent which they subsequently expanded. Our results reduce to theirs in the limit

$$\langle \mathbf{k}_L \mid T_a(\omega_0) \mid \mathbf{k}_R \rangle \approx \delta_{\mathbf{k}_{\parallel,L}, \mathbf{k}_{\parallel,R}} \int_{z_L}^{z_R} dz \int_{r_0}^{\infty} 2\pi\varrho\, d\varrho \times \frac{2e(z - z_L)\langle f \mid p_z \mid i \rangle}{[\varrho^2 + (z - z_L)^2]^{3/2}} = T(\omega_0, \bar{R}_\perp = z_L), \quad (17)$$

so that the matrix element is independent of $\mathbf{k}_L$ and $\mathbf{k}_R$ and is cut off a distance r_0 from the hydroxyl. In this limit we convert

$$\sum_{\mathbf{k}_i} \to \varrho_i \int d\xi_i, \qquad \xi_i = E_i - E_{FR}, \quad (18)$$

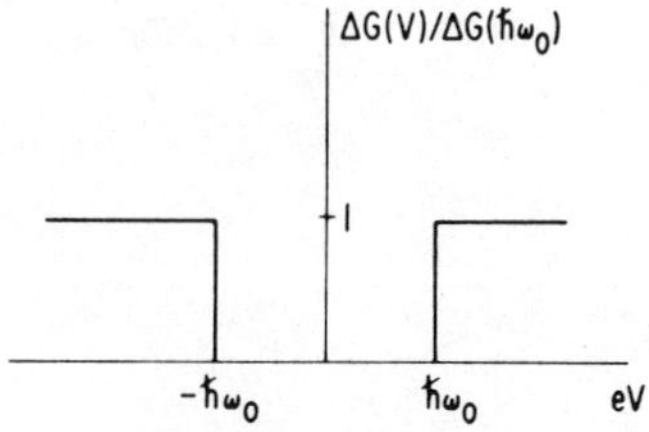

Fig. 4. Incremental conductance due to the onset of inelastic tunneling via the excitation of an excited state of energy $\hbar\omega_0$ of an impurity in the barrier.

with $\varrho_i = m_i k_{Fi}/2\pi^2\hbar^2$ being the density of states at the Fermi surface. At $T = 0$ for positive bias we find

$$j_{RL} = \frac{4\pi e N_i}{\hbar} \mid T(\omega_0) \mid^2 \varrho_L \varrho_R (eV - \hbar\omega_0) \theta(eV - \hbar\omega_0). \tag{19}$$

A similar result is valid for negative bias, so we get the results shown in Fig. 4. The sudden step in the conductance when $eV > \hbar\omega_0$ is due to the occurrence of a *new mechanism* for the electron to tunnel, as discussed in conjunction with Eqs. (1). Because of the sharp onset of the incremental conductance the approximations (17) merely renormalize the height of the step conductance. This is not always the case in the excitation of a band of collective modes, for which the neglect of the $\mathbf{k}_{\parallel}$ dependence of the analog of Eq. (16*b*) can alter the line shape as well.

In order to recover the Appelbaum–Anderson explanation of conductance maxima at zero bias, we must calculate $\mid \langle \mathbf{k}_L \mid M \mid \mathbf{k}_R \rangle \mid^2$ more accurately by incorporating the possibility of spin-flip excitations in the barrier by electrons which reflect back into the original metal rather than tunnel through the barrier. If there are paramagnetic impurities in the oxide, then in addition to the transition matrix elements (14) and (16) we also have

$$\langle \mathbf{k}_i \mid T^{Jd} \mid \mathbf{k}_j \rangle = \langle \mathbf{k}_i, m, M \mid \sigma^z S_n^{\,z} J_n[\mathbf{r} - \mathbf{R}_n^{(0)}] \mid \mathbf{k}_j, m, M \rangle, \tag{20a}$$

$$\langle \mathbf{k}_i \mid T^{\pm} \mid \mathbf{k}_j \rangle = \langle \mathbf{k}_i, m \pm 1, M \mp 1 \mid \tfrac{1}{2}\sigma^{\pm} S_n^{\,\mp} J_n[\mathbf{r} - \mathbf{R}_n^{(0)}] \mid \mathbf{k}_j, m, M \rangle, \tag{20b}$$

describing spin-flip ($T^{\pm}$) and no-spin-flip (T^{Jd}) tunneling as well as the analogous reflection terms $J^{d,\pm}$ describing reflection when $i = j$ in Eqs. (20). An explanation of the details of the calculation has been given in ([16]) so no effort is made to repeat them here.

To make the physics of the effect plausible, let us first consider the situation in the presence of a magnetic field. As $T \to 0$, the $g_n \mu_B S_n{}^z H$ term in Eq. (2*d*) reveals that the paramagnetic impurities in the junction will align antiparallel to the magnetic field, i.e., $\langle S^z \rangle \to -M$. For an electron to flip one of these spins, it must give to the impurity an energy

$$\hbar\omega_0 = g_n \mu_B H. \tag{21}$$

Thus the second-order terms give a conductance which is a superposition of Figs. 3 and 4, i.e., as shown in Fig. 5. Appelbaum gives the expressions using the same interaction $J_n(\mathbf{x}) = J(\mathbf{x})$ for all n (nsf = nospin-flip, sf = spin-flip):

$$\Delta G^{(2)}_{\text{nsf}}(eV) = (4\pi e/\hbar) N_i \varrho_R \varrho_L \{ | T_a |^2 + | T_J |^2 \langle M^2 \rangle \}, \tag{22a}$$

$$\Delta G^{(2)}_{\text{sf}}(eV) = \frac{4\pi e N_i}{\hbar} \varrho_L \varrho_R | T_J |^2$$
$$\times \{ S(S+1) - \langle M^2 \rangle + \langle M \rangle \, \theta(g\mu_B H - eV) \, \theta(g\mu_B H + eV) \}, \tag{22b}$$

where T_J is an average value of the integral

$$\langle \mathbf{k}_L \mid T_J \mid \mathbf{k}_R \rangle = \int d^3 r \exp[-i(\mathbf{k}_{\parallel,L} - \mathbf{k}_{\parallel,R}) \cdot \boldsymbol{\rho}] \, \bar{\chi}_{k_L}(z) \chi_{k_R}(z) J[\mathbf{r} - \mathbf{R}_n^{(0)}] \tag{22c}$$

$$z_L < R_\perp^{(0)} < z_R.$$

In a uniform field at $T = 0$ we use

$$\langle M \rangle \Rightarrow -S, \tag{23a}$$

$$\langle M^2 \rangle \Rightarrow +S^2. \tag{23b}$$

The results for $T \neq 0$ are given in Eqs. (40)–(44) of Appelbaum's paper ([16]).

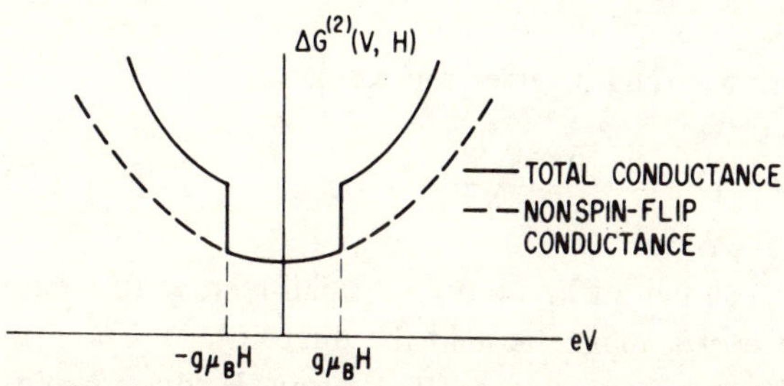

Fig. 5. Conductance obtained from the nonspin-flip and spin-flip terms of the *s–d* exchange interaction in second-order perturbation theory.

The resonance-scattering terms come from a third-order term for $|\langle f \mid M \mid i\rangle|^2$, which is the interference term between the above "single-step" process and the process in which an electron scatters from the impurity, making a virtual reflection back into the original metal prior to tunneling; i.e.,

$$\langle \mathbf{k}_L \mid M^{(2)} \mid \mathbf{k}_R\rangle = \sum_{\mathbf{k}'_R} \frac{\langle \mathbf{k}_L \mid T_J{}^{\pm} \mid \mathbf{k}_R{}'\rangle\langle \mathbf{k}_R{}' \mid J^{\mp} \mid \mathbf{k}_R\rangle}{E_R - E_R{}'}. \tag{24}$$

Mathematically, if one keeps track of the Fermi factors one finds that for ordinary impurity scattering $\mathbf{k}_R{}'$ runs over all values, but for spin-flip scattering there are processes restricted by $E_R{}' < E_{FR}$, so that Eq. (24) diverges as $\ln(E_R - E_{FR})$. This was first derived by Kondo in the context of bulk resistivity calculations. Its algebraic origin is seen by noting that the exclusion principle does not affect ordinary impurity scattering, so that one electron scatters independently of the presence of the others. However, for spin-flip scattering the commutation relations of the $\mathbf{S}_n(t)$ operators imply that the vacuum polarization processes induced by the impurity no longer exactly cancel the exclusion-principle restrictions on the direct scattering. A loose interpretation of this result is obtained by regarding the extra terms as a new scattering process becoming possible when $E \geq E_{FR}$. This scattering process in turn gives rise to resonance scattering at the Fermi energy. At zero bias both the initial and final energies of the electron are at $E = E_{FR}$, so this resonance is "tuned-in" and a great deal of current flows via (24). As eV increases, the initial and final energy are no longer both at the Fermi energies in their respective materials, so the resonance becomes detuned, and the current diminishes.

Appelbaum, using Eq. (22*c*) and a J which was an average of

$$\begin{aligned}\langle \mathbf{k}_R{}' \mid J \mid \mathbf{k}_R\rangle = {} & \int d^3r \exp[-i(\mathbf{k}'_{\parallel,R} - \mathbf{k}_{\parallel,R}) \cdot \boldsymbol{\rho}] \\ & \times \bar{\chi}_{k'_R}(z)\chi_{k_R}(z)J(\mathbf{r} - \mathbf{R}_\perp^{(0)}); \quad z_L < R_\perp < z_R \end{aligned} \tag{25}$$

for the reflection strength, gives the result

$$\Delta G^{(3)}(V) = \frac{16\pi e S(S+1)}{\hbar} N_i \varrho_i \varrho_R J \mid T_J \mid^2 [-F(eV)],$$

with $[-F(eV)]$ shown in Fig. 6 for eV small relative to a cutoff energy E_0. In the presence of a magnetic field the direct ($M^{(1)} = T_J$) tunneling terms give rise to three shifted peaks at $eV = \pm g\mu_B H$ (due to spin-flip and non-spin-flip direct transitions) and $eV = 0$ (due only to direct transitions with spin-flip). However, the two spin-flip $M^{(1)}$ terms are associated with differ-

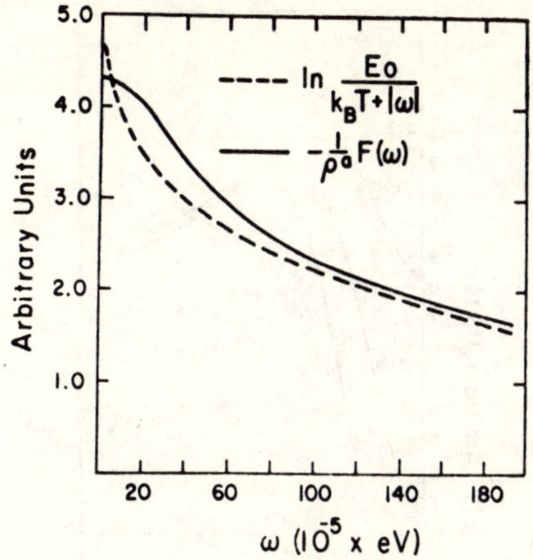

Fig. 6. The $H = 0$ resistance maximum form factor for $E_0 = 10$ meV and $\varkappa_B T = 0.1$ meV. This is Fig. 5 in ([16]).

ent $M^{(2)}$ which cancel at $T = 0$ to give zero for the coefficient of the eV peak for $|eV| < g\mu_B H$. Appelbaum finds [Eqs. (69)–(73)] at $T = 0$

$$\Delta G^{(3)}(eV) = \frac{8\pi e}{\hbar} \varrho_L \varrho_R N_i \,|\, T_J \,|^2 JS(S+1)$$

$$\times \left\{ \frac{-F(eV)}{S+1} [\theta(eV - g\mu_B H) + \theta(-eV - g\mu_B H)] \right.$$

$$+ \frac{-F(eV + g\mu_B H)}{S+1} \left[S + \frac{1}{2} - \frac{1}{2} \operatorname{sgn}(eV + g\mu_B H) \right]$$

$$\left. + \frac{-F(eV - g\mu_B H)}{S+1} \left[S + \frac{1}{2} - \frac{1}{2} \operatorname{sgn}(g\mu_B H - eV) \right] \right\}. \quad (26)$$

An important aspect of this result is the vanishing of the coefficient of the peak at $eV = 0$, so that in a magnetic field the maximum at $eV = 0$ splits into a doublet with maxima near $eV = \pm g\mu_B H$. Appelbaum made a numerical study of the line shapes as a function of the field H for a 10-meV cutoff (E_0) and $S = \frac{1}{2}, \frac{3}{2}, \frac{5}{2}$. The general line shape resulting from Eqs. (22) and (26) is sketched in Fig. 7.

We conclude with the observation that if resonance scattering is important enough to change Fig. 5 to Fig. 7 in a second-order calculation of $M^{(2)}$, one should sum higher-order terms in J as intermediate states in the calculation of M. The status of such calculations is still uncertain even for bulk properties of dilute magnetic alloys. Appelbaum *et al.* ([18]) argue from

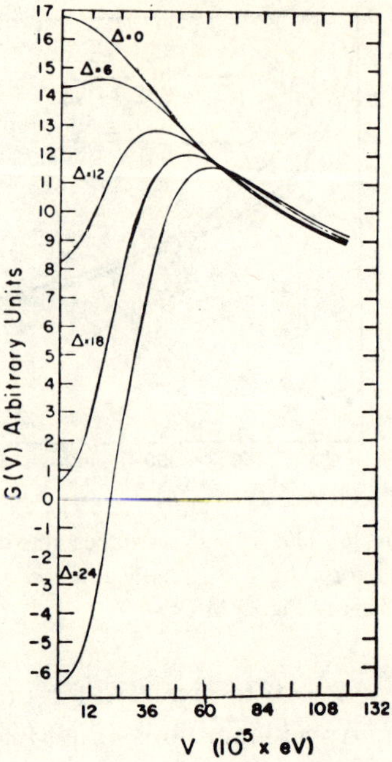

Fig. 7. The line shape for resonance scattering from paramagnetic impurities in a nonzero magnetic field $\Delta = g\mu_B H$. This is Fig. 7 in ([16]) for impurity spin $S = \frac{3}{2}$, $E_0 = 10$ meV, and $K_B T = 0.12$ meV. Δ is measured in units of 0.01 meV.

analogies to the bulk calculations that these higher-order terms form a bound state of binding energy $E_B = -E_0 \exp[1/(2J_\varrho)]$ for $J < 0$. Thus the conductivity would look like Fig. 5 even in zero field with $g\mu_B H \to E_B$. This interpretation is by no means universally agreed upon [see, e.g., ([19,20])].

The final topic of this chapter is the excitation of collective modes in the barrier by the tunneling electron. We recall that the coherent excitation of phonons by tunneling electrons is actually a very "old" topic, and was discussed as "phonon-assisted-tunneling" by Kleinman in Chapter 13. The coupling mechanism in this case is the "bulk" electron–phonon interaction represented by the terms $\vartheta[\mathbf{u}_n(t)]$ in Eq. (3*b*). In a similar way magnons are excited in a magnetically ordered barrier via the $\boldsymbol{\sigma} \cdot \mathbf{S}_n$ exchange coupling in Eq. (2*d*), with J expanded as was $\bar{V}$ in Eq. (3*b*). Magnon excitation can also occur via an impurity coupling mechanism like that discussed below for phonons.

We illustrate the application of the theory of inelastic tunneling to the excitation of collective modes by considering the incoherent excitation of phonons via collision of the electron with impurities in the barrier. The results of the complete calculation in this case follow closely the kinematical predictions made in Eqs. (1) for the conductance minimum. However, in a general calculation there are three sources of error of the simple kinematical arguments given in Eq. (1): (1) The dynamical form of the electron–phonon coupling introduces an extra function $| v(\varepsilon) |^2$ in the integrand of Eq. (1*a*). (2) The barrier penetraction factor of the electron depends on its energy, so that the electron in the initial and final states of the transition see different barriers. This effect is quite important in coherent phonon emission, in which case the transfer of an appreciable amount of parallel momentum to the phonon can increase the effective barrier height in the final state. (3) More than one phonon can be emitted by a tunneling electron. Phase space restrictions make such processes more slowly varying with bias near $eV = 0$ than single-phonon emission, but at larger bias values one in principle could see images of critical points in the multiple-phonon density of states.

For purposes of illustration let us consider a spin-independent impurity potential given by

$$V_{fn}[\mathbf{r} - \mathbf{R}_n(t)] = \sum_{\mathbf{q}} v(\mathbf{q}) \exp\{i\mathbf{q} \cdot [\mathbf{r} - \mathbf{R}_n(t)]\}. \tag{27}$$

The single-phonon emission term of the impurity potential is, in analogy with Eq. (3*b*), in the rigid-ion approximation

$$V_b^{\text{el-ph}}(\mathbf{r}, t) = \sum_{n,\mathbf{q}} v(\mathbf{q}) i\mathbf{q} \cdot \mathbf{u}_n(t) \exp\{i\mathbf{q} \cdot [\mathbf{r} - \mathbf{R}_n^{(0)}]\}. \tag{28}$$

The phonon creation–annihilation operators $a_{\mathbf{p}}, a_{\mathbf{p}}^+$ are introduced by expanding

$$\mathbf{u}_n(t) = \sum_{\mathbf{p}} \left(\frac{\hbar}{2MN\omega_{\mathbf{p}}} \right)^{1/2} \times \mathbf{e}_{\mathbf{p}} \{a_{\mathbf{p}} \exp(i[\mathbf{p} \cdot \mathbf{R}_n^{(0)} - \omega_{\mathbf{p}} t]) + a_{\mathbf{p}}^+ \exp(-i[\mathbf{p} \cdot \mathbf{R}_n^{(0)} - \omega_{\mathbf{p}} t])\}. \tag{29}$$

To calculate coherent phonon emission Eq. (29) is inserted into Eq. (28) and the sums over n performed over the barrier lattice. For incoherent emission one must first calculate the relevant matrix element of Eq. (28), then average its absolute square over a random distribution of impurities. We write

$$\begin{aligned}\langle \mathbf{k}_L \mid V_b^{\text{el-ph}}(t) \mid \mathbf{k}_R \rangle &= \sum_{n,\mathbf{q},\mathbf{p}} v(\mathbf{q}) i\mathbf{q} \cdot \mathbf{e}_\mathbf{p} (\hbar/2MN\omega_\mathbf{p})^{1/2} \\ &\times A^{-1} \int \exp[-i(\mathbf{k}_{\parallel,L} - \mathbf{k}_{\parallel,R}) \cdot \boldsymbol{\rho}] \exp(i\mathbf{q}_\parallel \cdot \boldsymbol{\rho}) d^2\varrho \\ &\times \int dz \exp(iq_\perp z) \bar{\chi}_{k_L}(z) \chi_{k_R}(z) \\ &\times \{a_\mathbf{p} \exp[i(\mathbf{p}-\mathbf{q}) \cdot \mathbf{R}_n^{(0)}] \exp(-i\omega_\mathbf{p} t) + a_\mathbf{p}^\dagger \exp[i(\mathbf{p}+\mathbf{q}) \cdot \mathbf{R}_n^{(0)}] \exp(i\omega_\mathbf{p} t)\}. \end{aligned} \tag{30}$$

Denoting the Fourier transform of the z wave function overlap integral by $M(q_\perp, k_L, k_R)$, we find that the terms in $[\langle V_b(t) V_b(0) \rangle]_{\text{AV}}$ which are diagonal in the phonon operators and averaged to lowest order in $c = N_i/N$ are

$$\begin{aligned}[\langle \mathbf{k}_L \mid V_b(t) \mid \mathbf{k}_R \rangle \langle \mathbf{k}_L \mid V_b(0) \mid \mathbf{k}_R \rangle]_{\text{av}} &= N_i \sum_{q_\perp} \frac{\hbar^2(\Delta^2 + q_\perp{}^2)}{2M} \\ &\times \Bigg(\mid v(\boldsymbol{\Delta} + q_\perp) M(q_\perp, k_L, k_R) \mid^2 \frac{1}{N} \sum_\mathbf{p} \frac{1}{\hbar\omega_\mathbf{p}} \\ &\times \{\exp(i\omega_\mathbf{p} t)[N(\omega_\mathbf{p})] + \exp(-i\omega_\mathbf{p} t)[N(\omega_\mathbf{p}) + 1]\} \Bigg), \end{aligned} \tag{31a}$$

$$\boldsymbol{\Delta} = \mathbf{k}_{\parallel,L} - \mathbf{k}_{\parallel,R}, \tag{31b}$$

$$N(\omega_\mathbf{p}) = [\exp(\hbar\omega_\mathbf{p}/\varkappa T) - 1]^{-1}. \tag{31c}$$

The important feature of Eq. (31*a*) is the factorization of the sum over $\mathbf{p}$ from the rest of the expression for $\langle V(t) V(0) \rangle$. This factorization occurs only for *incoherent* electron-collective-mode interactions and has the consequence that only energy conservation links the electrons and phonons. The sum over $q_\perp$ is the product of a recoil energy $R = \hbar^2(\Delta^2 + q_\perp{}^2)/2M$ of the impurity times an effective coupling matrix element which depends both on the impurity potential via $v(\mathbf{q})$ and on the barrier-penetration properties of the junction via $M(q_\perp, k_R, k_L)$. If we define in analogy to Eq. (17)

$$\begin{aligned}\bar{R} \mid T \mid^2 &\equiv N_i \sum_{q_\perp} \frac{\hbar^2(\Delta^2 + q_\perp{}^2)}{2M} \mid v(\boldsymbol{\Delta} + q_\perp) M(q_\perp, k_L, k_R) \mid^2 \\ &\approx \text{const}, \end{aligned} \tag{32}$$

then the second-order perturbation theory expression for j analogous to Eq. (16) turns out to be [22]

$$\begin{aligned} j(eV) &= \frac{4\pi e}{\hbar} \mid T \mid^2 \varrho_L \varrho_R \int d\xi_R \, d\xi_L [f(\xi_R) - f(\xi_R + eV)] \\ &\times \left\{ \frac{1}{N} \sum_\mathbf{p} \left(\frac{\bar{R}}{\hbar\omega_\mathbf{p}} \right) \{ \delta(\xi_L - \xi_R - eV + \hbar\omega_\mathbf{p}) [N(\omega_\mathbf{p}) + 1 - f(\xi_L)] \right. \\ &\left. + \delta(\xi_L - \xi_R - eV - \hbar\omega_\mathbf{p}) [N(\omega_\mathbf{p}) + f(\xi_L)] \} \right\} \\ &\xrightarrow{T \to 0} \frac{4\pi e}{\hbar} \mid T \mid^2 \varrho_L \varrho_R \frac{1}{N} \sum_\mathbf{p} \left(\frac{\bar{R}}{\hbar\omega_\mathbf{p}} \right) [\mid eV \mid - \hbar\omega_\mathbf{p}] \theta(\mid eV \mid - \hbar\omega_\mathbf{p}) \end{aligned} \tag{33a}$$

$$\xi_i = E_i - E_{FR}.$$

In the limit that $T \to 0$ the conductance is therefore given by

$$\Delta G^{(2)} = \frac{4\pi e}{\hbar} | T |^2 \varrho_L \varrho_R \frac{1}{N} \sum_{\mathbf{p}} \left(\frac{\bar{R}}{\hbar\omega_{\mathbf{p}}} \right) \theta(| eV | - \hbar\omega_p), \tag{33b}$$

which, except for the $(\bar{R}/\hbar\omega_{\mathbf{p}})$ weighting factor, is just Eq. (1*a*).

We emphasize that the result (33*b*) is a consequence only of the form of the expansion (29). Most of the dynamical features of the excitation process have been extracted by approximation (32). However, it is a careful consideration of precisely the form of the coupling and the temperature dependence of the line shape which permits the identification of the zero-bias minima in *p–n* diodes and M–S contacts with incoherent phonon

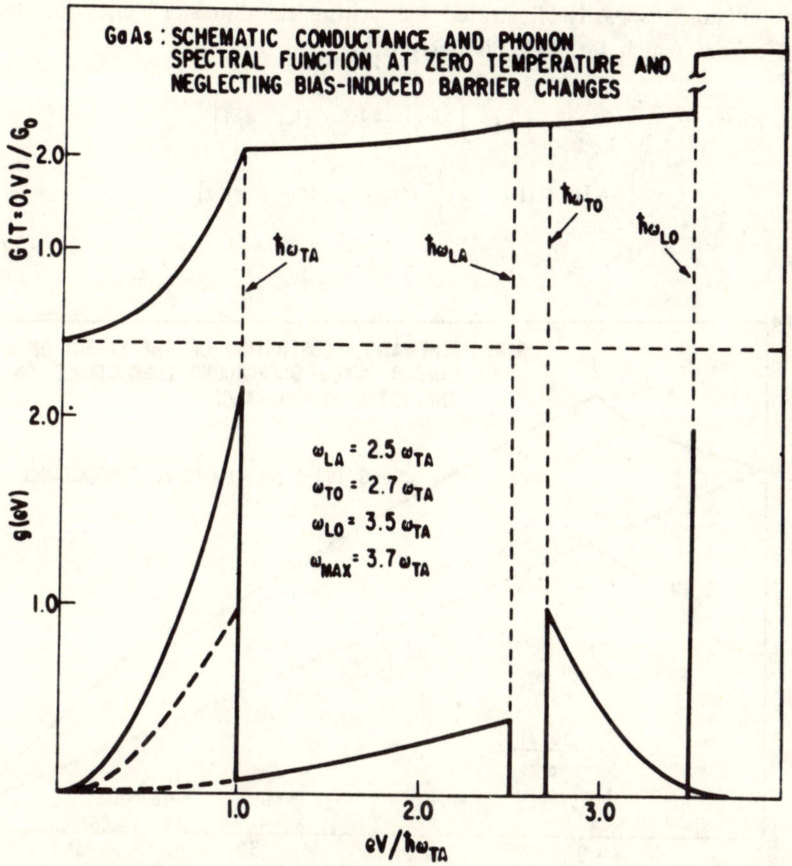

Fig. 8. The conductance line shape and model phonon density of states for GaAs *p–n* tunnel junctions in the constant $|T|^2$ approximation, assuming $E_F \gg \hbar\omega_{\mathrm{LO}}$ on both sides of the junction.

emission from impurities. Only a consideration of these aspects of the analysis permits one to investigate the satisfaction of the observed scaling laws of the conductance minimum on doping, type of material, temperature, etc. It is these laws which distinguish the various possible electron–excitation coupling mechanisms. In Fig. 8 a normalized form of Eq. (33*b*) is shown together with the phonon density of states used for a GaAs *p–n* junction or metal–semiconductor contact. The removal of the constant $|T|^2$ approximation is illustrated (for zero temperature) in Fig. 9. The temperature dependence of the line shapes is predicted by the model calculation and is compared with experimental data in Fig. 10.

The general relation between the contributions to the current due to energy-loss mechanisms and those due to self-energy many-body effects in the "metals" is easily illustrated by writing the "master" equation for $j(eV)$ from which Eq. (33*a*) is derived as:

$$j(eV) = \frac{4\pi e}{\hbar} \sum_{\mathbf{k}_L, \mathbf{k}_R, \mathbf{p}} \int d\varepsilon_L \int d\varepsilon_R \left[\frac{1}{\pi} \operatorname{Im} G(\mathbf{k}_L, \varepsilon_L)\right] \times \left[\frac{1}{\pi} \operatorname{Im} G(\mathbf{k}_R, \varepsilon_R)\right] [f(\varepsilon_R) - f(\varepsilon_R + eV)] \times T^2_{\mathbf{k}_L, \mathbf{k}_R}(\mathbf{p}, \varepsilon_L, \varepsilon_R). \tag{34}$$

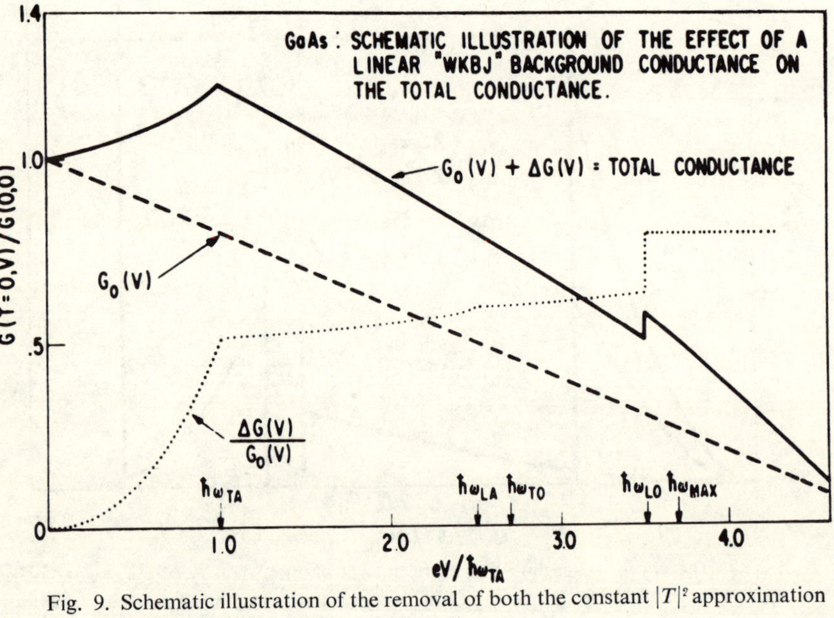

Fig. 9. Schematic illustration of the removal of both the constant $|T|^2$ approximation and the requirement that $E_F \gg \hbar\omega_{LO}$.

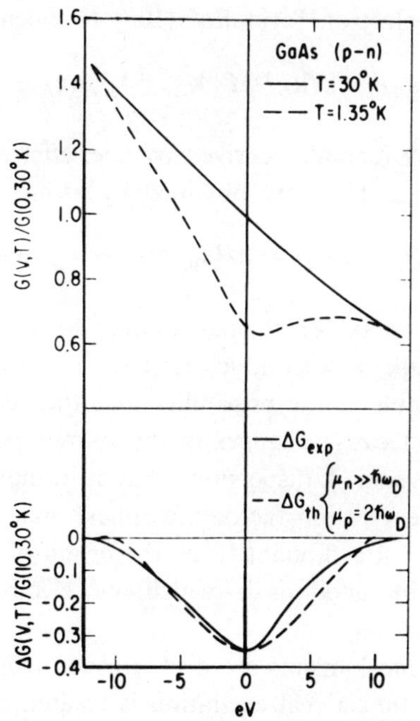

Fig. 10. Comparison of the temperature-dependent line shape with the data of Shewchun and Williams ([7]). The failure of the point of maximum inflection in the theoretical lineshape to occur at $eV_0 < \hbar\omega_D$ is a consequence of the use of a Debye phonon spectrum. The use of a more realistic model of the phonon spectrum removes this discrepancy between the model calculation and experimental data ([23]).

The self-energy effects are included in the propagators $G(\mathbf{k}, \varepsilon)$ as indicated in Chapter 21 by Schrieffer. For a free-electron model

$$G(\mathbf{k}, \varepsilon) = 1/(\varepsilon - \xi_\mathbf{k} + i\delta), \tag{35a}$$

$$\xi_\mathbf{k} = (\hbar^2 k^2/2m) - E_F, \tag{35b}$$

$$(1/\pi)\,\mathrm{Im}\, G(\mathbf{k}, \varepsilon) = \delta(\xi_\mathbf{k} - \varepsilon). \tag{35c}$$

A complete specification of a model calculation of $j(eV)$ requires a statement of the approximations used for $T^2_{\mathbf{k}_L, \mathbf{k}_R}$ and for the propagators $G(\mathbf{k}, \varepsilon)$. The free-electron model is obtained by using Eqs. (35) for $G(\mathbf{k}, \varepsilon)$ and using the appropriate one-electron form Eq. (10*b*) to calculate

$$T^2_{\mathbf{k}_L,\mathbf{k}_R}(\mathbf{p}, \varepsilon_L, \varepsilon_R) = |\langle \mathbf{k}_L | M | \mathbf{k}_R \rangle|^2 \delta_{\mathbf{p},0} \delta(\varepsilon_L - \varepsilon_R - eV). \tag{36}$$

The superconductivity formulas derived by Schrieffer contained only electron self-energy effects. They are obtained by using

$$T^2_{\mathbf{k}_L,\mathbf{k}_R}(\mathbf{q}, \varepsilon_L, \varepsilon_R) = M^2 \delta_{\mathbf{p},0} \delta(\varepsilon_L - \varepsilon_R - eV) \tag{37}$$

and the appropriate weak- or strong-coupling form for the one-electron propagators in a bulk superconductor. The constant-M approximation used in Eq. (37) eliminates the possibility that the self-energy effects described by Im $G(\mathbf{k}, \varepsilon)$ cause changes in the barrier penetration factor by introducing a new ε versus $\mathbf{k}$ dispersion relation in the barrier. This alteration in the barrier penetration factor plus phase space restrictions on the sums over $\mathbf{k}_R$ and $\mathbf{k}_L$, are thought to be the origin of the structure at the LO phonon energy in junctions in which one electrode is a degenerate semiconductor ([6,21]).

The energy-loss mechanisms are distinguished from self-energy effects by virtue of the fact that a real excitation is created during the tunneling process because of a time dependence of the one-electron potential in that barrier. Mathematically, this process is reflected in a contribution to T^2 which is inelastic, i.e.,

$$\begin{aligned} T^2_{\mathbf{k}_L,\mathbf{k}_R}(\mathbf{p}, \varepsilon_L, \varepsilon_R) = M^2(&\mathbf{k}_L, \mathbf{k}_R, \mathbf{p}) \\ &\times \{\delta(\varepsilon_R - \varepsilon_L + eV - \hbar\omega_{\mathbf{p}})[N(\omega_{\mathbf{p}}) + 1 - f(\varepsilon_L)] \\ &+ \delta(\varepsilon_R - \varepsilon_L + eV + \hbar\omega_{\mathbf{p}})[N(\omega_{\mathbf{p}}) + f(\varepsilon_L)]\}. \end{aligned} \tag{38}$$

The calculation of M^2 is described in Eqs. (31) and (32). In the approximation (32) it is given by $M^2 \approx |T|^2(\bar{R}/\hbar\omega_{\mathbf{p}})$. The full derivation of Eqs. (33) and (38) via Green's functions is outlined in ([14]). The incorporation of the possibility of energy-loss mechanisms "in the barrier" is simply an extension of the model calculation to incorporate many-body effects in the barrier into the theory with the same degree of precision as these effects have previously been incorporated in the analysis of the right- and left-hand "metals." Such an extension of the model calculation is referred to as the inclusion of "vertex corrections to the polarization diagram" in, e.g., calculations of bulk resistivity. We have seen above that it is this incorporation of energy-loss mechanisms into the model which permits the description of both conductance minima and maxima near zero bias as well as a variety of observed rapid variations in the conductance at higher bias. Therefore it appears that a detailed study of conductance "anomalies" is a potentially useful spectroscopic tool for studying the elementary excitation spectrum in the barrier of a tunnel junction.

REFERENCES

1. R. N. Hall, J. H. Racette, and H. Ehrenreich, Phys. Rev. Letters 4:456 (1960).
2. G. D. Mahan and C. B. Duke, Phys. Rev. 149:705 (1966).
3. B. M. Vul, E. I. Zavaritskaya, and N. V. Zavaritskii, Soviet Phys.—Solid State 8: 710 (1966).
4. R. A. and J. M. Rowell, Phys. Rev. Letters 13:404 (1964).
5. J. W. Conley and J. J. Tiemann, J. Appl. Phys. 38:2880 (1967).
6. J. W. Conley and G. D. Mahan, Phys. Rev. 169:681 (1967).
7. J. Shewchun and R. M. Williams, Phys. Rev. Letters 15:160 (1965).
8. J. M. Rowell and L. Y. L. Shen, Phys. Rev. Letters 17:15 (1966); J. M. Rowell and L. Y. L. Shen, Solid State Commun. 5:189 (1967); A. F. G. Wyatt, Phys. Rev. Letters 13:401 (1964).
9. J. E. Christopher, R. V. Coleman, A. Isin, and R. C. Morris, Phys. Rev. 172:485 (1968).
10. R. C. Jaklevic and J. Lambe, Phys. Rev. Letters 17:1139 (1966).
11. J. M. Rowell, unpublished.
12. J. Appelbaum, Phys. Rev. Letters 17:91 (1966).
13. P. W. Anderson, Phys. Rev. Letters 17:95 (1966).
14. C. B. Duke, S. D. Silverstein, and A. J. Bennett, Phys. Rev. Letters 19:315 (1967); Phys. Rev. (in press).
15. J. Bardeen, Phys. Rev. Letters 6:57 (1961).
16. J. Appelbaum, Phys. Rev. 154:633 (1967).
17. D. J. Scalapino and S. M. Marcus. Phys. Rev. Letters 18:459 (1967).
18. J. Appelbaum, J. C. Phillips, and G. Tzouras, Phys. Rev. (in press).
19. J. Kondo, Phys. Rev. 154:644 (1967).
20. C. B. Duke and S. D. Silverstein, Phys. Rev. 161:456; 161:470 (1967).
21. L. C. Davis and C. B. Duke, Solid State Commun. 6:193 (1968).

NOTES ADDED IN PROOF

Since the original writing of these lecture notes, this unified model description of tunneling has been extended substantially, and systematically developed in a general review of tunneling ([22]). In addition, considerable experimental work has been reported, including an observation of the predicted ([14]) phonon-emission lineshapes in palladium contacts on GaAs ([23]) and of phonon-emission in indium contacts on germanium ([24]) by a two-step "coherent" electron–phonon coupling mechanism rather than the incoherent impurity-interaction mechanism.

An additional major application of the model has occurred via the observations ([24–26]) and interpretation ([21,22]) of fine structure in the d^2I/dV^2 tunnel characteristics of metal–oxide–semiconductor ([25]) and metal–semiconductor ([24,26]) junctions due to the modifications of the electronic self-energy in the semiconductor electrode because of electron interactions with optical and local-mode phonons. The interpretations ([21,22]) follow from a direct application of Eq. (34). A comprehensive review of the experiments reported prior to September 1968, and their interpretation in terms of either self-energy or energy-loss mechanisms, has been given elsewhere ([22]).

ADDITIONAL REFERENCES

22. C. B. Duke, Tunneling in Solids, Academic Press (to be published).
23. P. Thomas and H. J. Queisser, Phys. Rev. (in press).
24. F. Steinrisser, L. C. Davis, and C. B. Duke, Phys. Rev. (in press).
25. E. L. Wolf, Phys. Rev. Letters 20: 204 (1968).
26. D. C. Tsui, Phys. Rev. Letters 21:994 (1968).

Chapter 29

Gapless Superconducting Tunneling—Theory

P. Fulde

Institut Max von Laue—Paul Langevin
Garching bei München

INTRODUCTION

This chapter is divided into two parts. In the first part we try to understand qualitatively the phenomenon of gapless superconducting tunneling by considering a very simple example. We study the case of a uniform current flowing in a superconducting thin film parallel to the tunneling junction assuming that the mean free path is long compared with the coherence distance. Although such a situation is difficult to achieve experimentally, it has the advantage that it can be studied fully without invoking any advanced mathematical methods. A computation of the density of states shows features which are characteristic for gapless superconductors and which are also present in more complex situations.

In the second part of the chapter we give a survey of the different pair-breaking mechanisms which have been studied until now. It turns out that the results for the density of states for various pair-breaking mechanisms can be grouped together quite conveniently and show many cross relations.

UNIFORM CURRENT PARALLEL TO A TUNNELING JUNCTION

Suppose we are dealing with an experimental arrangement as shown in Fig. 1. Let us assume that we are working at temperature $T = 0$ and that the mean free path in the sample is large compared with the coherence distance. Our aim is to calculate the density of states of the superconducting film in the presence of a current. For this purpose we determine the energy

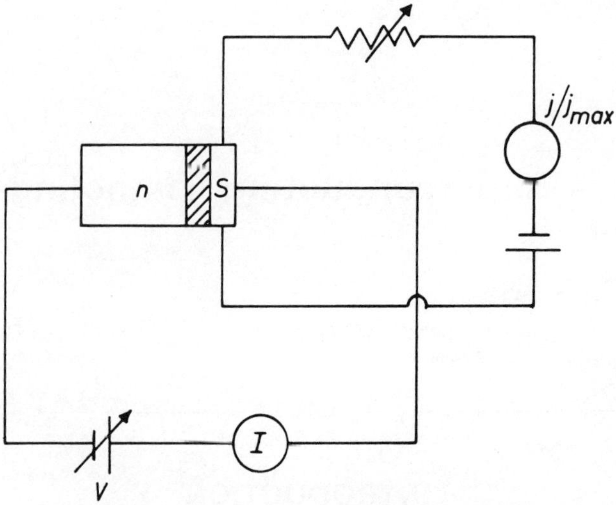

Fig. 1. Schematic representation of a tunneling experiment with a uniform current flowing parallel to the junction. From Fulde (8).

spectrum of an electron distribution shifted in momentum space by an amount $\mathbf{q}$ compared with the case where no current is flowing [see Fig. 2(*a*)]. In the rest frame of the pairs the excitation energy is $E_{\mathbf{p}} = (\varepsilon_p^2 + \Delta^2)^{1/2}$, where $\varepsilon_p = V_F(p - p_F)$ and Δ denotes the order parameter, which in general is $\mathbf{q}$-dependent. In the laboratory frame the excitation energy becomes

$$E_{\mathbf{p}} = (\varepsilon_p^2 + \Delta^2)^{1/2} + (\mathbf{q} \cdot \mathbf{p}/m). \tag{1}$$

If Δ_0 is the order parameter for $q = 0$, then for $q > q_s = \Delta_0/v_F$ the excitation energy may become negative for a certain region in a momentum space. However, $E_{\mathbf{p}} < 0$ implies a breakup of Cooper pairs [see Fig. (2*b*)], since for electrons in those momentum states it is energetically favorable to stay unpaired. Thus the surface of the region of unpaired electrons in momentum space is determined by the condition $E_{\mathbf{p}} = 0$. Since the unpaired electrons block the states which they occupy, for virtual pair scattering the order parameter Δ will decrease for $q > q_s$. We can calculate $\Delta(q)$ by solving the set of equations

$$\begin{aligned} E_{\mathbf{p}} &= 0 \\ \frac{1}{V} &= \sum_{\mathbf{p},\mathrm{excl}}^{\omega_0} 2^{-1}(\varepsilon_p^2 + \Delta^2)^{-1/2}. \end{aligned} \tag{2}$$

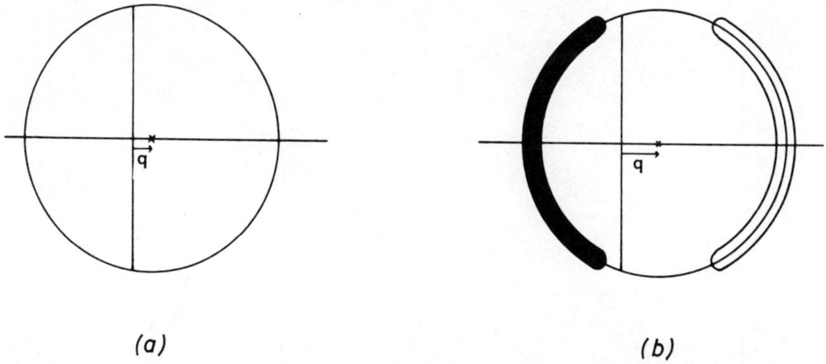

Fig. 2. Electron distribution for (*a*) $q < q_s$, (*b*) $q > q_s$. The shaded area denotes unpaired electrons.

Here V is the BCS interaction constant and "excl" indicates that all states with unpaired electrons have to be excluded from the sum.

Similarly, $j(q)$ can be calculated by just adding up the momenta of the occupied states belonging to a given ratio q/q_s. Numerical results for Δ/Δ_0 and j as functions of q/q_s are shown in Figs. 3 and 4.

Here we are primarily interested in calculating the density of states

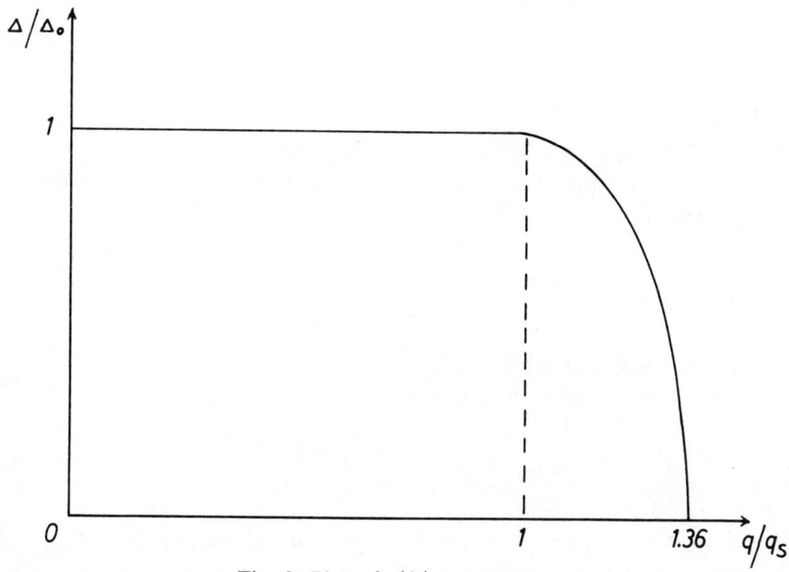

Fig. 3. Plot of Δ/Δ_0 versus q/q_s.

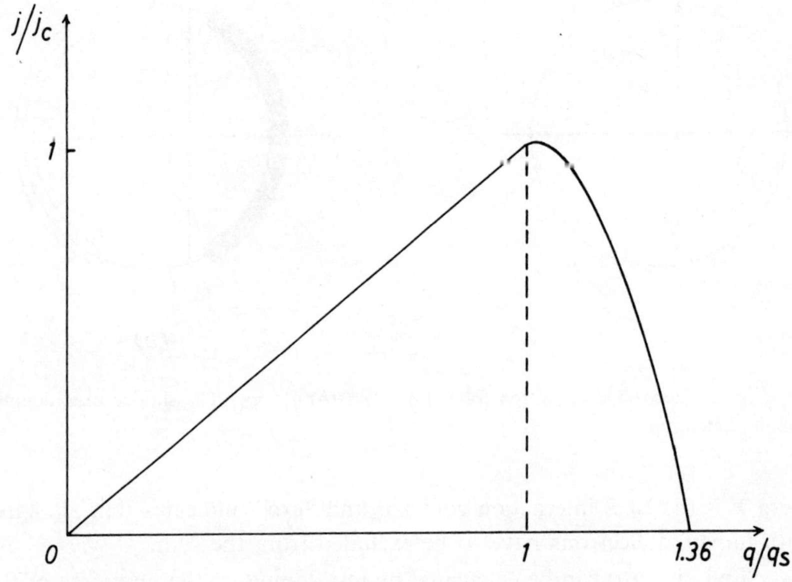

Fig. 4. Plot of j (in arbitrary units) versus q/q_s.

$N_s(\omega)$ from the given excitation spectrum (1). One finds that

$$N_s(\omega) = \frac{N(0)}{2qv_F} \operatorname{Re}\left\{ \frac{\omega + qv_F}{|\omega + qv_F|} [(\omega + qv_F)^2 - \Delta^2]^{1/2} - \frac{\omega - qv_F}{|\omega - qv_F|} [(\omega - qv_F)^2 - \Delta^2]^{1/2} \right\} \tag{3}$$

where $N(0) = mp_F/2\pi^2$. However, the shortest way to obtain Eq. (3) is by using the following relation between $N_s(\omega)$ and the single-particle Green's function $G(\mathbf{p}, \omega)$

$$N_s(\omega) = -\frac{1}{\pi} \operatorname{Im} \int \frac{d^3p}{(2\pi)^3} G(\mathbf{p}, \omega). \tag{4}$$

This relation is discussed in Chapter 21 of this volume. By using the appropriate Green's function for our situation, namely

$$G(\mathbf{p}, \omega) = \frac{\omega - (\mathbf{q} \cdot \mathbf{p}/m) + \varepsilon_p}{(\omega - \mathbf{q} \cdot \mathbf{p}/m)^2 - \varepsilon_p{}^2 - \Delta^2}, \tag{5}$$

one obtains Eq. (3) after performing the $d\varepsilon_p$ and $d\Omega$ integrations. Numerical results for $N_s(\omega)/N(0)$ are shown in Fig. 5. For $q > q_s$ the density of states

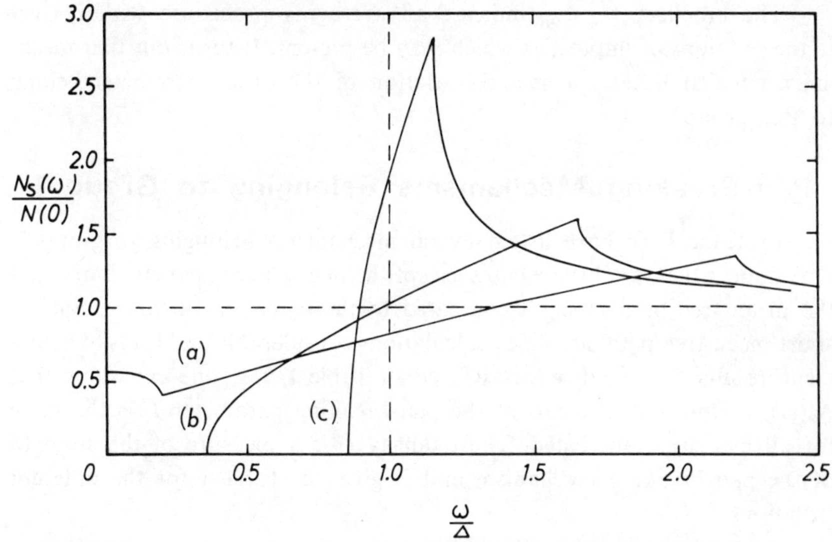

Fig. 5. Tunneling density of states for qv_F/Δ = (*a*) 1.20, (*b*) 0.646, (*c*) 0.146. From Fulde ([8]).

is nonvanishing for $\omega = 0$. This is a typical gapless situation. We note that a gapless density of states and the ability of the system to carry a persistent current do not exclude each other.

As will be seen later the effect of a finite mean free path is a smoothing out of the edges and cusps which appear in Fig. 5.

SURVEY OF THE THEORETICAL RESULTS ON GAPLESS SUPERCONDUCTING TUNNELING

We can divide the different mechanisms which lead to a breakup of Cooper pairs and hence to gapless superconductivity into two groups:

1. Pair-breaking mechanisms for which one is able to calculate $G(\mathbf{p}, \omega)$ and hence $N_s(\omega)$ for an arbitrary strength of the pair-breaking parameter.

2. Mechanisms for which $G(\mathbf{p}, \omega)$ and hence $N_s(\omega)$ can be calculated for small order parameter only (subcritical regime). In that case one expands $G(\mathbf{p}, \omega)$ in terms of the Green's function in the normal state, $G_0(\mathbf{p}, \omega)$, and the order parameter and keeps only the first two terms in powers of Δ. In a symbolic notation one has

$$G = \langle G_0 \rangle_{\text{im}} + \langle G_0 \Delta G_0^{-} \Delta^{+} G_0 \rangle_{\text{im}}. \tag{6}$$

The brackets $\langle \cdots \rangle_{\text{im}}$ indicate an averaging procedure with respect to the positions of impurities which may be present. It turns out that mechanisms which lead to a spatial variation of the order parameter belong to this group.

Pair-Breaking Mechanisms Belonging to Group 1

In Table I we have listed several mechanisms belonging to group 1. One notices that in all these cases except the one of paramagnetic impurities the mean free path is small compared to the coherence distance. Such a short mean free path simplifies calculations considerably and leads to equivalent results for the five cases listed in Table I, i.e., one can show that $N_s(\omega)$ if expressed in terms of the pair-breaking parameter Γ is the same for all five situations listed. The quantity Γ is a measure of the strength of the pair-breaking mechanism and is given in Table I for the different situations.

We now want to give a prescription for calculating the density of states for a given experimental situation. First we determine the ratio T_c/T_{c0} for that special case. Here T_c and T_{c0} are, respectively, the transition temperatures in the presence and in the absence of that pair-breaking mechanism. From the measured ratio T_c/T_{c0} we can determine Γ with the help of the following equation:

$$\ln \frac{T_{c0}}{T_c} = \psi\left(\frac{1}{2} + \frac{\Gamma}{2\pi T_c}\right) - \psi\left(\frac{1}{2}\right), \tag{7}$$

where $\psi(z)$ is the digamma function. With Γ determined in this way we obtain the density of states $N_s(\omega)$ as a function of Γ/Δ from

$$N_s(\omega) = N(0) \operatorname{Im} \frac{u}{(1 - u^2)^{1/2}}, \tag{8}$$

where the quantity u is connected with ω/Δ via

$$\frac{\omega}{\Delta} = u\left(1 - \frac{\Gamma}{\Delta(1 - u^2)^{1/2}}\right). \tag{9}$$

The only problem left is to compute Δ/Δ_0 as function of Γ and T. Here Δ_0 denotes the order parameter for vanishing Γ and T. Numerical results for $\Delta(\Gamma, T)/\Delta_0$ can be found in the literature. Here we show only numerical results for $N_s(\omega)/N(0)$ for various ratios Γ/Δ. One notes in Fig. 6 that for $\Gamma/\Delta > 1$ the density of states does not show a gap. Qualita-

TABLE I*

Physical situation	Pair-breaking mechanism	Pair-breaking parameter Γ/Δ_0	Remarks
Paramagnetic impurities	Exchange interaction $\sum_i J(\mathbf{r} - \mathbf{R}_i)\mathbf{s}\cdot\mathbf{S}_i$	$\dfrac{1}{\tau_s \Delta_0}$	$\dfrac{1}{\tau_s} = n_i \dfrac{mp_F}{\pi} \dfrac{S(S+1)}{4} \int \lvert J \rvert^2 \dfrac{d\Omega}{4\pi}$
Uniform current	Momentum $2\mathbf{q}$ of the Cooper pairs	$\dfrac{2}{3} \dfrac{\tau_{tr} v_F{}^2 q^2}{\Delta_0}$	$l \ll \xi_0$
Thin film in parallel field (local electrodynamics)	$\dfrac{e}{m}(\mathbf{p}\cdot\mathbf{A} + \mathbf{A}\cdot\mathbf{p})$	$\dfrac{\tau_{tr}(v_F e d H)^2}{18\Delta_0}$	$d \ll \delta_L$, $d \gg (l\xi_0)^{1/2}$ $l/\xi_0 \ll 1$, $d \gg l$
Thin film in parallel field (nonlocal electrodynamics)	$\dfrac{e}{m}(\mathbf{p}\cdot\mathbf{A} + \mathbf{A}\cdot\mathbf{p})$	$\dfrac{\tau_{tr}(v_F e d H)^2}{18\Delta_0} g\left(\dfrac{\pi}{d} l\right)$	$d \ll \delta_L$, $d \gg (l\xi_0)^{1/2}$, $l/\xi_0 \ll 1$
Pauli paramagnetism	$-\mu_B \mathbf{H}\cdot\mathbf{s}$	$\dfrac{\tau_{so}(\mu H)^2}{2\Delta_0}$	$l_{so} \ll \xi_0$

* $\mathbf{R}_i$ is the position of the ith impurity, n_i is the impurity concentration, τ_{tr} is the transport mean free time, l and l_{so} are the mean free path and the spin orbital mean free path, respectively, δ_L is the London penetration depth, d is the film thickness, and H is the external field. Furthermore, $g(x) = (3/2x^3)\{(1 + x^2) \arctan x - x\}$.

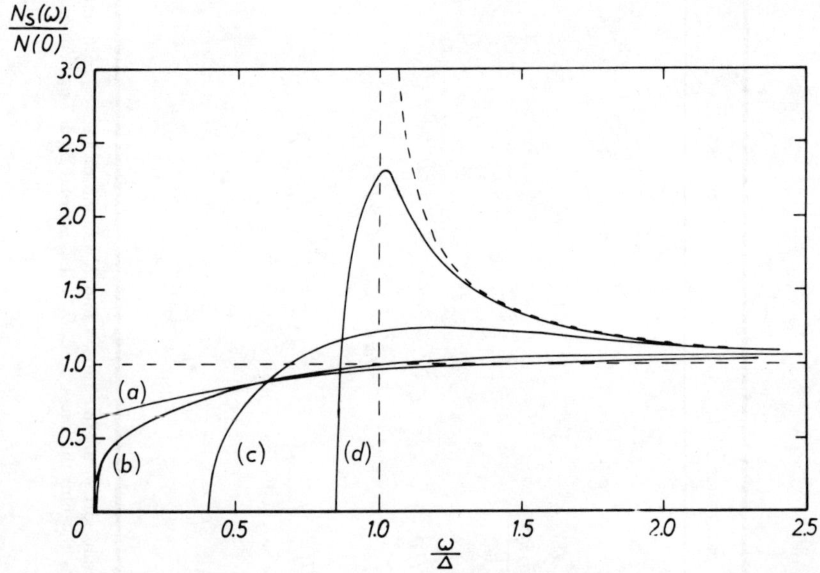

Fig. 6. Tunneling density of states for Γ/Δ = (*a*) 1.31, (*b*) 1, (*c*) 0.30, (*d*) 0.032. From Fulde ([8]).

tively, the curves resemble the ones of Fig. 5. It should be noted that in the subcritical regime, where Δ is small, the density of states simplifies to ([2])

$$N_s(\omega) = N(0)\left[1 + \frac{\Delta^2}{2}\,\frac{\omega^2 - \Gamma^2}{(\omega^2 + \Gamma^2)^2}\right], \tag{10}$$

which demonstrates explicitly the vanishing of the gap in $N_s(\omega)$. From what has been said it is clear that the pair-breaking mechanisms listed in Table I are additive in the sense that if more than one of them act at the same time on the superconducting electrons, their effects can be obtained by adding up the different values of Γ.

It turns out that there are mechanisms other than those listed in Table I for which the density of states can be calculated for an arbitrary strength of the pair-breaking parameter, but which do not show an equivalence to the previous situations. For example, Strässler and Wyder ([13]) have investigated the case of thin film in a parallel field assuming that the film consists of an array of close-packed spheres with specular boundary conditions at the surface. For that case those authors were able to perform the calculations for arbitrary mean free path. They obtained the result that the density of states shows appreciable deviations from the dirty limit for ratios l/ξ_0

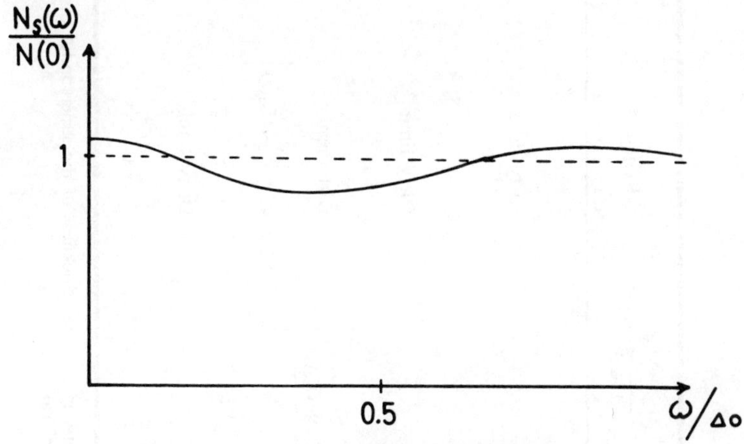

Fig. 7. Density of states as it may arise if the effect of the field on the electron spins becomes important [for details see ([20])].

even as small as 1/10. Furthermore, they found that for a given ratio H/H_c the pair-breaking effect is larger the larger the mean free path is.

Another situation which can be treated completely is that of a thin film in a magnetic field which is so large that the effect of the field not only on the electronic orbits but also on the electron spins has to be taken into account. While the case of short spin-orbital mean free path is contained in Table I, the case of long spin-orbital mean free path leads to new features. If the influence of Pauli paramagnetism is not too great, so that we still have second-order phase transitions, then we expect the density of states in the gapless regime to consist of a superposition of two curves of the form of Eq. (10) but shifted by an amount $\pm\mu H$. It can be shown that this may lead, for example, to a density of states as shown in Fig. 7. Thus for $\omega = 0$ the density of states of the stable superconducting state may even exceed that for the normal state.

Pair-Breaking Mechanisms Belonging to Group 2

Table II contains a list of pair-breaking mechanisms belonging to group 2 with the exception of the thin film case, which we have written down for comparision only. One notes that the other situations have in common that the order parameter is a space-dependent function. A measure of this spatial variation is the quantity $\beta = \langle|\Delta|^4\rangle/\langle|\Delta|^2\rangle^2$ which now is different from one. In calculating $N_s(\omega)$ one restricts oneself to the subcritical regime where Δ is small. Equation (6) is used to calculate $G(\mathbf{p}, \omega)$.

TABLE II*

Physical situation	$\varrho = \Gamma/2\pi T$	$f_1(\varrho)$	$\beta = \frac{\langle \lvert\Delta\rvert^4 \rangle}{\langle \lvert\Delta\rvert^2 \rangle^2}$	Remarks
Thin film in parallel magnetic field	$\frac{\tau_{tr}(v_F edH)^2}{36\pi T}$	$-\frac{1}{2}\psi^{(2)}(\frac{1}{2}+\varrho) - \frac{1}{6}\varrho\psi^{(3)}(\frac{1}{2}+\varrho)$	1	Dirty limit
Type II	$\frac{\tau_{tr}v_F^2 eH}{6\pi T}$	$-\frac{1}{2}\psi^{(2)}(\frac{1}{2}+\varrho)$	1.16	Dirty limit
Surface sheath	$\frac{0.59\tau_{tr}v_F^2 eH}{6\pi T_B}$	$-\frac{1}{2}\psi^{(2)}(\frac{1}{2}+\varrho)$	$\frac{d}{2}\left[\frac{0.59eH_{c3}(t)}{\pi}\right]^{1/2}$	Dirty limit; $d \gg \left[\frac{e}{2}H_{c2}(t)\right]^{-1/2}$
Contact between superconductor and paramagnetic film	$\frac{\tau_{tr}v_F^2}{12\pi T}\left(\frac{\pi}{2d_s}\right)^2$	$-\frac{1}{2}\psi^{(2)}(\frac{1}{2}+\varrho) + \frac{1}{18}\varrho\psi^{(3)}(\frac{1}{2}+\varrho)$	1.50	Dirty limit; $d_s > (l\xi_0)^{1/2}$

* $H_{c2}(t)$ and $H_{c3}(t)$ are the upper critical field and the nucleation field of the surface state respectively, d_s is the thickness of the superconducting film, $\psi^{(n)}(z)$ is the nth derivative of the digamma function. All other symbols are the same as in Table I.

This way one obtains up to terms of the order $|\Delta(\mathbf{r})|^2$

$$N_s(\omega, \mathbf{r}) = N(0)\left[1 + \frac{|\Delta(\mathbf{r})|^2}{2} \frac{\omega^2 - \Gamma^2}{(\omega^2 + \Gamma^2)^2}\right]. \tag{11}$$

This equation looks very similar to Eq. (10). One notices a local relationship between $N_s(\omega, \mathbf{r})$ and the order parameter $\Delta(\mathbf{r})$.

Differences for the different situations appear if one computes the average $\langle|\Delta(\mathbf{r})|^2\rangle_{\text{av}}$ over the junction area as function of $[1 - (T/T_c)]$. In that case the generalized Ginzburg–Landau equation is applied for the computation of the normalization factor. The latter can be written as

$$\left[\psi\left(\frac{1}{2} + \varrho\right) - \psi\left(\frac{1}{2}\right) - \ln\frac{T_{c0}}{T}\right]\langle|\Delta(\mathbf{r})|^2\rangle + \frac{1}{2(2\pi T)^2} f_1(\varrho)\langle|\Delta(\mathbf{r})|^4\rangle = 0, \tag{12}$$

where the solutions of the linearized equation are used to calculate the averages $\langle|\Delta(\mathbf{r})|^{2,4}\rangle$. The function $f_1(\varrho)$ with $\varrho = \Gamma/2\pi T$ is different for different physical situations and is listed in Table II. From Eq. (12) one finds by expansion

$$\langle|\Delta(\mathbf{r})|^2\rangle = 2(2\pi T_c)^2 \frac{1 - \varrho\psi^{(1)}(\frac{1}{2} + \varrho)}{\beta f_1(\varrho)}\left(1 - \frac{T}{T_2}\right), \tag{13}$$

where $\beta = \langle\Delta^4\rangle/\langle\Delta^2\rangle^2$ is also listed in Table II. The quantities ϱ and T_c are connected via Eq. (7). Thus because of the different normalization factor of $\langle|\Delta(\mathbf{r})|^2\rangle$ for different physical situations one is able to discriminate between the different mechanisms. A simple measurable quantity into which the normalization of the order parameter enters is the tunneling conductance. The ratio of the zero bias tunneling conductance in the superconducting and normal states is of special of interest. The latter can be shown to be of the form

$$1 - \left.\frac{dI_s/dV}{dI_n/dV}\right|_{V=0} = -\frac{\langle|\Delta(\mathbf{r})|^2\rangle_{\text{av}}}{2(2\pi T)^2}\psi^{(2)}\left(\frac{1}{2} + \varrho\right) = -\frac{\langle|\Delta(\mathbf{r})|^2\rangle_{\text{av}}}{\langle|\Delta(\mathbf{r})|^2\rangle}\left(1 - \frac{T}{T_c}\right)\frac{1 - \varrho\psi^{(1)}(\varrho)}{\beta f_1(\varrho)}\psi^{(2)}\left(\frac{1}{2} + \varrho\right). \tag{14}$$

As an example, consider a superconducting film in contact with a paramagnetic film. In that case $\langle|\Delta(\mathbf{r})|^2\rangle_{\text{av}} = 2\langle|\Delta(\mathbf{r})|^2\rangle$ and $\beta = \frac{3}{2}$. Thus we write

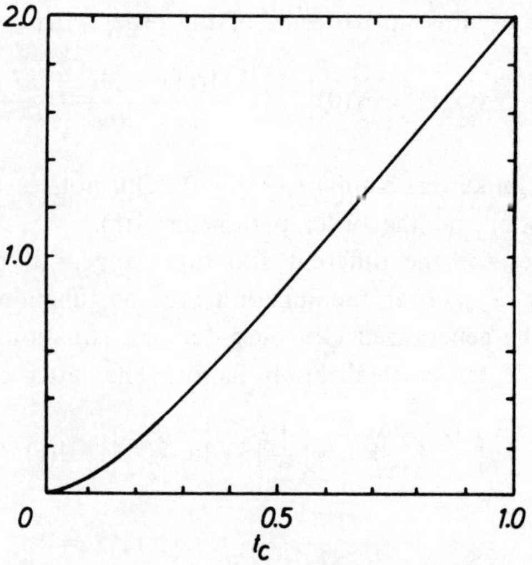

Fig. 8. Plot of $C(t_c)$ as function of $t_c = T_c/T_{c0}$. From Fulde and Maki ([28]).

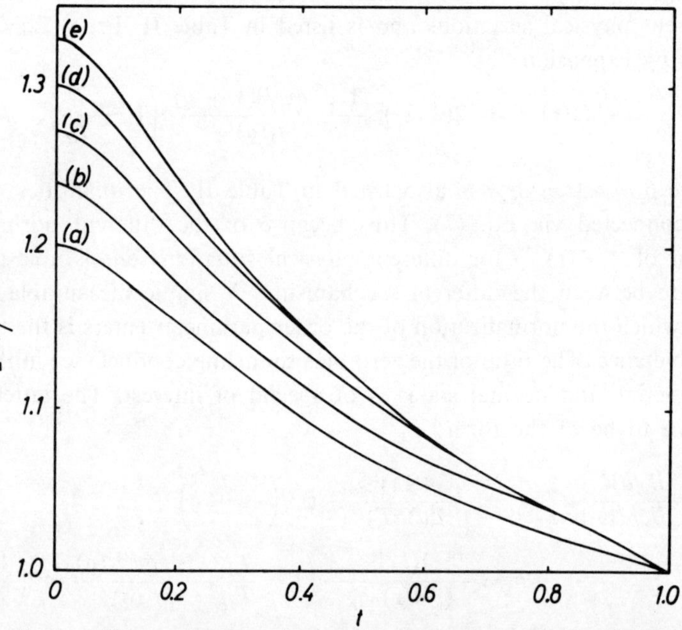

Fig. 9. $\varkappa_2(t)/\varkappa$ as function of t for the simultaneous pair-breaking influence of a contact between a superconducting and a paramagnetic film and a magnetic field perpendicular to the contact (type II). Curves (a)–(e) belong to situations with decreasing superconducting film thickness. For details see ([28]).

$$1 - \frac{dI_s/dV}{dI_n/dV}\bigg|_{V=0} = \frac{4}{3}\left(1 - \frac{T}{T_2}\right)C(t_c), \tag{15}$$

where $C(t_c)$ is shown in Fig. 8. On the other hand, for the type II case we have $\langle|\varDelta(\mathbf{r})|^2\rangle = \langle|\varDelta(\mathbf{r})|^2\rangle_{\text{av}}$ and $\beta = 1.16$, and because of a different functional dependence of $f_1(\varrho)$ a slightly different function $C(t_c)$. From this it is clear that pair-breaking mechanisms are no longer additive if they act simultaneously. In order to study such situations, one would have to compute $C(t_c)$ for such cases. Instead the quantity

$$\frac{\varkappa}{\varkappa_2(t)} = \frac{2}{\pi^2}\,\psi^{(1)}\left(\frac{1}{2} + \varrho\right)\left[\frac{7\zeta(3)}{f_1(\varrho)}\right]^{1/2} \tag{16}$$

is often calculated, which is suitable to describe changes in the functional dependence of $f_1(\varrho)$. In Figs. 9 and 10 we show series of curves for $\varkappa_2(t)/\varkappa$

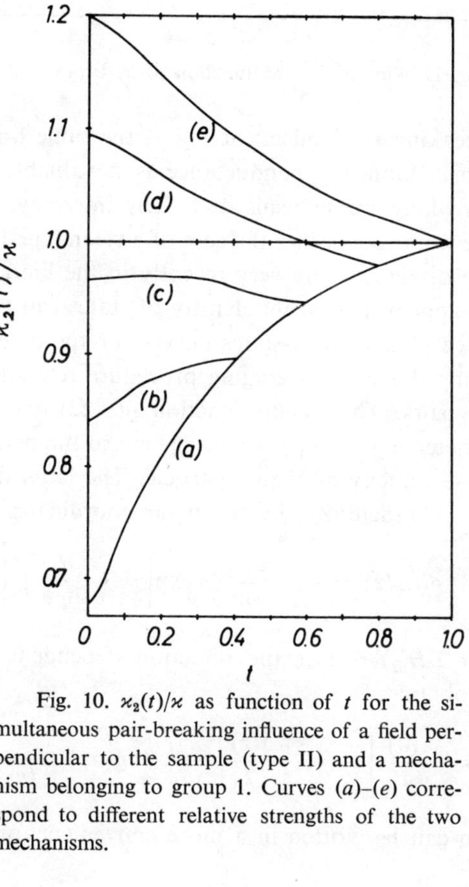

Fig. 10. $\varkappa_2(t)/\varkappa$ as function of t for the simultaneous pair-breaking influence of a field perpendicular to the sample (type II) and a mechanism belonging to group 1. Curves (a)–(e) correspond to different relative strengths of the two mechanisms.

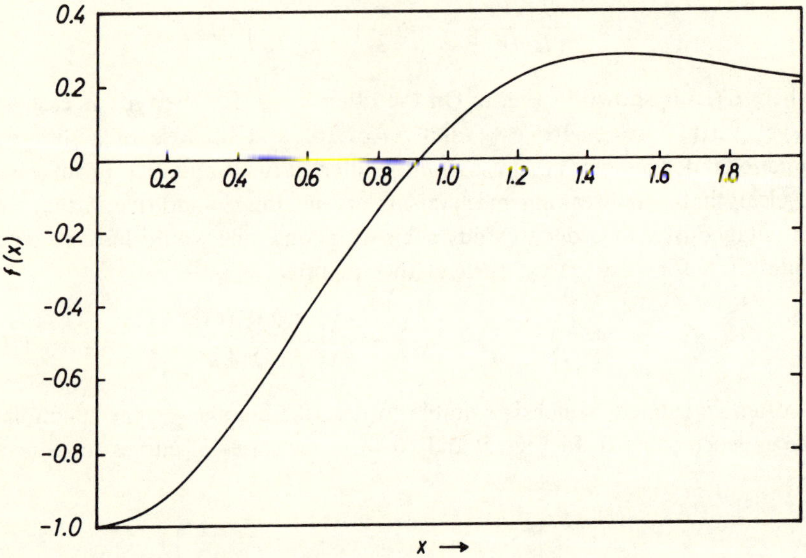

Fig. 11. Plot of $f(x)$ as function of x. From Maki ([21]).

for two pair-breaking mechanisms acting at the same time. Thus it seems that the zero-bias tunneling conductance is a valuable tool in studying different pairbreaking mechanisms and their interplay.

Finally, we want to consider the case of a clean type-II superconductor, which has been discussed only very recently in the literature. It turns out that to a good approximation the density of states can be considered as a weighted average of density-of-states curves for the case when a uniform current is flowing. Here the averaging procedure refers to the momentum of the Cooper pairs. The weight function $\varrho(\alpha, \Omega)$ depends on the pair-breaking parameter α $(= 2\mathbf{v}\mathbf{q}_s)$ where $2\mathbf{q}_s$ is the momentum of a Cooper pair and $\mathbf{v}$ is the velocity of a quasiparticle. The latter defines a direction Ω which can be characterized by the polar coordinates (ϑ, φ). One finds

$$\varrho(\alpha, \Omega) = \frac{1}{\pi^{1/2}\varepsilon \sin\vartheta} \exp\left[-\left(\frac{\alpha}{\sin\vartheta}\right)^2\right], \tag{17}$$

where $\varepsilon = v_F(e/2\, H_{c2})^{1/2}$. For the direction dependent density of states $N_s(\omega, \Omega)$ one obtains

$$N_s(\omega, \Omega) = \frac{N(0)}{4\pi}\left[1 + \frac{\langle|\Delta(\mathbf{r})|^2\rangle}{2}\int_{-\infty}^{+\infty} \varrho(\alpha, \Omega)\frac{d\alpha}{(\omega-\alpha)^2}\right]. \tag{18}$$

This expression can be written in a more convenient form as

$$N_s(\omega, \Omega) = \frac{N(0)}{2}\left\{1 + \frac{\langle|\Delta(\mathbf{r})|^2\rangle}{(\varepsilon \sin \vartheta)^2} f\left(\frac{\omega}{\varepsilon \sin \vartheta}\right)\right\}, \tag{19}$$

where the function

$$f(x) = 2xe^{-x^2} \int_0^x e^{y^2}\, dy - 1$$

is shown in Fig. 11.

REFERENCES

Textbooks

1. J. R. Schrieffer, Theory of Superconductivity, W. A. Benjamin, Inc., New York, 1964.
2. P. G. de Gennes, Superconductivity of Metals and Alloys, W. A. Benjamin, Inc., New York, 1966.
3. K. Maki, Gapless Superconductivity, in: Treatise on Superconductivity, (by R. D. Parks, ed.), Marcel Dekker, New York, 1968. This reference contains a rather complete list of contributions to the subject.

The following list of references is not intended to be complete, and should serve only as a starting point for further information.

Uniform Current Case

4. J. Bardeen, Rev. Mod. Phys. 34:667 (1964).
5. R. H. Parmenter, RCA Rev. 26:323 (1962).
6. K. Maki, Progr. Theor. Phys. (Kyoto) 29:10 (1963); 29:333 (1963).
7. P. Fulde and R. A. Ferrell, Phys. Rev. 131:2459 (1963).
8. P. Fulde, Phys. Rev. 137:A783 (1965).

Thin Film in Parallel Field (Local Electrodyn)

9. K. Maki, Progr. Theor. Phys. (Kyoto) 31:731 (1964).
10. P. G. de Gennes and M. Tinkham, Physics 1:107 (1964).

Thin Film in Parallel Field (Nonlocal Electrodyn)

11. R. S. Thompson and A. Baratoff, Phys. Rev. Lett. 15:471 (1965).
12. E. Guyon, F. Meunier, and R. S. Thompson, Phys. Rev. 156:452 (1967).

Thin Film—Arbitrary Mean Free Path

13. S. Strässler and P. Wyder, Phys. Rev. 158:319 (1967).

Paramagnetic Impurities

14. A. A. Abriksov and L. P. Gorkov, Zh. Eksperim. i Teor. Fiz. 39:1781 (1960) [English translation: Soviet Phys.—JETP 12:1243 (1961)].
15. S. Skalski, O. Betbeder-Matibet, and P. R. Weiss, Phys. Rev. 136:A1500 (1964).

Pauli Paramagnetism

16. G. Sarma, J. Phys. Chem. Solids 24:1029 (1963).
17. K. Maki and T. Tsuneto, Progr. Theor. Phys. (Kyoto) 31:945 (1964).
18. L. P. Gorkov and A. J. Rusinov, Zh. Eksperim. i Teor. Fiz. 46:1361 (1964) [English translation: Soviet Phys.—JETP 19:922 (1964)].
19. P. Fulde and K. Maki, Phys. Rev. 141:215 (1966); Phys. Rev. (Errata) 147:414 (1966).
20. P. Fulde, Solid State Commun. 5:181 (1967).

Type II (Dirty Limit)

21. K. Maki, Physics 1:21 (1964); 1:127 (1964).
22. P. G. de Gennes, Physik kondensierten Materie 3:79 (1964).

Type II (Clean Limit)

23. K. Maki, Phys. Rev. Letters 18:835 (1967).
24. U. Brandt, W. Pesch, and L. Tewordt, Z. Phys. 201:209 (1967).

Surface Sheath

25. D. Saint-James and P. G. de Gennes, Phys. Rev. Letters 9:211 (1664).

Proximity Effect

26. P. G. de Gennes and E. Guyon, Phys. Letters 3:168 (1963).
27. N. R. Werthammer, Phys. Rev. 132:2440 (1963).
28. P. Fulde and K. Maki, Phys. Rev. Letters 18:675 (1965); Physik kondensierten Materie 5:380 (1966).
29. P. G. de Gennes and S. Mauro, Solid State Commun. 3:381 (1965).

Chapter 30

Gapless Superconductor Tunneling—Experiment

T. Claeson

Department of Physics
Chalmers University of Technology
Göteborg, Sweden

In Chapter 29 of this volume Fulde described the different mechanisms leading to gapless superconductivity. Here we will make no attempt to give a comprehensive review of this phenomenon but briefly discuss three different pair-breaking mechanisms that have been observed.

Pair-breaking in a superconductor depresses the superconducting order. In Fig. 1 we see the order parameter versus temperature for several degrees of pair scattering. This curve was calculated by Skalski *et al.* ([1]) for the paramagnetic impurity case. As long as the superconducting order parameter is larger than the pair-breaking parameter a forbidden energy region of excited states is obtained. However, as the pair-breaking strength increases, more and more states appear in the previously forbidden region, until the gap no longer exists. The gapless state is obtained when the excitation energy breadth caused by the pair scattering equals the order parameter. The boundary between the gapless and the gap regions is shown as a dashed curve in Fig. 1. The gapless region is the one below the boundary curve. We see that for a slightly depressed transition temperature T_c gaplessness exists only in a small temperature region close to T_c. At lower temperatures a gap appears, although it is smaller than the one expected from the T_c value.

The density of excited states versus energy is shown in Fig. 2. The curves are normalized to the density of electron states at the Fermi energy in the normal state, N_0. Different pair-breaking strengths are used as parameters. At an infinite collision time we get the typical BCS piled-up density of states. As the pair-breaking is increased, the gap decreases and the densities of states are rounded off considerably.

The tunneling technique is very well suited for studies of gapless

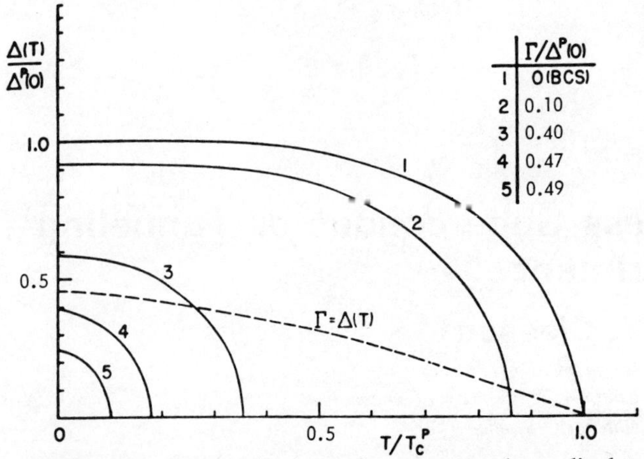

Fig. 1. The superconducting order parameter (normalized to the value found with no pair breaking, the energy gap) versus temperature (T_c^P is the transition temperature of the undisturbed superconductor). Curves with different degrees of pair breaking, Γ, are plotted, as is the function defined by order parameter = pair breaking strength. Below this latter boundary gaplessness is obtained. The critical temperature connected to each pair-breaking strength is given by $\Delta(T_c) = 0$. Note that when the pair breaking is not very strong gaplessness exists only in a limited temperature region close to T_c. From Skalski *et al.* (1).

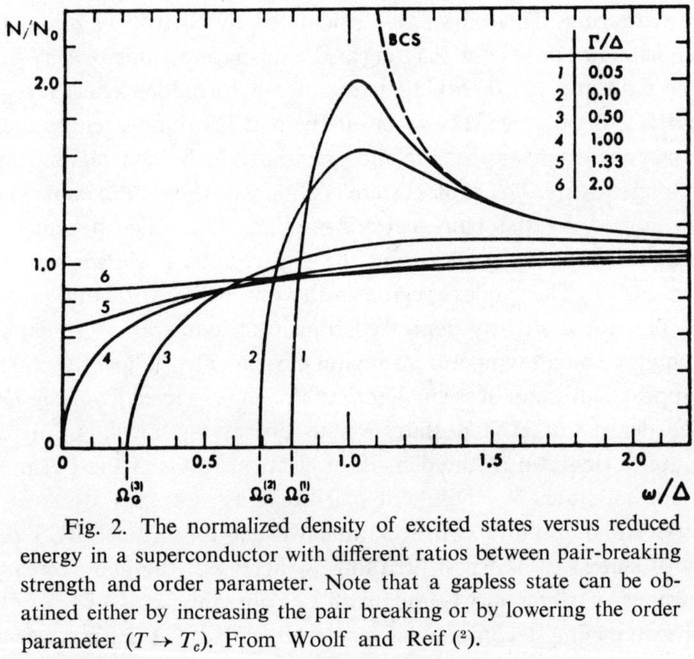

Fig. 2. The normalized density of excited states versus reduced energy in a superconductor with different ratios between pair-breaking strength and order parameter. Note that a gapless state can be obtained either by increasing the pair breaking or by lowering the order parameter ($T \to T_c$). From Woolf and Reif (2).

superconductivity. We can measure the energy gap directly and compare it with the superconducting transition temperature (determined resistively or by noting the complete disappearance of structure due to superconductivity in tunneling curves) as the pair-breaking strength is varied. We know that at 0°K the derivative of the current with respect to the applied voltage in a tunneling experiment is a direct measure of the electron density of states in a superconductor. However, one must be careful with thermal broadening at finite temperatures, a phenomenon which causes rounded-off tunneling characteristics, as the pair-breaking does. If we first assume a normal BCS gap and density of states, then Fig. 3 shows the normalized derivative of current with respect to voltage (dI/dV) versus applied energy (in eV) for several reduced temperature ratios. It is apparent that sharp curves, reminiscent of the BCS density of states, are obtained only at temperatures considerably below the critical temperature.

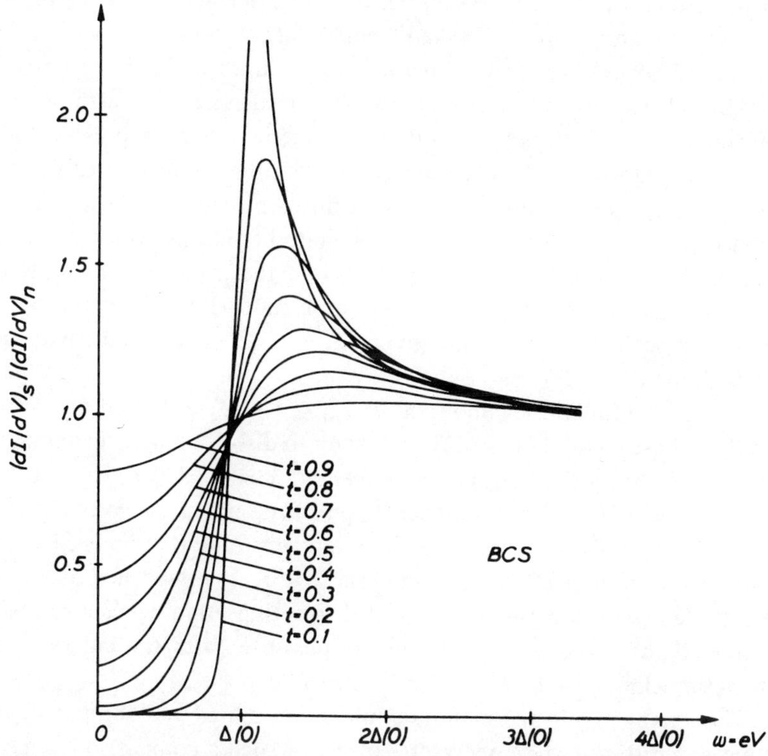

Fig. 3. Normalized conductance versus voltage curves with temperature as a parameter. The curves were calculated from a BCS density of states with a temperature-dependent gap and with thermal smearing. From Skalski *et al.* (1).

In order to observe pair breaking (and gaplessness) we can determine several different entities from our tunneling curves. For example, we can measure a property which is easy to determine, like dI/dV at zero bias voltage, and note how this ratio changes with temperature and pair-breaking strength. In Chapter 29 Fulde showed how these properties are related to each other. We can also try to determine the density of states directly by taking dI/dV at low temperatures. In the latter case for good observation of appreciable pair breaking we want to fulfill two conditions: (1) $T_c/T_{c0} \lesssim 0.5$ (where T_c and T_{c0} are the transition temperatures of the disturbed and undisturbed metal respectively), and (2) $T/T_c \ll 1$.

PAIR-BREAKING INDUCED BY PARAMAGNETIC IMPURITIES

Tunneling into superconductors with paramagnetic impurities was first performed by Reif and Woolf ([2]). It had earlier been observed in bulk alloys that magnetic impurities had a much larger influence upon T_c than nonmagnetic ones. The essential feature of the superconductive state is the correlation between electron pairs in time-reversed states (i.e., with opposite momenta and spins). Collisions with nonmagnetic impurities preserve time-reversal invariance and the pairs are not broken. In a collision with a magnetic impurity, on the other hand, spin flip is possible. Time reversal invariance is broken and the pairs have a finite lifetime τ, which causes an energy spread $\Gamma = \hbar/\tau$ introducing states into the gap. Here T_c is rapidly depressed, but the energy gap decreases even more. At 0.91 of the critical impurity concentration (the one giving $T_c = 0°K$) superconductivity persists without a gap ([3]).

In order to fullfill the conditions $T_c/T_{c0} < 1$, $T/T_c \ll 1$, Reif and Woolf used Pb as base metal. However, it is difficult to dissolve magnetic impurities in Pb. Therefore a flash evaporation method was used, i.e., small pellets of a mixture of the two components were evaporated to give a film of about 10 Å (i.e., four atomic layers) on the substrate at a time. Since the substrate was held at a temperature close to 1°K, the deposited atoms are not expected to diffuse but remain in their positions, and hence a solubility is obtained that otherwise would not be possible. With the aid of a He^3 system tunneling could be performed down to 0.4°K.

The energy gap in these alloys could be estimated and compared with the measured transition temperature. In Fig. 4 such a comparison is made. We see that the energy gap decreases faster than the critical temperature does. The concentration at which the experimental gap is zero is somewhat

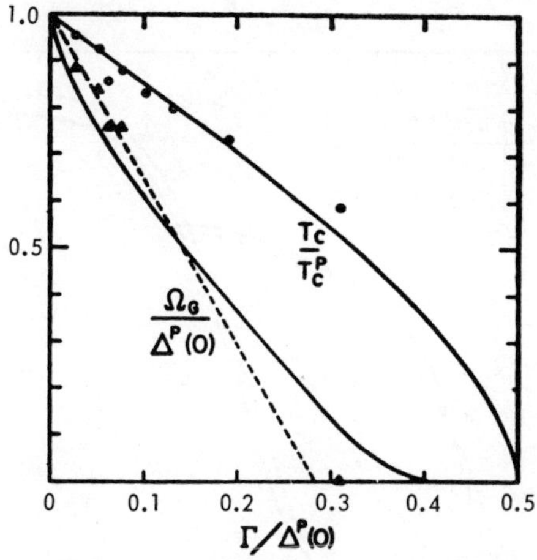

Fig. 4. The superconducting transition temperature and the half-energy gap at $T/T_c = 0.25$ versus pair-breaking strength for a metal with paramagnetic impurities. (Superscript P denotes pure metal.) The open circles and triangles are the experimental points for the critical temperature and the half-energy gap, respectively, as measured by Reif and Woolf ([2]). The dashed curve is their linear extrapolation of the low-impurity concentration points. The full lines are calculated by Skalski *et al.* ([1]). Note that the energy gap goes to zero faster than T_c does.

uncertain. The corresponding theoretical curves from the work of Skalski *et al.* ([1]) are also included in the figure, and we see that the agreement between the experimental and the theoretical curves is rather good.

Figure 5 shows the normalized dI/dV (or conductance) versus energy curves obtained for two concentrations in the Pd–Gd system. The corresponding theoretically predicted curves are indicated by the dashed lines. As can be seen the agreement between the theoretical and experimental curves is not too bad when using rare-earth impurities. Transition-metal impurities, however, gave experimental curves which were smeared considerably more.

TUNNELING IN MAGNETIC FIELDS

A magnetic field applied to a dirty superconductor breaks time-reversal invariance and hence pairs, and the energy gap will disappear at

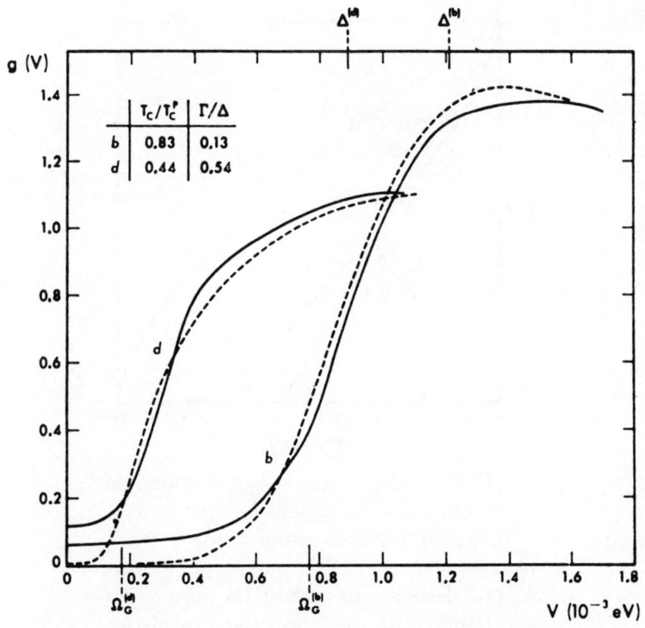

Fig. 5. Plot of dI/dV versus V for two concentrations of Gd in Pb. Solid lines are experimental curves, while the dashed lines were obtained theoretically. From Woolf and Reif ([2]).

a field related to the critical field by $(H/H_c)^2 = 0.91$. Guyon and co-workers ([4]) have performed tunneling on "dirty" superconductors (superconductors with a small mean free path, e.g., metals with a high impurity content or thin films). The measurements were done in high magnetic fields parallel or perpendicular to the film surface (i.e., in fields lower than but close to H_{c3} and H_{c2}, respectively). Figure 6 shows an example of their results. The points are measured conductance values at different applied voltages. Two theoretical BCS curves are also shown, one with a gap chosen such that dI/dV at zero bias voltage reproduces the experimental value and the other such that the energy, where the maximum in the conductance curve appears, is the same as in the experimental case. Neither of these two curves coincides with the experimental values. On the other hand, a good fit is obtained with a theoretical curve where pair breaking is included. In the gapless region the conductance curves had an energy scale independent of the applied magnetic field, whereas any BCS-like approach with a gap parameter decreased by field would lead to a field-dependent energy scale.

Similar experiments have been performed by Levine ([5]) and by Mill-

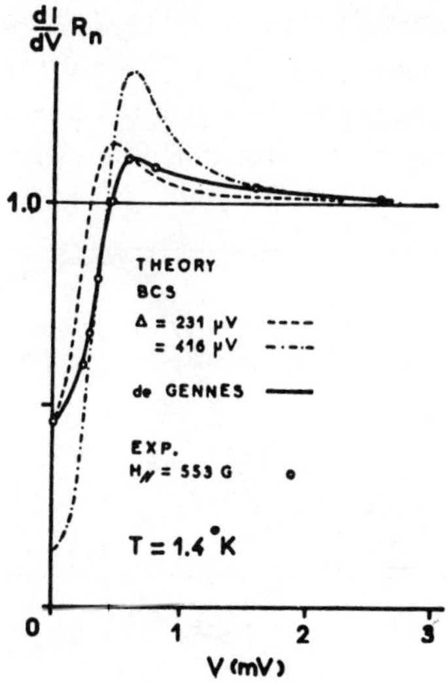

Fig. 6. Plot of dI/dV versus V for a dirty superconductor ($Sn_{0.95}In_{0.05}$) in a magnetic field parallel to the junction and close to the critical field value. Experimental values are given by open circles. The solid line passing through the points is calculated with a pair breaking. The dashed lines represent conductance curves calculated from the BCS theory. There does not exist any gap value for which the BCS characteristic is in reasonable agreement with experiment. From Guyon *et al.* ([4]).

stein and Tinkham ([6]). Some conductance curves taken by the latter are shown in Fig. 7. A He^3 cryostat was used to obtain low temperatures. Manganese was alloyed into Al to prevent the latter film becoming superconducting. Edges were trimmed to sharpen the resistive transition in magnetic fields. It was found that the experimental curves deviated slightly from theoretical curves calculated from the Abrikosov–Gorkov ([3]) theory as applied to the magnetic field problem by Maki and Maki and Fulde ([7]). The latter curves were calculated in the dirty limit, i.e., when $1/\xi_0 = 0$, where l is the electronic mean free path and ξ_0 is the coherence length. One would imagine that this limit would be applicable to the experimental situation, as $1/\xi_0$ was about 0.55 for a tin film and about 0.14 for a Sn–In alloy film. However, Strässler and Wyder ([8]) have recently shown that the degree of pair breaking is dependent on the mean free path. Even for such a low ratio of $1/\xi_0$ as 0.1 the density of excited states is appreciably changed compared to the $l = 0$ limit. We see from Fig. 7 that a very good agreement between experimental points and theoretical curves is obtained when a finite mean free path is assumed. In Fig. 8 the normalized conductance at zero bias voltage for two films with different mean free paths is shown.

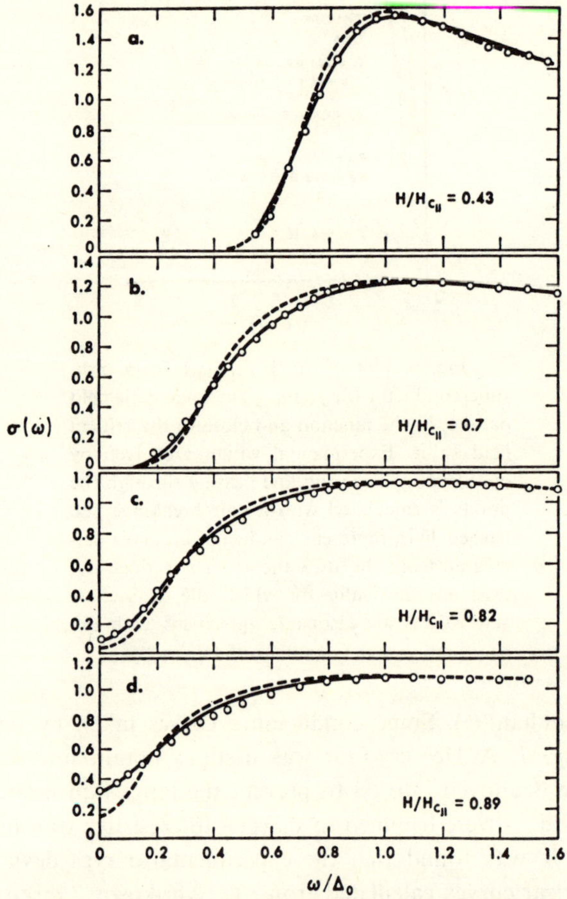

Fig. 7. Normalized dI/dV versus V $[\sigma(\omega) = \{(dI/dV)_s/(dI/dV)_n\}_\omega;\ \omega = eV]$ values for a Sn–5% In alloy film in magnetic fields are given by the open circles. The measured ratio of l/ξ_0 in the film was 0.14. The magnetic field was applied parallel to the film, and the data were taken at $T = 0.36$°K. The dashed and the solid curves are computed with l/ξ_0 equal to 0 and 0.31, respectively. A very good agreement between experimental and theoretical curves is obtained with a finite mean free path. From Millstein and Tinkham ([6]).

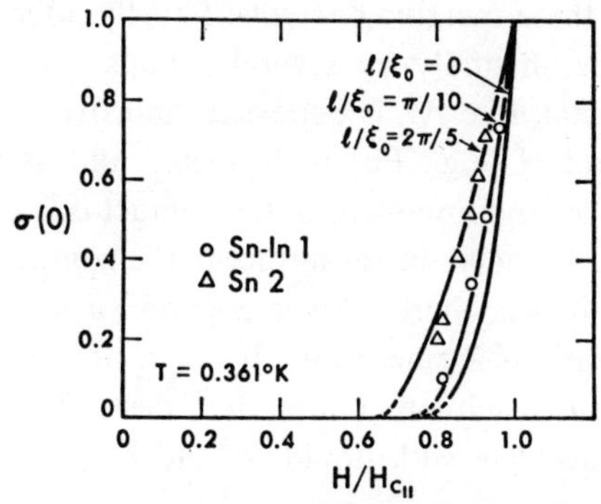

Fig. 8. Magnetic field dependence of the normalized conductance value at zero bias voltage in parallel field. The experimental points were obtained by Millstein and Tinkham ([6]), and the solid lines were calculated using the Strässler-Wyder ([8]) theory with the values of l/ξ_0 indicated. Independent estimates suggested $l/\xi_0 \approx 0.14$ for the Sn–In alloy film and $l/\xi_0 \approx 0.55$ for the tin film. Note that these are in a ratio of 1:4, just as are the values for the theoretical curves.

Theoretical lines calculated from the Strässler–Wyder theory are also given.

When taking characteristics with magnetic fields one has to be careful with trapped flux. Curves taken at zero field before any field has been applied are BCS-like, while conductance curves at $H = 0$, registered after the superconducting film has been driven normal, can have some of the pair-broken characteristics left due to flux trapped in the film.

PROXIMITY EFFECT

The third case of pair breaking or order-suppressing confirmed by tunneling that will be mentioned in this brief review is the case of superimposed films in a metallic contact. Due to the proximity of a nonsuperconducting metal in contact with a thin film of superconducting material the superconducting transition temperature of the latter can be considerably depressed if the film is thinner than a coherence length. In addition, superconductivity can be induced in a metal film in contact with a superconductor. Pairs of superelectrons "leak" into the "nonsuperconducting" metal where they tend to be broken, while "normal" electrons "leak" into the superconductor. Fulde and Maki ([9]) predicted the existence of pair breaking and gaplessness in these metallic contacts. Shortly afterwards the situation was confirmed experimentally by several groups ([10]).

A reference metal, like Al, is deposited and oxidized. Then a layer of the metal to be studied (e.g., Pb) is deposited and as soon as possible a second layer of the other component of the contact is laid upon the first one. In order to get a clean metal–metal surface in the contact, one tries to avoid an absorbed layer of gas there. Hence a good vacuum and a short time between the evaporations are desirable. If one is afraid that the superposed films will diffuse into each other and thus destroy the sharp boundary, one can cool the substrate with liquid N_2 and keep it at that temperature until the test.

Experiments by Hauser ([11]) have shown conclusively that the ratio between the zero-bias conductances in the normal and the superconducting states approaches unity linearly as $T \to T_c$. Data were taken on Pb films

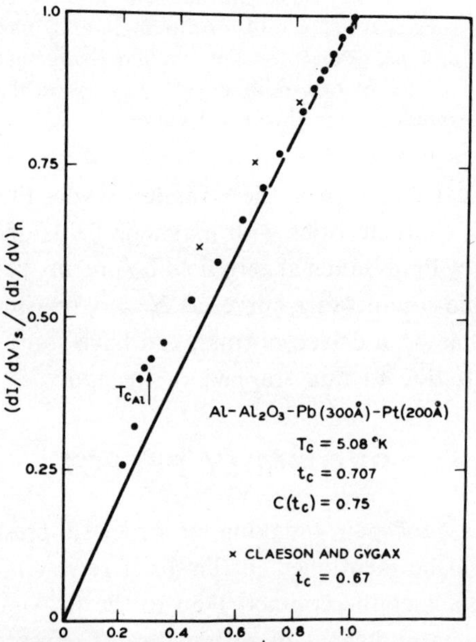

Fig. 9. The normalized conductance at zero bias voltage as a function of $t = T/T_c$ for an $Al/Al_2O_3/Pb/Pt$ junction [measured by J. J. Hauser ([11])]. Note that the conductance approaches the value 1 linearly as $T \to T_c$. The slope of the linear portion is denoted by $1.33C(t_c)$. $t_c = T_c/T_{c0}$.

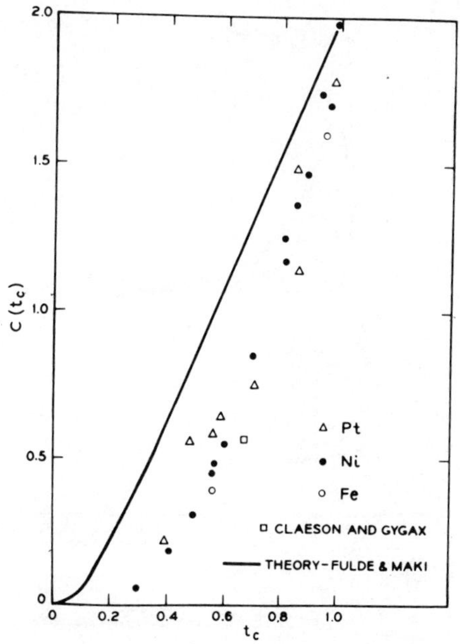

Fig. 10. The slope of the zero-bias conductance versus reduced temperature curve, $C(t_c)$, as a function of $t_c = T_c/T_{c0}$ for a great number of double layers. $C(t_c)$ is a measure of the degree of gaplessness, and we can note that sandwiches with either a magnetic (Fe, Ni) or nonmagnetic (Pt, Ag) "normal" layer having the same T_c dipslay the same degree of "gaplessness." The theoretically computed curve for $C(t_c)$ is also given. Both theoretical and experimental values vary from 0 to 2 as T_c varies from 0 to T_{c0}, but we note that the experimental points systematically fall below the theoretical curve, a fact which indicates that the pair breaking is stronger in the metallic contact than theoretically predicted. From Hauser ([11]).

backed by magnetic or nonmagnetic "normal" films. A typical set of data is given in Fig. 9. The slope of the linear portion is proportional to the function $C(T_c/T_{c0})$, as discussed by Fulde in Chapter 29. This function is not only dependent upon the strength of the pair breaking (measured by the T_c/T_{c0} ratio) but also slightly upon the kind of pair breaking considered. Experimental values of $C(t_c)$ as a function of $t_c = T_c/T_{c0}$ are given in Fig. 10.

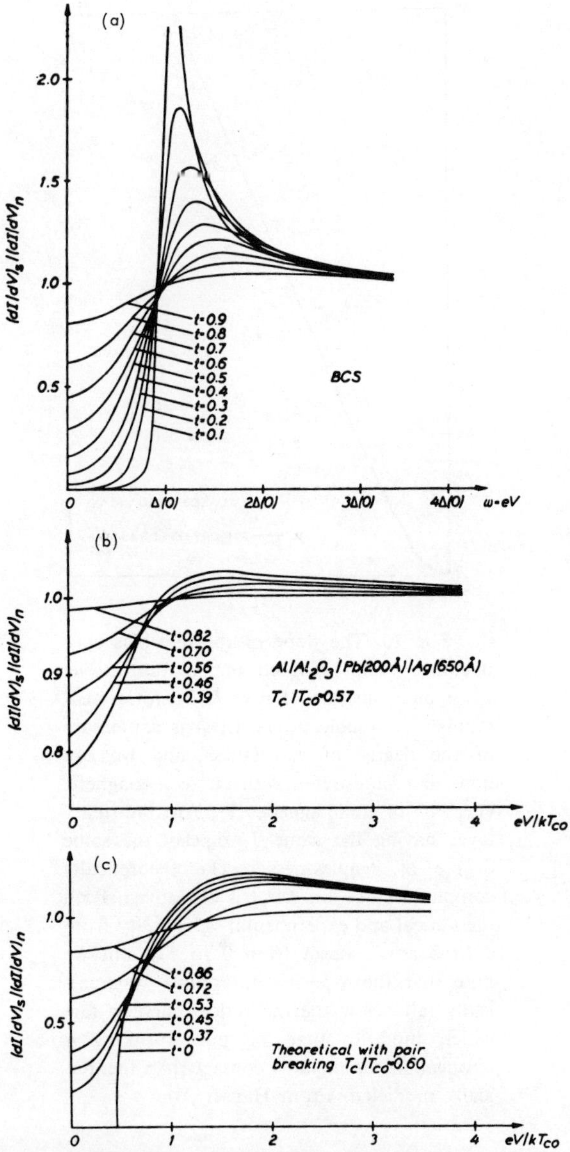

Fig. 11. (*a*) Theoretical normalized conductance versus energy curves calculated from the BCS theory for several T/T_c ratios. $\Delta(0) = 1.76\, kT_c$ was used. (*b*) Experimental dI/dV versus eV/kT_{c0} curves for an $Al/Al_2O_3/Pb/Ag$ junction. $t = T/T_c$. The offset zero might be caused by leakage currents. Note the flat appearance of curves in 11(*b*) compared to those in 11(*a*). (*c*) dI/dV versus eV/kT_{c0} calculated for a pair-breaking parameter corresponding to $T_c/T_{c0} \approx 0.60$ and for several $t = T/T_c$ ratios. The calculation is based on an analogy with the paramagnetic impurity case. From Claeson *et al.* ([10]).

Leakage currents can disturb the tunneling results. Presumably, areas of the reference metal can come into contact with the superconductor at edges or in holes of the insulating barrier. Thus we might get $I = I$ (tunneling into superconductor) $+ I$ (leakage). In a derivative plot we get a constant contribution to dI/dV independent of applied voltage. If this problem is serious, we can try to solve it in different ways. We can try to estimate and subtract the constant contribution from dI/dV versus V curves. If a fit to theoretical curves is obtained at one temperature, the fit should also be good at other temperatures in order to have a valid agreement. We can also normalize experimental and theoretical curves to an easily distinguishable value.

In Figs. 11(a) and (b) we can compare theoretical BCS curves with experimental dI/dV versus V curves for tunneling into Pb backed by Ag. Even if we subtract a large "leakage" current, which probably is taken to be larger than it really is, the experimental curves are much more smeared than the theoretical curves. This, of course, is an indication of pair breaking.

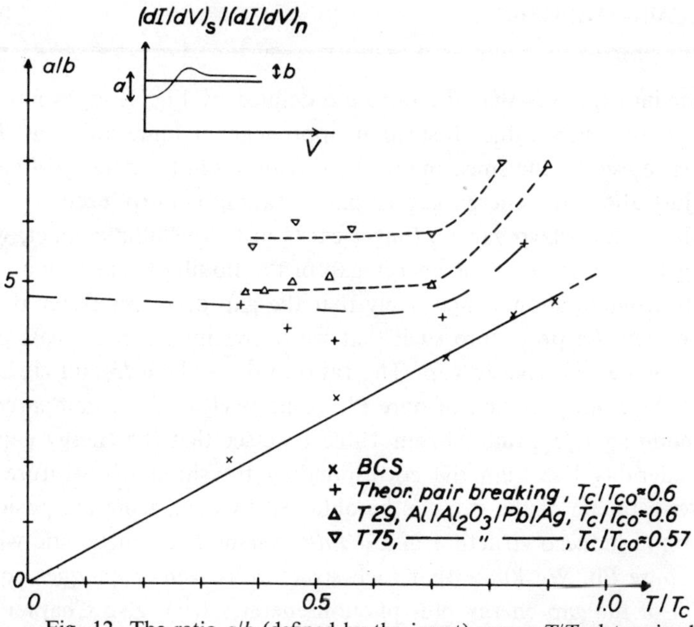

Fig. 12. The ratio a/b (defined by the insert) versus T/T_c determined by tunneling into the Pb side of a Pb–Ag sandwich, and the same ratio found theoretically with a pure BCS gap (full line) and with pair breaking introduced (long dashed line). Note that the experimental curves seem to reach a finite value at low temperatures, while the BCS ratio goes to zero. From Claeson *et al.* ([10]).

TABLE I

Comparison of Estimated Energy Gap and Superconducting Transition Temperature Ratios between Superimposed Film Sandwiches and Pure Lead

Junction	Ω/Δ_0 upper limit	T_c/T_{c0} lower limit
Al/Al_2O_3/Pb(350 Å)/Ag(1000 Å)	0.44	0.67
Al/Al_2O_3/Pb(180 Å)/Ag(600 Å)	0.46	0.65
Al/Al_2O_3/Pb(300 Å)/Ag(650 Å)	0.41	0.60
Al/Al_2O_3/Pb(160 Å)/Ag(800 Å)	0.30	0.45
Al/Al_2O_3/Pb(300 Å)/Ag(1200 Å)	0.22	0.31
Al/Al_2O_3/Pb(1120 Å)/Al(3100 Å)	0.37	0.60
Al/Al_2O_3/Pb(105 Å)/Al(1600 Å)	0.30	0.63
Al/Al_2O_3/Ag(520 Å)/Pb(300 Å)	0.32	0.62
Al/Al_2O_3/Ag(630 Å)/Pb(250 Å)	0.29	0.59
Al/Al_2O_3/Ag(650 Å)/Pb(220 Å)	0.33	0.58
Al/Al_2O_3/Al(1480 Å)/Pb(105 Å)	0.37	0.56

Another indication is that the ratio a/b defined by Fig. 12 does not extrapolate to the BCS value. Instead it approaches a finite value as $T \to 0$, which is caused by the smearing of the singularity in the density of electron states just above the energy gap as pair breaking is introduced.

Although a sharp gap no longer exists in the metallic contact, we can try to get an estimate of the energy gap of the Pb film backed by a normal film. In order to show conclusively that the gap decreases faster than T_c, let us design the procedure such that we overestimate, or at least do not underestimate, the energy gap. The ratio Ω/Δ_0, (where Ω and Δ_0 are the gaps of the composite and of pure Pb, respectively) will be compared with corresponding T_c/T_{c0} ratio. From Table I we see that the energy gap ratio is considerably less than the corresponding transition temperature ratio. A better estimate of the gap can be obtained by comparing the position of the phonon-induced structure in a d^2I/dV^2 versus V characteristic with the one of pure Pb. We know that such structure is seen at energies equal to the sum of the gap energy plus phonon energies [([12]); also Chapter 20 by Rowell in this volume].* Hence if we assume that the phonon spectrum is unchanged, we can measure the gap by noting the position of the singula-

* With pair breaking, the infinite density of states just above the energy gap is smeared. Still, we have an abrupt rise from the zero value at the gap.

rities and subtracting corresponding phonon energies. With $T_c/T_{c0} \approx 0.6$ it was found that $0.25 < \Omega/\Delta_0 < 0.32$. If we had taken the corresponding value from the theory of pair-breaking caused by magnetic impurities [1], we would have obtained $\Omega/\Delta_0 \approx 0.28$ with $T_c/T_{c0} = 0.6$.

The order parameter $\Delta(r)$ is spatially varying in our double layer. In Chapter 29 it was shown that the metallic contact belongs to a group of pair-breaking mechanisms other than the paramagnetic impurity case. In the latter case Δ is not space-dependent, and theoretical results are well

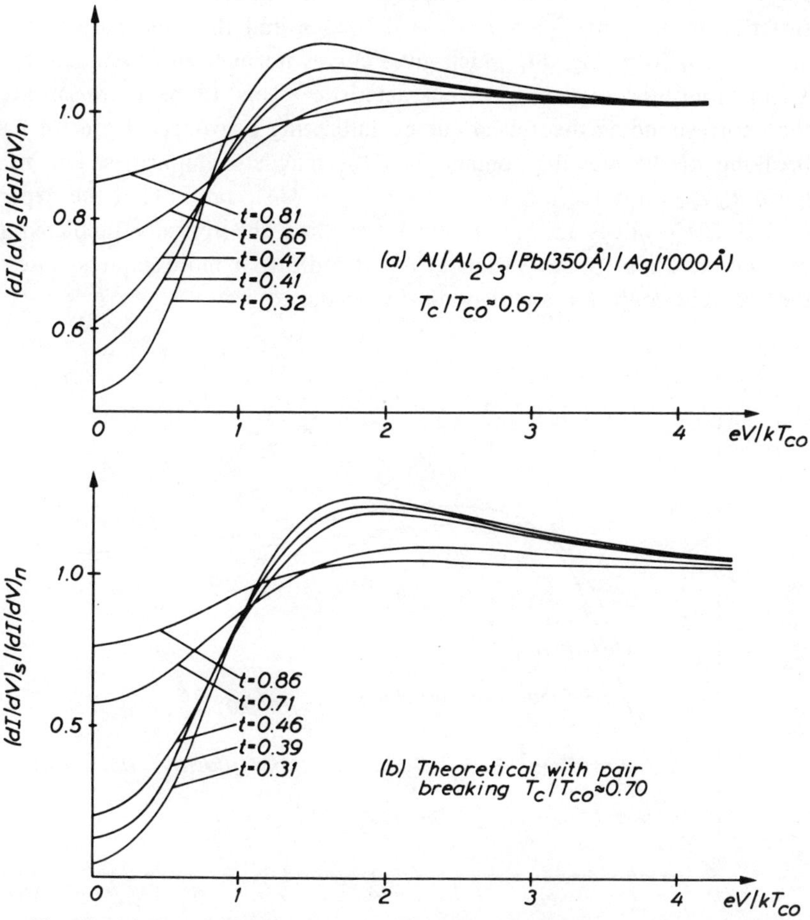

Fig. 13. (*a*) Experimental conductance versus energy curves for an $Al/Al_2O_3/Pb/Ag$ junction. $t = T/T_c$. (*b*) Theoretical curves computed in analogy with the paramagnetic impurity case for a pair-breaking parameter corresponding to $T_c/T_{c0} \approx 0.70$. From Claeson *et al.* [10].

established. In order to use these results, let us assume that $\Delta(r) = \Delta = \text{const}$ in the metallic contact. From a knowledge of the T_c/T_{c0} ratio we can determine the order parameter and the pair-breaking parameter at each temperature considered, and once these are known density-of-states versus energy curves can be computed as shown in Chapter 29. After adding the thermal smearing contribution we obtain theoretical dI/dV versus V curves, and these can be compared with experimental ones. In Figs. 11(*b*) and (*c*) we compare the two types of curves for a T_c/T_{c0} ratio of 0.6 and we see that the experimental plots agree much better with these pair-breaking curves than they did with the simple BCS ones. Figure 13 shows that the situation is the same for a $T_c/T_{c0} \approx 0.7$ ratio, and the close agreement is also evident from Fig. 14, which gives curves normalized in another way. Throughout, the experimental characteristics seem to be more smeared than corresponding theoretical curves, indicating a stronger degree of pair breaking in the metallic contact than for magnetic impurities, i.e., to a given T_c/T_{c0} value there corresponds a lower Ω/Δ_0 ratio. That the experimental $C(t_c)$ values in Fig. 10 lie lower than theoretically predicted (a calculation using spatial variance) might indicate a more gapless state in the metallic contact than theoretically predicted.

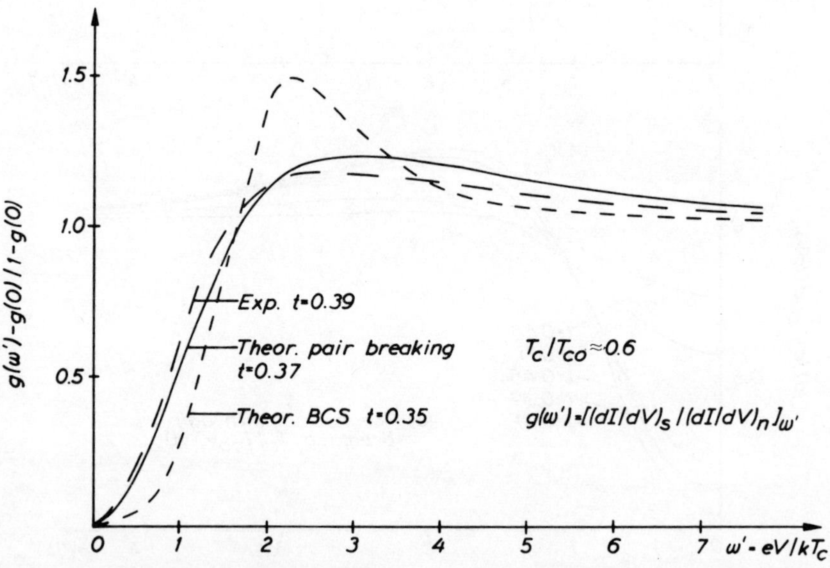

Fig. 14. Theoretical and experimental conductance curves normalized in such a way as to get rid of any "leakage" contribution to the tunneling current. The experimental characteristic is from an $Al/Al_2O_3/Pb/Ag$ junction with $T_c/T_{c0} \approx 0.6$. Here $t = T/T_c$ falls in the range 0.35–0.39. From Claeson *et al.* ([10]).

CONCLUSION

We have seen that pair breaking and gaplessness have been shown to exist experimentally. Three examples have been given here. With paramagnetic impurities the superconducting dI/dV versus V characteristic, which depends upon the density of excited states, was smeared, and the energy gap measured was smaller than would be expected from the BCS theory when T_c was known. "Dirty" superconductors in high magnetic fields also gave characteristics in accord with a theory where pair breaking had been included. The metallic contact gave very smeared curves too. The pair breaking seemed to be larger in that case than in the case with paramagnetic impurities.

REFERENCES

1. S. Skalski, O. Betbeder-Matibet, and P. R. Weiss, Phys. Rev. 136:A1500 (1964).
2. F. Reif and M. A. Woolf, Phys. Rev. Letters 9:315 (1962); M. A. Woolf and F. Reif, Phys. Rev. 137:A557 (1965).
3. A. A. Abrikosov and L. P. Gorkov, Zh. Eksperim. i Teor. Fiz. 39:1781 (1960) [English traslation: Soviet Phys.—JETP 12:1243 (1961)].
4. E. Guyon, A. Martinet, J. Matricon, and P. Pincus, Phys. Rev. 138:A746 (1965); E. Guyon, Adv. Physics 15:417 (1966).
5. J. L. Levine, Phys. Rev. 155:373 (1967).
6. J. Millstein and M. Tinkham, Phys. Rev. 158:325 (1967).
7. K. Maki, Progr. Theoret. Phys. (Kyoto) 29:10 (1963); 29:333 (1963); 29:603 (1963); 31:731 (1964); K. Maki and P. Fulde, Phys. Rev. 140:A1586 (1965).
8. S. Strässler and P. Wyder, Phys. Rev. 158:319 (1967).
9. P. Fulde and K. Maki, Phys. Rev. Letters 15:675 (1965); Physik kondensierten Materie 5:380 (1966).
10. T. Claeson and S. Gygax, Solid State Commun. 4:385 (1966); J. J. Hauser, Physics 2:247 (1966); E. Guyon, A. Martinet, S. Mauro, and F. Meunier, Physik kondensierten Materie 5:123 (1966); H. Tsuya, J. Phys. Soc. Japan 21:1011 (1966); T. Claeson, S. Gygax, and K. Maki, Physik kondensierten Materie 6:23 (1967).
11. J. J. Hauser, Phys. Rev. 164:558 (1967).
12. D. J. Scalapino, J. R. Schrieffer, and J. W. Wilkins, Phys. Rev. 148:263 (1966); W. L. McMillan and J. M. Rowell, "Tunneling and the Strong Coupling Superconductor," in: Treatise on Superconductivity, (R. D. Parks, ed.), Marcel Dekker, New York, 1968.

Chapter 31

DC Josephson Effect

J. E. Mercereau

Ford Scientific Laboratory
Newport Beach, California
and California Institute of Technology
Pasadena, California

INTRODUCTION

At the outset it should be stressed that the Josephson effect ([1]) is not *necessarily* a tunneling phenomenon but is the result of far more fundamental physical processes. It is true historically that tunnel junctions have been used to study this effect, but the tunneling is not essential to the phenomenon. Primarily, tunnel junctions have been used because this is the only physical situation for which detailed calculations of certain amplitude effects can be made.

In this chapter we will consider Josephson's effect as arising from some sort of weak coupling between two macroscopic quantum states. For the purpose of illustrating Josephson effects it is only necessary to assume that the macroscopic quantum state exists, and it becomes entirely possible to neglect exactly what the $\sim 10^{22}$ electrons had to do in order to get into this state. In order to derive the Josephson equations, all that is required is the simplest quantum mechanics and the requirement of gauge invariance. As has been the usual practice, the Josephson equations will be derived and then modified to make them gauge invariant.

THE MACROSCOPIC QUANTUM STATE ([2])

In a superconductor all electron pairs may be considered to have condensed into the same quantum state which we write as:

$$\Psi = \sqrt{\varrho}\, e^{i\gamma},$$

where ϱ is the Cooper-pair particle density and γ is the common phase of all the pairs. All superelectron pairs will also have the same energy μ. The time dependence of this wave function is of course given by $\partial\Psi/\partial t = (-i/\hbar)$ (energy)Ψ. Furthermore, since all the electrons must do exactly the same thing, the time rate for the macroscopic state Ψ must be the same as for a single pair, or, $\partial\Psi/\partial t = (-i/\hbar)\mu\Psi$.

In the Josephson effect two superconductors are required (Fig. 1). The two superconductors are in some way weakly connected, so that there is a small probability that the wave function in superconductor 1 will change by transfering an electron to superconductor 2—by tunneling or some other process.

Mathematically, this is written as

$$\partial\Psi_1/\partial t = (-i/\hbar)[\mu_1\Psi_1 + \xi\Psi_2].$$

A similar equation exists for Ψ_2,

$$\partial\Psi_2/\partial t = (-i/\hbar)[\mu_2\Psi_2 + \xi\Psi_1].$$

These equations can be solved by substituting $\Psi = \sqrt{\varrho}\, e^{i\gamma}$ and equating real and imaginary parts. The result (if ϱ_1 is assumed equal to ϱ_2) is

$$\dot{\varrho}_1 = \dot{\varrho}_2 = (\xi/\hbar)\varrho_1 \sin(\gamma_1 - \gamma_2),$$

$$(\partial/\partial t)(\gamma_1 - \gamma_2) = (\mu_1 - \mu_2)\hbar.$$

From the continuity equation it is possible to relate the density change to a current and show that the Josephson current density is

$$j = \frac{2e}{\hbar}\mathscr{E}\sin(\gamma_1 - \gamma_2) = j_1 \sin(\gamma_1 - \gamma_2).$$

Here $\mathscr{E}$ is the energy per unit area ("binding energy" arising from ξ), where the area involved is the contact area between the two superconductors.

In the case of a tunnel junction a calculation (1) of maximum total binding energy W yields an expression involving the BCS energy gap Δ:

$$W = \frac{h}{2e}I_0 = \frac{\pi}{2}\left(\frac{h}{2e}\right)\frac{\Delta}{R},$$

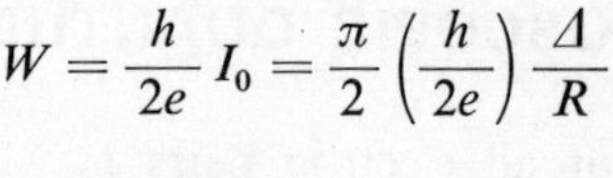

Fig. 1.

where $2\Delta = 3.5kT_c$ and I_0 is the maximum total Josephson current. I_0 is about 10 mA for a 0.1-ohm junction. In order for the thermal fluctuation effects to be small it is necessary to have $W \gg kT$. This implies that the resistance of the tunnel junction must be less than about an ohm. Typically,

$$W = I_0(h/2e) \approx 10^{-2} \times 10^{-15} \approx 10^{-17}\,j.$$

Consider a metal bridge (Fig. 2) and ask how large a region (111) must be "normal" in order for there to be Josephson effects. Here I am defining "normal" to be a region of possibly finite superconducting density ϱ but lacking phase coherence. For order of magnitude effects I will presume that phase *in*coherence will generate a condensation energy and that this condensation energy is then $\sim(1/2\mu_0)B_c^2(\text{Vol})$, where μ is the permeability of free space, B_c the critical field, and (Vol) the volume of the "normal" region. Using the energy of $10^{-17}\,j$, the volume for niobium is on the order of 10^{-21} m^3, or a region $\sim$1000 Å on each side. Numbers like this appear to be approximately correct for point-contact junctions ([3]).

Unfortunately, there are no "phase meters" available to check these effects directly, so it is now necessary to find how this phase depends on electromagnetic fields. Equations describing any measurable quantities must be gauge invariant, and it will be found convenient to specifically require gauge invariance in the Josephson equations.

From E + M theory

$$\mathbf{B} = \nabla \times \mathbf{A}, \qquad \mathbf{E} = -\nabla\phi - (\partial \mathbf{A}/\partial t),$$

where $\mathbf{A}$ and ϕ are the vector and scalar potentials. Gauge invariance means that the physics of any situation is unchanged if $\mathbf{A} \to \mathbf{A}' + \nabla\chi$ and $\phi \to \phi' - (\partial\chi/\partial t)$ where χ is an arbitrary scalar quantity. We must now find how the phase transforms under such a transformation.

The velocity operator for electron pairs is

$$v = (1/2m)\{i\hbar\nabla + 2e\mathbf{A}\},$$

and when applied to the macroscopic wave function, $\sqrt{\varrho}e^{i\gamma}$, gives as the expectation value for velocity $\langle v\rangle = (1/2m)\{\hbar\nabla\gamma - 2eA\}$, as a real physical observable. Now, the phase must change in such a way that $\langle v\rangle$ is unaltered

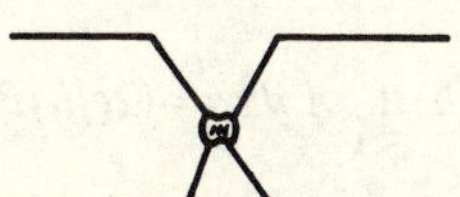

Fig. 2.

by a gauge transform on **A**. This can be effected by replacing by γ by γ', where

$$\gamma \rightarrow \gamma' + (2e/\hbar)\chi$$

Therefore $(\gamma_1 - \gamma_2)$ is *not* gauge invariant, and must be replaced by

$$(\gamma_1 - \gamma_2) \Rightarrow \left\{\gamma_1 - \gamma_2 + (2e/\hbar) \int_1^2 A\, dx\right\}.$$

Hence,

$$j = j_1 \sin\left(\gamma_1 - \gamma_2 + (2e/\hbar) \int_1^2 A\, dx\right)$$

and

$$\frac{\partial}{\partial t}\left(\gamma_1 - \gamma_2 + \frac{2e}{\hbar}\int_1^2 A\, dx\right) = (\mu_1 - \mu_2)\hbar^{-1} = \frac{\Delta\mu(\text{thermo})}{\hbar} + \frac{2e}{\hbar}\int_1^2 E\, dx.$$

Since $\mathbf{E} = -\nabla\phi - (\partial\mathbf{A}/\partial t)$, the final result, illustrating the explicit electromagnetic effects and assigning $s\,\Delta t$ as a thermal energy, is

$$j = j_1 \sin\left(\gamma_1 - \gamma_2 + \frac{2e}{\hbar}\int_1^2 A(t)\, dx\right),$$

$$\frac{\partial}{\partial t}(\gamma_1 - \gamma_2) = \frac{2e}{\hbar}\int_1^2 \nabla\phi \cdot dx + \frac{s}{\hbar}\int_1^2 \Delta T\, dx.$$

Most experiments are done at constant temperature, so that the time rate of change of the phase difference, $(\gamma_1 - \gamma_2)$, will be determined by the electrostatic potential, ϕ. It will be of considerable interest to show the entropy effect $(s\,\Delta T)$; however, the only experiment ([4]) that I know of has thus far not had sufficient sensitivity.

JOSEPHSON EFFECTS AT ZERO CURRENT

All the above results are quite general and are independent of the actual mechanism of charge transfer from superconductor 1 to superconductor 2. We now apply these equations to some simple cases of zero net current to determine the effects of applied fields on rectangular, uniform tunnel junctions (Fig. 3). Let **B** be a uniform magnetic field, $\mathbf{B} = -\hat{z}B_0$. We choose the gauge $\mathbf{A} = B_0 y\hat{x}$. Then this field **B** requires the quantum phase to vary as

$$(2e/\hbar)\int_1^2 A\, dx = (2e/\hbar)B_0 ty.$$

Now, since the net current is defined zero, $(\gamma_1 - \gamma_2)_0$ must also be zero

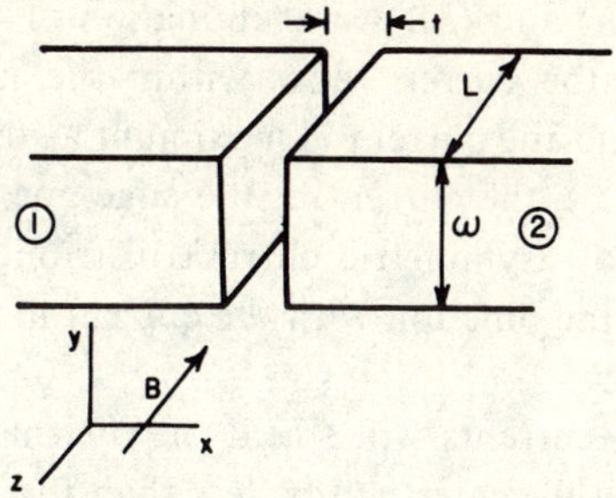

Fig. 3. Rectangular tunnel junction. Here t is the thickness of the barrier in the x direction, L is the length of the junction in the z direction, and w is the length in the y direction.

in this gauge. Actually, $(\gamma_1 - \gamma_2)$ is a "boundary condition" determined by the required net current. Therefore here j reduces to

$$j = j_1 \sin\left(\frac{2e}{\hbar} B_0 t y\right). \tag{1}$$

So far we are neglecting phase shifts due to surface currents in the superconductors. If the junction is not connected into an external circuit, no net current can flow. This requires that the current density be symmetric about the center of the junction. Two current distributions from Eq. (1)

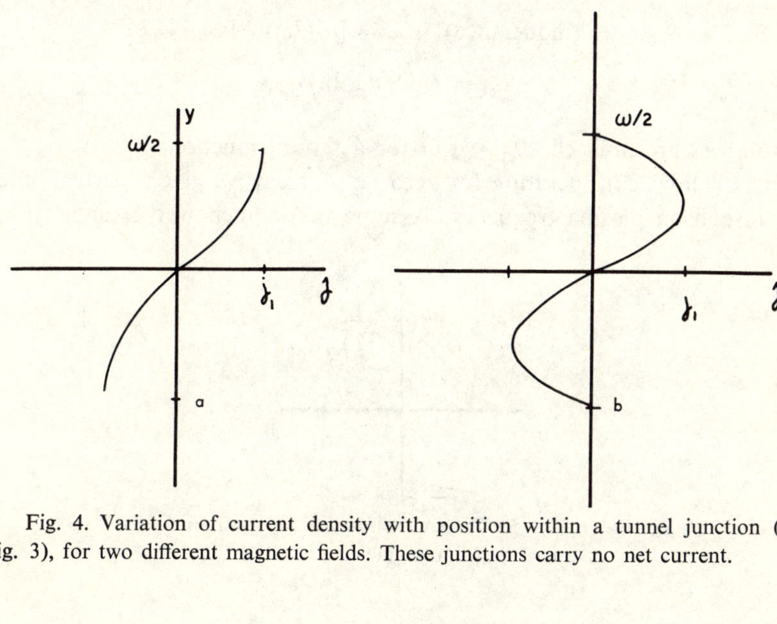

Fig. 4. Variation of current density with position within a tunnel junction (see Fig. 3), for two different magnetic fields. These junctions carry no net current.

are shown in Figs. 4(a) and (b). An examination of these current patterns shows the details of the current flow within the junction. In Fig. 4(a), $B_0 = (1/2wt)(h/2e)$, and the current is maximum at the edge. If B increases to $B_0 = (1/wt) \times (h/2e)$, the current at the edge goes to zero, as shown in Fig. 4(b). This type of asymmetric current distribution corresponds to a circulating current in the junction with the current loops closed in the body of the superconductor.

These circulating currents will shield the internal volume of the junctions, and the field in the center will be less than the applied field. To keep these shielding effects small, it is required that the induced flux be small compared to the flux unit. Thus very approximately, Induced Flux = inductance × induced current $\ll h/2e$, or $(\mu_0 w\lambda/L) \times (j_1 wL) \ll (h/2e)$, where λ is the magnetic field penetration depth. Thus there is a critical width, $w^2 = (h/2e)(1/j_1\lambda\mu_0)$, below which these shielding effects are unimportant. This width is called the "Josephson penetration depth" λ_J and for the usual junction is about 1 mm.

If $w \gg \lambda_J$, in low fields the current can flow only at λ_J from the edges as in Fig. 5. (Again no net current). However, as the applied field is increased, the induced current will also increase by Faraday induction and shield the internal area of the junction. Since the current density is limited to a maximum of j_1, above a certain critical field the magnetic field can no longer be shielded and will begin to penetrate into the junction. This critical field, B_{c1}, is reached when the induced currents are maximum, or

$$B_{c1}\lambda\omega = (\text{inductance}) \times I_c = [\mu_0(\lambda\omega)/L] \times \lambda_J j_1 L,$$

$$B_{c1} = (h/2e)(1/\lambda\lambda_J);$$

if numbers are inserted, $B_{c1} \approx 1$ G for a typical junction.

As a final consideration for zero net current we give a derivation of the Josephson plasma frequency. Assume that somehow the tunnel junc-

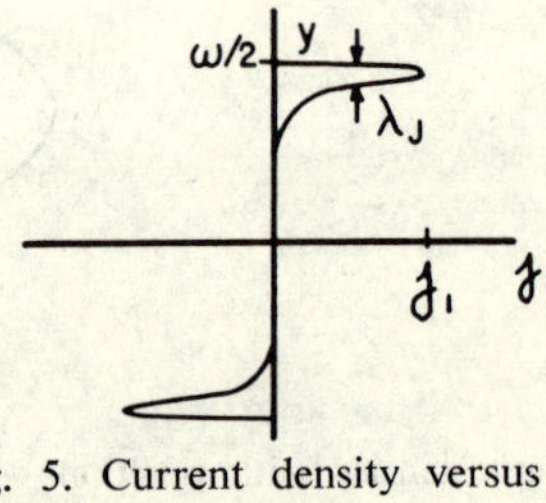

Fig. 5. Current density versus position for a wide junction at the critical field, B_{c1}.

tion, considered as a capacitor, is charged. This charge can then flow across the junction as Josephson current, reversing the polarity of the capacitor. As current flows across the junction, the charge transferred is $\Delta Q = I_J/\omega$, where ω is the frequency of phase precession, or $\omega = 2eV/\hbar$. If the junction has capacity C then the frequency can also be written as $\omega = (2e/\hbar)(Q/C)$. This leads to

$$\omega^2 = \frac{(2e/\hbar)I_J}{C(\Delta Q/Q)}.$$

The lowest frequency is given when all of the charge is transferred each cycle, or $\Delta Q = Q$ and $\omega_0{}^2 = (I_J 2e/\hbar)/C$; ω_0 is called "Josephson plasma frequency". By analogy to an electrical circuit, $(h/2e)I_J^{-1}$ corresponds formally to an inductance and has been called the Josephson inductance. Actually, however, the Josephson oscillation exchanges energy between the electrostatic energy of the capacitor and the quantum mechanical binding energy of the junction rather than between electrostatic energy and magnetic energy as in a real LC circuit.

JOSEPHSON EFFECTS AT FINITE CURRENT

As we noted before, the phase difference $(\gamma_1 - \gamma_2)$ can be thought of as a boundary condition on the current. Previously we had taken $(\gamma_1 - \gamma_2)$ at the center of the junction, $(\gamma_1 - \gamma_2)_0$, to be zero, and relied on symmetry to determine the variation of current with position. Following along with this procedure finite current will require that $(\gamma_1 - \gamma_2)_0 \neq 0$, and maximum current will be achieved when $(\gamma_1 - \gamma_2)_0 = \pi/2$. For this case of maximum current Fig. 1(*a*) [in which the magnetic field $B = (1/2wt)(h/2e)$] transforms into Fig. 6(*a*).

Current density is zero at the edges but has its maximum value, j_1, at the center. The total current flowing to the right in this figure is then just the area under the curve. Similarly, Figure 1(*b*) [where B has been increased to $(1/wt)(h/2e)$] transforms into Fig. 6(*b*). At this field the current reverses direction twice within the junction in such a way that there can be *no* net flow, and the junction has become an infinite impedance for net supercurrent transport.

Increasing the field still further, to $B = (3/2)(1/wt)(h/2e)$ for example, gives Fig. 6(*c*). The current reverses twice in this junction also. However, now the net flow to the right is again finite but is smaller than before because of internal cancellation, and the junction again has zero impedance to supercurrents.

The capability of carrying supercurrent thus oscillates and decreases

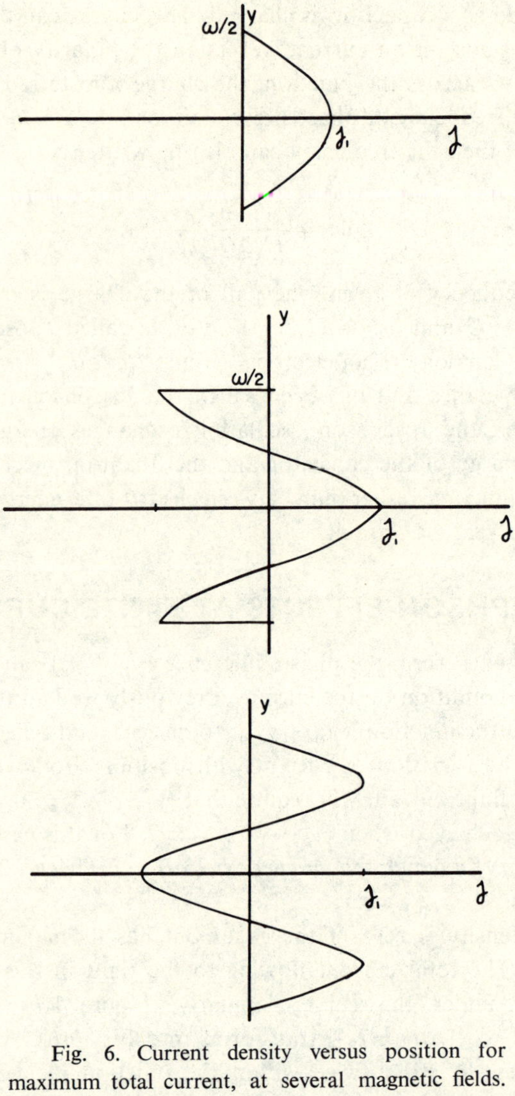

Fig. 6. Current density versus position for maximum total current, at several magnetic fields.

with increasing magnetic field. A mathematical derivation of this effect will now be given for a junction sufficiently small that internal shielding effects are negligible, i.e., $w \ll \lambda_J$. We will use the same coordinate system as before and assume no variation in the z direction. The total Josephson current is given by

$$I_J = j_1 L \int_{-w/2}^{w/2} \sin\left(\gamma_1 - \gamma_2 + \frac{2e}{\hbar} Byt\right) dy.$$

The integration will be done at the surface of the superconductors separated by an insulating tunnel barrier. However, in a magnetic field a surface current must flow in the superconductor—the Meissner effect. The effect of these currents has been explicitly ignored in the discussion so far and now must be accounted for in the calculation.

We have shown that the electron drift velocity can be written in terms of $\nabla\gamma$ and the vector potential, $\langle v\rangle = (1/2m)\{\hbar\nabla\gamma - 2eA\}$. In addition, we may define supercurrent within a single superconductor as $j = \varrho e\langle v\rangle$. Therefore the phase gradient in terms of current density and vector potential is

$$\nabla\gamma = (2e/\hbar)\{(m/\varrho e^2)j + A\}.$$

This expression can then tell us how the phase varies at the surface of the superconductor if we know the current and potential. Previously we chose $A_y \equiv 0$, so that in this gauge the phase at y relative to the center is

$$\gamma(y) = \gamma_0 + (2e/\hbar)(m/\varrho e^2)j_s y,$$

where j_s is the surface current. The factor $(m/\mu_0\varrho e^2)^{1/2}$ is the penetration depth, λ, of magnetic field into the material, so that

$$\gamma(y) = \gamma_0 + \frac{2e}{\hbar}\mu_0\lambda^2 j_s y.$$

Shielding currents will flow in opposite directions in the two superconductors, so that finally substituting all of this back into the expression for current, we get

$$I_J = j_1 L \int_{-w/2}^{w/2} \sin\{\gamma_1(0) - \gamma_2(0) + (2e/\hbar)t(B + 2\mu_0\lambda^2 j_s)y\}\, dy.$$

However, since $\nabla\times B = \mu_0 j$, the boundary condition on the field is just $B = \mu_0\lambda j_s$ and the integral becomes

$$I_J = j_1 L \int_{-w/2}^{w/2} \sin\{\gamma_1(0) - \gamma_2(0) + (2e/\hbar)B(t + 2\lambda)y\}\, dy.$$

Thus the effect of the surface currents is to make the effective aperture of the junction wider than the actual spacing t and become $t + 2\lambda$, which, for sufficiently thin film structures, may even be wider than the junction itself. The integration above is simple, and the total Josephson current becomes

$$I_J = I_0 \frac{\sin\phi}{\phi} \sin\{\Delta\gamma(0)\}, \tag{2}$$

where $I_0 = j_1 L w$, $\phi = (2e/\hbar)B(t + 2\lambda)$, and $\Delta\gamma(0) = \gamma_1(0) - \gamma_2(0)$.

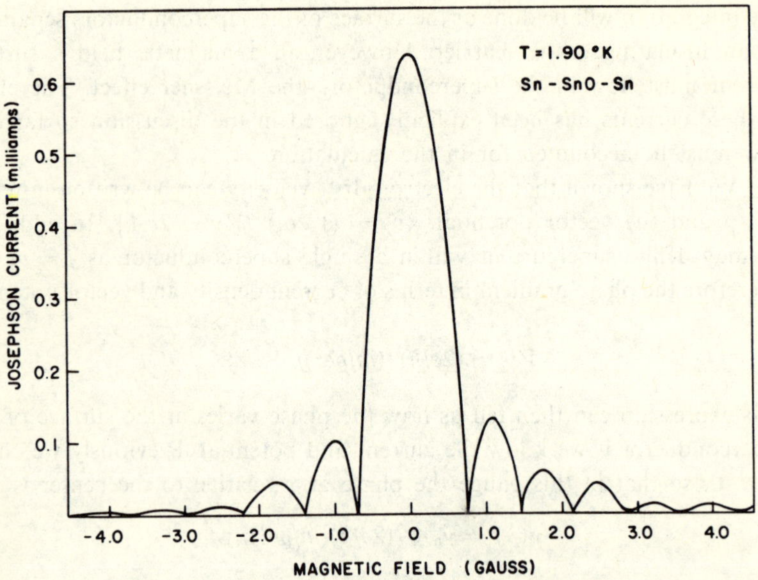

Fig. 7. Josephson current as a function of magnetic field, showing the diffraction effects anticipated.

As it must, this expression for Josephson current has all the features discussed above. The relative phase, $\Delta\gamma(0)$, is a boundary condition determined by the current impressed on the junction and adjusts to meet the requirements. The maximum zero-voltage current is

$$I_{\max} = I_0(\sin\phi)/\phi\,,$$

and has the periodicity and amplitude dependence noted before. Actual experimental data are shown in Fig. 7 and very nicely confirm ([5]) the anticipated results.

JOSEPHSON EFFECTS AT FINITE VOLTAGE

The data shown above were taken by connecting a current source to the junction and determining the maximum zero-voltage current as a function of magnetic field. However, connecting the junction to a voltage source immediately produces alternating currents. This is the subject of Chapter 32 by Scalapino; however, I would like to indicate here the transition to these AC effects.

Under an applied voltage the relative phase precesses, as we noted previously. To the extent that the phase variation of the *electric* field can be neglected in the junction, a voltage then simply time-modulates the previously derived Josephson current, or

$$I_J = I_0 \frac{\sin\phi}{\phi} \sin\frac{2e}{\hbar}\left(\int V\,dt + \alpha\right).$$

A steady voltage V_0 then time-modulates the Josephson current at frequency $2eV_0/\hbar$ but generates *no* net supercurrent. The junction is thus somewhat anomalous, having a zero DC impedance for supercurrents at zero voltage but an infinite DC impedance for supercurrent at a finite voltage. No DC current flows upon applying a DC voltage, only internal alternating currents with oscillation frequency $2eV_0/\hbar$, i.e., with a time-dependent phase $\gamma = (2eV_0/h)t$. Even with a time-dependent phase it may still be possible to have DC current if the time average of $\sin\gamma$ is constant and nonzero.

This can be achieved by adding an alternating voltage to the steady voltage V_0. The resulting phase is

$$\gamma = \frac{2e}{\hbar}\int (V_0 + v\sin\Omega t)\,dt.$$

In this circumstance the period τ of the phase interval 0–2π is itself normally time dependent. However, if the frequency of the alternating voltage is chosen as $(2e/\hbar)V_0 = N\Omega$, inspection of the expression for γ above shows that this period becomes time-invariant. But with such a time-invariant period the half periods are clearly *not* equal. The symmetry of the sine is thus upset and the time average, $(1/\tau_0)\int_0^\tau \sin\gamma$, is finite [and when $(2e/\hbar)V_0 = N\Omega$, also constant] satisfying the requirement for a DC current. A direct expansion of $\sin\gamma$ also shows these "frequency modulation" effects, which were first shown experimentally by Shapiro ([6]).

THE JOSEPHSON QUANTUM PHASE DETECTOR

In a superconductor tunnel junction the pair tunneling current depends on the quantum phase as $j = j_1\sin(\gamma_1 - \gamma_2)$. A knowledge of the actual mechanism responsible for charge transfer is needed to calculate j_1, but variations in the phase can be determined using such general considerations as gauge invariance and energy conservation. It is for this reason that experiments involving the quantum phase have been so precise and successful. The time dependence of the relative phase

$$\frac{\partial}{\partial t}(\gamma_1 - \gamma_2) = \frac{\Delta\mu}{\hbar} = \frac{2eV}{\hbar} \tag{3}$$

is independent of any tunneling mechanism. In addition, the phase variation within a superconductor was found to be given by

$$\nabla\gamma = (2/\hbar)\{m\langle v\rangle + eA\}. \tag{4}$$

Thus by virtue of the macroscopic quantum nature of superconductivity an experimental study of the phase can avoid most of the very difficult materials problems of solid-state physics. In this chapter the observable effects of the above two equations will be discussed. Josephson junctions will be used as "phase detectors" to investigate how the phase varies within a solid superconductor.

Equation (3) has been examined with extreme precision by Langenberg *et al.* (7) and is the subject of Chapter 33 of this volume. I will concentrate on an experimental study of Eq. (4) and its implications. To do this it will be necessary to consider two Josephson junctions (symbolized as capacitors) connected in parallel (Fig. 8). Such a superconducting device we call a superconducting interferometer.

Total Josephson current through the interferometer is, of course, $I_T = I_1 + I_2$. If we assume the junctions are identical, then from Equation (2)

$$I_T = I_0 \frac{\sin\phi}{\phi}\{\sin(\gamma_a - \gamma_b) + \sin(\gamma_c - \gamma_d)\}, \tag{5}$$

where ϕ and γ are defined as before. As the current is increased through the device the phases $(\gamma_a - \gamma_b)$ and $(\gamma_c - \gamma_d)$ adjust to accomodate the current until they reach their maximum value, $\pi/2$. If the phases can adjust independently, then the maximum supercurrent is just $I_T(\max) = 2I_0$.

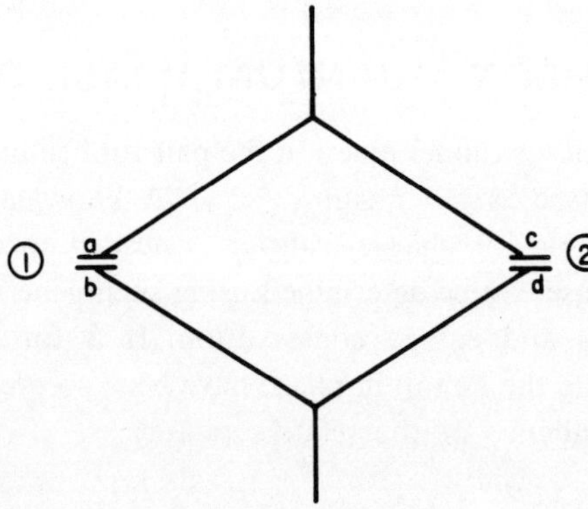

Fig. 8.

However, if *a–c* and *b–d* are connected by superconducting links [which preserve the phase as in Eq. (4)] then a separate and adjustable constraint appears on $(\gamma_a - \gamma_b)$ and $(\gamma_c - \gamma_d)$.

If *a–c* is a superconductor, integrating Eq. (4) will display this constraint,

$$\gamma_a - \gamma_c = \frac{2}{\hbar} \int_c^a (m\langle v\rangle + eA)\, dl,$$

and similarly for $(\gamma_b - \gamma_d)$. If we rewrite Eq. (5) in terms of products of sines and cosines, we find

$$I_T = 2I_0 \frac{\sin\phi}{\phi} \cos \tfrac{1}{2}(\gamma_a - \gamma_c + \gamma_d - \gamma_b) \sin \tfrac{1}{2}(\gamma_a + \gamma_c - \gamma_d - \gamma_b).$$

Substituting the above integrals for phase differences in the superconductor, we get

$$I_T = 2I_0 \frac{\sin\phi}{\phi} \left[\cos \frac{1}{\hbar} \oint (m\langle v\rangle + eA)\, dl\right] \sin \delta\gamma_0, \tag{6}$$

where

$$\delta\gamma_0 = \{(\gamma_a + \gamma_c) - (\gamma_b + \gamma_d)\}.$$

As in the single junction, the maximum current occurs when $\delta\gamma_0 = \pi/2$, and at sufficiently low frequencies $\overline{\delta\dot{\gamma}_0} = (2e/\hbar)V$. We will limit this discussion to zero voltage ($V = 0$) and consider the maximum zero-voltage current $I_{\max}$,

$$I_{\max} = 2I_0 \frac{\sin\phi}{\phi} \cos\,(1/\hbar) \oint (m\langle v\rangle + eA)\, dl. \tag{7}$$

Thus the result of the phase constraint introduced by the superconducting links is to impose an "interference modulation" on the maximum Josephson current. The form of this current is a diffraction-modulated, $(\sin\phi)/\phi$, interference effect, $\cos 1/\hbar \oint (m\langle v\rangle + eA)\, dl$. The diffraction is due to flux (ϕ) within the individual junctions and the interference is determined by the integral around the inteferometer. This integral is *not* quantized as would be expected if the loop were entirely superconducting ([8]). The superconducting path is broken by two insulating barriers, removing the requirement of quantization, but preserving current flow by the Josephson effect.

The initial experiments ([9]) with these devices were conducted simply to determine whether Eq. (4) actually applied in a real superconductor. If it did not apply—if long-range phase coherence were not achieved—then there would be no interference. In the first experiments modulation was by means of a magnetic field. If we can neglect the velocity modulation (for example, a symmetric device) then Eq. (7) becomes

$$I_{max} = 2I_0 \left| \frac{\sin \phi}{\phi} \right| \left| \cos \frac{e}{\hbar} \phi_T \right|,$$

where ϕ is the normalized flux in the junction and $\phi_T = \oint A\, dl$ is the flux enclosed by the interferometer. To examine this prediction, thin film devices were fabricated as shown in Fig. 9.

The maximum current at zero voltage was recorded as a function of magnetic field. Experimental results ([10]) are shown in Fig. 10. This figure shows only the central diffraction peak of the junctions (extending about ± 350 mG) but does include many periods of interference modulation. This particular interferometer displays almost ideal behavior—from the interference pattern the ratio of junction area to interferometer aperture is about 1/20. The periodicity of current with flux has always been found to be $h/2e$ regardless of material, demonstrating the true universal nature of the phenomena. Interferometers have been constructed with apertures up to about 3 cm^2 and, by constructing special low-inductance circuits, phase coherence has been demonstrated at distances of over a meter ([11]).

The velocity term in the phase modulation has also been investigated ([12]). In these experiments an additional velocity was added to the electrons by means of a controllable "drift current." This drift current was intro-

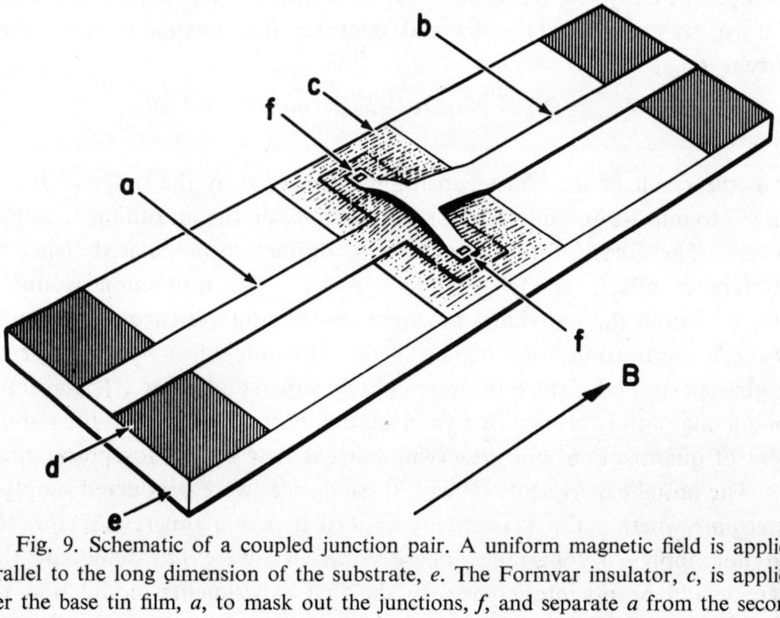

Fig. 9. Schematic of a coupled junction pair. A uniform magnetic field is applied parallel to the long dimension of the substrate, *e*. The Formvar insulator, *c*, is applied over the base tin film, *a*, to mask out the junctions, *f*, and separate *a* from the second tin film, *b*.

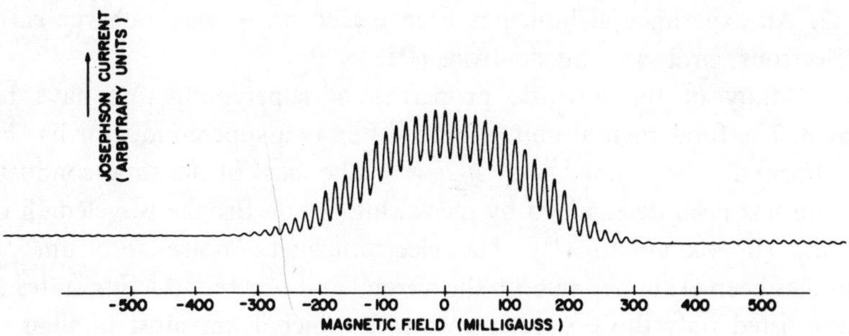

Fig. 10. Experimental trace of I_T versus magnetic field showing both interference and diffraction effects. The interference has a field periodicity of about 16 mG. Approximate maximum Josephson current is 0.5 mA; the junction separation is about 3 mm and junction width $w = 0.5$ mm. The small magnetic offset is due to an incompletely compensated background field.

duced into the interferometer circuit in such a way that the magnetic fields generated by the current were negligible compared to the velocity effects. It was found, as expected, that velocity modulation did occur, and the de Broglie wavelength for the supercurrent has been determined. The experimental results for velocity modulation are entirely similar to those shown in Fig. 10, replacing the magnetic field by a drift current.

Such experiments as these have been used to confirm all aspects of the phase relation, Eq. (4), and therefore also Eq. (7). The interferometer is well enough understood now for it to be used as a measuring instrument in other experiments. Devices of various types utilizing these interference techniques have actually been engineered into operating instruments.

SUPERCONDUCTING INTERFEROMETRY

These superconducting instruments are, of course, exceptionally sensitive to electromagnetic fields. In our laboratories we have operating magnetometers whose noise level is less than 2×10^{-10} G per root cycle (or 10^{-10} G with a 1 sec time constant) and voltmeters whose effective noise is less than 10^{-14} V at a 1 msec time constant.

With instruments such as these many unique experiments have been possible. Some of them are noted here.

1. Since these instruments respond fundamentally to the vector potential (**A**), they have been used to observe effects directly attributed to the vector potential *itself* even in the absence of both electric and magnetic fields ([13]).

2. An experimental limit has been placed on *magnetic* charge carried by electrons, protons, and neutrons ([14]).

3. Many of the intrinsic properties of superconductors have been studied. The fundamental unit of pinned flux in a superconductor has been experimentally determined to be $h/2e$ ([15]). The mass of the superconducting electron has been determined by measuring the de Broglie wavelength of a rotating superconductor ([16]). The electromagnetic noise spectrum of a metal has been examined in both the normal and superconducting states ([17]). I have listed only those experiments with which I am most familiar. No doubt many more experiments exist and I apologize for the omissions.

ACKNOWLEDGMENTS

I wish to thank Prof. W. Parker for his careful notes of my lectures at the Advanced Study Institute at Risø, from which this chapter has been written.

REFERENCES

1. B. D. Josephson, Phys. Letters 1:251 (1962); P. W. Anderson, Lectures on the Many-Body Problem, Ravello, 1963, Vol. II (E. R. Caianello, ed.), Academic Press, New York, p. 113.
2. R. P. Feynman, Lectures on Physics, Vol. III, Addison-Wesley Publishing Co., Massachusetts, 1965; J. E. Mercereau, in: Treatise on Superconductivity (R. Parks, ed.), Marcel Dekker, New York, 1968, Chapter 9.
3. J. E. Zimmerman and A. H. Silver, Phys. Rev. 141:367 (1966).
4. H. A. Notarys, private communication.
5. P. W. Anderson and J. M. Rowell, Phys. Rev. Letters 10:230 (1963).
6. S. Shapiro, Phys. Rev. Letters 11:80 (1963).
7. D. N. Langenberg, D. J. Scalapino, B. N. Taylor, and R. E. Eck, Phys. Rev. Letters 15:294 (1965).
8. B. S. Deaver and W. M. Fairbank, Phys. Rev. Letters 7:43 (1961).
9. R. C. Jaklevic, J. Lambe, A. H. Silver, and J. E. Mercereau, Phys. Rev. Letters 12:159 (1964).
10. R. C. Jaklevic, J. Lambe, J. E. Mercereau, A. H. Silver, Phys. Rev. 140A:1628 (1965).
11. L. L. Vant-Hull and J. E. Mercereau, Phys. Rev. Letters 17:629 (1966).
12. R. C. Jaklevic, J. J. Lambe, A. H. Silver, and J. E. Mercereau, in: Low Temperature Physics—LT9 (Part A), Plenum Press, New York, 1965, p. 446.
13. R. C. Jaklevic, J. J. Lambe, A. H. Silver, and J. E. Mercereau, Phys. Rev. Letters 12:274 (1964).
14. L. L. Vant-Hull, Bull. Am. Phys. Soc. 12:1136 (1967).
15. J. E. Zimmerman and J. E. Mercereau, Phys. Rev. Letters 13:125 (1964).
16. J. E. Zimmerman and J. E. Mercereau, Phys. Rev. Letters 14:887 (1965).
17. L. L. Vant-Hull, R. A. Simpkins, and J. T. Harding, Physics Letters 24A:736 (1967); B. T. Ulrich, Phys. Rev. Letters 20:381 (1968).

Chapter 32

The Theory of Josephson Tunneling

D. J. Scalapino*

Department of Physics and
Laboratory for Research on the Structure of Matter
University of Pennsylvania
Philadelphia, Pennsylvania

INTRODUCTION

I'll begin by making an analogy between the long-range spatial order of a solid and the long-range phase order of a superconductor. The first question which one might ask is, how does a short-range force produce long-range order? In a solid the typical range of the interatomic forces are of the order of the interatomic spacings. However, once a part of the crystal forms, the continued propagation of the order is established in a step-wise manner with the arrangement of the occupied sites legislating the arrangement of the subsequent level through the short-range interactions. In the case of a superconductor we are dealing with an interaction whose "range" is associated with the retarded nature of the phonon-exchange interaction and may "extend"† on the order of $v_F/\omega_D \approx 10^{-5}$ cm. The attractive nature of this interaction leads to the formation of pairs which are spread out over a coherence length ξ of order 10^{-4} cm. Now, the center-of-mass coordinates of some 10^6 pairs lie in a sphere of diameter ξ. In order for each of the pairs to have the optimum space-time correlations necessary to take advantage of the attractive phonon exchange interaction, it is necessary that they all have the same center-of-mass momentum. More generally, in the presence of electromagnetic fields we may extend this to the statement

* A. P. Sloan Foundation Fellow.

† The actual spatial extent of the interaction is highly localized, probably on the order of interatomic spacings. However, since the important lattice frequencies are of the order of the Debye frequency, an electron with velocity v_F can come from a distance v_F/ω_D to feel the lattice disturbance produced by another electron.

that the pairs must have the same center-of-mass phase. To the extent that they do not, the condensation energy is lost. Therefore, just as atoms of a solid control the local spatial order around them, pairs control the local phase order within a coherence length. With this in mind the extension of the long-range order in a superconductor can be understood as a sort of "poor man's analytic continuation." In Fig. 1 the circles represent spheres of diameter ξ. We have just argued that the pair center-of-mass phase must be locked over a distance ξ. Starting from sphere 1, we note that in the region of overlap with sphere 2 the phases must be the same. Now if sphere 2 locks phases within it, the phase locking is continued from the basic length to the entire metal. This phase continuation is *forced* by the *condensation energy* associated with the overlap regions.

In the solid the order parameter is the atomic position, while in the superconductor it is the pair center-of-mass phase. Now let us consider some special cases of nonuniform order. In a solid one might imagine a screw dislocation in which the order advanced by one lattice spacing in going around the dislocation line. In the superconductor the analog is a vortex line in which the pair phase advances by 2π in going around the vortex core. Next consider a slip plane in the solid where the atomic spacing is shifted by Δa. The analog in this case is a superconducting tunnel junction in which the pair phase difference between the two superconductors is $\Delta\varphi$.

It is possible to go further with this analogy and argue that the ordering of the solid breaks the translational invariance of the original Hamiltonian while the phase ordering of the superconductor breaks the "local" gauge invariance in the superconducting case. Anderson [1], Nambu [2], and Goldstone [3] have discussed the fact that such broken symmetries imply

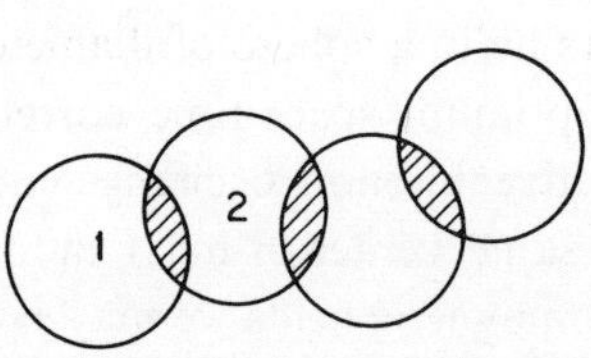

Fig. 1. Schematic illustration of the establishment of long-range phase order in the superconducting state. Here the phase locking over a coherence length ξ is continued from one sphere of diameter ξ to another by virtue of the condensation energy associated with the overlap between the two spheres.

the existence of elementary (order-fluctuation) excitations. In the solid the interatomic forces are short range, so the arguments, which "prove" that these low-lying order-fluctuation modes are phonons, *work*. As Anderson ([1]) has discussed, the corresponding fluctuations of the phase in the superconducting state involve fluctuations in the charge density, and the long-range Coulomb forces give them the plasmon energy in a bulk superconductor. In systems with boundaries additional phase fluctuation modes may appear. As emphasized by Ferrell ([4]), the nature of the long-wavelength phase fluctuation excitations plays a critical role in determinig the stability of the long-range phase ordering. In one and two dimensions the superconducting state is unstable with respect to these phase excitations and their zero-point motion destroys the long-range order ([4–6]). In certain three-dimensional geometries the thermal excitation of phase fluctuation modes can lead to the disintegration of the phase ordering below the "bulk" transition temperature determined by the quasiparticle excitations ([7–9]). The analogy here is simply to a solid near its melting point.

In this chapter we will treat the theory of Josephson tunneling ([10]) in a way which illustrates a number of these general features of superconducting phase ordering. Besides offering an especially simple case in which a number of results can be worked out in detail, these phenomena are all directly observable in the tunneling I–V characteristics of the junction. This latter circumstance is a result of two unique features of Josephson tunneling: (1) the pair part of the tunneling current is a periodic function of the relative pair phase shift across the junction, so that one can probe the relative pair phase by measuring this current and, (2) as Anderson ([11]) has emphasized, a Josephson junction is a "weak superconductor," so that the characteristic junction penetration length and plasma frequency are on the experimentally convenient scale of millimeters and gigahertz rather than the London or Pippard penetration depths ($\sim$ hundreds of Å) and metallic plasma frequencies ($\sim 10^{15}$–10^{16} cps) characteristic of bulk superconductors.

In Section II the "condensation" energy associated with the establishment of phase coherence across the barrier is calculated and the resulting phase-coherent state displayed. The constitutive equations which relate the tunneling current to the relative pair phase and this phase to the electromagnetic fields in the barrier are then derived. Using these relations, the simplest examples of the DC and AC Josephson effects are then reviewed to illustrate the manner in which the junction I–V characteristics allow us to probe the relative pair phase shift across the junction.

In Section III Maxwell's equations are combined with the Josephson constitutive relations to obtain a description of the electrodynamics of the junction. Using this formalism, we discuss nonuniform phase ordering in which vortices are produced in the barrier by an external magnetic field. The structure of the vortices is reflected in the behavior of the critical current as a function of magnetic field. Next we determine the elementary excitations associated with fluctuations of the phase ordering and show how they give rise to structure in the *I–V* characteristics. The final section, Section IV, is devoted to reviewing the manner in which the relative pair phase order across the junction is destroyed near the transition temperature.

There are a number of excellent review articles ([12]) on the Josephson effect, to which we refer the reader for complementary views and further information (also see Chapters 31 and 33 of this volume).

THE JOSEPHSON EQUATIONS

Near the Fermi surface the tunneling matrix element T decreases exponentially with the oxide thickness measured on a scale of angstroms. This suggests that one should be able to treat the tunneling interaction by perturbation theory. Indeed, using lowest-order perturbation theory one finds a quasiparticle tunneling current proportional to $|T|^2$, which fits a number of features of the observed *I–V* characteristics as discussed in Chapters 19 and 21. The signature of this quasiparticle current is a step discontinuity at a voltage $2\Delta/e$. Proceeding to fourth order in T, one finds two-quasiparticle processes, which have a step current discontinuity at a voltage Δ/e ([13]). In this same order there is a zero-bias δ-function singularity for pair transfer. This singularity arises because the state with one pair transferred is degenerate with the initial state, and is a warning that we have not started the perturbation calculation with the correct linear combination of the degenerate states. There are, in fact, a large number of essentially degenerate states having n pairs transferred from one superconductor to the other which differ only by an electrostatic energy $(2en)^2/2C$, where C is the junction capacitance. The resolution of this degeneracy leads in a natural way to the Josephson equations describing pair tunneling.

Figure 2 shows two pieces of superconductor with equal chemical potentials separated by a macroscopic distance. The ground state of this system is given by antisymmetrizing the product of the ground state $\psi_0^{(1)}(N_1)$ of metal 1 containing N_1 electrons times the ground state $\psi_0^{(2)}(N_2)$ of metal 2 containing N_2 electrons:

$$\Psi_0 = A\psi_0^{(1)}(N_1)\psi_0^{(2)}(N_2).$$

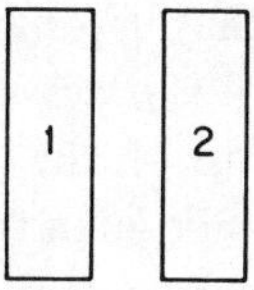

Fig. 2. Two pieces of superconductor with equal chemical potentials separated by a macroscopic distance. As they are brought together, the tunneling interaction acts to lift the pair transfer degeneracy and form a state coherent in the pair phases.

The energy is just the sum of the ground-state energies of the two superconductors:

$$E_0 = E_0^{(1)}(N_1) + E_0^{(2)}(N_2).$$

There is a large class of states obtained by transferring n electron pairs,

$$\Psi_n = A\psi_0^{(1)}(N_1 - 2n)\psi_0^{(2)}(N_2 + 2n), \tag{1}$$

which are essentially degenerate with Ψ_0. For example, the energy of the state Ψ_n is

$$E_0^{(1)}(N_1 - 2n) + E_0^{(2)}(N + 2n) + \frac{(2en)^2}{2C} = E_0^{(1)}(N_1) + E_0^{(2)}(N_2) + \frac{(2en)^2}{2C} + 2n(\mu_2 - \mu_1). \tag{2}$$

Here C is the relative capacitance of the two pieces of superconductor and $\mu_1 - \mu_2$ is the difference in chemical potential between them. Since we are assuming μ_1 equals μ_2, it follows that Ψ_n differs in energy from w_0 only by the small electrostatic term $(2en)^2/2C$. As the metals are moved closer and their separation becomes microscopic, the tunneling interaction dominates the electrostatic energy and legislates the resolution of the near degeneracy of the Ψ_n states ($n = 0, \pm 1, \pm 2, \ldots$).

Here we take the point of view of Ferrell and Prange ([14]) and treat this problem using degenerate-state perturbation theory ([15]). The tunneling interaction is assumed to be adequately described by the Hamiltonian

$$H_T = \sum_{kq} T_{kq}(C_{k\uparrow}^+ C_{q\uparrow} + C_{q\downarrow}^+ C_{k\downarrow}) + \text{h.c.}$$

The operator $C_{k\uparrow}^+(C_{-q\downarrow}^+)$ creates an electron in a Bloch state of momentum k $(-q)$ and spin $\uparrow(\downarrow)$ on side 1 (2), and the equality

$$T_{kq} = T_{k\uparrow q\uparrow} = T_{-q\downarrow -k\downarrow}$$

follows from the assumption that H_T has time-reversal symmetry. Since H_T transfers only one electron, it is necessary in order to lift the degeneracy to consider the matrix elements of the second-order interaction $H_T(E_0 - H_0)^{-1}H_T$. Here H_0 is the Hamiltonian for the two noninteracting metals. The second-order interaction has matrix elements connecting states with Δn equal to 0 or ± 1. Evaluating these matrix elements, one finds

$$\begin{aligned} \langle \Psi_m \mid H_T \frac{1}{E_0 - H_0} H_T \mid \Psi_n \rangle &= -E_1, \qquad n = m, \\ &= -\tfrac{1}{2}E_1, \qquad n = m \pm 1, \end{aligned} \tag{3}$$

with

$$E_1 = \sum_{qk} \mid T_{qk} \mid^2 \frac{\Delta_q}{E_q} \frac{\Delta_k}{E_k} \frac{1 - f_q - f_k}{E_q + E_k}. \tag{4}$$

Here Δ_q is the gap, $E_q = \sqrt{(\varepsilon_q{}^2 + \Delta_q{}^2)^{1/2}}$ is the quasiparticle excitation energy, and $f_q = f(E_q)$ is the Fermi factor. For identical superconductors the expression for E_1 reduces to

$$E_1 = \pi^2 N^2(0) \mid T \mid^2 \Delta, \tag{5}$$

where $\mid T \mid^2$ is an average tunneling matrix element for electron transfer near the Fermi surface and $N(0)$ is the single-particle Bloch density of states at the Fermi surface.

According to degenerate-state perturbation theory, the correct linear combination of the Ψ_n states is determined by diagonalizing the matrix $\langle \Psi_m \mid H_T(E_0 - H_0)^{-1}H_T \mid \Psi_n \rangle$ shown in Fig. 3. This is straightforward and the resulting states are

$$\Psi_\varphi = \sum_n e^{in\varphi} \Psi_n \tag{6}$$

with energies

$$E_\varphi = (E_0 - E_1) - E_1 \cos\varphi. \tag{7}$$

The contribution $-E_1$ comes from the diagonal term and would be present

$$\langle n | H_T(E_0 - H_0)^{-1} H_T | m \rangle = -E_1 \begin{pmatrix} \ddots & & & & & \\ \frac{1}{2} & 1 & \frac{1}{2} & & & 0 \\ & \frac{1}{2} & 1 & \frac{1}{2} & & \\ & & \frac{1}{2} & 1 & \frac{1}{2} & \\ 0 & & & \frac{1}{2} & 1 & \frac{1}{2} \\ & & & & & \ddots \end{pmatrix}$$

Fig. 3. The matrix $\langle \Psi_m \mid H_T(E_0 - H_0)^{-1}H_T \mid \Psi_n \rangle$. The linear combinations of Ψ_n states which diagonalize this form the phase-coherent state associated with the Josephson phenomena.

regardless of whether a single state Ψ_n or the linear combination given by Eq. (6) were used. The term $-E_1 \cos\varphi$ represents the "condensation" or binding energy associated with the phase-coherent linear combination given by Eq. (6). Since $\cos\varphi$ can vary between 1 and -1, the original set of degenerate states are spread out into a band of width $2E_1$ with the lowest energy state having $\varphi = 0 \bmod 2\pi$.

In order to identify the parameter φ, consider what happens when the phase of the pairs on side 2 is shifted by $\Delta\varphi$ relative to those on side 1. The Ψ_n wave functions transform as

$$\Psi_n \to e^{i\Delta\varphi n}\Psi_n,$$

so that

$$\Psi_\varphi \to \Psi_{\varphi+\Delta\varphi}.$$

Therefore φ is the relative pair phase difference between superconductor 2 and superconductor 1. When this relative phase is zero (mod 2π), pairs from one superconductor "fit" into the other, and the system lowers its energy. As the relative pair phase shifts from zero toward $\pi/2$, the binding energy decreases toward zero due to the interference of the different pair transfer contributions. A junction forms a particularly simple illustration of the relative pair phase locking which we mentioned earlier.

In the Ψ_φ basis the operator which measures the number of pairs transferred is

$$n_{\text{op}} = \frac{1}{i}\frac{\partial}{\partial\varphi}. \tag{8}$$

The time rate of change of this is

$$\dot{n}_{\text{op}} = \frac{i}{\hbar}[H, n_{\text{op}}],$$

so that the current operator associated with pair transfer can be written as

$$I_{\text{op}} = 2e\dot{n}_{\text{op}} = \frac{2e}{\hbar}\left[H, \frac{\partial}{\partial\varphi}\right].$$

The expectation value of the pair current in the state Ψ_φ is, therefore*

$$I = \langle\Psi_\varphi \mid I_{\text{op}} \mid \Psi_\varphi\rangle = -\frac{2e}{\hbar}\frac{\partial}{\partial\varphi} E_1 \cos\varphi = I_1 \sin\varphi \tag{9}$$

* This result is valid to order $[N(0) \mid T \mid]^2$. The next higher order contribution of H_T to the "condensation" energy has the form $\Delta[N(0) \mid T \mid]^4 \cos(2\varphi)$ and gives a correction to I of the form $I_1[N(0) \mid T \mid]^2 \cos(2\varphi)$.

with the pair current amplitude $I_1 = 2eE_1/h$. For identical superconductors it follows from Eq. (5) that

$$I_1 = (4\pi e^2/\hbar)N^2(0) \mid T \mid^2 (\pi\Delta/2e). \tag{10}$$

This is just equal to the *normal*-state current which flows at a voltage $\pi\Delta/2e$.

Ambegaokar and Baratoff ([16]) have determined the effect of thermally excited quasiparticles* on the Josephson current amplitude. They find that for identical superconductors

$$I_1(T) = I_1(0)\frac{\Delta(T)}{\Delta(0)}\tanh[\beta\Delta(T)/2], \tag{11}$$

which decreased like $\Delta^2(T)$ near the transition temperature. The excited quasiparticles block the pair tunneling processes and thereby reduce the binding energy and hence the current amplitude associated with pair transfers. This result is in agreement with measurements of $I_1(T)$ by Fiske ([17]). In this case the transition matrix elements, Eq. (3), which enter in determining the "condensation" energy are taken between finite temperature states which have the *same* quasiparticle excitation but differ by 1 in the number of ground-state pairs transferred. It follows that the energy difference between these two states is still given by Eq. (2) with $\mu_2 - \mu_1$, the difference in chemical potential between the two superconductors. The chemical potential continues to enter at finite temperature because the energy change of interest is at fixed entropy.

It is also worth noting the effect of a bias voltage (and high-frequency electromagnetic fields) on the amplitude I_1. We will see shortly that this amplitude determines the size of the Josephson AC tunneling current which exists when the junction is biased. Riedel ([18]) showed that $I_1(V)$ has a logarithmic singularity as the bias voltage V approaches $2\Delta/e$ if quasiparticle lifetime effects are neglected. This arises because the energy $E_k + E_q - eV$ of the intermediate state which appears in expression Eq. (4) for the coupling energy vanishes when $E_k = E_q = \Delta$ and $eV = 2\Delta$. In a similar manner the Josephson current response to high-frequency radiation can be shown to have a logarithmic singularity for $nh\nu = 2\Delta$. In both these cases it is important to recognize that the pair tunneling goes via a two-step process which involves a single-particle transfer across a potential or with the absorption of quanta at an intermediate stage. The logarithmic singu-

* In Section IV we discuss the effect of phase fluctuation excitations on $I_1(T)$.

larity is washed out by finite lifetime effects ([19]), but the fact that I_1 peaks for voltages and frequencies near 2Δ suggests that the AC Josephson effect may be observable at voltages and frequencies above 2Δ. Without this resonance once might well have expected I_1 to have fallen off for these energies and response times. Grimes *et al.* ([20]) have reported high-frequency response measurements which lend support to the idea that I_1 remains large in the frequency region 2Δ.

Having resolved to second order the degeneracy and determined the pair current state Ψ_φ, we will pause in this development of the Josephson equation to look at the simplest cases of the DC and AC Josephson effects. In the ground state φ is 0 (mod 2π), the "condensation" energy $-E_1 \cos\varphi$ is a maximum, and no current flows. Suppose a battery in series with a resistor is attached across the junction. Then current flows, charging the capacitor formed by the junction, so that the chemical potentials of the two superconductors are no longer equal. From Eq. (2) it follows that the energy of Ψ_n changes at a rate

$$-2n[\mu_2(t) - \mu_1(t)] = 2neV(t)$$

relative to the state with no pairs transferred. Here $V(t)$ is the potential across the junction, $V_1 - V_2$. This means that the state evolves within an overall phase factor, as

$$\sum_n \exp\left[i\frac{2en}{\hbar}\int_0^t V(t')\,dt'\right]\Psi_n.$$

If the resistor is such that the potential across the junction vanishes at a current I less than I_1, then this state evolves to Ψ_{φ^*} with φ^* determined by

$$\varphi^* = \sin^{-1}(I/I_1).$$

In this case rather than allowing a voltage to develop, the system carries a DC supercurrent and preserves a part of the binding energy, $-E_1\cos\varphi^*$. The rate at which the phase changes, $2eV/h$, is approximately 500 MHz/μV. This means, for example, that 1 μV for $\frac{1}{2}$ nsec is sufficient to shift the phase from 0 to $\pi/2$. Experimentally, the Josephson DC current appears as a zero-bias vertical current step (Fig. 4).

As φ goes to $\pi/2$, the binding energy vanishes and, beyond the critical current I_1 a potential is developed across the junction. If this change is adiabatic, the state of the system is given by $\Psi_{\varphi(t)}$ where

$$\partial\varphi/\partial t = -2[\mu_2(t) - \mu_1(t)]/\hbar. \tag{12}$$

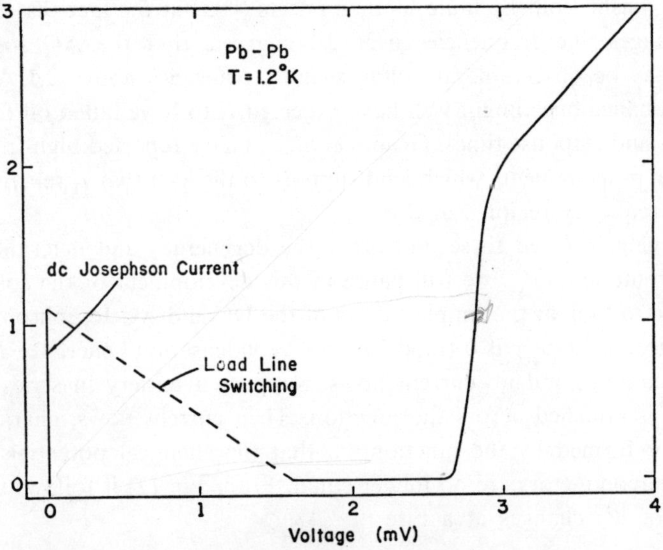

Fig. 4. The I–V curve for a Pb–Pb oxide–Pb junction, displaying the DC Josephson current. When the maximum DC Josephson current is reached, the junction switches along the circuit load line to the normal I–V curve. From Langenberg *et al.* ([40]).

For the case of a constant DC bias, $eV = \mu_1 - \mu_2$, this integrates to

$$\varphi(t) = (2eVt/\hbar) + \varphi_0,$$

and an AC current

$$I = I_1 \sin\left(\frac{2eVt}{\hbar} + \varphi_0\right)$$

with a frequency of approximately 500 MHz/μV is produced across the junction.* This is the AC Josephson effect. Since the average binding energy $-E_1 \cos \varphi(t)$ is zero, this state is much more subject to fluctuations than the DC state. In practice, these oscillating current states are often stabilized by a synchronously applied or self-generated electromagnetic fields.

As we have seen, the parameter φ is the relative pair phase difference between superconductor 2 and superconductor 1. Returning to our program of developing the Josephson equations, let us examine some properties

* Recent measurements by Parker *et al.* ([21]) showed that this frequency is $2e/h = 483.5916 \pm 0.0026$ MHz/μV. This value is nearly 40 ppm smaller than the presently accepted value given by Cohen and DuMond. This has a number of interesting implications [see ([21])].

of the pair phase in more detail. In the superconducting state the electrons are paired in time-reversed states, so that the pair phase is determined by the center-of-mass motion of the pair. Since the magnetic field strengths in which we will be interested are such that the cyclotron radius of the pair center-of-mass motion is much larger than the coherence length, a semi-classical approximation for this motion is adequate, and the pair phase difference between two points in a *bulk* superconductor is given by

$$\frac{2}{\hbar}\int_1^2 \left(m\mathbf{v} + \frac{e}{c}\mathbf{A}\right)\cdot d\mathbf{x} - \frac{2}{\hbar}\int_0^t \Delta\mu_{21}(t')\,dt' + \text{const.} \tag{13}$$

Here $\mathbf{v}$ is the pair center-of-mass velocity, $\mathbf{A}$ is the vector potential, e is the electron charge, and $\Delta\mu_{21}(t')$ is the difference in chemical potential between the points at time t'. Mercereau and his co-workers have carried out a number of beautiful experiments which illustrate the validity of Eq. (13), and some of these are discussed in by Mercereau in Chapter 31. Within the framework of the tunneling Hamiltonian model of the junction which we are using the relative pair phase difference between two points on opposite sides of the junction (see Fig. 5) is given by

$$\varphi_2 - \varphi_1 = \frac{2e}{\hbar c}\int_1^2 \mathbf{A}(x, t)\cdot d\mathbf{x} - \frac{2}{\hbar}\int_0^t \Delta\mu_{21}(t')\,dt' + \text{const.} \tag{14}$$

Here $\mathbf{A}$ is the vector potential and $\Delta\mu_{21}$ is the difference in chemical potential between points 2 and 1. The absence of the velocity term here as compared with Eq. (13) is consistent with the tunneling Hamiltonian formalism, since the transfer part of the phase shift can be absorbed in the constant.

The time derivative of Eq. (14) can be written in the form

$$\frac{\partial}{\partial t}\left(\varphi_2 - \varphi_1 - \frac{2e}{\hbar c}\int_1^2 \mathbf{A}(x, t)\cdot d\mathbf{x}\right) = \frac{-2\Delta\mu_{21}(t)}{\hbar}. \tag{15}$$

This generalizes our earlier expression, Eq. (12), to the case in which there is a vector potential. Since this vector potential may vary in space, the combination

$$\varphi_2 - \varphi_1 - \frac{2e}{\hbar c}\int_1^2 \mathbf{A}(x, t)\cdot d\mathbf{x} \equiv \varphi(y, z, t) \tag{16}$$

depends upon the y, z coordinates of the insulating layer and defines a local phase which we will simply write as $\varphi(y, z, t)$. Note that Eq. (16) is invariant under a gauge transformation in which

$$\mathbf{A} \to A + \nabla\chi, \qquad \varphi_1 \to \varphi_1 + (2e/\hbar c)\chi(1), \qquad \varphi_2 \to \varphi_2 + (2e/\hbar c)\chi(2);$$

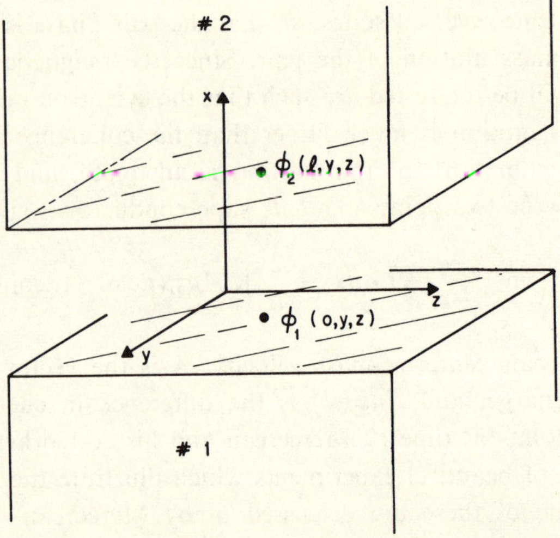

Fig. 5. Junction geometry showing two typical points (l, y, z) in superconductor 2 and $(0, y, z)$ in superconductor 1 between which the relative pair phase difference $\varphi_2 - \varphi_1 = \varphi_2(l, y, z) - \varphi_1(0, y, z)$ is computed.

$\varphi(y, z, t)$ is just the gauge-invariant generalization of the relative pair phase parameter φ. The effect of the vector potential **A** on the local condensation energy is simply to introduce $\varphi(y, z, t)$ in place of φ. Assuming a uniform oxide layer, the local current density is normal to the barrier and has the magnitude

$$j(y, z, t) = j_1 \sin \varphi(y, z, t). \tag{17}$$

Here j_1 is I_1 divided by the junction area. If the oxide layer is nonuniform, then the current density amplitude varies with position in the y–z plane. The introduction of a local current density and generalized relative pair phase assumes that the spatial scale of variation is large compared to the coherence length.

Before concluding this section it is worthwhile developing one further relationship which relates the gradient of the phase to the DC magnetic field. This will be needed in the analysis of the junction electrodynamics which follows. A junction composed of superconducting films which are thick compared to a penetration depth is shown in Fig. 6. The phase differences in the bulk superconductors are determined by Eq. (13). For superconducting films which are thick compared to the penetration depth

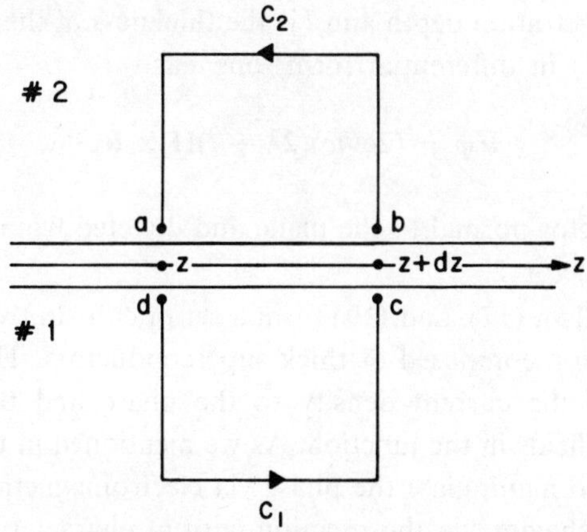

Fig. 6. Contours C_1 and C_2 used with Eq. (13) to relate $\nabla\varphi$ to the magnetic field H_{dc}. The contours extend into the superconductors for several penetration depths.

the parts of the contours parallel to the barrier can be taken sufficiently deep that the contribution of the pair center-of-mass velocity is negligible. Since the dominant part of the current flows parallel to the barrier surface, shielding the field from the interior of the bulk superconductors, the contribution of the velocity to the parts of the contour normal to the junction can also be neglected. Therefore for the contours illustrated in Fig. 6 Eq. (13) gives the phase differences

$$\varphi_a(z) - \varphi_b(z + dz) = \frac{2e}{\hbar c} \int_{C_1} \mathbf{A} \cdot d\mathbf{x}$$

and

$$\varphi_c(z + dz) - \varphi_d(z) = \frac{2e}{\hbar c} \int_{C_2} \mathbf{A} \cdot d\mathbf{x}.$$

Combining these, we have

$$[\varphi_a(z) - \varphi_d(z)] - [\varphi_b(z + dz) - \varphi_c(z + dz)] = \frac{2e}{\hbar c} \int_{C_1+C_2} \mathbf{A} \cdot d\mathbf{x}.$$

Now, adding the line integrals from a to d and c to b to both sides and replacing the line integral of A around the closed loop by the surface integral of H, we obtain

$$\varphi(z) - \varphi(z + dz) = (2e/\hbar c)(2\lambda + l)dz\, H(z). \tag{18}$$

Here λ is the penetration depth and l is the thickness of the insulating layer. Writing Eq. (18) in differential form, one has

$$\nabla\varphi = (2e/\hbar c)(2\lambda + l)\mathbf{H} \times \hat{\mathbf{n}}, \tag{19}$$

with $\hat{\mathbf{n}}$ a unit vector normal to the plane and directed from superconductor 1 to superconductor 2.

Equations (15), (17), and (19) form a set of constitutive relations for a Josephson junction composed of thick superconductors. These constitutive equations relate the current density to the phase and the phase to the electromagnetic fields in the junction. As we mentioned in the introduction, it is the ability to manipulate the phase via electromagnetic fields and then to observe its behavior via the junction current characteristics that allows one to explore the superconducting phase ordering associated with such junctions.

THE ELECTRODYNAMICS OF JOSEPHSON JUNCTION

The electrodynamics of Josephson junctions are obtained by combining Maxwell's equations with the constitutive current–phase–field relations discussed in the previous section. For the case of thick superconductors the derivation is simple. When the thicknesses of the superconductors are less than or of order the penetration depth the behavior of the fields outside the junction must be taken into account by solving the detailed boundary value problem. While this is interesting, particularly with respect to the coupling of electromagnetic energy into and out of the junction ([22]), we will not treat it here.

For the case of thick superconductors it is useful to integrate the Maxwell equation $\nabla \times E = -(1/c)\partial H/\partial t$ over a surface $abcd$ (Fig. 7) which extends into the superconductors well beyond the penetration depth. Since E_x is screened within a Fermi–Thomas screening depth

$$\int \nabla \times E \cdot ds = \int_{(abcd)} E \cdot dx = l[E_x^{(\mathrm{o})}(z + dz) - E_x^{(\mathrm{o})}(z)],$$

where $E_x^{(\mathrm{o})}$ is the electric field in the oxide and l is the effective oxide thickness. The magnetic field integration gives

$$\frac{1}{c}\int \frac{\partial H}{\partial t} \cdot ds = \frac{1}{c}\,\frac{\partial H_y^{(\mathrm{o})}}{\partial t}\,(2\lambda + l)\,dz.$$

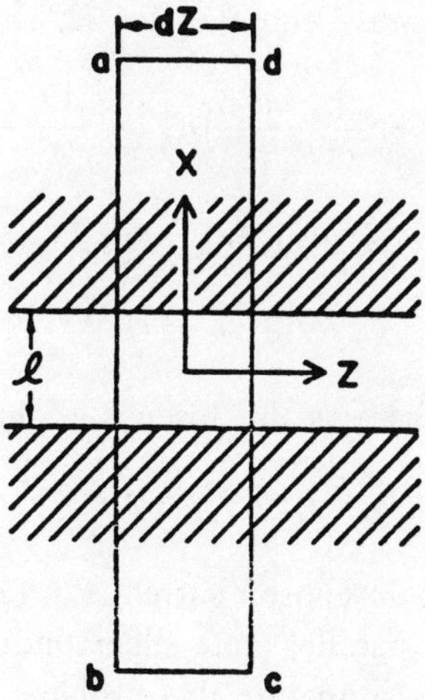

Fig. 7. Infinite plane geometry used to determine the coupling of the AC Josephson current to the electromagnetic field in the junction.

Here $H_y^{(0)}$ is the y component of the magnetic field in the oxide and the penetration depth λ is defined by

$$\lambda = \int_{l/2}^{\infty} dx \, \frac{H_y(x)}{H_y^{(0)}}.$$

Combining these equations, we find

$$l \frac{\partial E_x^{(0)}}{\partial z} = -\frac{2\lambda + l}{c} \frac{\partial H_y^{(0)}}{\partial t}. \tag{20a}$$

In a similar way, taking the normal of the surface along the z axis, one obtains

$$l \frac{\partial E_x^{(0)}}{\partial y} = \frac{2\lambda + l}{c} \frac{\partial H_z^{(0)}}{\partial t}. \tag{20b}$$

The x component of $\nabla \times H = (4\pi/c)j + (\varepsilon/c)\partial E/\partial t$ in the oxide layer is

$$\frac{\partial H_z^{(0)}}{\partial y} - \frac{\partial H_y^{(0)}}{\partial z} = \frac{4\pi}{c} j_x + \frac{\varepsilon}{c} \frac{\partial E_x^{(0)}}{\sigma t}, \tag{21}$$

where ε is the effective dielectric constant of the oxide layer. Now, combining

Eq. (20a, b) and (21), a wave equation for $E_x^{(0)}$ is constructed:

$$\left(\frac{\partial^2}{\partial y^2} + \frac{\partial^2}{\partial z^2} - \frac{1}{\bar{c}^2}\frac{\partial^2}{\partial t^2}\right)E_x^{(0)} = \frac{4\pi}{\varepsilon}\frac{1}{\bar{c}^2}\frac{\partial}{\partial t}j_x. \tag{22}$$

Here the phase velocity $\bar{c}$ is given by

$$\bar{c} = c[l/(2\lambda + l)\varepsilon]^{1/2}$$

and the drive current density is the Josephson current density

$$j_x = j_1 \sin\varphi$$

In the absence of the Josephson current Eq. (22) describes the electromagnetic properties of a parallel plate superconducting transmission line. Swihart ([23]) was the first to analyze this problem and show the existence of slow TM waves with phase (and group) velocity $\bar{c}$. In practice, for Josephson junctions $\bar{c}$ is of order 1/20 the speed of light. Physically, the reduced $\bar{c}/c$ ratio reflects the large disparity between the spatial extent of the electric and magnetic parts of the field. As we have noted, the electric field is essentially confined to the oxide layer of thickness l, while the magnetic field extends an additional two penetration depths $2\lambda + l$. Since the average stored energy in these fields are approximately equal

$$\varepsilon E_0{}^2 l/8\pi \approx (H_0{}^2/8\pi)(2\lambda + l),$$

we find

$$\frac{H_0}{E_0} = \left(\frac{\varepsilon l}{2\lambda + l}\right)^{1/2}.$$

Now the group velocity v_g times the average stored energy per unit width of the junction is equal to the power transmitted per unit width

$$v_g\frac{\varepsilon E_0{}^2 l}{4\pi} = c\int\frac{(\mathbf{E}\times\mathbf{H})\cdot ds}{4\pi} = \frac{clE_0H_0}{4\pi},$$

so that

$$v_g = \frac{c}{\varepsilon}\frac{H_0}{E_0} = c\left(\frac{l}{\varepsilon(2\lambda + l)}\right)^{1/2}.$$

This is just the $\bar{c}$ which appears in Swihart's dispersion relation. In the presence of the Josephson coupling, energy can be stored as "condensation" energy $-E_1\cos\varphi$. In the long-wavelength limit we will find that the energy oscillates back and forth between the electric field and this condensation energy. In this case H_0 goes to zero and the group velocity vanishes.

The excitation is a plasma-like mode in which pairs oscillate back and forth across the junction.

The derivation we have given of Eq. (22) which governs the electrodynamics of a Josephson junction follows that of ([24]). Josephson ([25]) has written this equation in a more compact form by using the relation [our Eq. (15)]

$$\frac{\partial \varphi}{\partial t} = -\frac{2\Delta\mu_{21}}{\hbar} = \frac{2eE_x^{(0)}l}{\hbar}$$

to eliminate $E_x{}^0$ in favor of φ:

$$\left(\frac{\partial^2}{\partial y^2} + \frac{\partial^2}{\partial z^2} - \frac{1}{\bar{c}^2}\frac{\partial^2}{\partial t^2}\right)\varphi = \frac{1}{\lambda_J{}^2}\sin\varphi. \tag{23}$$

Here λ_J, which has the dimensions of length, is the Josephson penetration depth

$$\lambda_J = \left(\frac{\hbar c^2}{8\pi e(2\lambda + l)j_1}\right)^{1/2}.$$

We will find that it is analogous to the London penetration depth in a bulk superconductor. For typical junctions λ_J is of the order of millimeters.

The electrodynamic properties of Josephson junctions are determined from the solutions of Eq. (22) or Eq. (23) with the appropriate boundary conditions. Since these are nonlinear equations, there are a rich variety of solutions. We now proceed to review some results for three special cases. The first of these illustrates the nature of nonuniform phase ordering, and the second and third provide some insight into the nature of the phase fluctuation excitations.

Vortex Structure and Critical Currents

For a junction with dimensions small compared to λ_J (which is of the order of millimeters for a typical junction), the magnetic field generated by the Josephson current can be neglected and

$$\partial\varphi/\partial z = (2ed/\hbar c)H_e.$$

Here H_e is the magnitude of the *externally* applied field in the $-y$ direction and $d = 2\lambda + l$. Integrating this equation for a uniform field, the expression for the current density can be written as

$$j = j_1 \sin\left(\frac{2ed}{\hbar c}H_e z + \varphi_0\right).$$

Maximizing the current through the junction by setting φ_0 equal to $-edH_eL/\hbar c$, one finds that the maximum (or *critical*) current which the junction can carry without developing a voltage is given by

$$I_c = wj_1 \int_0^L dz \sin\left[\frac{2edH_e}{\hbar c}\left(z - \frac{L}{2}\right)\right]$$
$$= I_1 \left| \frac{\sin[(ed/\hbar c)H_eL]}{(ed/\hbar c)H_eL} \right|. \tag{24}$$

Here w is the width of the junction and I_1 is the current density j_1 times the area of the junction wL. This Fraunhofer diffraction of I_c versus H_e was

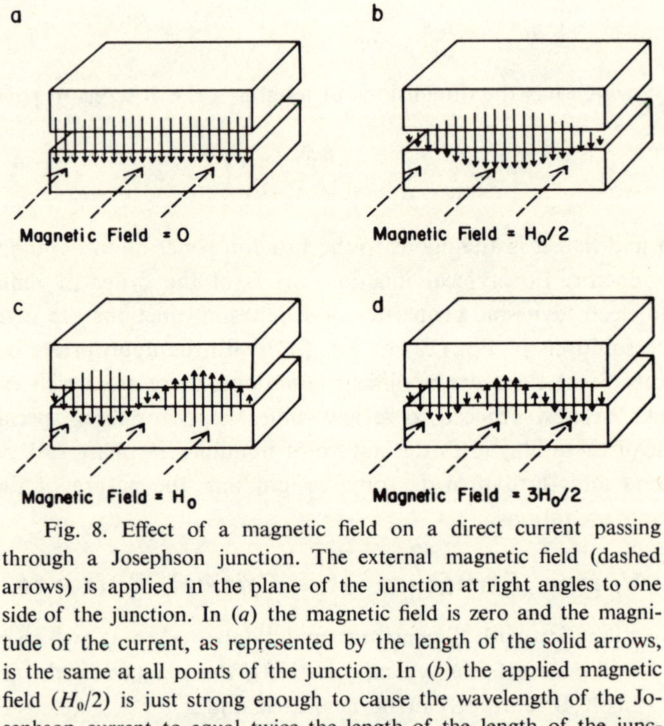

Fig. 8. Effect of a magnetic field on a direct current passing through a Josephson junction. The external magnetic field (dashed arrows) is applied in the plane of the junction at right angles to one side of the junction. In (*a*) the magnetic field is zero and the magnitude of the current, as represented by the length of the solid arrows, is the same at all points of the junction. In (*b*) the applied magnetic field ($H_0/2$) is just strong enough to cause the wavelength of the Josephson current to equal twice the length of the length of the junction. In (*c*) the magnetic field has been increased to H_0, so that a full wavelength fits into the junction. The current now reverses direction once; equal currents flow in both directions and the total junction current is zero. In (*d*) the field has been increased to $3H_0/2$; three half-wavelengths fit into the junction, the current reverses direction twice, and again there is a net current across the junction. The total current is not as large as it was in (*a*) or (*b*). Figure 9 shows how the junction critical current depends on the strength of the magnetic field. From Langenberg *et al.* ([41]).

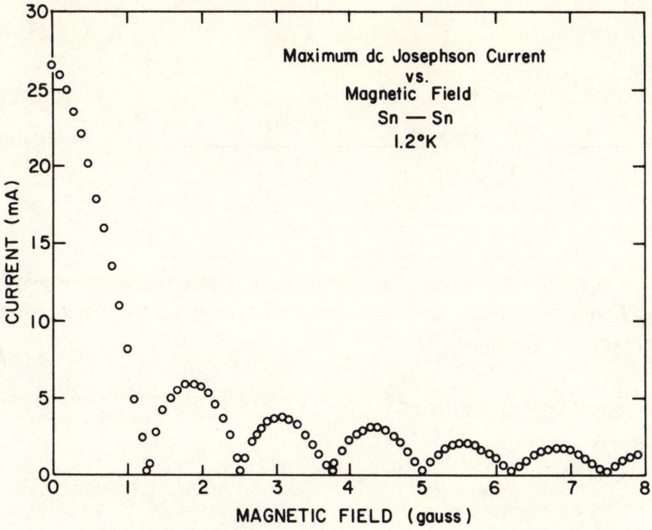

Fig. 9. Magnetic field dependence of the maximum DC Josephson current of a junction. The data is for a Sn–Sn oxide–Sn junction at 1.2°K, with the magnetic field oriented along the y axis (see Fig. 5). For this sample L was 0.25 mm and the period of the Fraunhofer pattern is 1.25 G. From Langenberg *et al.* ([41]).

first reported by Rowell ([26]) and provided striking evidence for the existence of the DC Josephson effect. The current density profiles for the critical current condition at various field values are illustrated in Fig. 8 and the observed critical current versus magnetic field for a short junction is shown in Fig. 9. Each time the junction contains an integral number of flux units, $dLH_e = n(hc/2e)$, the critical current goes to zero. The application of such fields effectively decouples the phase, since the binding energy, $-E_1 \int \cos \varphi(z) dz$, goes to zero. Near these points the effects of phase fluctuations become important. These will be discussed in Section IV.

For a junction with dimensions large compared to λ_J the self magnetic field produced by the Josephson current must be taken into account. In this case it is possible for vortex-like phase ordering to occur. Owen and Scalapino ([27]) have studied the critical current behavior for the symmetric junction geometry shown in Fig. 10. The external field H_e is in the $-y$ direction and the current flows from left to right. When $L \gtrsim \lambda_J$ the phase must be determined from the static limit of Eq. (22),

$$\partial^2\varphi/\partial z^2 = (1/\lambda_J^2) \sin \varphi$$

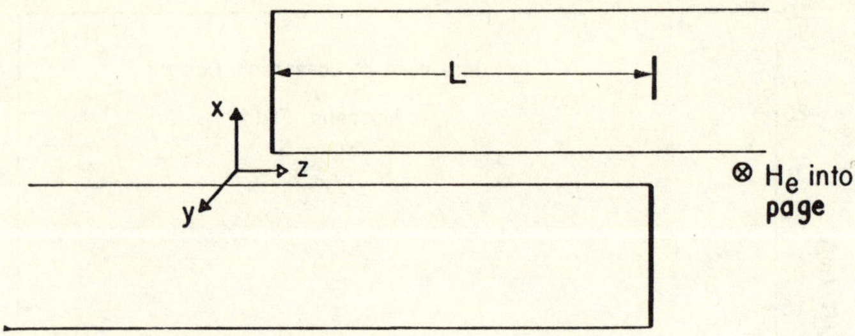

Fig. 10. The tunnel junction geometry. Here the junction length is L and the applied H field is into the plane of the page. From Owen and Scalapino ([22]).

with the boundary conditions

$$H_y^{(0)} = \frac{\hbar c}{2ed} \frac{\partial \varphi}{\partial z}\bigg|_{z=0}, \qquad H_y^{(L)} = \frac{\hbar c}{2ed} \frac{\partial \varphi}{\partial z}\bigg|_{z=L}.$$

The values of $H_y(0)$ and $H_y(L)$ are related to the externally applied magnetic field H_e and the current flowing through the junction I by Ampere's law

$$H_y(L) - H_y(0) = 4\pi I/c, \qquad H_y(L) + H_y(0) = 2H_e.$$

The general solution of the differential equations leads to Jacobian elliptic functions, and the resulting current density and magnetic field distributions are given by

$$j(z) = -\frac{2j_1}{k}\, sn\left(\frac{z-z_0}{\lambda_J}\,\bigg|\,\frac{1}{k^2}\right) dn\left(\frac{z-z_0}{\lambda_J}\,\bigg|\,\frac{1}{k^2}\right),$$

$$H(z) = \frac{\hbar c}{ed\lambda_J k}\, cn\left(\frac{z-z_0}{\lambda_J}\,\bigg|\,\frac{1}{k^2}\right).$$

The parameters k and z_0 are used to fit the boundary conditions. The wavelength and the detailed shape of the current distribution depend on k. In the limit of small k the current distribution has a sinusoidal shape with a wavelength $\pi k \lambda_J$. As k increases this sinusoidal shape becomes distorted and the period is given by $2k\lambda_J K(k^2)$, where K is the complete elliptic integral

$$K(k^2) = \int_0^{\pi/2} \frac{d\theta}{[1 - k^2 \sin^2\theta]^{1/2}}.$$

In Fig. 11, the current density for maximum current with zero applied external magnetic field is shown for junctions with different L/λ_J ratios. For the short junctions the current distribution is almost uniform. However, as the L/λ_J ratio increases the we can see that the current is carried within a distance of the order of several λ_J of the surface. For an infinite junction the maximum current per unit length along the edge is $2\lambda_J I_1$. The case of an infinite junction was originally discussed by Ferrell and Prange ([14]). It is interesting to contrast the peaking of the current density at a depth λ_J and its vanishing at the surface with the peaking of the current at the surface of a bulk superconductor in the London theory. This distinction arises because of the nonlinearity of the Josephson constitutive relation as opposed to the linear London relation.

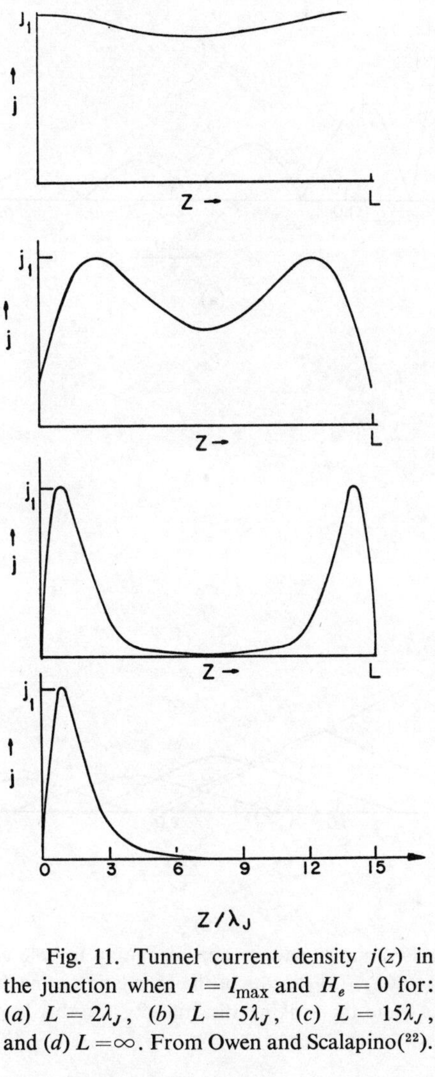

Fig. 11. Tunnel current density $j(z)$ in the junction when $I = I_{max}$ and $H_e = 0$ for: (a) $L = 2\lambda_J$, (b) $L = 5\lambda_J$, (c) $L = 15\lambda_J$, and (d) $L = \infty$. From Owen and Scalapino([22]).

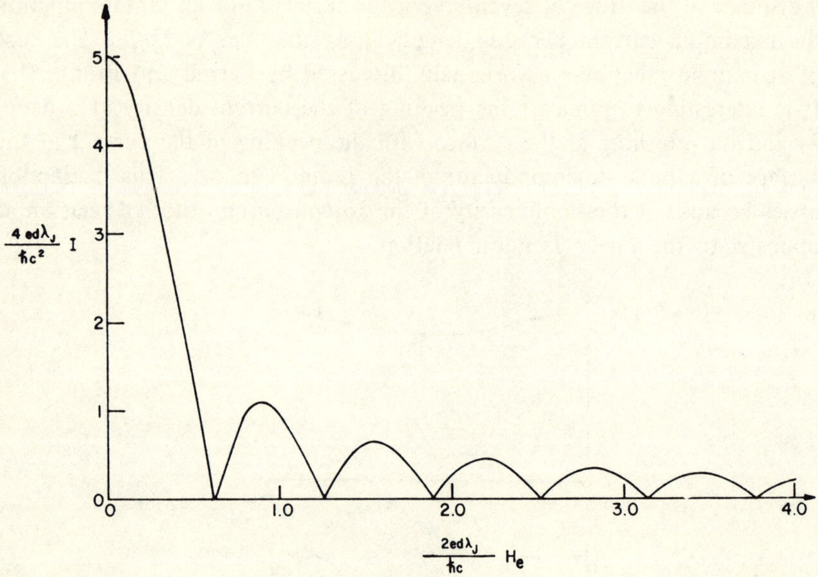

(a)

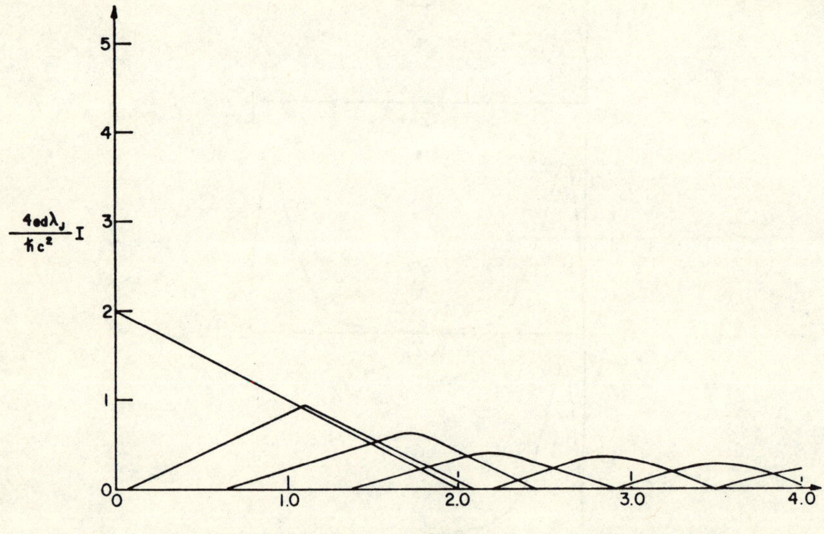

(b)

Fig. 12. (*a*) The Fraunhofer pattern approximation for the critical current versus applied field when $L = 10\lambda_J$. (*b*) Results of the calculation using Jacobian elliptic functions when $L = 10\lambda_J$. Note the striking dissimilarity for this large L/λ_J ratio. From Owen and Scalapino ([22]).

In the presence of a weak external magnetic field Meissner-like currents flow within several λ_J of the junction edge, shielding the interior of the junction and preserving most of the barrier "condensation" energy. As the field is increased these shielding currents are no longer capable of preventing the field from entering, and as a result phase variations and current oscillations appear in the junction interior. The resulting vortex array is reminiscent of the type-II superconducting phase. However, because this is basically a one-dimensional system (and, furthermore, since in practice the junctions are finite in length) there is no sharp phase transition between the Meissner and type-II vortex phases of a Josephson junction. The structure of these non uniform phase states can be probed by measuring the critical current versus magnetic field, and in Fig. 12 we have plotted I_c versus H_e for a junction with an L/λ_J ratio of 10. The envelope of the maxi-

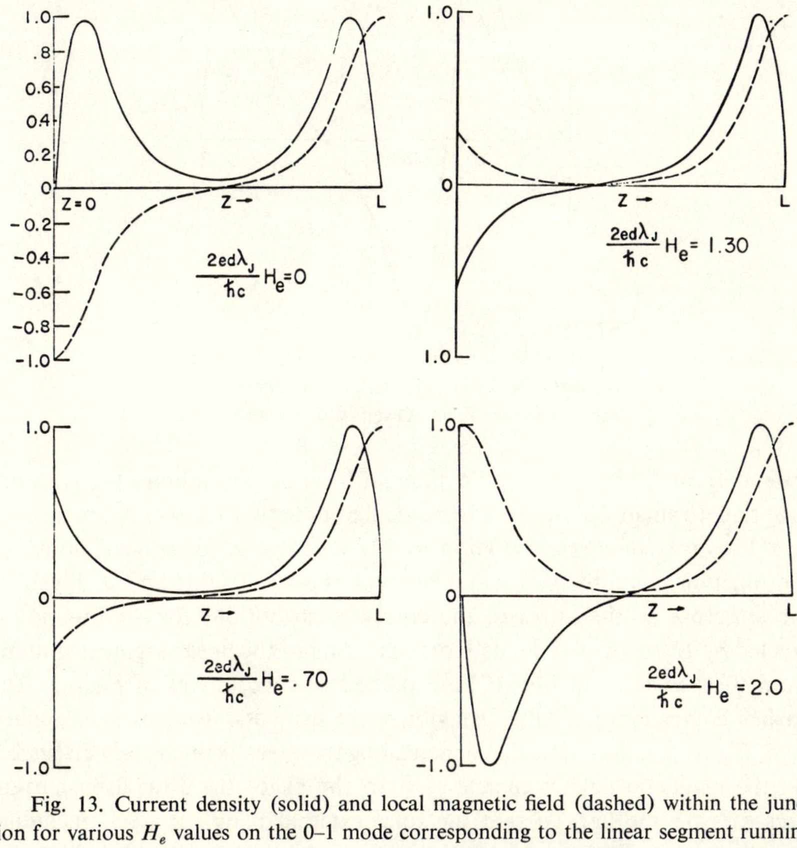

Fig. 13. Current density (solid) and local magnetic field (dashed) within the junction for various H_e values on the 0–1 mode corresponding to the linear segment running from (0, 2) to (2, 0) in Fig. 12 (*b*). From Owen and Scalapino ([22]).

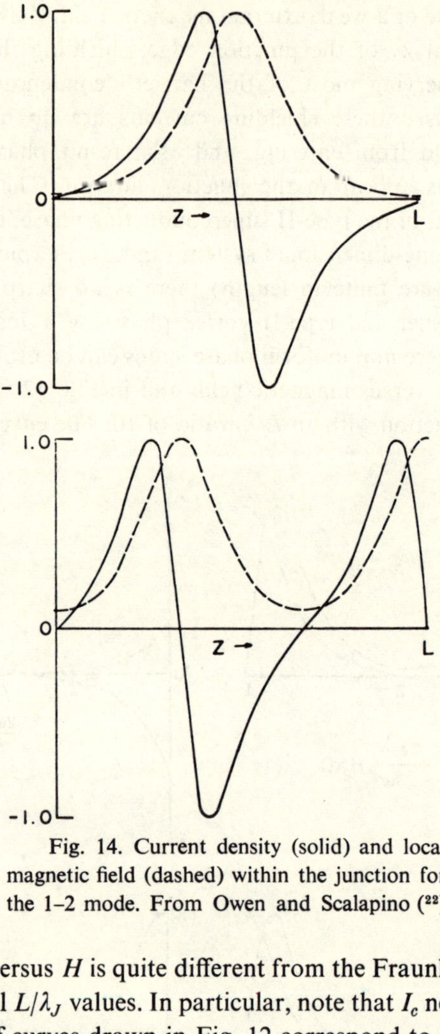

Fig. 14. Current density (solid) and local magnetic field (dashed) within the junction for the 1–2 mode. From Owen and Scalapino ([22]).

mum current I_c versus H is quite different from the Fraunhofer behavior appropriate to small L/λ_J values. In particular, note that I_c never drops to zero.

The array of curves drawn in Fig. 12 correspond to "modes" in which the junction contains n to $n + 1$ vortices. To clarify this and to illustrate the structure of the vortices, the current distributions for various points labeled by H_e on the 0–1 mode [corresponding to the linear segment running from (0,2) to (2,0) in Fig. 12] are plotted as solid curves in Fig. 13. The dashed curves represent the behavior of the magnetic field. At zero applied field, Fig. 13(a), the current is carried along the edges as previously discussed. As the magnetic field is increased (into the page) the Josephson current decreases on the left, so that the total external magnetic field (including that due to the DC current flowing in the circuit) is screened from the

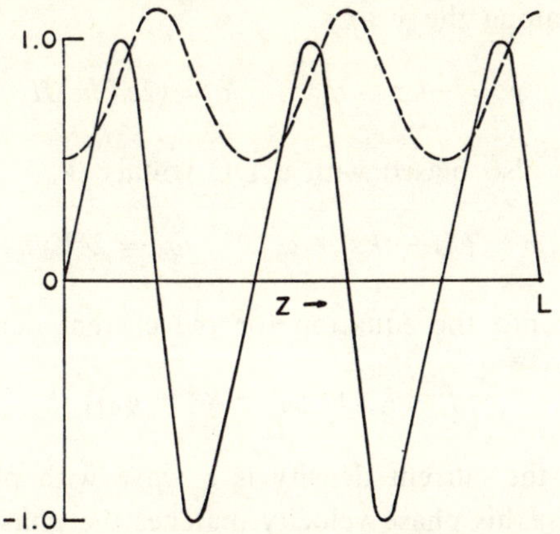

Fig. 15. Current density (solid) and local magnetic field (dashed) within the junction for the 2–3 mode with $H_e = 1.8(2ed\lambda_J/\hbar c)$. From Owen and Scalapino ([22]).

interior. This process continues until at a field $H_e = 2$ no net current can be carried. The magnetic field is measured in units of $(\hbar c/2e)/d\lambda_J$. During this process the junction has gone from a state with no vortices, Fig. 13(*a*), to a state containing one vortex, Fig. 13(*d*).

Another "one-vortex" solution is illustrated in Fig. 14(*a*). For the $L/\lambda_J = 10$ case the external field H_e appropriate to *this* one-vortex solution is 0.06. Note that this current density carries zero net current across the barrier. At this field it is a state of higher free energy than the 0–1 solution which was just discussed. However, as H_e is further increased the current density of this mode evolves from a one- to a two-vortex array, as shown in Fig. 14(*b*). For $H_e = 1.1$ the critical current carried by this 1–2 mode is, in fact, greater than that carried by the 0–1 mode. From $H_e \approx 1.7$ to 2.1 the 2–3 mode array, illustrated in Fig. 15 for $H_e = 1.8$, carries the maximum critical current. As the field increases the current distribution becomes more sinusoidal and the I_c curve approaches the Fraunhofer pattern. Thus, in principle, an analysis of the critical current versus magnetic field offers the possibility of giving detailed information on vortex structure in Josephson junctions ([28]).

Excitation of Electromagnetic Modes of the Junction by the Josephson Current

For a junction with dimensions less than λ_J we have seen that the spatial variation of the phase is proportional to the externally applied magnetic field H. For H along the y axis

$$\varphi = -kz + \varphi_0; \qquad k = (2ed/\hbar c)H.$$

If the junction is also biased with a DC voltage V_0

$$\varphi = \omega_0 t - kz + \varphi_0, \qquad \omega_0 = 2eV_0/\hbar.$$

Substituting this into the equation for the current density,

$$j = j_1 \sin(\omega_0 t - kz + \varphi_0), \tag{25}$$

and we see that the current density is a wave with phase velocity $\omega_0/k = (V/dH)c$. When this phase velocity matches the phase velocity $\bar{c}$ of the Swihart mode it is natural to expect that the Josephson current will transfer energy to the TM electromagnetic mode, which then dissipates it by losses in the barrier region and by external radiation. This power is supplied by the battery, so that for a fixed magnetic field one expects that the junction I–V characteristic will exhibit a peak when V is such that

$$(V/dH)c = \bar{c}. \tag{26}$$

In Fig. 16 results reported by Eck *et al.* ([24]) for just this situation are illustrated. Fig. 17 shows the relationship between the peak voltage and the applied magnetic field. At voltages such that ω_0 is greater than 2Δ losses due to pair breaking occur. These can account for the broadening of the peak shown in the upper curve in Fig. 16, as well as the deviation of the V H curve from a straight line at higher voltages.

For a short junction with well-defined edges the boundary conditions determine a set of allowed electromagnetic field modes. Assuming the surface currents along the z direction of the junction vanish at the ends, the normal modes are $\cos(n\pi z/L)$ and the allowed frequencies are given by $\omega_n = n\pi\bar{c}/2L$. Corresponding to these allowed frequencies there is a set of voltages

$$V_n = n(\pi\hbar\bar{c}/2eL) \tag{27}$$

for which the Josephson current oscillates at a frequency of an allowed electromagnetic mode of the junction. Expanding the electric field in these modes, the solution of Eq. (20) with a driving current density given by Eq. (25) is

$$E_x(z, t) = \frac{4\pi j_1}{\varepsilon\omega} \sum_{n=0}^{\infty} \cos\left(\frac{n\pi z}{L}\right) \frac{a_n \cos(\omega t + \theta_n) + b_n \sin(\omega t + \theta_n)}{\{[1 - (n\pi\bar{c}/\omega L)^2]^2 + (1/Q_n^2)\}^{1/2}},$$

where

$$\theta_n = \tan^{-1} \frac{1/Q_n}{1 - (n\pi\bar{c}/\omega L)^2}$$

and the current-field overlap coefficients a_n and b_n are

$$a_n = \frac{2kL \sin(kL - n\pi)}{(kL)^2 - (n\pi)^2} \times \begin{cases} \frac{1}{2} & n = 0 \\ 1 & n > 0 \end{cases},$$

$$b_n = \frac{2kL[1 - \cos(kL - n\pi)]}{(kL)^2 - (n\pi)^2} \times \begin{cases} \frac{1}{2} & n = 0 \\ 1 & n > 0 \end{cases}.$$

Here only field modes were considered which have no spatial variation

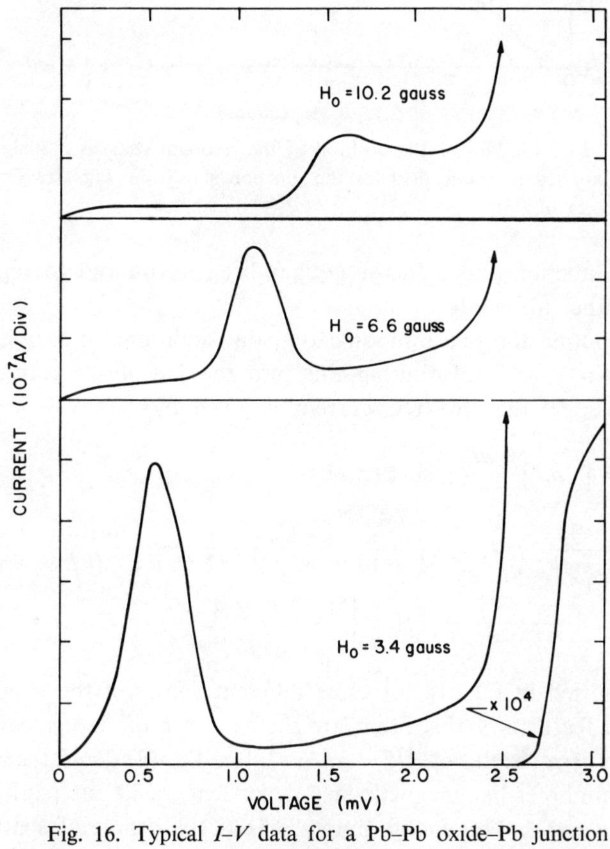

Fig. 16. Typical I–V data for a Pb–Pb oxide–Pb junction at 1.2°K, showing the resonant peak and its dependence on magnetic field. From Eck *et al.* ([24]).

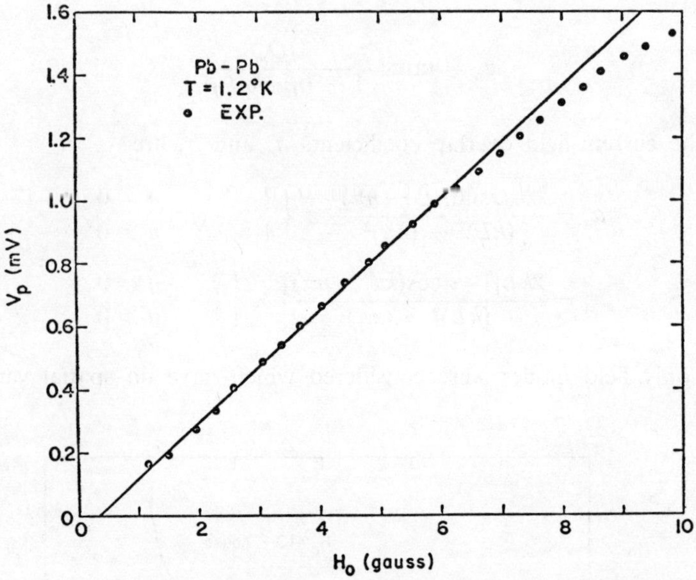

Fig. 17. Plot of the position of the resonant peak as a function of applied magnetic field for the junction shown in Fig. 16. From Eck *et al.* ([24]).

in the y direction and a factor Q_n has been introduced to represent the losses in the nth mode.

Computing the phase modification this additional contribution to E_x produces in φ and substituting this into the Josephson current density, one finds ([29,30]) that the DC current is given by

$$\begin{aligned} I_{dc} &= \int_0^L dz \int_0^\tau \frac{dt}{\tau} j_1 \sin\varphi(z, t) \\ &= \frac{4\pi l}{\varepsilon\omega V_0} j_1^2 L \sum_{n=0}^{\infty} \frac{1/Q_n}{[1-(n\pi c/\omega L)^2]^2 + (1/Q_n)^2} \left[\frac{\sin[(kL-n\pi)/2]}{(kL-n\pi)/2}\right]^2 \\ &\quad \times \frac{1}{(1+n\pi/kL)^2} \times \begin{cases} \frac{1}{2} & n = 0 \\ 1 & n > 0 \end{cases}. \end{aligned}$$

This is plotted in Fig. 18 for three different values of the magnetic field. The main features of this result are the existence of resonances separated in voltage by multiples of $\Delta V = \pi h\bar{c}/2eL$ [see Eq. (27)] and the appearance of the Fraunhofer-like magnetic-field-dependent coefficient $\{\sin[(kL - n\pi)/2]/\frac{1}{2}(kL - n\pi)\}^2$. This factor simply reflects the degree of spatial overlap of the Josephson current density and the electric field mode. If H is adjusted for peak coupling to the nth mode the $n \pm 2m$ modes are not excited. Under this condition there should be three modes consecutive excited and then only every other mode as one moves outward from these three. The additional factors in the expression for the current tend to modify the exact size of the resonances as shown in Fig. 18, but the selection rule determined by the Fraunhofer factor is obeyed. For a general setting of the magnetic field all resonances can be excited. With a current source feeding the junction the I–V trace follows only the positive dI/dV part of the I–V characteristic, as shown in Fig. 19. Step structures of this type were first reported by Fiske ([31]) and the details of the characteristic predicted by Eq. (28)

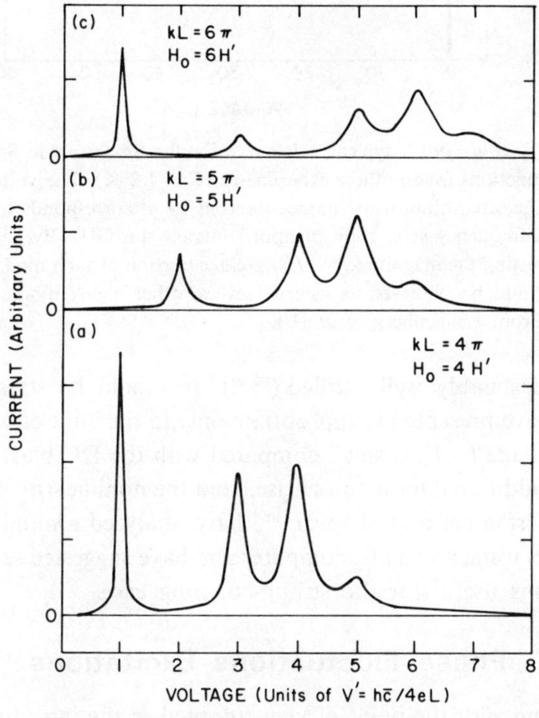

Fig. 18. Schematic diagram showing how the magnetic field determines the coupling to the junction modes. The I–V curves are plotted using Eq. (28) for three different values of applied magnetic field. The current scale is the same for each H_0 but is otherwise arbitrary. The voltage scale is in units of the fundamental mode spacing $V = h\bar{c}/4eL$. For all three curves $Q_n = 10$. In (a) the field is chosen for optimum coupling to the 4th mode, in (b), to the 5th mode, and in (c), to the 6th mode. Thus $H_0 = 4H'$, $5H'$, $6H'$ respectively, where $H' = hc/4edL$. From Langenberg *et al.* ([40]).

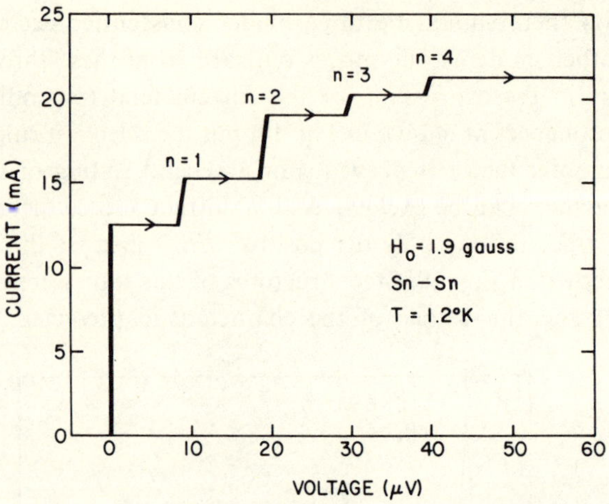

Fig. 19. A typical *I–V* curve for the Sn–Sn oxide–Sn junctions used in these experiments ($T = 1.2°K$). The voltage separation of the modes (labeled by n) corresponds to a frequency separation of approximately 4.6 GHz. By adjusting the magnitude of H_0 a greater portion of each mode could be observed as well as several other higher modes. From Langenberg *et al.* ([40]).

have been reasonably well verified ([29,30]). It should be stressed that the analysis we have presented is appropriate only in the limit of weak coupling where the voltage $l \cdot E_x$ is small compared with the DC bias. When this is not the case additional harmonics arise from the nonlinearity of the electrodynamics. Werthamer and Shapiro ([32]) have analyzed a number of strong-coupling cases using an analog computer and have suggested several analytic approximations useful for the strong-coupling case.

Phase Fluctuations Excitations

In keeping with the point of view adopted in the introduction we will conclude this section with a discussion of the elementary excitations associated with a Josephson junction. The equation of motion for these excitations has already been given, Eq. (23), but we reproduce it here for convenience

$$\left(\nabla_{y,z}^2 - \frac{1}{\bar{c}^2}\frac{\partial^2}{\partial t^2}\right)\varphi = \frac{1}{{\lambda_J}^2}\sin\varphi .$$

If the right-hand side is neglected, the phase fluctuations are just the Swihart modes discussed above in Section III,B. However, the inclusion of the barrier "condensation" energy term leads to a modification of the excitation spectrum. Here we will treat the problem of small fluctuations about a *static* phase $\varphi_0(y, z)$.

First, suppose $\varphi_0(y, z)$ is zero. Then, linearizing the equation of motion, one obtains the dispersion relation ([25])

$$\omega = [\omega_p{}^2 + (k\bar{c})^2]^{1/2}. \tag{29}$$

Here ω_p is the Josephson plasma frequency

$$\omega_p = \frac{\bar{c}}{\lambda_J} = \sqrt{\frac{(2e)^2 E_1}{\hbar^2 C}}$$

which is of order 10^9–10^{10} cps for a typical junction. The first person to discuss this excitation was P. W. Anderson ([11]), and it has recently been observed by Dahm *et al.* ([39]). The dispersion relation, Eq. (29), is plotted in Fig. 20 and contrasted with the dispersion of the Swihart excitation (shown as a dotted curve). The characteristic excitations of the unbiased junction differ in the long-wavelength limit from the TM strip-line mode by the appearance of a cutoff plasma frequency associated with pair transfer. From a circuit point of view the existence of ω_p is simply a reflection of the junction inductance L which arises from the Josephson constitutive equation

$$\frac{dI}{dt} = I_1 \cos\varphi\dot{\varphi} = I_1 \cos\varphi \frac{2eV}{\hbar} \equiv \frac{1}{L} V.$$

For small phase fluctuations about φ equal to zero, $\cos\varphi$ is unity, so that the effective junction inductance is given by $L = \hbar/2eI_1$. This inductance in parallel with the junction capacitance C makes the junction respond as a tank circuit with a resonant frequency given by $\omega_p{}^2 = 1/LC$. It is interesting to note that as φ is varied toward $\pi/2$ the inductance $\hbar/(2eI_1 \cos\varphi)$ becomes

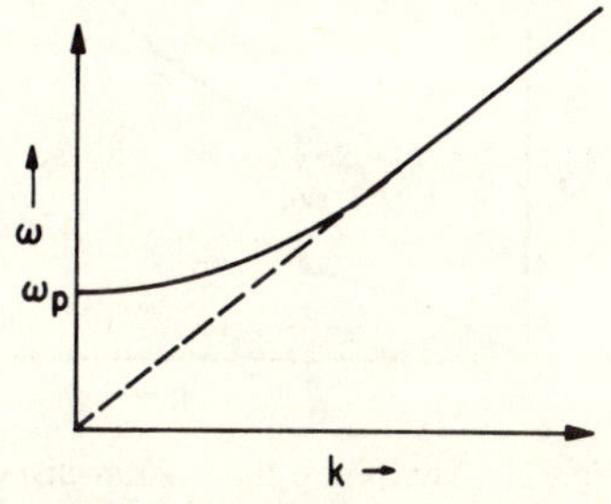

Fig. 20. Plot of the Josephson (solid) and Swihart (dashed) dispersion relations.

infinity, driving the plasma frequency to zero. This is reasonable, since, as we have seen, the phase coupling vanishes when φ is $\pi/2$.

In Section III,A we saw that an external magnetic field could produce vortices in the junction. This was described by solutions of the equation

$$\nabla^2_{y,z}\varphi_0(y, z) = \frac{1}{\lambda_J{}^2} \sin \varphi_0(y, z)$$

with the appropriate boundary conditions. If we set

$$\varphi = \varphi_0(y, z) + \delta\varphi,$$

then the linearized wave equation for phase fluctuations is

$$\left(\nabla^2_{y,z} - \frac{1}{\bar{c}^2}\frac{\partial^2}{\partial t^2}\right)\delta\varphi = \frac{1}{\lambda_J{}^2} \cos \varphi_0(y, z)\, \delta\varphi.$$

Lebwohl and Stephen ([33]) have solved this equation and obtained a dispersion relation for the case where vortices are present, which has the form sketched in Fig. 21. Here Bragg scattering of the phase fluctuations from the periodic vortex array produces a band gap in the spectrum when k is equal to π divided by the vortex spacing. There is no gap at zero wave vector because an infinite junction geometry was assumed. For a finite junction the gap at zero wave vector will be given by $\omega_p(\int dy\, dz \cos \varphi(y, z)/\int dy\, dz)^{1/2}$. In the limit where L/λ_J is of order unity or less this has the Fraunhofer form

$$\omega_p \left| \sin\left(\frac{ed}{\hbar c} HL\right) \bigg/ \frac{ed}{\hbar c} HL \right|^{1/2}.$$

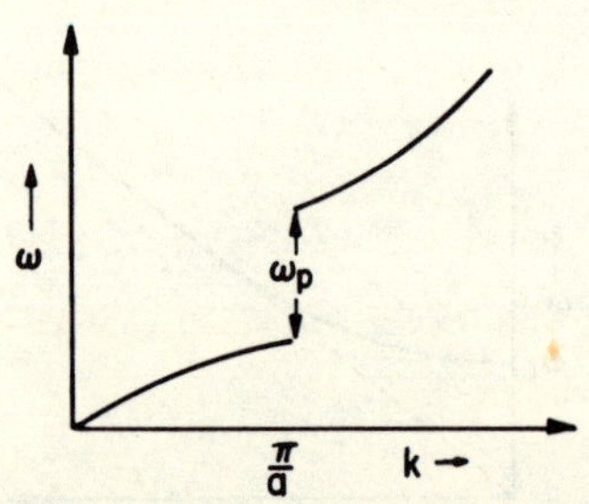

Fig. 21. The phase fluctuation dispersion relation in the presence of a vortex array with spacing a.

FLUCTUATIONS IN JOSEPHSON JUNCTIONS

In calculating the coupling energy and hence the Josephson current we assumed a definite choice of phase φ. Since the pair transfer number n is conjugate to φ, the choice of a definite value for φ implies, via the uncertainty relation, infinite fluctuations in n. Now, although the electrostatic energy is small, it will eventually act to suppress large fluctuations in n. This in turn implies a variation in φ. These zero-point oscillations of φ reduce the binding energy and hence the Josephson current amplitude I_1. The amount of this suppression depends upon the ratio of Coulomb energy $(2e)^2/2C$ to the binding energy E_1 and is usually negligible for practical junctions. However, near the critical current or for magnetic fields near the nulls in the Fraunhofer pattern these fluctuations can eventually destroy the phase coherence. In addition, at finite temperatures thermal phase fluctuations are present which reduce I_1. As one would imagine, this effect depends upon the ratio $kT/E_1(T)$, so that as T approaches T_c and $E_1(T)$ goes to zero it becomes important.

Phase fluctuations are a source of junction noise, and along with quasiparticle excitations contribute to the noise power spectrum of the junction. This latter quantity is of interest when one considers devices based on Josephson junctions. At present various superconducting devices based upon the Josephson phenomena are sensitive and stable enough so that one can measure voltages of 10^{-14} to 10^{-15} V and magnetic fields of 10^{-10} G and detect millimeter radiation at a level of 10^{-14} W with a one-cycle bandpass ([34]). One would like to understand the limits on such systems. In addition, the Josephson radiation linewidth has been measured as a function of temperature and operating point ([35]). In Sn–Sn oxide–Sn junctions emitting at 10^{10} Hz, this linewidth was observed to increase from approximately 2 kHz to 10 MHz as the temperature increased from $T/T_c = 0.3$ to $T/T_c = 0.98$. This provides a direct proble of the low-lying fluctuations. Finally, questions concerning the lifetime of superconducting current-carrying metastable states are of at least conceptual interest. All of these things are closely related to the simple idea of phase fluctuations, and we will briefly discuss some features of this theory here.

Recognizing that φ and n are conjugate variables, the Hamiltonian describing the phase fluctuations of the junction is

$$H = -\int \varepsilon_1 \cos \varphi(s)\, ds + \frac{1}{8\pi} \int (\varepsilon E^2 + H^2)\, dV. \tag{30}$$

Here the first term is simply the binding energy of the junction, with $\varphi(s)$

the relative pair phase at a point s in the plane of the barrier and ε_1 the energy density

$$\varepsilon_1 = E_1/A = \hbar j_1/2e.$$

The energy density ε_1 could also be a function of position in the plane of the barrier. The last term in Eq. (30) is the electromagnetic field energy. From the discussion of the electromagnetic modes we know that the bulk of the electric field energy is confined in the insulating region. This means that this part of the field energy can be expressed in terms of the local pair transfer density $n(s)$ and the junction capacitance per unit area C.

$$\frac{1}{8\pi}\int \varepsilon E^2\, dV = \int ds \frac{2e^2}{C} n^2(s).$$

Finally, using the Josephson relation

$$\nabla\varphi = (2ed/\hbar c)\mathbf{H} \times \hat{\mathbf{n}}$$

which relates the gradient of the phase to be transverse magnetic field, the magnetic energy term becomes

$$\int dV \frac{H^2}{8\pi} = \int ds \frac{\varepsilon_1}{2} \lambda_J{}^2 \,|\, \nabla\varphi \,|^2.$$

The full Hamiltonian can therefore be written in the form

$$H = \int ds\left(-\varepsilon_1 \cos\varphi + \frac{2e^2}{G} n^2 + \frac{\varepsilon_1}{2} \lambda_J{}^2 \,|\, \Delta\varphi \,|^2\right) \tag{31}$$

with the commutation relations

$$[\varphi(s), n\,(s')] = i\delta(s - s').$$

The Heisenberg equations of motion for the phase field operator φ lead to the nonlinear wave equation

$$\left(\nabla^2_{y,z} - \frac{1}{\bar{c}^2}\frac{\partial^2}{\partial t^2}\right)\varphi = \frac{1}{\lambda_J{}^2} \sin\varphi,$$

which was previously used to describe the classical c-number phase. This is now to be interpreted as an operator equation. The linearized form of this equation has the same dispersion relation discussed in Section III,C,

$$\omega = [\omega_p{}^2 + (k\bar{c})^2]^{1/2}.$$

The effect of quantization is simply to introduce the zero-point fluctuations of these excitations into the ground-state properties. In the absence of the phase locking ω_p vanishes and the dispersion relation for these phase fluctuations reduce to the Swihart form. In this case the phase fluctuations are related to the zero-point blackbody electromagnetic radiation in the junction by $\dot{\varphi} = 2elE_x/\hbar$.

For a short junction the normal mode expansion of the φ-field operators can be approximated by keeping only the uniform spatial part, and the Hamiltonian, Eq. (31), reduces to a form originally introduced by Anderson ([11])

$$H = -E_1 \cos\varphi + [(2en)^2/2C]. \tag{32}$$

This will be adequate to give an idea of the type of phenomena associated with phase fluctuations.* Furthermore, in calculating quantities such as the binding energy one is interested in the *difference* between the phase-coupled Josephson modes and the Swihart modes. The energy difference between these excitations for a junction of length L goes as

$$\Delta\omega = \left[\omega_p{}^2 + \bar{c}^2\left(\frac{\pi n}{L}\right)^2\right]^{1/2} - \frac{\bar{c}\pi n}{L} \cong \left(\frac{\omega_p L}{2\bar{c}\pi n}\right)\omega_p$$

for $n > 0$ and $\omega_p L/\bar{c}\pi n \ll 1$. For typical junctions $\omega_p L/\bar{c}\pi n \lesssim 10^{-1}$, so that only the uniform $n = 0$ mode differs in a significant way from the Swihart modes.

Phase Fluctuations and the Critical Current

The effect of the quasiparticle excitations can be taken into account by setting

$$E_1 = E_1(0)\frac{\Delta(T)}{\Delta(0)}\tanh\left[\frac{\beta\Delta(T)}{2}\right].$$

Here, we address our attention to the effect of phase fluctuation excitations on the critical current. If, as we argued in Section II, the effect of the external circuit is to fix the average phase across the junction, then the Hamiltonian for the junction and the external drive can be approximated by†

* Recently Fetter and Stephen ([36]) have discussed fluctuations in an infinite junction using the Hamiltonian given in Eq. (31).

† This form was used by Anderson ([12]), and we believe it is the relevant one for this problem. One might argue that the external circuit fixed the average current through the system rather than the average phase. In this case an additional term $-\lambda\sin\varphi$

$$-E_1 \cos\varphi + \frac{(2en)^2}{2C} - \lambda\varphi$$

with λ adjusted so that $\langle\varphi\rangle$ equals a specified phase φ_0.

Setting $\varphi = \varphi_0 + \delta\varphi$, the Hamiltonian can be written in the form

$$H = -E_1(\cos\varphi_0 \cos\delta\varphi - \sin\varphi_0 \sin\delta\varphi) + \frac{(2en)^2}{2c} - \lambda(\varphi_0 + \delta\varphi),$$

taking

$$\lambda = E_1 \sin\varphi_0$$

and expanding to second order in $\delta\varphi$,

$$H \approx \frac{E_1 \cos\varphi_0}{2}\,\delta\varphi^2 + \frac{(2en)^2}{2C} = \tilde{\omega}_p(a^+a + \tfrac{1}{2}). \tag{33}$$

Here, $\tilde{\omega}_p$ is related to the Josephson plasma frequency by

$$\tilde{\omega}_p = \omega_p(\cos\varphi_0)^{1/2}.$$

Expanding the current operator in the same way, one finds

$$I = I_1 \sin\varphi \approx I_1[\sin\varphi_0\,(1 - \tfrac{1}{2}\delta\varphi^2) + \cos\varphi_0\,\delta\varphi]. \tag{34}$$

Taking a thermal average of Eq. (34) with respect to H, Eq. (33), we obtain an approximate expression for the current in the presence of phase fluctuations:

$$\langle I\rangle \approx I_1 \sin\varphi_0(1 - \tfrac{1}{2}\langle\delta\varphi^2\rangle) \approx I_1 \sin\varphi_0 \exp(-\tfrac{1}{2}\langle\delta\varphi^2\rangle). \tag{35}$$

Here

$$\langle\delta\varphi^2\rangle = \frac{\omega_p}{2E_1}\coth\frac{\beta\tilde{\omega}_p}{2} = \left(\frac{e^2}{CE_1}\right)^{1/2}\coth\frac{\beta\tilde{\omega}_p}{2} \tag{36}$$

and the phase fluctuation reduction of $\langle I\rangle$ enters as a Debye–Waller-like factor.

For low temperatures such that $\omega_p \gg kT$,

$$\langle I\rangle \sim I_1 \sin\varphi_0 \exp[-(e^2/2CE_1)^{1/2}].$$

---

should be added to the junction Hamiltonian, Eq. (32). A straightforward calculation shows that this leads to an excitation energy $\omega_p(\cos\varphi_0)^{-\frac{1}{2}}$, which diverges as φ_0 approaches $\pi/2$. Formally, this would lead to an absence of phase fluctuations at the critical current. This is, of course, an artifact of the model and would not occur if we allowed the possibility of a normal current component.

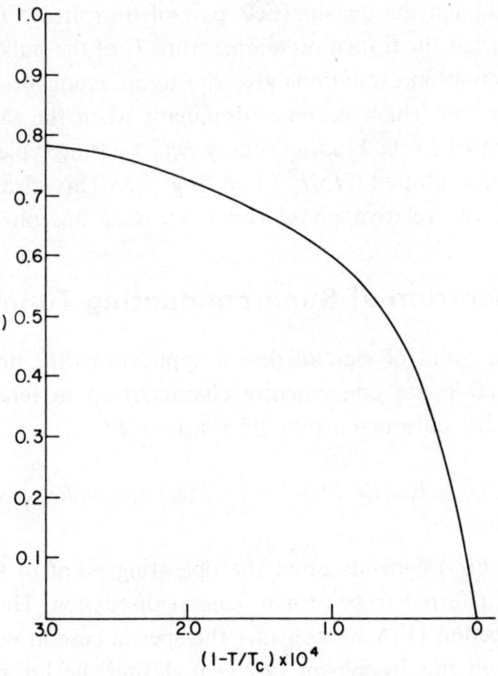

Fig. 22. The ratio of the junction critical current $\langle I \rangle$ including phase fluctuations to $I_1(T)$ (which includes only the effects of quasi particle excitations) versus $(1 - T/T_c)$. Very near the transition temperature the phase fluctuation excitations dominate and the long-range phase ordering "melts."

Here we can see that the reduction of $\langle I \rangle$ due to the zero-point phase motion is small as long as E_1 is large compared to the electrostatic energy $e^2/2C$. When the temperature is high enough that the thermal energy kT is large compared to ω_p, the current becomes

$$\langle I \rangle = I_1(T) \sin \varphi_0 \exp(-kT/2E_1(T) \cos \varphi_0).$$

Maximizing this with respect to φ_0 one finds that the critical current occurs when

$$\cos \varphi_0 = \tfrac{1}{2}(kT/E_1) \tan^2 \varphi_0.$$

Since $\sin \varphi_0 \sim 1$, the critical current for T near T_c becomes

$$\langle I \rangle = I_1(T) \exp[-(kT/2E_1(T))^{2/3}]. \tag{37}$$

Therefore although the quasiparticle part of the current $I_1(T)$ extrapolates linearly to zero at the transition temperature T_c of the bulk superconductor, the phase fluctuation excitations give rise to an exponential decrease in the Josephson current which becomes dominant when the thermal energy kT is large compared to the binding energy $E_1(T)$. Using typical junction parameters we have plotted $\langle I\rangle/I_1(T)$ in Fig. 22. This clearly indicates the "melting" of the relative phase order across a Josephson junction.

Power Spectrum of Superconducting Tunnel Junctions

From the point of view of device applications the noise properties of a tunnel junction are conveniently characterized in terms of the power spectrum of the current–current fluctuations $P(\omega)$:

$$\mathrm{Re}\langle I(t)I(0)\rangle = \int_0^\infty P(\omega)\cos\omega t\, d\omega. \tag{38}$$

The form of $P(\omega)$ depends upon the operating point of the junction, and the reader is referred to ([31]) for a general discussion. Here, following the analysis in Section IV,A we consider the special case in which the junction is operating on the Josephson DC step. Using the linearized form of H and I given in the previous section the power spectrum $P(\omega)$ (neglecting the quasiparticle part) is simply evaluated and we find

$$P(\omega) = A\delta(\omega) + B\frac{\Gamma/\pi}{(\omega - \tilde{\omega}_p)^2 + \Gamma^2}, \tag{39}$$

with

$$\begin{aligned} A &= (I_1\sin\varphi_0)^2(1 - \langle\delta\varphi^2\rangle) \approx \left[I_1\sin\varphi_0\exp\left(-\frac{\langle\delta\varphi^2\rangle}{2}\right)\right]^2 \\ B &= (I_1\cos\varphi_0)^2\langle\delta\varphi^2\rangle. \end{aligned} \tag{40}$$

Here $\langle\delta\varphi^2\rangle$ is given by Eq. (36) and Γ is a width introduced to take into account the junction losses at a frequency ω_p. The presence of the δ-function at zero frequency simply represents the memory of the DC Josephson supercurrent response. The DC Josephson current is, in fact, equal to $\sqrt{A}$ [see Eq. (35)].

This form for $P(\omega)$ was arrived at by linearizing about a given operating point. One of the consequences of this procedure is the delta-function at zero frequency. In order to understand the nature of the power spectrum in the very low frequency region it is necessary to solve the problem in more detail. We will see that it is, in fact, closely related to the metastability of persistent current-carrying states in a loop of superconductor containing a Josephson junction. In rough terms, the width of the "delta-function" is just 1 over the lifetime of the metastable state.

To formulate these ideas, let us consider the simplest case, in which no DC current is flowing, so that $\lambda = 0$. Then, setting $n = (1/i)\,\partial/\partial\varphi$ in Eq. (32) we find that the junction Hamiltonian describes a one-dimensional band problem. The linearization procedure used in Section IV,A corresponded to approximating the periodic potential by a single harmonic well with discrete levels. The higher levels were assumed to broaden because of additional interactions, but the lowest level was stable and gave rise to the δ-function in $P(\omega)$. Now we can see that the solution of the actual problem will have a sequence of bands. The true power spectrum will therefore have the zero-frequency δ-function replaced by a peak with the width appropriate to that of the lowest band.

Before estimating this width it is useful to look at the related problem of metastability of persistent currents. For a loop containing a Josephson junction Goldman *et al.* ([38]) found that the metastable states could be determined by considering the energy form

$$-E_1 \cos\varphi + \tfrac{1}{2}LI^2. \tag{41}$$

Here L is the inductance of the loop and the second term is just the magnetic energy associated with a current I. Now the relative phase φ is related to the current by integrating the vector potential A around the loop as illustrated in Fig. 23. For the thick loop of interest this integration gives a simple, linear relationship between the phase change across the junction and the current:

$$\varphi = \oint \frac{2eA \cdot dl}{\hbar c} = \frac{2e}{\hbar c}\int H \cdot ds = \frac{LI}{\hbar/2e}. \tag{42}$$

This means that the energy is given by $E = -E_1 \cos(LI/\hbar/2e) + \tfrac{1}{2}LI^2$,

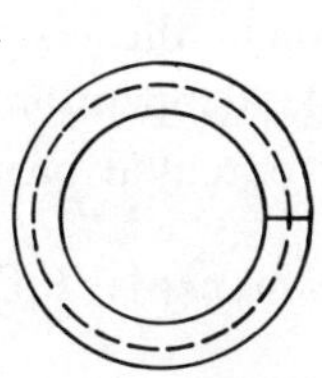

Fig. 23. Idealized geometry of a thick superconducting loop containing a Josephson junction. The dashed line represents the integration path used in Eq. (42).

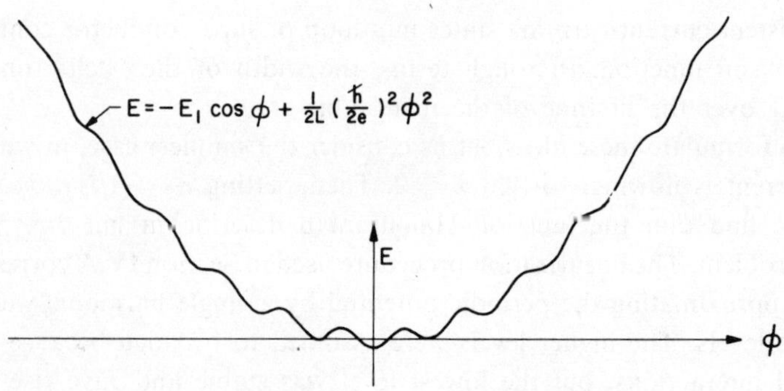

Fig. 24. The φ-space potential $-E_1 \cos\varphi + (1/2L)(\hbar/2e)^2\varphi^2$ appropriate to junctions of the type used in ([38]). The local minima represent the current-carrying metastable states of a loop with a Josephson junction. From Goldman *et al.* ([38]).

which is plotted in Fig. 24. This shows a certain number of metastable current-carrying states. Now replacing $\frac{1}{2}LI^2$ by $(\hbar/2e)^2\varphi^2/2L$ and adding the electrostatic energy $-(2e\,\partial/\partial\varphi)^2/2C$, we have a Hamiltonian for a one-dimensional particle moving in the potential shown in Fig. 24. At zero temperature it can tunnel from one minimum to another. The time associated with this determines the lifetime of the current-carrying metastable states. For a junction with a small inductance the time associated with the decay of the low-lying states is inversely proportional to the bandwidth of the power spectrum $P(\omega)$ we have just discussed.

It seems appropriate to conclude by estimating the lifetime (or bandwidth) from the *tunneling* rate. In the low-temperature region the spontaneous decay lifetime is

$$1/\tau_s \approx \omega_p \exp\{-\pi[E_1C/(2e)^2]^{1/2}\}. \tag{43}$$

Here ω_p represents the rate at which the "phase" coordinate strikes the barrier, and the exponential is just the tunneling probability for a transition which carries the phase through the classically forbidden region of the potential $-E_1 \cos\varphi$. At finite temperature the phase coordinate can be thermally activated over the potential barrier, so that

$$1/\tau \approx \omega_p \exp[-E_1(T)/kT]. \tag{44}$$

Estimating τ_s by taking ω_p equal to 10^{10}, E_1 of order 10^{-17}–10^{-19} J, and C of order 10^{-8} F, one finds

$$1/\tau_s \approx 10^{10} \exp(-10^5).$$

Thus such metastable current-carrying states are *quite* persistent.* However, since E_1 is proportional to $|T|^2$, which varies exponentially on a scale of angstroms, changing the width of the oxide layer by, say 30 Å decreases E_1 to the point where $[E_1C/(2e)^2]^{1/2}$ is of order unity and $1/\tau_s \approx 10^{10}$. In this way one can understand the rapid change between Josephson and non-Josephson tunneling with oxide thickness (or normal junction resistance).

ACKNOWLEDGMENTS

These notes represent a composite of talks I have given at the Symposium on the Physics of Superconducting Devices, University of Virginia, April 1967, the NATO Summer School on Tunneling, Risø, June 1967, and the Gordon Conference on the Dynamics of Quantum Liquids and Solids, Crystal Mountain, July 1967. This work was supported by the Advanced Research Projects Agency, the National Science Foundation, and the Alfred P. Sloan Foundation.

REFERENCES

1. P. W. Anderson, Phys. Rev. 112:1900 (1958).
2. Y. Nambu, Phys. Rev. 117:648 (1960).
3. J. Goldstone, Nuovo Cimento 19:154 (1961).
4. R. Ferrell, Phys. Rev. Letters 14:330 (1964).
5. R. M. Rice, Phys. Rev. 140:A1889 (1965).
6. P. C. Hohenberg, Phys. Rev. 156:396 (1967).
7. R. D. Parks and R. P. Groff, Phys. Rev. Letters 18:342 (1967).
8. T. K. Hunt and J. E. Mercereau, Phys. Rev. Letters 18:551 (1967).
9. R. P. Groff, S. Marcelja, W. E. Masker, and R. D. Parks, Phys. Rev. Letters 19: 1328 (1967).
10. B. D. Josephson, Phys. Letters 1:251 (1962).
11. P. W. Anderson, Lectures on the Many-Body Problem, Ravello 1963, Vol. II, (E. R. Caianello, ed.), Academic Press, New York, 1964, p. 113.
12. B. D. Josephson, Adv. Physics 14:419 (1965); P. W. Anderson, Progress in Low Temperature Physics, Vol. 5, (C. J. Gorter, ed.), North-Holland Publishing Co., Amsterdam; J. E. Mercereau, in: *Superconductivity*, (R. D. Parks, ed.), Marcel Dekker, New York, Chapter 8, 1968.
13. J. R. Schrieffer and J. W. Wilkins, Phys. Rev. Letters 10:17 (1963).

* Because of the inductance, the higher-lying metastable states become exponentially easier to tunnel out of. Furthermore, as the temperature increases and $E_1(T)$ decreases transitions become more probable. Finally, when a current is driven through a junction the bandwidth of the δ-function in $P(\omega)$ is increased.

14. R. A. Ferrell and R. E. Prange, Phys. Rev. Letters 10:479 (1963).
15. L. I. Schiff, Quantum Mechanics, 2nd ed., McGraw-Hill, New York, 1955, p. 156.
16. V. Ambegaokar and A. Baratoff, Phys. Rev. Letters 10:486 (1963); Errata, Phys. Rev. Letters 11:104 (1963).
17. M. D. Fiske, Rev. Mod. Phys. 36:221 (1964).
18. V. E. Riedel, A. Naturforsch. 19A:1634 (1964).
19. D. J. Scalapino and T. M. Wu, Phys. Rev. Letters 17:315 (1966).
20. C. C. Grimes, P. L. Richards, and S. Shapiro, Phys. Rev. Letters 17:431 (1966).
21. W. H. Parker, B. N. Taylor, and D. N. Langenberg, Phys. Rev. Letters 18:287 (1967).
22. C. S. Owen and D. J. Scalapino, Phys. Rev. 164: 538 (1967).
23. J. C. Swihart, J. Appl. Phys. 32:461 (1961).
24. R. E. Eck, D. J. Scalapino, and B. N. Taylor, Phys. Rev. Letters 13:15 (1963).
25. B. D. Josephson, Adv. Physics 14:419 (1965).
26. J. M. Rowell, Phys. Rev. Letters 11:200 (1963).
27. C. S. Owen and D. J. Scalapino, Phys. Rev. 164:538 (1967).
28. A. M. Goldman and P. J. Kreisman, Phys. Rev. 164:544 (1967).
29. R. E. Eck, D. J. Scalapino, and B. N. Taylor, in: Low Temperature Physics IX, Proceedings of the 9th International Conference (J. G. Daunt *et al.*, eds.) Plenum Press, New York, 1965, p. 415.
30. D. D. Coon and M. D. Fiske, Phys. Rev. 138:A744 (1965).
31. Milan D. Fiske, Rev. Mod. Phys. 36:221 (1964).
32. N. R. Werthamer and S. Shapiro, Phys. Rev. 164:523 (1967).
33. P. Lebwohl and M. J. Stephen, Phys. Rev., to be published.
34. Proceedings of the Symposium on the Physics of Superconducting Devices, Univ. of Virginia, 1967, p. G1.
35. W. H. Parker, A. J. Dahm, D. N. Langenberg, and D. J. Scalapino, in: Proceedings of 2nd International Conference on Tunneling, Solid State Commun., Vol. 5, No. 5, p. vii.
36. A. L. Fetter and M. J. Stephen, Phys. Rev., to be published.
37. D. J. Scalapino, in: Proceedings of the Symposium on the Physics of Superconducting Devices, Univ. of Virginia, 1967, p. G1.
38. A. M. Goldman, P. J. Kreisman, and D. J. Scalapino, Phys. Rev. Letters 15:495 (1965).
39. A. J. Dahm, A. Denenstein, T. F. Finnegan, D. N. Langenberg, and D. J. Scalapino, Phys. Rev. Letters 20:859 (1968).
40. D. N. Langenberg, D. J. Scalapino, and B. N. Taylor, Proc. IEEE 54:560 (1966).
41. D. N. Langenberg, D. J. Scalapino, and B. N. Taylor, Sc. Am. Vol. 214, No. 5, p. 30 (1966).

Chapter 33

AC Josephson Tunneling—Experiment

D. N. Langenberg

Laboratoire de Physique
École Normale Supérieure
Paris, France
and Department of Physics
*University of Pennsylvania**
Philadelphia, Pennsylvania

In Chapter 32 of this volume Scalapino discussed some of the consequences of phase coherence for the behavior of superconducting systems in which two or more pieces of superconductor are weakly coupled. In Chapter 31 Mercereau described remarkable experimental manifestations of this phase coherence in situations where the phase is time independent, i.e., DC Josephson effects. In this chapter I will describe some of the experimental effects which are observed in situations where the phase is time dependent, i.e., the AC Josephson effect. Before I begin I would like to note that while this volume is devoted to tunneling, most of the phenomena discussed in the present chapter and in Chapters 31, 32, and 34 have no necessary connection with tunneling. The key ideas are *phase coherence* and *weakly coupled superconductors*. The tunneling process is one way to couple two superconductors weakly, but it is not the only way. Figure 1 shows schematically some of the different types of systems which display Josephson effects. All of them are conventionally called "Josephson junctions." Fig. 1(*a*) shows a junction which consists of two superconducting evaporated thin films separated by an insulating oxide film. It is an example of a general class of junctions in which two superconductors are separated by an insulating layer. This layer is usually an oxide of one of the superconductors and may be intentionally created or may simply be the natural oxide film which exists on the surface of most metals exposed to the atmosphere. In this type of junction the two superconductors really are coupled by a tunneling

* Permanent address.

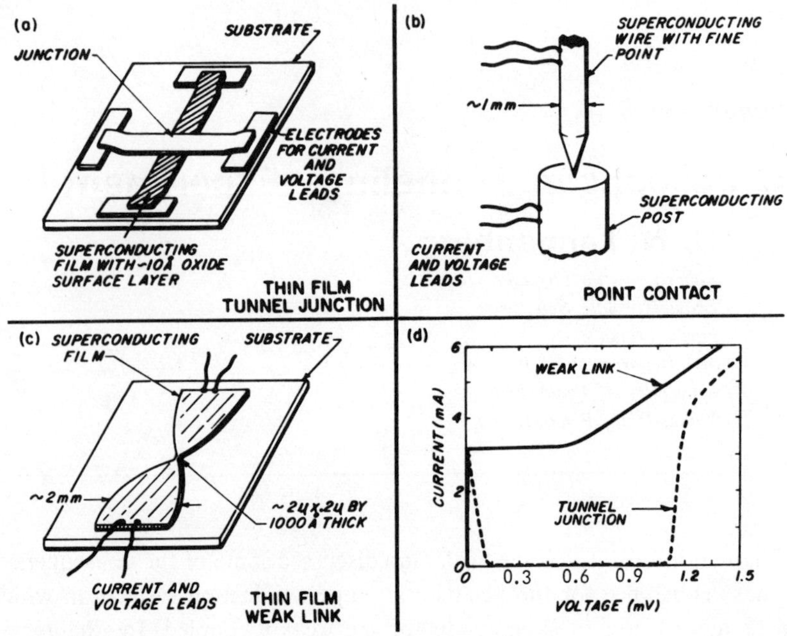

Fig. 1. Some types of Josephson junctions, with typical DC current–voltage characteristics.

process, the tunneling of Cooper pairs discussed by Scalapino. A rather different sort of junction is shown in Fig. 1(*c*). This is essentially two "bulk" pieces of superconductor connected by a very small bridge with dimensions of the order of a micron. It is usually an evaporated thin film with the bridge formed by evaporating through an appropriate mask or by photo-etch techniques. The mechanism by which the two superconductors are coupled via the "weak link" is somewhat obscure, but it is certainly not simply Cooper pair tunneling of the kind which occurs in thin-film tunnel junctions. The point contact junction shown in Fig. 1(*b*) is simply a small area contact between a micron-size sharp point on one piece of superconductor and another piece of superconductor. It is of particular interest because it is extremely simple to fabricate. It can behave like a weak link or like a tunnel junction depending on the contact pressure and the nature of the oxide films on the superconductors at the contact.

Typical current–voltage (*I–V*) characteristics for these types of junction are shown in Fig. 1(*d*). The *I–V* characteristic for a thin-film tunnel Josephson junction displays a zero-voltage current (the DC Josephson current) followed by switching along the circuit load line to the single-particle tunneling curve discussed by Giaever in Chapter 19. If the two superconductors are identical, this consists of a very small current at voltages below the voltage $2\Delta/e$ which corresponds to the superconducting energy gap 2Δ. At the gap voltage there is a sharp rise in the current followed by an asymptotic approach to the linear normal-state I–V curve. The I–V characteristic for a weak link junction displays a zero-voltage current followed by a smooth transition to the normal-state current, as shown. The essential characteristics of the AC Josephson effect observed in all these types of junction are the same.

As Scalapino has demonstrated, the time dependence of the relative pair phase difference φ between the two weakly coupled superconductors in a Josephson junction is given by

$$\partial\varphi/\partial t = 2\Delta\mu/\hbar, \tag{1}$$

where $\Delta\mu$ is the difference between the electrochemical potentials of the two superconductors ([1]). If there is a DC voltage V_0 across the junction, $\Delta\mu = eV_0$, and Eq. (1) then implies that φ varies linearly in time like $\varphi = \omega_0 t + \varphi_0$, where $\omega_0 = 2eV_0/\hbar$ is the (angular) Josephson frequency corresponding to the voltage V_0. The current density in the junction is in general a periodic function of this phase difference. For thin-film tunnel junctions, at least, the current density is to a good approximation given by ([1])

$$j = j_1 \sin\varphi, \tag{2}$$

where j_1 is a parameter which depends on the strength of the junction coupling and on the temperature ([2]). We will treat it as a constant, although actually it is also voltage dependent ([3,4]). Equations (1) and (2) together imply that in the presence of a nonzero DC voltage across the junction the current density oscillates at the Josephson frequency $\nu_0 = \omega_0/2\pi = 2eV_0/h$. What is the order of magnitude of this frequency? In a general way, the characteristic superconducting properties of a junction are restricted to voltages below several times the gap voltage. Typical gap voltages are 1 or 2 mV and the constant $2e/h$ is of order 500 MHz/μV. The frequencies of interest therefore range up to perhaps a THz.

If one has a system of charged particles which has some characteristic frequency, one can usually find out something about the system by probing it with an electromagnetic field at that frequency. Just such an experiment, the first to demonstrate the existence of the oscillating AC Josephson supercurrent, was suggested by Josephson ([1]) and done by Shapiro ([5]). The idea was to expose a Josephson junction to microwave radiation and to observe the effect on the DC *I–V* characteristic. A typical experimental set up for doing such an experiment is shown in Fig. 2. The junction, here a thin-film tunnel junction, is placed in a waveguide which is immersed in liquid helium. An adjustable short-circuit plunger in the waveguide below the junction provides some control over the coupling of the microwave power into the junction. Electrical leads for observing the junction *I–V* characteristic are brought out through the waveguide walls to an amplifying and recording system which plots the DC *I–V* characteristic. Since the Josephson effects are very sensitive to magnetic fields, a magnetic shield is placed outside the dewar system to eliminate the earth's magnetic field and stray magnetic fields generated in the laboratory. Helmholtz coils for producing a controllable magnetic field are placed in the liquid nitrogen Dewar and arranged to produce a uniform magnetic field parallel to the junction plane.

The experiment consists of applying microwave power to the junction via the waveguide and plotting the *I–V* characteristic. Figure 3 shows

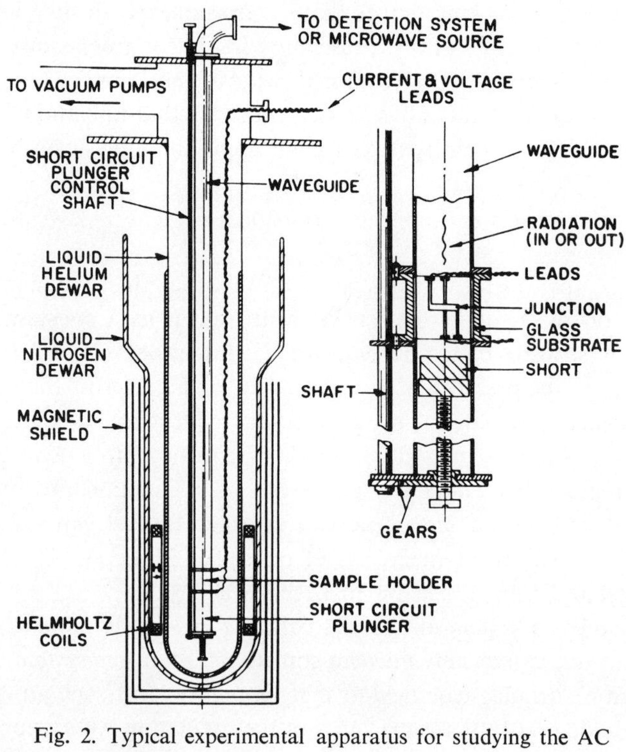

Fig. 2. Typical experimental apparatus for studying the AC Josephson effect.

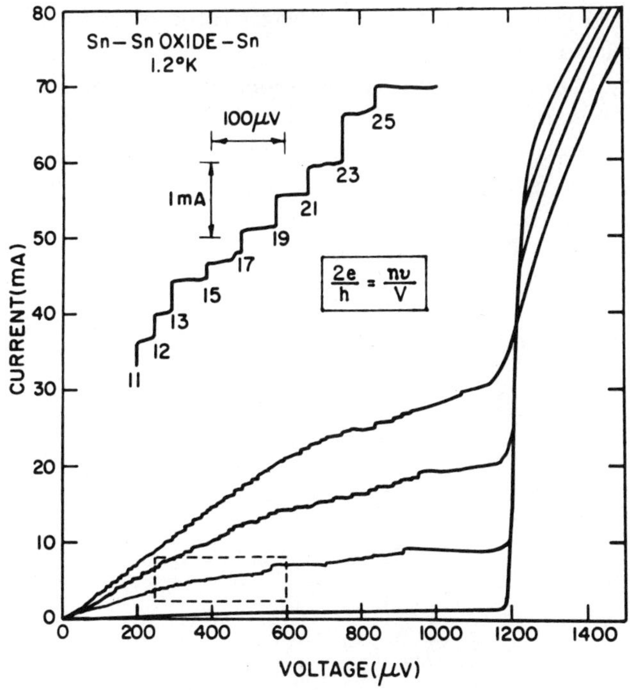

Fig. 3. Microwave-induced steps in the current–voltage characteristic of a thin-film tunnel junction. The curve observed in the absence of microwave power is the lowest one at voltages below the gap voltage (1200 μV) and the highest one above the gap voltage. The DC Josephson current has been suppressed by applying a small magnetic field. With increasing microwave power the general current level increases below the gap voltage and decreases above the gap voltage.

typical experimental results observed on a tin–tin oxide–tin junction. The curves show the characteristic at different levels of applied microwave power. As the microwave power is increased the general level of current increases in the region below the gap voltage and decreases in the region above. This can be understood as a result of rectification of the microwave signal by the nonlinear junction characteristic.* Of more interest here is the appearance of a series of "steps" in the current. A portion of one of the curves is shown on an enlarged scale in the inset. It is found that the

* This was originally suggested by P. W. Anderson in private conversation and has been verified in detail by A. Dahm, A. Denenstein, and W. H. Parker at the University of Pennsylvania.

steps occur at voltages $V_n = nh\nu/2e$, where ν is the applied microwave frequency and n is an integer.

The appearance of these microwave-induced steps can be explained using a simple argument given by Shapiro ([5]). We start with the equation for the time variation of the phase and we suppose that the effect of the microwave power is to add an AC component to the junction voltage, so that the total voltage across the junction is

$$V = V_0 + v_0 \cos(\omega t + \theta). \tag{3}$$

The time-dependent junction phase then has the form

$$\varphi = \omega_0 t + \frac{2ev_0}{\hbar\omega} \sin(\omega t + \theta) + \varphi_0, \tag{4}$$

and the Josephson current density is

$$\begin{aligned} j &= j_1 \sin\left[\omega_0 t + \frac{2ev_0}{\hbar\omega} \sin(\omega t + \theta) + \varphi_0\right] \\ &= j_1 \sum_{s=-\infty}^{\infty} J_s\left(\frac{2ev_0}{\hbar\omega}\right) \sin[(\omega_0 + s\omega)t + \varphi_0 + s\theta], \end{aligned} \tag{5}$$

where J_s is the ordinary Bessel function of order s. We see that the current is phase modulated. If the Josephson frequency ω_0 is equal to an integer multiple $n\omega$ of the microwave frequency, the Josephson current density contains a zero-frequency component given by

$$j_{DC} = j_1(-1)nJ_n\left(\frac{2ev_0}{\hbar\omega}\right) \sin(\varphi_0 - n\theta). \tag{6}$$

The condition $\omega_0 = n\omega$ is just the condition $V_n = nh\nu/2e$ experimentally observed.

The magnitude and sign of j_{DC} depends on the relative phase $\varphi_0 - n\theta$. The variable which is actually under the control of the experimenter is of course j_{DC}, not this relative phase. Equation (6) should therefore be viewed as predicting that the current density on the nth step can be varied continuously over the range between $\pm \mid j_1 J_n(2ev_0/\hbar\omega) \mid$ while the voltage remains constant at $nh\nu/2e$ and the relative phase varies from $-\pi/2$ to $+\pi/2$. The junction can actually display a real, negative DC resistance, with a negative current flowing at positive voltage. (The coupling to the microwave source saves us from awkward confrontations with the laws of thermodynamics. The junction can either absorb or emit radiation, depending on the relative pair-microwave phase). It should be noted that the simple theory we have presented takes no account of spatial variation of the microwave field in the junction. The Bessel function dependence of the step amplitude on the amplitude of the microwave field predicted by this theory is observable in small junctions, but is somewhat washed out in junctions with dimensions comparable to the microwave wavelength in the junction. It remains true, however, that the order n of the steps induced with appreciable amplitude increases with microwave power.

These microwave-induced steps are not restricted to tunnel junctions. They are also observed in weak links and in point contacts. Figure 4 shows

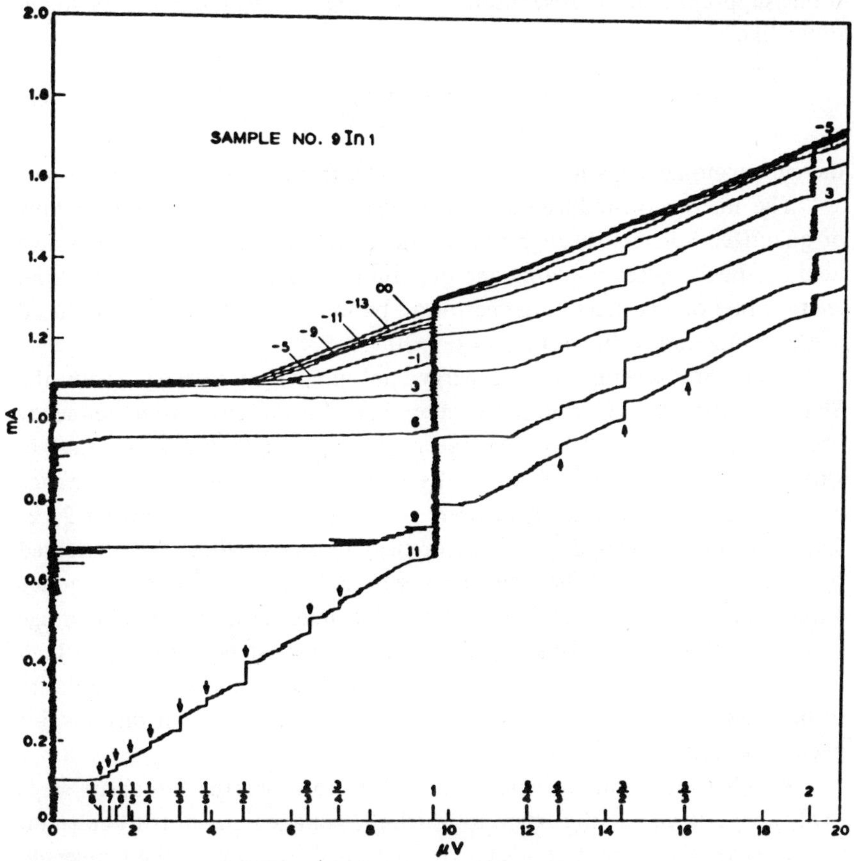

Fig. 4. Microwave-induced steps in the current–voltage characteristic of a weak link junction. The numbers attached to the traces indicate the input power level in dBm. The frequency was about 5 GHz, so that the $n = 1$ step occurred near $10\,\mu$V. From Dayem and Wiegand ([6]).

some results of Dayem and Wiegand ([6]) taken with weak links, in which such steps are clearly apparent. These weak link structures are discussed in detail by Wyatt in Chapter 34, so here we will simply point out the appearance of subharmonics, i.e., steps occurring at voltages corresponding to fractional harmonics of the applied microwave frequency. Such subharmonic steps are commonly seen in weak links and point contacts but seldom if ever seen in tunnel junctions. One can explain this using the simple picture we have presented if one assumes that the relation between the Josephson current and the phase in a weak link or point contact is not simply sinusoidal. [There is other experimental evidence for this ([7]), and the theory gives the sinusoidal relation only in the limit of weak coupling]. If one supposes that the Josephson current is related to the phase by something like

$$j = \sum_{s=1}^{\infty} j_s \sin s\varphi , \tag{7}$$

the subharmonic steps follow quite naturally from the theory given above.

The microwave-induced steps can also be viewed in terms of a flow of quantized vortices through the junction synchronized by the microwave field ([8]). Each vortex contains one quantum of flux, $hc/2e$. A flow with an average rate of n vortices in m microwave periods corresponds to an induced average junction voltage $V_{n/m} = (n/m)(h\nu/2e)$.

There are several other experimental points of interest about the Shapiro steps. At DC voltages greater than Δ/e the Josephson frequency becomes greater than $2\Delta/h$, and one might expect pair breaking to set in with some attendant modification of the steps. In fact, no marked change is observed. Steps can often be seen at voltages up to and beyond $2\Delta/e$. This is probably related to the theoretical results of Reidel ([3]) discussed by Scalapino. These predict that the Josephson current amplitude contains a logarithmic singularity at $2\Delta/e$ and remains finite for a considerable range of voltage above $2\Delta/e$. This would explain the persistence of steps to high voltages, but it raises another question. One might expect the singularity to be reflected in an increase of step amplitude near $2\Delta/e$, but this has not been observed.

Another interesting question is the extent to which the Josephson steps really are vertical, i.e., the extent to which the voltage remains truly constant as one varies the current on a step. We have had occasion at the University of Pennsylvania to look at this rather closely and have found that in most cases the steps are remarkably vertical. In a typical tunnel junction with a normal-state resistance of a few milliohms the voltage is constant to within a fraction of a nanovolt over the full range of current on the step—the voltage is constant to better than one part per million. However, as is often the case, if you look at a wall hard enough, you begin to see some cracks. Tunnel junctions and (even more so) point contacts with normal-state junction resistances on the order of 100 milliohms or more seem to have step widths of the order of 5 to 10 nV. The step widths seem to be independent of the order of the step for a given junction and increase with the normal-state junction resistance, occasionally reaching several hundred nanovolts in point contacts with normal-state resistances of the order of an ohm. In all cases the mid-current point on the step is the point which corresponds precisely to $nh\nu/2e$. This finite width is probably related to fluctuations of the relative pair phase, but the connection is not yet clear.

There is another amusing variation of experiments of this sort. If one applies microwaves to a junction, leaving the junction essentially open circuited, the measured DC voltage across it exhibits a variety of structure which is a function of microwave power and applied magnetic field. One of the things that can happen in this situation is that the junction can sit at one of zero crossing points of the microwave-induced steps; the junction can act as a sort of quantum rectifier for which the output voltage is $V_n = nh\nu/2e$, where ν is the microwave frequency and n is an integer. If the space variation of the relative pair phase in the junction is ignored, the pair phase can be shown to satisfy the equation of motion ([9])

$$\frac{\partial^2\varphi}{\partial t^2} + \omega_J{}^2 \sin\varphi = \frac{I_0}{I_1}\,\omega_J{}^2 \sin\omega t, \tag{8}$$

where $\omega_J{}^2 = 2eI_1/\hbar C$ is the square of the Josephson plasma frequency ([1]) (C is the junction capacitance), I_1 is the Josephson current density parameter j_1 multiplied by the junction area, and I_0 is the amplitude of the microwave driving current. This is the equation of motion of a simple pendulum driven by a periodic torque. The quantum voltage V_n corresponds to rotation of the pendulum about its pivot with average frequency $n\nu$. Surprisingly, no detailed treatment of the nature and stability of the solutions of this equation seems to exist. However, simple physical arguments based on the pendulum model suggest that stable solutions corresponding to the observed quantum voltages do exist.

One can also find interesting magnetic field dependences of the junction voltage in this situation. Figure 5 shows an example. The sharp spikes represent momentary switching to quantized voltages as the magnetic field is swept. Note the remarkably precise antisymmetry of the curve.

We come now to experiments in which the presence of the Josephson

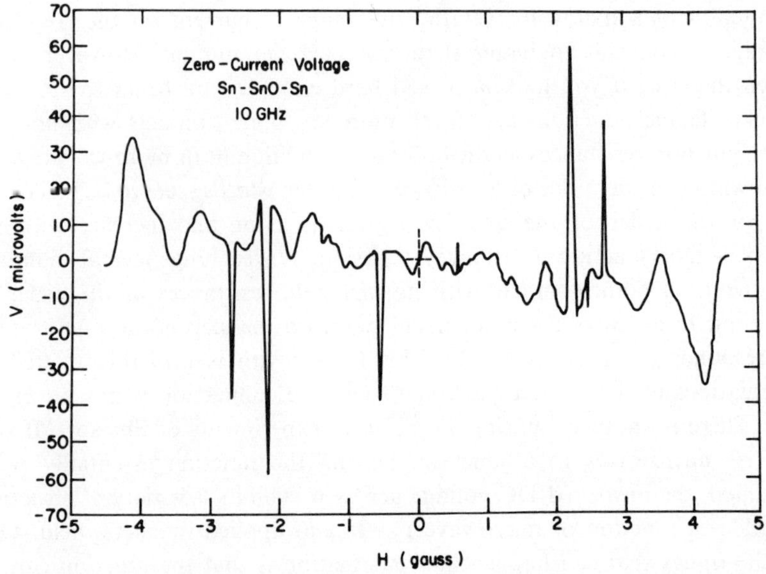

Fig. 5. Magnetic field dependence of the DC open-circuit voltage across a thin-film tunnel junction in the presence of applied microwave power. The sharp spikes represent momentary switching to the quantum voltages discussed in the text. From unpublished work of A. Dahm, A. Denenstein, and W. H. Parker.

current is inferred from structure which appears in the I–V characteristic of the junction even in the absence of externally applied microwave power. We have already seen that in the presence of a DC voltage bias on the junction the junction phase varies linearly in time. In the presence of a uniform magnetic field $\mathbf{H}_0$ in the plane of the junction, the Josephson phase also has a space dependence determined by the equation

$$\nabla\varphi = (2ed/\hbar c)\mathbf{H}_0 \times \hat{\mathbf{n}}, \tag{9}$$

where $d = l + 2\lambda$, l is the thickness of the junction oxide layer, λ is the superconducting penetration depth, and $\hat{\mathbf{n}}$ is a unit vector normal to the junction plane. As Scalapino showed (Chapter 32), this equation is a consequence of the relation between the Josephson phase and the magnetic vector potential and is intimately connected with the quantization of flux. If a DC bias voltage and a magnetic field are applied together, the Josephson current density is given by

$$j = j_1 \sin(\omega_0 t - kz + \varphi_0), \tag{10}$$

where the wave vector $k = 2edH_0/\hbar c$. There is thus a traveling current density wave with phase velocity

$$\frac{\omega_0}{k} = \left(\frac{V_0}{dH_0}\right)c. \tag{11}$$

Now a Josephson tunnel junction consists of a pair of conducting parallel plates separated by an insulating oxide layer, and such a structure can of course support propagating electromagnetic waves. In the particular case where the parallel plates are superconducting but the Josephson current is neglected, the propagation characteristics have been worked out by Swihart ([10]). It turns out that the phase velocity for such waves is given by

$$\bar{c} = \left(\frac{l}{\varepsilon d}\right)^{1/2} c, \tag{12}$$

where ε is the dielectric constant of the oxide. It is thus reasonable to suppose that if we make the phase velocity of the Josephson current density wave equal to the phase velocity for electromagnetic wave propagation in the junction, there will be a strong coupling between the two. We would expect that the Josephson current would then be able to generate a relatively large electromagnetic field in the junction and that this electromagnetic field could then react back on the Josephson current in much the same way as did the applied microwave radiation in the Shapiro experiment, causing some sort of singularity or anomaly in the I–V characteristic. Figure 6 shows experimental data of Eck *et al.* ([11]) which show this effect. What appears is a resonance-like "bump" in the I–V characteristic. The bump appears at a voltage which is linearly proportional to the applied field, just as Eqs. (11) and (12) predict.

The Josephson tunnel junction, however, is not an infinite transmission line, but an open-circuited line of finite length. It should therefore have a series of resonant electromagnetic modes at frequencies determined by equating the length of the junction to integer multiples of a half wavelength. From the formulas we have given we see that in a junction with dimensions of the order of 1 mm the modes occur at multiples of about 10 GHz. When the Josephson frequency coincides with the frequency of one of these modes the Josephson current can drive the mode resonantly, and structure should appear in the I–V characteristic. Structure of this type was first reported by Fiske ([12]); an experimental curve showing "Fiske steps" is shown in Fig. 7. The detailed theory of this effect ([13]) has been discussed by Scalapino in Chapter 32. The theory predicts a series of peaks in the dc current at voltages for which the Josephson frequency matches a junction mode

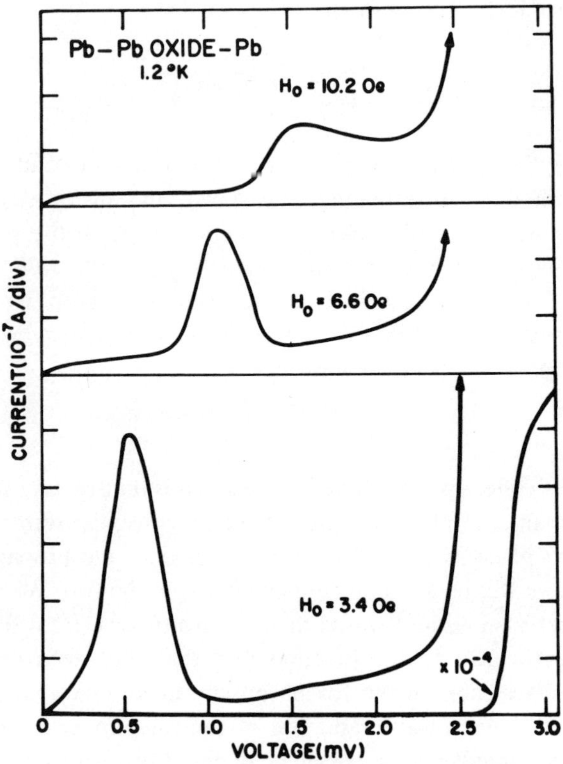

Fig. 6. Resonance bump observed in the *I–V* characteristic of a thin-film tunnel junction in the presence of an applied magnetic field. From Eck *et al.* ([11]).

frequency. What is ordinarily observed experimentally is not resonances but steps. The reason for this is that the impedance of the Josephson junction is very small and the normal experimental bias conditions correspond to constant current bias. One therefore observes only portions of these resonance curves. The steps have a finite width corresponding to the finite Q of the electromagnetic modes. Experimentally, this Q typically lies between ten and several hundred. The role of the external magnetic field in this case is to control the coupling of the Josephson current to the electromagnetic modes. If the external field is adjusted so that the space variation of the Josephson current coincides with the space variation of the electromagnetic fields in one of the resonant modes, the coupling to that mode is optimized.

It should be noted that the junction modes simply provide a resonant feedback path by which the electromagnetic fields generated by the Joseph-

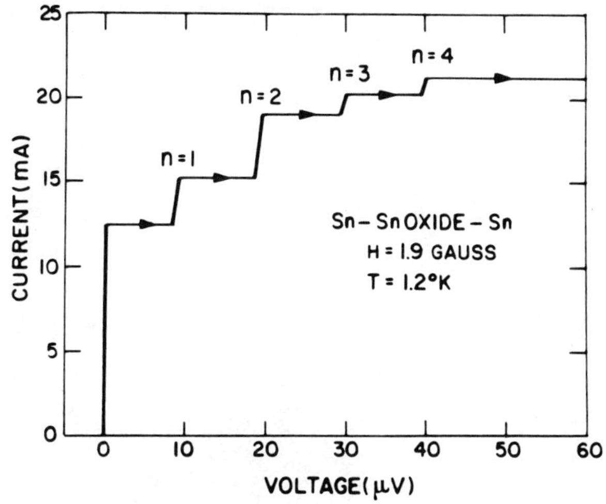

Fig. 7. Self-induced steps in the I–V characteristic of a thin-film tunnel junction. The steps occur at voltages corresponding to the frequencies of the resonant electromagnetic modes of the junction.

son current can react back on the current. The resonator need not be the junction itself. It can be an external resonator. Figure 8 shows an I–V curve for a point-contact Josephson junction coupled to a conventional coaxial microwave resonator ([14]). The point contact has no resonant modes itself, but its I–V curve shows structure at voltages corresponding to the three lowest-order modes of the external coaxial resonator.

After Josephson's original prediction of the AC Josephson effect there were repeated but unsuccessful efforts by several workers to detect externally the radiation emitted by the oscillating Josephson supercurrent. The Fiske modes are the key to efficient excitation of radiation inside a tunnel junction. A practical problem arises in the coupling of the radiation out of the junction. If one supposes that the radiation is emitted at the open ends of the junction, one can estimate the power emitted by considering the characteristic impedance for electromagnetic wave propagation in the junction. This impedance is

$$Z_J = \frac{c}{\bar{c}}\left(\frac{l}{\varepsilon w}\right)Z_0, \tag{13}$$

where w is the junction width and Z_0 is the impedance of free space. If we imagine that this impedance is coupled discontinuously at the edge of the junction to an exterior impedance Z_G which is of the order of the impedance

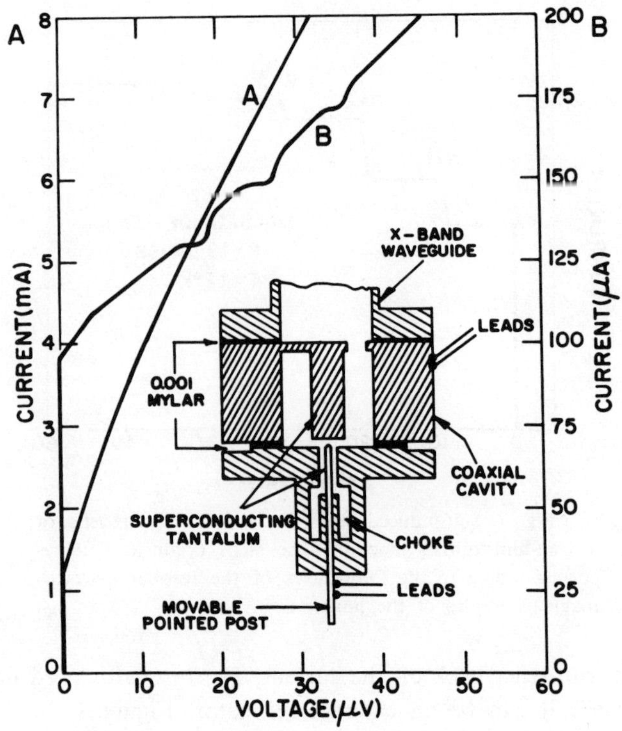

Fig. 8. Self-induced steps in the I–V characteristic of a point contact. The inset shows how the point contact is coupled to an external coaxial resonator. Curve B is the I–V characteristic for relatively light contact pressure and curve A is the characteristic for heavier contact pressure; note the different current scales for the two curves. Curve B shows steps at voltages corresponding to the frequencies of the three lowest modes of the resonator. From Dayem and Grimes ([14]).

of free space, the power transmission coefficient through this discontinuity is given by transmission-line theory as

$$T = \frac{4Z_G Z_J}{(Z_G + Z_J)^2} \approx 4\,\frac{Z_G}{Z_0}. \tag{14}$$

Taking typical values for all the parameters involved, one finds a transmission coefficient of the order of 10^{-5}. This corresponds to a junction Q of 10^5 against external radiation. If all of the DC power supplied to the junction is converted to electromagnetic radiation inside the junction, one will typically have a power generated inside the junction of the order of 10^{-7} W.

If the total junction Q is 10–100, we would expect to have only 10^{-3}–10^{-4} of this power radiated, i.e., 10^{-11}–10^{-10} W. There is some evidence that this picture of the manner in which radiation is coupled into and out of the junction is oversimplified, and that significant coupling can occur directly through the superconducting films. Nevertheless, the power radiated is expected to be rather small, and successful external detection requires a proper understanding of the role of the Fiske modes together with sensitive microwave superheterodyne detection techniques.

The need for the latter was circumvented in an ingenious experiment by Giaever ([15]) using the double junction structure shown in Fig. 9. Radiation generated in one junction propagated in the shielded oxide layer (or perhaps through the superconducting film common to the two junctions) into the second junction. There it was detected by observing structure in the I–V curve due to photon-assisted normal particle tunneling of the type first observed by Dayem and Martin ([16]).

External detection of the radiation was first reported by Yanson *et al.* ([17]) and shortly thereafter by Langenberg *et al.* ([18,19]). Figure 10 shows an example of the sort of signal one sees by connecting a superheterodyne receiver to the waveguide of Fig. 2 and biasing the thin-film tunnel junction on an appropriate Fiske mode. Under good conditions one can see of the order of 10^{-11} W (we have observed as much as 10^{-10} W radiated by a

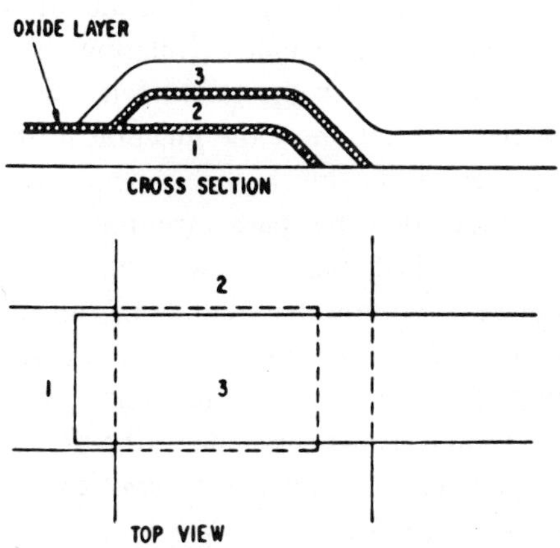

Fig. 9. The double thin-film tunnel junction device used by Giaever to detect radiation emitted by the Josephson current. The junction between films 2 and 3 was the generator and the junction between films 1 and 2 served as a detector.

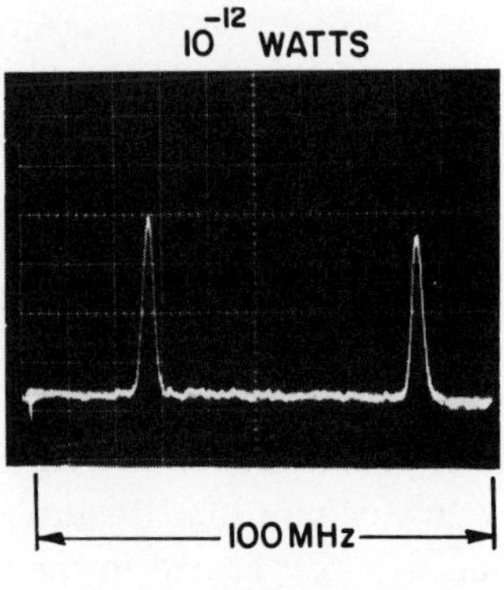

Fig. 10. Josephson radiation from a thin-film tunnel junction as detected by a conventional microwave superheterodyne receiver at 10 GHz. The local oscillator frequency was swept over a range of 100 MHz and the junction was biased at constant DC voltage. The output signal peaks occurred when the frequency difference between the Josephson and local oscillator frequencies passed through the pass band of the 30 MHz i.f. amplifier.

thin-film tunnel junction at 70 GHz). This is about what we estimated earlier from our impedance mismatch calculation and suggests that the picture we used there may have some truth in it. On the other hand, we have observed that the orientation of the junction in the waveguide really doesn't affect the power output much. The junction orientation shown in Fig. 2 is based on the idea that the space variation of the fields in the junction is normal to the applied magnetic field and in the direction of the waveguide axis, and that the power is radiated from the upper and lower ends of the junction. In fact, this junction axis can be oriented in either of the other two mutually perpendicular directions and one still sees roughly the same amount of power. This is just one of the pieces of evidence which indicate that the coupling of radiation in and out of a junction is a complicated and ill-understood business.

Point contacts and weak links should of course radiate too. Radiation from point contacts has been reported by Dayem and Grimes ([14]) and by Zimmerman *et al.* ([20]). In these experiments the resonant feedback was provided by coupling the point contact to an external resonant circuit (Fig. 8 shows the method used by Dayem and Grimes). Much higher efficiencies were found, presumably because the open structure of a point contact makes it easier to couple power out. The net output power was of the same order as observed in thin-film tunnel junction experiments because the input powers are typically lower for point contacts than for tunnel junctions.

According to the simple theory the radiation emitted by a junction biased at constant voltage is monochromatic and has a frequency determined solely by the bias voltage. A number of interesting deviations from this behavior are observed. Suppose, first of all, that one chooses the dimensions of a thin-film tunnel junction so that the frequencies of the Fiske modes are multiples of 5 GHz (they then occur approximately at multiples of 10 μV). If one then looks at the junction with a 10-GHz receiver and biases the junction on the second Fiske mode, one observes 10-GHz radiation, as expected. However, 10-GHz radiation is *also* usually observed when the junction is biased on any of the other modes, i.e., the output spectrum is very rich in harmonics and subharmonics of the fundamental Josephson frequency. Similar behavior has been reported for point contacts. ([14]). This is not really surprising if one remembers how nonlinear a Josephson junction really is, even though one tends to think in terms of linearized small-signal models of its behavior. It is rewarding to estimate the magnitude of the rf field in a thin-film tunnel junction. We can write the stored energy in a Fiske mode as

$$(1/8\pi)E_{\mathrm{rf}}^2 V = PQ/\omega\,, \tag{15}$$

where E_{rf} is the amplitude of the rf electric field, V is the volume of the junction oxide layer, P is the power dissipated in the junction, Q is the junction Q, and ω is the mode frequency. Taking $P = 10^{-7}\ W = 1$ erg-sec^{-1}, $Q = 10^2$, $\omega = 10^{11}$ sec^{-1}, and $V = 10^{-9}$ cm^3, one finds $E_{\mathrm{rf}} \approx 3$ statvolt-cm$^{-1} \approx 10^3$ V-cm^{-1}. This corresponds to about 100 μV of rf voltage across the junction, a value comparable to typical DC bias voltages. It is not surprising then that the junction behaves in a highly nonlinear manner. A complete theory of this nonlinear behavior has not been (and may never be) given, but some aspects have been treated by Langenberg *et al.* ([18]) and by Werthamer ([4]).

Still more complicated radiation effects have been observed in thin-film tunnel junctions ([21]). If such a junction is biased on one of the low-order Fiske steps, up to, say, $n = 5$, the frequency of the observed radiation is accurately (within one part in 10^5) a harmonic or subharmonic of the fundamental Josephson frequency which corresponds to the bias voltage.

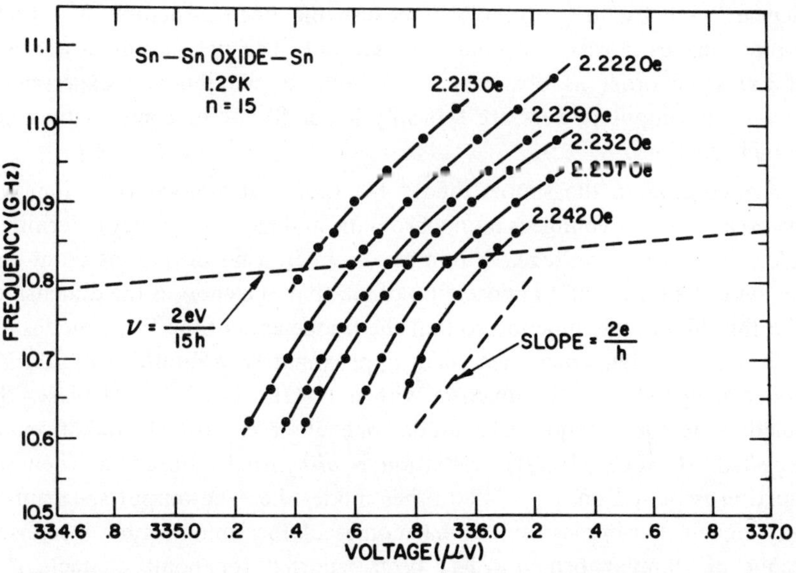

Fig. 11. The voltage and magnetic field dependence of the frequency of non-Josephson radiation from a thin-film tunnel junction. The junction was biased on the 15th step at a voltage corresponding to a Josephson frequency of about 150 GHz. The component of the emitted radiation with a frequency near 10 GHz was detected. The frequency of this component is shown as a function of bias voltage and applied magnetic field. The dotted line shows the voltage dependence expected for the 15th subharmonic of the Josephson frequency; this frequency should be field-independent.

If the junction is biased on higher order steps, this is no longer true. The observed frequency can depend in a complicated way on bias voltage and applied magnetic field. Figure 11 shows an example. The origin of this "non-Josephson" radiation is not understood. It suggests the appearance of another form of oscillation, perhaps some sort of parametric excitation of the junction, accompanied by nonlinear mixing effects.

All this indicates a rather complex frequency spectrum for the Josephson radiation. What about the linewidth of an individual component of the spectrum? The simple theory we have discussed is concerned with average values of the superconducting pair phase and says nothing about possible fluctuations in the phase. Scalapino discussed this problem in Chapter 32. From the experimentalist's point of view it is sufficient to note that no absorption or emission line is ever perfectly sharp and that considerable information about a system can usually be obtained by investigating the mechanisms which contribute to a nonzero linewidth. Some recent work on this question has been performed by Parker *et al.* ([28]). The purpose of these experiments is to see if the linewidth can be made to give fundamental information about fluctuations of the pair phase. The experimental situation is still somewhat obscure, but the following features are emerging: Linewidths as small as 2 kHz are observed at 10 GHz in thin-film tunnel junctions at temperatures well below the superconducting transition temperature T_c. This is a fractional linewidth of two parts in 10^7. The linewidth increases with the junction dynamic resistance on the Fiske step, and increases rapidly as T_c is approached. Such experiments are a bit tricky, as can be seen by observing that a 2-kH_z linewidth can be produced by a fluctuation of the junction bias voltage of a few picovolts!

With such a variety of nonlinear phenomena available in the radio frequency, microwave, and submillimeter regions it is not surprising that Josephson junctions are attracting attention in device applications. We have discussed some of their properties as radiation sources and detectors. Silver and Zimmerman ([22]) have demonstrated that a point-contact junction can be used as an active detector in an rf spectrometer and have used such a spectrometer to detect nuclear resonance. Grimes *et al.* ([23]) have shown that point contacts are good infrared detectors. Figure 12 is taken from their paper and shows the response of a Nb–Nb point contact as a function of frequency. Some of the prominent structure may be related to the Riedel current singularity. Shapiro ([24]) has studied microwave harmonic

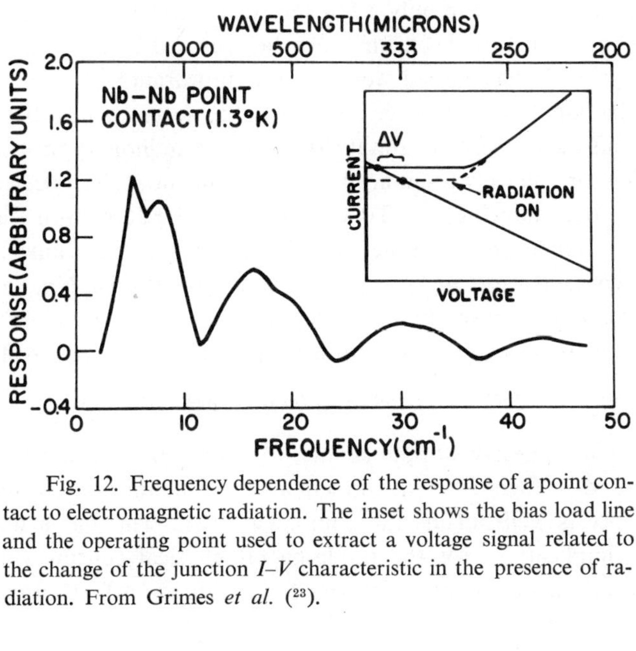

Fig. 12. Frequency dependence of the response of a point contact to electromagnetic radiation. The inset shows the bias load line and the operating point used to extract a voltage signal related to the change of the junction *I–V* characteristic in the presence of radiation. From Grimes *et al.* ([23]).

generation in Josephson junctions. Further results along these lines will undoubtedly appear in the future.

I would like to close with a brief discussion of recent experiments ([25,26]) at the University of Pennsylvania which center around the quantitative accuracy of the Josephson frequency–voltage relation $\nu_0 = 2eV_0/h$. The apparent simplicity and generality of this relation make it a tempting target for the experimentalist for two reasons. The first is that it is alleged to describe a fundamental property of a rather complicated interacting many-body system, the Josephson junction. One might well ask whether it is really the whole story or whether sufficiently precise probing might disclose discrepancies necessitating interesting corrections. The second reason is that it provides a means of measuring the fundamental physical constant e/h if the frequency–voltage relation really is exactly correct.

The experimental problem is to measure accurately a frequency and a voltage. This might be done either by irradiating a junction with microwaves with accurately known frequency and measuring the voltages at which Shapiro steps occur, or by biasing a junction with an accurately known voltage and measuring the frequency of the emitted radiation. A microwave frequency can be measured to perhaps one part in 10^8 with simple conventional techniques. The voltage measurement is less easy. The maximum voltage available is limited to something like superconducting energy gap voltages, a few millivolts at most. The voltages corresponding to convenient microwave frequencies are only a few tens of microvolts, but the nonlinear harmonic effects in Josephson junctions can be used to get voltages in the millivolt range. It has turned out that the appearance of non-Josephson radiation has limited the voltage and hence the accuracy possible with the radiation-out experiment to about 10 parts per million (ppm). With the radiation-in or microwave-induced-step experiment it has been possible to establish the following: (1) To within 2 ppm the Josephson frequency–voltage ratio is independent of the type of junction (thin-film tunnel junctions and point contacts were used), superconductor (Sn, Pb, Ta, Nb, and Nb_3Sn were tried), temperature, induced step number, microwave power, and applied magnetic field. (2) If the ratio is taken to be $2e/h$, the results yield

$$2e/h = 483.5916 \pm 0.0026 \text{ MHz}/\mu\text{V}.$$

This value is nearly 40 ppm smaller than the value obtained by Cohen and DuMond in their 1963 adjustment of the fundamental constants ([27]); the difference is well outside the combined errors. This has a number of important implications for the fundamental physical constants and even for quantum electrodynamics [cf. ([26]) for a discussion]. A discussion of these implications would take me too far afield; let me simply say that here is further evidence that underlying the apparent complexity of the Josephson junction is a general and fundamental property of the superconducting state, long-range phase coherence. Despite the richness of phenomena it displays, the Josephson junction is basically a very simple two-body quantum mechanical system, indeed, perhaps simpler than the hydrogen atom.

REFERENCES

1. B. D. Josephson, Phys. Letters 1:251 (1962); Rev. Mod. Phys. 36:216 (1964); Adv. Phys. 14:419 (1965).
2. V. Ambegaokar and A. Baratoff, Phys. Rev. Letters 10:486 (1963); Erratum, Phys. Rev. Letters 11:104 (1963).
3. E. Riedel, Z. Naturforsch. 19A:1634 (1964).
4. N. R. Werthamer, Phys. Rev. 147:255 (1966).
5. S. Shapiro, Phys. Rev. Letters 11:80 (1963); S. Shapiro, A. R. Janus, and S. Holly, Rev. Mod. Phys. 36:223 (1964).
6. A. H. Dayem and J. J. Wiegand, Phys. Rev. 155:419 (1967).
7. J. E. Zimmerman and A. H. Silver, Solid State Commun. 4:133 (1966).
8. P. W. Anderson and A. H. Dayem, Phys. Rev. Letters 13:195 (1964); P. W. Anderson, Rev. Mod. Phys. 38:298 (1966).
9. D. N. Langenberg, D. J. Scalapino, B. N. Taylor, and R. E. Eck, Phys. Letters 20: 563 (1966).
10. J. C. Swihart, J. Appl. Phys. 32:461 (1961).
11. R. E. Eck, D. J. Scalapino, and B. N. Taylor, Phys. Rev. Letters 13:15 (1964).
12. M. D. Fiske, Rev. Mod. Phys. 36:221 (1964).
13. R. E. Eck, D. J. Scalapino, and B. N. Taylor, in: Low Temperature Physics IX, Proceedings of the Ninth International Conference (J. G. Daunt *et al.*, eds.), Plenum Press, New York, 1965, p. 415.
14. A. H. Dayem and C. C. Grimes, Appl. Phys. Letters 9:47 (1966).
15. I. Giaever, Phys. Rev. Letters 14:904 (1965).
16. A. H. Dayem and R. J. Martin, Phys. Rev. Letters 8:246 (1962).
17. I. K. Yanson, V. M. Svistunov, and I. M. Dmitrenko, Zh. Eksperim. i Teor. Fiz. 48:976 (1965) [English translation: Soviet Physics—JETP 21:650 (1965)].
18. D. N. Langenberg, D. J. Scalapino, B. N. Taylor, and R. E. Eck, Phys. Rev. Letters 15:294 (1965); Erratum, Phys. Rev. Letters 15:842 (1965).
19. D. N. Langenberg, D. J. Scalapino, and B. N. Taylor, Proc. IEEE 54:560 (1966).
20. J. E. Zimmerman, J. A. Cowan, and A. H. Silver, Appl. Phys. Letters 9:353 (1966).
21. D. N. Langenberg, W. H. Parker, and B. N. Taylor, Phys. Letters 22:259 (1966).
22. A. H. Silver and J. E. Zimmerman, Appl. Phys. Letters 10:142 (1967).
23. C. C. Grimes, P. L. Richards, and S. Shapiro, Phys. Rev. Letters 17:431 (1966).
24. S. Shapiro, J. Appl. Phys. 38:1879 (1967).
25. D. N. Langenberg, W. H. Parker, and B. N. Taylor, Phys. Rev. 150:186 (1966).
26. W. H. Parker, B. N. Taylor, and D. N. Langenberg, Phys. Rev. Letters 18:287 (1967).
27. E. R. Cohen and J. W. M. DuMond, Rev. Mod Phys. 37:537 (1965).
28. W. H. Parker, A. J. Dahm, D. N. Langenberg, and D. J. Scalapino, Solid State Commun. 5:7 (1967).

Chapter 34

Weakly Coupled Superconductors

A. F. G. Wyatt

University of Nottingham
Nottingham, England

Weakly coupled superconductors exhibit the AC Josephson effect ([1]), and in fact this might be taken as a rough definition of a superconducting weak link, although they also have several other interesting properties. Josephson junctions are the most well understood form of weak link and their properties have been discussed in previous chapters. Here we shall discuss other forms of weak coupling and indicate their similarities as well as the properties which are peculiar to them individually.

There are some general requirements for all weak links between two bulk superconductors. First, the link must be superconducting, so that the bulk regions remain correlated with each other. A measure of the correlation is the phase of the wave equation which describes the superconducting state. Mercereau in Chapter 31 has shown how the phase changes under the influence of electric and magnetic fields.

The rate of change of the phase between two points 1 and 2 depends linearly on the voltage between them:

$$\partial \Delta\phi_{12}/\partial t = 2eV_{12}/\hbar. \tag{1}$$

in addition, the phase difference between two points depends on the line integral of the canonical momentum between them

$$\Delta\phi_{12} = \frac{2}{\hbar}\int_1^2 \left(mv - \frac{eA}{c}\right)\cdot dl, \tag{2}$$

and if the velocity (v) of the supercurrent is small then this depends only on the vector potential A.

If we can get these potentials into a superconductor, we can change the phase differences and then observe the effect of this on, for example, its current-carrying capacity.

The second requirement is concerned with the application of magnetic and electric fields in such a way that the superconductivity is preserved. For a magnetic field to penetrate appreciably into the link, then either the dimensions of the link must be small or the penetration depth λ must be large. As λ is inversely proportional to the square root of the superfluid density, it is clear why it is advantageous for the link to be only weakly superconducting. If the link is strong, there is also difficulty in trying to apply an electric field, because usually we should exceed the critical current before we get a voltage. The link must have a dissipative mechanism in the superconducting state if we are to have an electric field and a current flowing simultaneously. Energy dissipation can be by one or a combination of different processes, such as normal current flow, radiation, and creation and motion of flux lines.

The third requirement is that the phase difference must be reasonably well localized in the superconductor for the properties to be observed. This point can be exemplified by considering why a strip of type-II material held in a magnetic field between H_{c1} and H_{c2} would not show the AC Josephson effect, although we can get magnetic and electric fields in it while it is superconducting. In the mixed state dissipation is by the flow of vortices predominantly in a direction normal to the current flow. When a vortex flows between two points the phase changes by 2π because this is the change in phase in going once around a vortex. However, the motion of vortices are not exactly correlated with each other and the phase changes add incoherently from different parts of the strip. The radiation therefore has a noise spectrum. This has been observed by Van Gurp ([2]). If the total phase change is well localized, then the effects will be coherent, and consequently we get a line spectrum.

Weak links which satisfy these requirements fall into two groups; those which are large but very weakly superconducting and those which have a small physical size. The first type rely on the proximity effect to make the link, which would usually be normal, superconducting. The Josephson junction is the well-known example of this group, where the thin oxide tunnel barrier is weakly superconducting. The oxide thickness is of the order of 10 Å. A normal metal can also be sandwiched between two superconductors ([3]) but because a metal has a much stronger proximity effect the thickness has to be of the order of 10^4 Å for it to be sufficiently weak.

The second type of weak links are the constricted thin films, which are of the same material throughout but with the width of the film constricted to a few microns at one point ([4]). A weak link can also be made by pressing a sharply pointed superconductor onto another flat piece [5]. It is not clear whether the link is through an oxide or a metal bridge or a combination of both. Point contacts do have a number of practical advantages in that they are robust and strongly coupled to radiation; however, they are rather unstable, since they are so sensitive to pressure changes.

Much of the work on weakly coupled superconductors has been concentrated on Josephson junctions. Thin film constrictions are much less well understood and some of the ideas that will now be discussed may have to be revised in the light of further experiments.

One aspect of constrictions is clear, and that is their physical construction. The material is continuous and uniform and the only dimension that varies is the width, and that even at the constriction is greater than ξ_0 and λ_0.

The constriction conducts a DC supercurrent, but this is not surprising —as it was for the tunnel junction. The constriction has not shown a conductance which varies periodically with applied field. This also is not surprising, as λ is small compared with the width of the constriction ($\sim 3\ \mu$), so we would expect the critical field to be the same at the constriction as at other parts of the film. Once a sufficiently large field has been applied, to stop the supercurrent, the film is normal.

It is the AC effects which first demonstrated the similarity between constrictions and Josephson junctions. This was the experiment of Anderson and Dayem [4]. They saw current steps in the I–V characteristic when the constriction was in a microwave field. The steps were at voltages $V = n\hbar\omega/m2e$ where m and n are integers; the steps at $m = 1$ were much larger than the steps at submultiples of the microwave frequency, i.e., $m > 1$.

We repeated this experiment and found that observing the steps is not as easy as it might seem because they only exist over a small range of microwave power and this varies with temperature, as shown in Fig. 1.

At temperatures $T < T_c - 0.5°\mathrm{K}$ we do not see any steps at any microwave power, most likely because of induced AC supercurrents and the heating effect of the microwaves. These effects are appreciable and can change the I–V characteristic and in the extreme case drive the sample normal. In the Josephson junction the steps are seen at all temperatures. The constriction steps are seen on a resistive background, and this raises the question of what dissipation mechanisms are involved. Apparently, radiation losses can be ruled out, as no one has detected radiation from a constriction and it should be reasonably well coupled to a waveguide or cavity. This is in contrast to a Josephson junction, which is not too well coupled to a waveguide, yet gives radiation which has been detected.

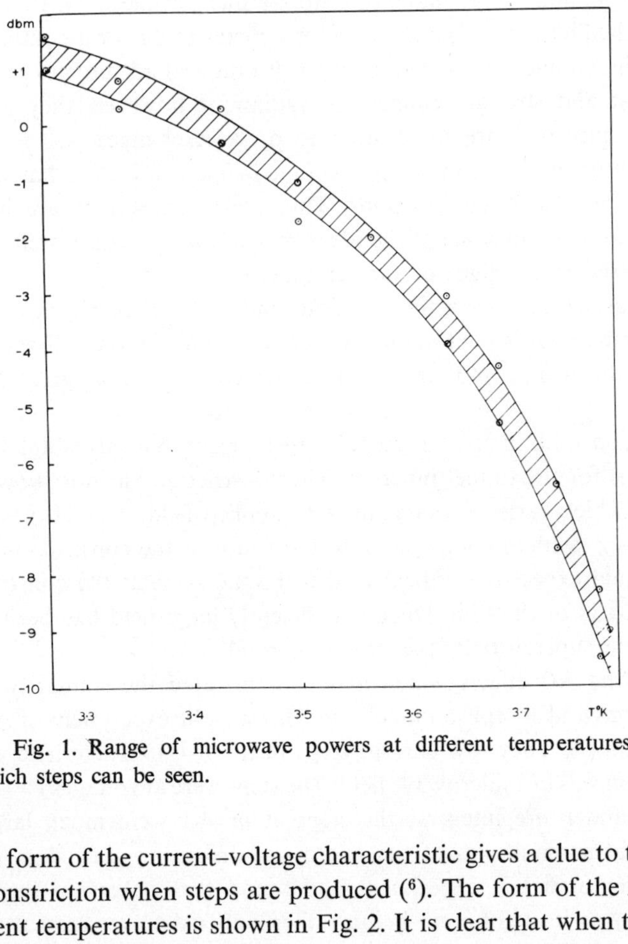

Fig. 1. Range of microwave powers at different temperatures for which steps can be seen.

The form of the current–voltage characteristic gives a clue to the nature of the constriction when steps are produced ([6]). The form of the *I–V* curve at different temperatures is shown in Fig. 2. It is clear that when the critical current is exceeded the *I–V* curves are not linear, i.e., ohmic. However, we would only get an ohmic behavior if the whole sample went normal; what is far more likely is that the center of the sample, where it is narrow and the current density high, goes normal and the outer parts of the film stay superconducting. There will be a phase boundary as shown in Fig. 3, where the current density along the boundary is just equal to the critical current density. The calculation of the position of the boundary is not simple because the critical current density used must be the one corresponding to the temperature of the boundary, and this is not the temperature of the bath because of the Joule heating in the normal region. The general form of the *I–V* for this model can be simply seen. If at some current we increase the current, then not only does the voltage across the normal region increase

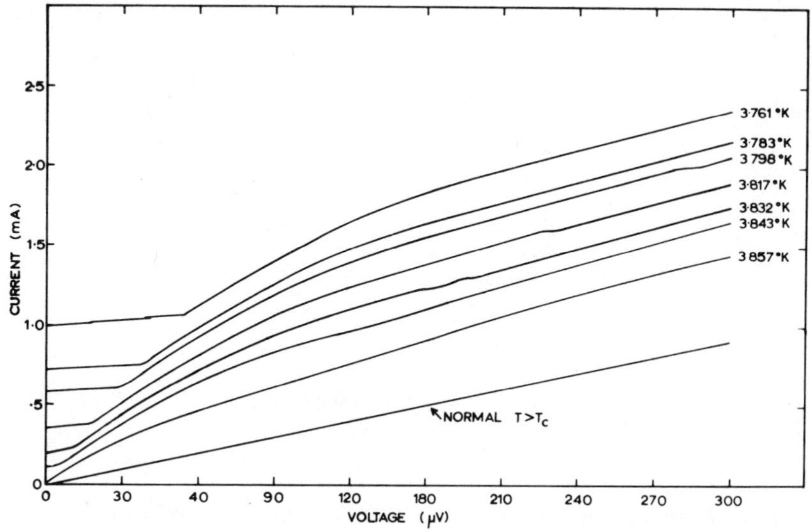

Fig. 2. Current–voltage curves at different temperatures for a thin-film constriction. No microwaves are applied. From Proceedings of the Tenth International Conference on Low Temperature Physics, Moscow, 1966.

but the region itself grows, and so the voltage increases faster than the current.

The model was confirmed by a fortuitous scratch on one of the constrictions we made. There was a piece out of the side of the film at $\sim 30\ \mu$ from the constriction. When the phase boundary reached this point the voltage increased faster and caused a kink in the I–V curve. At all temperatures this should occur at the same sample resistance, and indeed we find that the kinks on the different I–V curves do lie on a straight line.

This model probably explains why a parallel-sided strip does not act as a weak link: because such a strip cannot support a phase boundary, and either the whole strip is normal or it is superconducting, and when it is normal the separation between the two superconductors is too great.

It is not certain what the exact nature of the resistive material is. Most likely it is just normal metal, but the possibility of some complicated mixed state has not been excluded.

The picture is, however, very akin to the S–N–S junction, which is similar to Josephson junctions and probably can be explained in the same way. The separation of the superconducting regions turns out to be quite

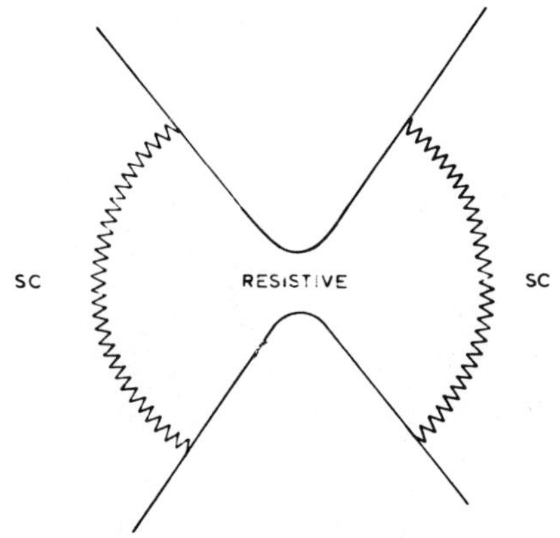

Fig. 3. Schematic picture of the two phases in a constriction with a voltage across it.

large in the region where steps are seen: it is of the order of 10 μ and possibly larger. There remains the dilemma of a material with a DC current in it greater than the critical current acting as a superconducting path for AC.

The current steps do not vary periodically with microwave power, as the Josephson junction does. This may be due to one of two reasons. If the rf voltage is not uniform across the normal region, then different points of the constriction will be on different parts of the step-height versus power curve. As this is an oscillatory function, there would be destructive interference. More likely, however, the normal region grows as the microwave power is increased, and so the coupling between the two superconducting regions becomes smaller and hence the size of the steps smaller.

Besides producing supercurrents at finite voltages, microwaves also increase the size of the critical supercurrent (I_c) that can flow at zero voltage ([6]). The effect is more pronounced near the transition temperature (T_c), as shown in Fig. 4. As the power is increased the I_c goes through a maximum and then decreases in the expected way. Another property of constrictions is that they have two critical temperatures, (Fig. 5) which we associate with different regions going superconducting at different temperatures. The large regions on each side of the constriction have a transition at a higher temperature than the narrow constriction. Hunt and Mercereau ([7]) have shown how the difference in transition temperature ΔT_c depends on the dimensions of the narrow region and offer an interesting idea of a connection between the two T_c's and the increase in I_c with microwave power.

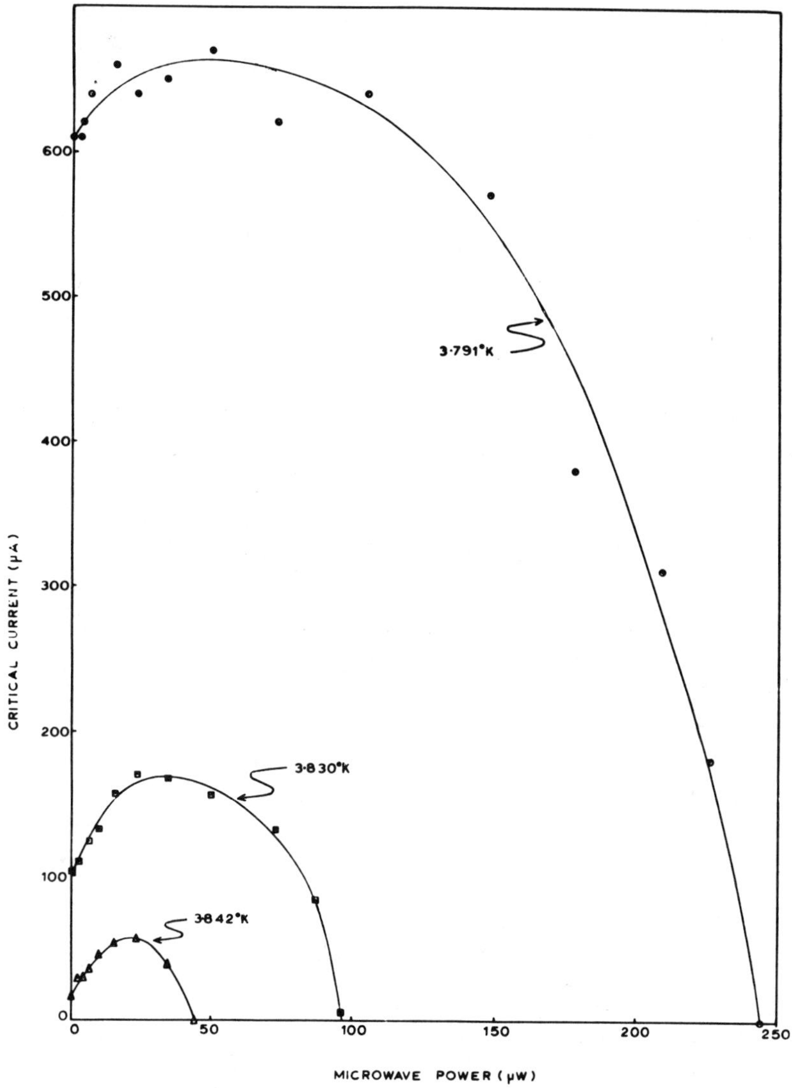

Fig. 4. Critical current as a function of microwave power at different temperatures. From Phys. Rev. Letters 16:1166 (1966).

They suggest that microwaves raise the transition temperature of the constricted region to that of the rest of the sample. This gives the increase in I_c because before the microwaves are applied I_c is a function of $[T_c$ (constriction) $- T] = \Delta T$, and with microwaves, the same function but

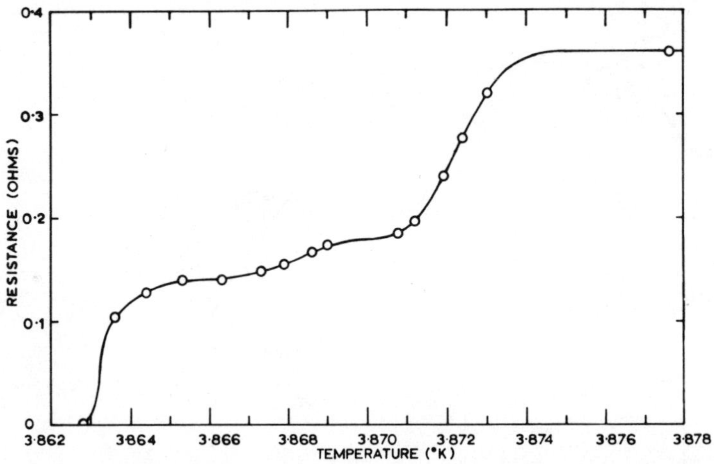

Fig. 5. Zero-current resistance of a constriction as a function of temperature in the transition region. From Proceedings of the Tenth International Conference on Low Temperature Physics, Moscow, 1966.

now of $[T_c(\text{bulk}) - T] = \Delta T + \Delta T_c$, and the larger the argument, the larger the current. This accounts for the largest percentage increase occurring when ΔT is small. They have calculated the increase in I_c for the ΔT_c of one of our samples and the agreement is very good.

Weakly coupled superconductors have provided us with many interesting phenomena and have emphasized the importance of the phase of the order parameters. However, not all the interesting questions have been answered, and perhaps not all of their highly individual characteristics found.

REFERENCES

1. B. D. Josephson, Adv. Phys. 14:419 (1965).
2. G. J. Van Gurp, Phys. Rev. 166:436 (1968).
3. J. Clarke, in: Proceedings of the Tenth International Conference on Low Temperature Physics, Moscow, 1966, to be published.
4. P. W. Anderson and A. H. Dayem, Phys. Rev. Letters 13:195 (1964). A. H. Dayem and J. J. Wigand, Phys. Rev. 155:419 (1967).
5. J. Lambe, A. H. Silver, J. E. Mercereau, and R. C. Jaklevic, Phys. Letters 11:16 (1964); A. H. Silver and J. E. Zimmerman, Phys. Rev. Letters 15:888 (1965). J. Clarke, Phil. Mag. 13:115 (1966); J. E. Zimmerman and A. H. Silver, Phys. Rev. 141:367 (1966); D. N. Langenberg, W. H. Parker, and B. N. Taylor, Phys. Rev. 150:186 (1966); C. C. Grimes, P. L. Richards, and S. Shapiro, Phys. Rev. Letters 17:431 (1966); A. H. Dayem and C. C. Grimes, Appl. Phys. Letters 9:47 (1966); J. E. Zimmerman, J. A. Cohen, and A. H. Silver, Appl. Phys. Letters 9:353 (1966).

Chapter 35

Atomic Tunneling in Solids*

J. A. Krumhansl

Laboratory of Atomic and Solid State Physics
Cornell University
Ithaca, New York

INTRODUCTION

We will discuss some aspects of the motion of atoms by way of tunneling between two or more equivalent potential wells in a crystal. This motion should not be confused with ordinary diffusion, because the energies involved are much smaller, of the order 10^{-4}–10^{-2} eV. Tunneling of atoms is of interest because the spectrum of elementary excitations associated therewith may span both the microwave range and the thermal range, i.e., tunneling frequencies are in the range 10^{10}–10^{12} Hz. Examples of systems where atomic tunneling occurs are dilute solutions of light ions in ionic crystals, e.g., Li:KCl, KCl:OH, and KH_2PO_4. We will discuss atomic tunneling in connection with optical, thermal, and inelastic effects and the symmetry of defects.

ISOLATED TUNNELING SITES

Tunneling versus Diffusion

The potential of a Li^- ion in KCl looks typically as in Fig. 1, where the energies and distances are indicated.

Tunneling Potentials

The radius of a Li^+ ion, 0.85 Å, is considerably smaller than that of the K^+ ion, 1.33 Å. Lombardo and Pohl (1) and Sack and co-workers (2) have suggested that the small Li^+ ion might not be stable at the lattice

* Based on work supported by the United States Atomic Energy Commission under Contract AT-(30-1)-3699, Technical Report NYO-3699-22.

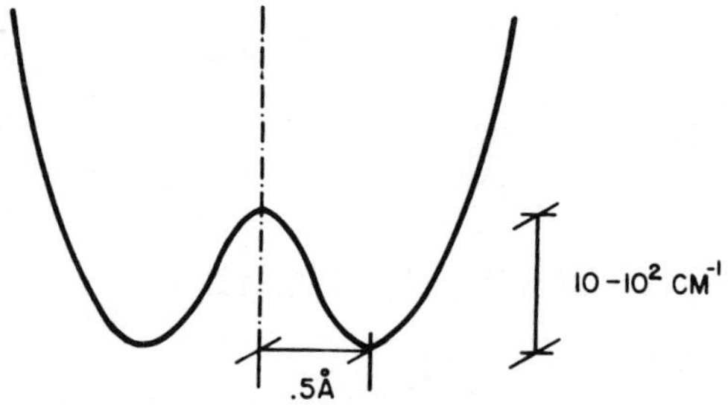

Fig. 1. Typical parameters for an atomic tunneling potential.

site. This suggestion has been confirmed in many different experiments. We can understand this theoretically from the following arguments ([3]).

The potential energy φ for a Li^+ ion consists of three parts, $\varphi = \varphi$ (Coulomb) + φ(repulsion) + φ(polarization). In a cubic crystal the Coulombic term is of the order x^4 for small deviations x from the lattice site. There is thus no Coulombic contribution to a harmonic oscillator potential. For a potassium ion in KCl the positive term φ(repulsion) is by far the dominant contribution at small deviations. In the case of a small ion (e.g., Li^+ in KCl) φ(repulsion) is so much reduced that we get a potential maximum. For large displacements of the small ion the overlap becomes greater and the change in energy ultimately becomes positive. Thus there is a possibility of low potential extrema off the lattice site as in Fig. 1. The sensitivity to the parameters involved is illustrated by the system Li^+:KBr, where there is no manywell potential but a flat potential, which can give rise to a very-low-frequency localized mode ([4]).

The effects of the weak coupling of the Li^+ ion to its environment has been observed in a variety of experiments:

1. Anomalous dielectric constant at low temperatures ([5]).
2. Electrocaloric effect (cooling on adiabatic depolarization at low temperatures) ([6]).
3. Thermal conductivity ([7]).
4. Specific heat ([8]).
5. Inelastic properties ([9]).
6. Microwave absorption ([10]).

Tunnel States–Statics

First we treat a simple one-dimensional model in which we have tunneling between two wells. The impurity is assumed to move in a double-well truncated harmonic oscillator potential (Fig. 2),

$$V = V_A = \tfrac{1}{2}m\omega^2(\mathbf{x} + \mathbf{x}_0)^2, \qquad \mathbf{x} < 0,$$
$$= V_B = \tfrac{1}{2}m\omega^2(\mathbf{x} - \mathbf{x}_0)^2, \qquad \mathbf{x} > 0,$$

assuming the rest of the lattice rigid. We express the lowest states as linear combinations of the ground states $|\,a\rangle$ and $|\,b\rangle$ of the unperturbed harmonic oscillators. The energies are given by the secular equation

$$\begin{vmatrix} E_0 - E & -SE \\ -SE & E_0 - E \end{vmatrix} = 0,$$

where $E_0 = \langle a \mid H \mid a\rangle$, and $H_1 = \langle a \mid H \mid b\rangle$ and $S = \langle a \mid b\rangle$. We get one symmetric and one antisymmetric solution,

$$E_\pm = \frac{E_0 \pm H_1}{1 \pm S}, \qquad \Psi_\pm = \frac{|\,a\rangle \pm |\,b\rangle}{[2(1 \pm S)]^{1/2}}.$$

There are of course many higher states of the system, and with increasing energy, these approach the states of a single harmonic oscillator. Typical tunnel splitting in these systems is $\delta = E_+ - E_- = 1\ \mathrm{cm}^{-1}$.

We now extend the model to three dimensions ([11]). For the cubic symmetry of the impurity environment there are three multiwell potentials which are consistent with the octahedral symmetry. These are:

XY_6 with 6 minima in the $\langle 100\rangle$ directions

XY_8 with 8 minima in the $\langle 111\rangle$ directions

XY_{12} with 12 minima in the $\langle 110\rangle$ directions

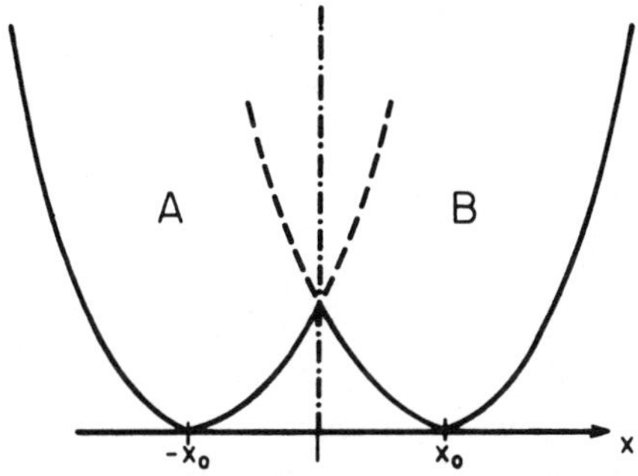

Fig. 2. Model potential for tunneling calculation.

Inelastic measurements ([2]) suggest that for Li:KCl the proper symmetry is XY_8. We make the assumption that the wells are symmetric, since this gives simple results,

$$V_i = \frac{m\omega^2}{2} [(|\,\mathbf{x}\,| - \mathbf{x}_i)^2 + (|\,y\,| - y_i)^2 + (|\,z\,| - z_i)^2], \qquad i = a, b, \ldots,$$

with the unperturbed oscillator functions

$$|\,a\rangle = \left(\frac{m\omega}{\pi\hbar}\right)^{3/4} \exp\left\{-\frac{m\omega}{2\hbar}[(\mathbf{x} - \mathbf{x}_a)^2 + (y - y_a)^2 + (z - z_a)^2]\right\}.$$

This system has an eightfold degeneracy. The correct linear combinations of basis states are easily determined by group-theoretical methods.

The energy splitting is indicated in Fig. 3, where the symmetry and degeneracy are given. It will be noted that the original eightfold degeneracy is accounted for in the total number of states for the four levels. One can show that the overlap and tunneling along the cube edges dominates for isotropic wells ([12]). Thus the ion may be thought of as shuttling around the surface of the cube. This kind of motion is equivalent for these low-lying tunneling states to a quasirotational motion associated with rigid rotators in an octahedral field with large barriers. In Fig. 4 the energy level structure compatible with the symmetry considerations is given for two extremes, corresponding to the single three-dimensional harmonic oscillator well on the extreme left and to deep isolated XY_8 wells on the extreme right.

Recently Walton ([13]) devised a method for phonon spectroscopy based on the magnetic-field tunable spin-phonon interaction and has confirmed this tunneling structure completely in Li^+:KCl.

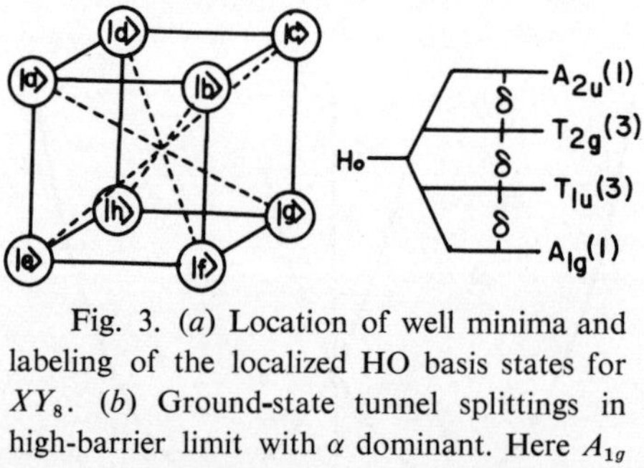

Fig. 3. (*a*) Location of well minima and labeling of the localized HO basis states for XY_8. (*b*) Ground-state tunnel splittings in high-barrier limit with α dominant. Here A_{1g} and A_{2u} are singly degenerate; T_{1u} and T_{2g} are triply degenerate. $\delta = 2(\alpha - \frac{1}{2}\hbar\omega S)$. From Phys. Rev. Letters 16:1105 (1966).

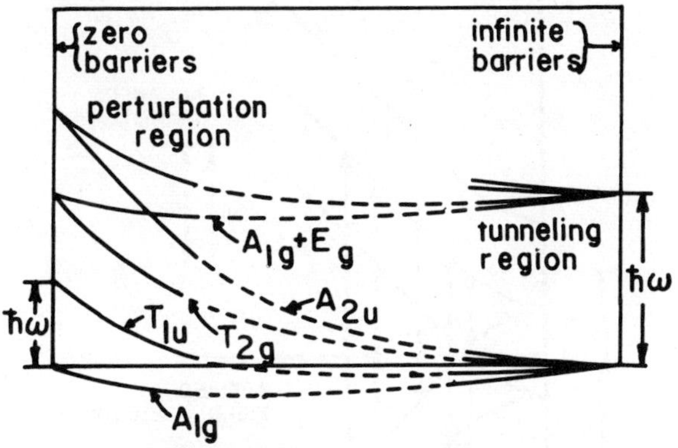

Fig. 4. The energy-level structure compatible with symmetry considerations is given for two extremes, corresponding to the single three-dimensional well HO on the extreme left and to deep isolated XY_8 wells on the extreme right. When tunneling between deep wells is introduced the levels are modified and may be classified according to symmetry types as shown by the solid lines near the right-hand side of the figure; the structure shown in Fig. 1(*b*) corresponds to the lowest set of levels A_{1g}, T_{1u}, T_{2g}, A_{2u}. In the other limit the introduction of a weak multiwell perturbation splits the HO levels as shown near the left-hand side. Between these limits the levels connect as indicated by the dotted lines. From Phys. Rev. Letters 16:1105 (1966).

Effects of Static Field

The introduction of an electric field reduces the symmetry of the problem, e.g., the O_h symmetry becomes replaced by C_{4v} when a field is applied along the $\langle 100 \rangle$ direction. Figure 5 shows the nature of the splitting of the energy levels.

We see that by applying an electric field one can tune the system to get absorption for a fixed microwave frequency.

Typical Results-Isotope Dependence

Additional confirmation of the fundamental role of tunneling as the basic mechanism can be found from the isotope dependence of the energy level structure. Harrison *et al.* ([14]) have used specific heat measurements on Li^6- and Li^7-doped KCl to show how marked this dependence is. This is to be expected from the typical exponential dependence of tunneling, i.e., $\omega \propto \exp[-(2V_0 m/\hbar^2)^{1/2}]$.

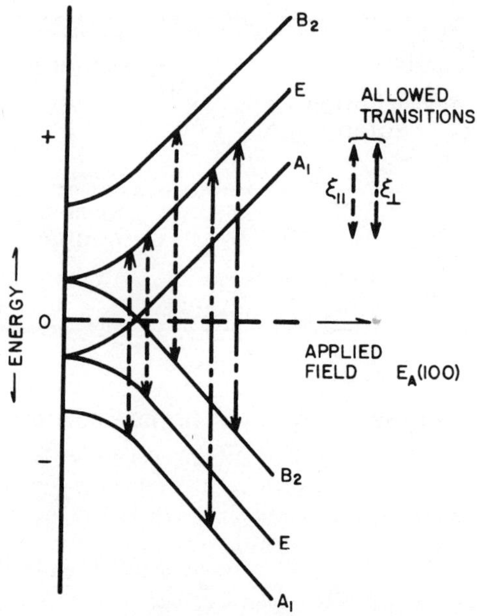

Fig. 5. Electric field splitting of levels shown in Fig. 3. From Phys. Rev. 153:1009 (1967).

Specific Heat–Equilibrium Susceptibility

The level systems shown in Figs. 3 and 5 represent a basis set for thermal or electric field excitation. Thus the low-temperature susceptibility and dielectric properties of a material like KCl can be strongly affected by a small amount of Li. The specific heat follows a generalized Schottky behavior, and the susceptibility can easily be investigated from the quantum mechanical limit $\delta \gg kT$ to the opposite limit. When a static electric field (or uniaxial stress) is applied the dielectric and optical (microwave) properties become quite anisotropic; the anisotropy has been used to determine the defect symmetry.

INTERACTING TUNNELING SYSTEMS

Interaction with Phonons*

The idealization of holding the lattice rigid is obviously not valid. This may be corrected by considering the coupled tunneling site–lattice

* Based on ([17]).

system and anharmonicity. The complete Hamiltonian is

$$H = H_{\text{imp}} + H_{\text{latt}} + H_{\text{int}}.$$

The rest of the lattice is treated as harmonic,

$$V = V(\mathbf{R}_{\text{imp}}, \mathbf{R}_l) = V(\mathbf{R}_{\text{imp}}, \mathbf{x}_l) + \frac{\partial V(\mathbf{R}_{\text{imp}}, \mathbf{x}_l)}{\partial \mathbf{x}_l} \cdot \hat{\mathbf{u}}_l + \ldots,$$

where

$$\mathbf{R}_l = \mathbf{x}_l + \hat{\mathbf{u}}_l$$

and

$$\hat{\mathbf{u}}_l = \sum_s (\hbar/2m\omega_s)^{1/2}(a_s^{\dagger} + \text{h.c.}).$$

Denoting the noninteracting tunneling levels by ε_n, then in the usual second quantization language the contributions to the Hamiltonian may be written

$$H_{\text{imp}} = \sum_n \varepsilon_n \hat{b}_n^{\dagger} \hat{b}_n,$$

$$H_{\text{latt}} = \sum_s \hbar\omega_s(a_s^{\dagger} a_s + \tfrac{1}{2}).$$

Due to the interaction $V(\mathbf{R}_{\text{imp}}, \mathbf{R}_l)$ there is a coupling to various energy states within the tunneling system,

$$\frac{\partial V(\mathbf{R}_{\text{imp}}, \mathbf{x}_l)}{\partial \mathbf{x}_l} = \sum_{nn'} \langle n \mid \frac{\sigma V}{\partial \mathbf{x}_l} \mid n' \rangle \hat{b}_n^{\dagger} \hat{b}_{n'},$$

so

$$\hat{H}_{\text{int}} = \sum_{nn's} \Phi\begin{pmatrix} n, n' \\ s \end{pmatrix} \hat{b}_n^{\dagger} \hat{b}_{n'} (a_s^{\dagger} + a_s).$$

The equations are treated in the usual many-body way, and the idea is to compute the excitation spectrum of the interacting system. The spectrum character may be typically seen in response functions.

Of particular interest is the electric susceptibility of the impurity system. The dipole moment operator is

$$\hat{\mathbf{M}} = \sum_{nn'} \mathbf{M}_{nn'} \hat{b}_n^{\dagger} \hat{b}_{n'},$$

and

$$\hat{H}_{\text{applied}} = \hat{\mathbf{M}} \cdot \mathscr{E}.$$

With a thermodynamic Green's function defined as

$$G(n_4 n_2 n_1 n_3; \tau) = -i\langle T\hat{b}_{n_1}^{\dagger}(\tau)\hat{b}_{n_2}(\tau)\hat{b}_{n_3}^{\dagger}(0)b_{n_4}(0)\rangle,$$

it is well known that the susceptibility is

$$\chi_{\hat{M},\hat{M}} = \lim_{\delta\to 0+} \frac{N_i}{V} \sum_{n_1 n_2 n_3 n_4} M_{n_1 n_2} M_{n_3 n_4} (-1) G(n_4 n_2 n_1 n_3; \omega + i\delta).$$

The equation of motion is

$$i \frac{\partial}{\partial\tau} G(n_4 n_2 n_1 n_3; \tau) = i\delta(\tau) \langle \hat{b}^\dagger_{n_3} \hat{b}_{n_4} \hat{b}^\dagger_{n_1} \hat{b}_{n_2} - \hat{b}^\dagger_{n_1} \hat{b}_{n_2} \hat{b}^\dagger_{n_3} \hat{b}_{n_4} \rangle$$
$$- i \langle T[\hat{b}_{n_1}(\tau) \hat{b}_{n_2}(\tau), H] \hat{b}^\dagger_{n_3}(0) \hat{b}_{n_4}(0) \rangle .$$

For the noninteracting case we get

$$-iG(n_4 n_2 n_1 n_3; \omega) = \frac{\delta_{n_4 n_1} \delta_{n_2 n_3} (\bar{N}_2{}^T - \bar{N}_1{}^T)}{\omega - (\varepsilon_{n_2} - \varepsilon_{n_1})}$$

and

$$\chi(\omega) = \lim_{\delta\to 0+} \frac{N_i}{V} \sum_{n_1 n_2} M_{n_1 n_2} M^*_{n_2 n_1} \frac{\bar{N}_2{}^T - \bar{N}_1{}^T}{\omega - (\varepsilon_{n_2} - \varepsilon_{n_1}) + i\delta} .$$

The real part of the susceptibility is the principal value of this integral. For instance, for a two-level system we have

$$N_a^{(0)} - N_b^{(0)} = \tanh \beta(\varepsilon_b - \varepsilon_a)/2,$$

and accordingly

$$\chi' \propto \frac{N_i}{V} 2 \mid M_{ab} \mid^2 \frac{(\varepsilon_b - \varepsilon_a)}{\omega^2 - (\varepsilon_b - \varepsilon_a)^2} \tanh \beta(\varepsilon_b - \varepsilon_a)/2.$$

In this case the imaginary part is

$$\chi'' \propto \delta[\omega - (\varepsilon_b - \varepsilon_a)] \tanh \beta(\varepsilon_b - \varepsilon_a)/2,$$

with peaks as in Fig. 6.

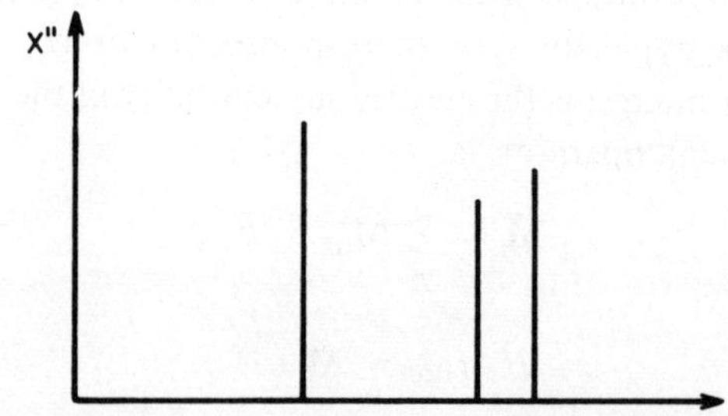

Fig. 6. Imaginary part of susceptibility (absorption spectral density) for tunneling system alone.

For the tunneling system interacting with the lattice one may treat the coupled Green's functions in the usual way and thus get self-energy corrections, i.e.,

$$G \sim \frac{1}{\omega - (\varepsilon_{n_2} - \varepsilon_{n_1}) - \Sigma(\omega) - i\Gamma(\omega)}.$$

The overall spectrum looks typically as in Fig. 7, where the tunneling modes lie in the region of the ordinary phonon spectrum. The nonzero imaginary part causes damping of the tunneling modes by phonon radiation, which is responsible for the finite width in the real tunneling system indicated in Fig. 8. The phonon radiation damping rate depends on the density of phonon states at the frequency of the tunneling level splitting; since this will become small at low frequencies, it may be seen that the original idealization of treating the lattice as rigid may have some rationalization in the limiting case of slow tunneling.

In this section we have given primary attention to the interaction between a single tunneling impurity and the lattice phonons, a useful low concentration limit. At higher concentrations impurities will interact with each other.

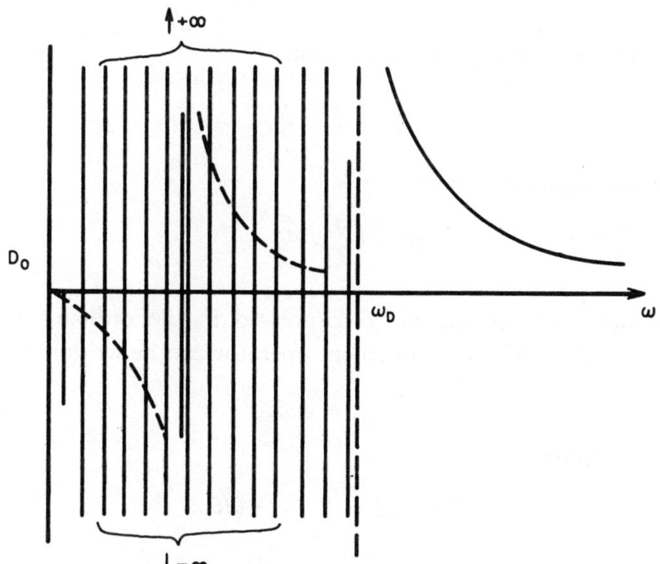

Fig. 7. Response for excitation at tunneling site, lying in "continuum" of lattice response D_0.

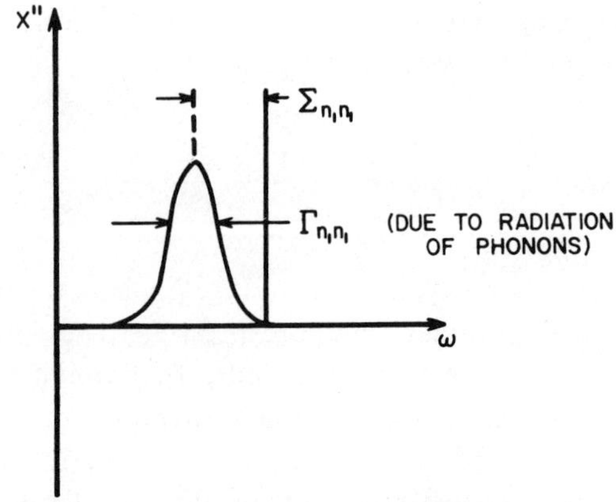

Fig. 8. Broadening of absorption line in Fig. 6 due to lattice interaction.

Spin Model for Two-Well Systems

We will now look at a method for treating mutually interacting tunneling systems ([15,16]). For simplicity we use a two-well system like that in Fig. 1. The wave function can be expressed as a sum,

$$\Psi = c_1\Psi_L + c_2\Psi_R.$$

For this we introduce a matrix notation,

$$\Psi = \begin{pmatrix} c_1 \\ c_2 \end{pmatrix},$$

in which the Hamiltonian is

$$H = \begin{pmatrix} E_0 & \Omega \\ \Omega & E_0 \end{pmatrix},$$

where Ω describes the tunneling. Expressed in the ordinary Pauli spin matrices σ_x, σ_y, σ_z the dipole moment operator can be written as

$$\hat{p}_z = ea\hat{\sigma}_z,$$

and the Hamiltonian

$$\hat{H}_0 = \hat{H} - E_0 = -\Omega\hat{\sigma}_x.$$

The equation of motion becomes

$$\ddot{\sigma}_z = -(2\Omega)^2\sigma_z.$$

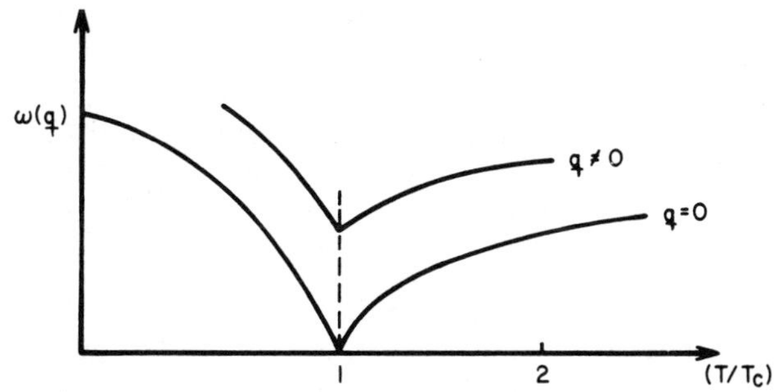

Fig. 9. Temperature dependence of a cooperative tunneling mode; at (T/T_c) the system would spontaneously polarize in the $q = 0$ mode.

For interacting wells we postulate an exchange interaction ([15,16])

$$\hat{H} = -\Omega \sum_i \sigma_x^{(i)} - \sum_{ij} J_{ij} \sigma_z^{(i)} \sigma_z^{(j)},$$

where the sums are over different lattice sites. We try to diagonalize this Hamiltonian by assuming a periodic array and polarization waves of the type

$$\sigma_z{}^{\mathbf{q}} = \frac{1}{\sqrt{N}} \sum_i (\exp -i\mathbf{q} \cdot \boldsymbol{\chi}_i) \sigma_z^{(i)}.$$

The (ω, q) dispersion behavior is shown in Figure 9. In general we get a coupled set of nonlinear equations as in the ferromagnetic spin wave case. But if we linearize about thermal equilibrium and employ the methods used in the latter case, the dynamical and collective properties may be studied. For certain values of the exchange integral J_{ij} it is possible to get a spontaneous polarization ferroelectricity ([16]).

We have only sketched the approach here, because as yet the details for a real system remain to be worked out.

SUMMARY

To sum up, we have seen that:

There are a number of excitations in solids which are tunneling excitations. They are particularly important for light ionic impurities.

The simplest theory for the tunneling excitations is a MO-like calculation.

It is in principle possible to dress the tunneling motion due to lattice dynamics.

The tunneling systems can couple to each other, and we can get spin waves.

REFERENCES

1. G. Lombardo and R. O. Pohl, Phys. Rev. Letters 15:291 (1965).
2. H. S. Sack and M. S. Moriarty, Solid State Comm. 3:93 (1965). N. E. Byer and H. S. Sack, Phys. Rev. Letters 17:72 (1966). F. Baumann, Bull. Am. Phys. Soc. 9: 644 (1964). A. I. Lakatos and H. S. Sack, Bull. Am. Phys. Soc. 11:229 (1966). H. Bogardus and H. S. Sack, Bull. Am. Phys. Soc. 11:229 (1966).
3. J. A. D. Matthew, Solid State Commun. 3:365 (1965).
4. A. J. Sievers and J. S. Takeno, Phys. Rev. 140:1030 (1965).
5. H. S. Sack and M. S. Moriarty, Solid State Commun. 3:93 (1965).
6. G. Lombardo and R. O. Pohl, Phys. Rev. Letters 15:291 (1965).
7. F. Baumann, Bull. Am. Phys. Soc. 9:644 (1964).
8. Harrison and R. O. Pohl, private communication.
9. N. E. Byer and H. S. Sack, Phys. Rev. Letters 17:72 (1966).
10. A. I. Lakatos and H. S. Sack, Bull. Am. Phys. Soc. 11:229 (1966).
11. S. P. Bowen, M. Gomez, J. A. Krumhansl, and J. A. D. Matthew, Phys. Rev. Letters 16:1105 (1966). M. Gomez, S. P. Bowen, and J. A. Krumhansl, Phys. Rev. 153: 1009 (1967).
12. R. J. Quigley and T. P. Das, Solid State Commun. 5:407 (1967).
13. D. Walton, Phys. Rev. Letters 19:305 (1967).
14. J. P. Harrison, P. P. Peressini, and R. O. Pohl, in: Proceedings of the First International Conference on Localized Excitation in Solids, Plenum Press, New York, 1968.
15. P. G. DeGennes, Solid State Commun. 1:132 (1963).
16. R. Brout, K. A. Muller, and H. Thomas, Solid State Commun. 4:507 (1966).
17. S. Bowen, Ph. D. thesis, Cornell University (1967).

Chapter 36

The Detection of Atomic Tunneling in Solids

Y. Imry

Israel Atomic Energy Commission
Soreq Nuclear Research Center
Yavne, Israel

INTRODUCTION

In this chapter we shall examine theoretically the interaction of a tunneling atom with external radiation. We shall be interested in emission, scattering, and absorption processes, the radiation being light, neutrons, γ rays, etc. We shall also discuss relevant experiments. Our purpose will be to demonstrate how the tunneling phenomenon is *directly* detectable through the following effects: (1) splitting of energy levels, (2) interesting interference effects, and (3) changes in transition probabilities.

The systems with which we shall be mainly concerned are:

1. Hydrogen bonds, where a proton might tunnel between two minima, especially in crystals of the KH_2PO_4 type ([1]).

2. The tunneling of hydrogen through many minima in V ([2,3]) and possibly in other metals, and its relation to the band picture.

3. Heavier impurities such as Kr in hydroquinone clathrates and perhaps Fe in KCl and KBr. These are of interest in Mössbauer studies.

This is by no means a complete list of cases where atomic tunneling is possible. In these systems the experimental situation is better known, especially with regard to the processes that we wish to discuss. One should remark that for hydrogen atoms a special convenience exists. By isotopic substitution to deuterium one can reduce tunneling probabilities (due to the larger mass of D) without changing other properties too much. This is extremely helpful for the isolation of tunneling effects.

In Chapter 35 of this volume Krumhansl described how a multiple-minima situation might arise for an impurity in an ionic crystal when the size of the impurity atom is smaller enough than the volume available to it.

A somewhat similar situation exists in the linear A—H—B system, say an O—H—O bond. For fixed A and B, H sees two molecular potentials, A—H and H—B. If the A—B distance is large enough, the superposition of these two potentials will exhibit two minima. For example, if two O—H Morse-type potentials are superposed, it is easily seen that for O—H—O distances larger than about 2.45 Å, two minima are obtained for the potential acting on the proton. The distance between the minima and the height of the barrier between them increase further with O—O distance ([1]).

Similar results are obtained with other types of potentials, as well as in more complicated three-dimensional cases. The simple one-dimensional case is important for H bonds, and since the mathematical treatment is easy, we shall use it extensively. We shall not be interested in quantitative details of high accuracy, such as the exact shape of the potential barrier. This does not contain much new physics and it is not justified by the present experimental accuracy.

REVIEW OF THE QUANTUM MECHANICS OF MULTIPLE-WELL POTENTIALS

We consider a particle of mass m in the simple asymmetric double-minimum potential of Fig. 1. ([1]) This potential consists of two harmonic

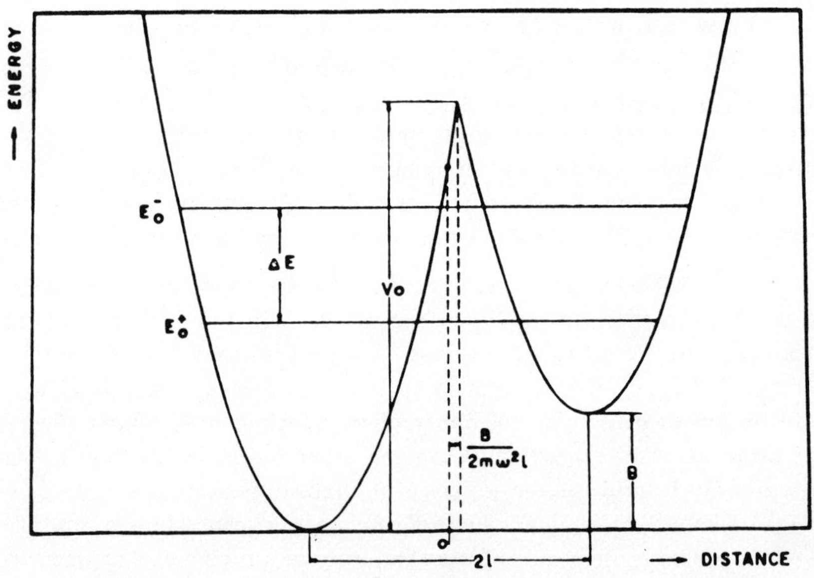

Fig. 1. Asymmetric double-minimum potential. From Imry *et al.* ([1]).

oscillator parabolas having the same curvature k and $\omega = (k/m)^{1/2}$, and separated by a distance of $2x_0$. In addition, we assume the bottom of the right-hand oscillator to be higher by $B\hbar\omega$ than the bottom of the left-hand oscillator. We assume $x_0 \gg 2\sigma = [\hbar/(m\omega)]^{1/2}$ (2σ is the zero-point amplitude of each oscillator) and $B \ll 1$.

One would like to do a perturbative approximation of the problem, but it is not clear what should be taken as the perturbation. It is simpler to resort to a variational method, calculate the Hamiltonian matrix in some convenient basis, diagonalize it, and hope that the correct wave function can be approximated by a linear combination of the basis functions. The accuracy of this method can be determined by the study of its convergence when the basis is made larger. For $B \ll 1$ one feels that the strongest interaction is between close levels in the left- and right-hand oscillators, so that each such pair of levels $|\, nl\rangle$ and $|\, nr\rangle$ (respectively, the nth oscillator level on the left- and right-hand sides) will combine to a pair of levels of our problem. Thus $|\, 0l\rangle$ and $|\, 0r\rangle$ seem to be good basis for the two lowest levels. [This expectation is borne out by numerical calculations using another method ([5]) and by comparison to the exact solution for $B = 0$ ([6])]. The H-matrix is easily calculated to be ([4])

$$H = \text{const} + \begin{pmatrix} -B/2 & \Omega_0 \\ \Omega_0 & B/2 \end{pmatrix} \hbar\omega ,$$

where

$$\Omega_0 = (2/\sqrt{\pi})q \exp(-q^2), \qquad q = x_0/\sigma \gg 1 , \tag{1}$$

and the overlap $\langle 0l \,|\, 0r\rangle$ is neglected. The energy levels are

$$E_{1,2} = \text{const} \pm [(B/2)^2 + \Omega_0{}^2]^{1/2} \hbar\omega \tag{2}$$

and the wave functions

$$\begin{gathered} |\, 1\rangle = \alpha \,|\, 0l\rangle + \beta \,|\, 0r\rangle , \qquad |\, 2\rangle = \alpha \,|\, 0r\rangle - \beta \,|\, 0l\rangle , \\ \alpha, \beta > 0; \qquad \alpha^2 + \beta^2 = 1; \qquad \alpha/\beta = [(B/2)^2 + \Omega_0{}^2]^{1/2}/\Omega_0 . \end{gathered} \tag{3}$$

For the symmetric case ($B \ll \Omega_0$): $E_2 - E_1 = 2\Omega_0\hbar\omega$, $\alpha = \beta$. For the extreme asymmetric case ($1 \gg B \gg \Omega_0$): $E_2 - E_1 = B\hbar\omega$, $\beta \to 0$. The factor $2\Omega_0\hbar\omega$ is called "the tunneling splitting." We note the sharp decrease of Ω_0, connected with the tunneling probability, with increasing m (isotope effect). The dependence of Ω_0 on x_0/σ is also qualitatively clear.

The equality $\alpha = \beta$ of the weights of the wave functions in the two wells in the symmetric case also follows from general symmetry arguments.

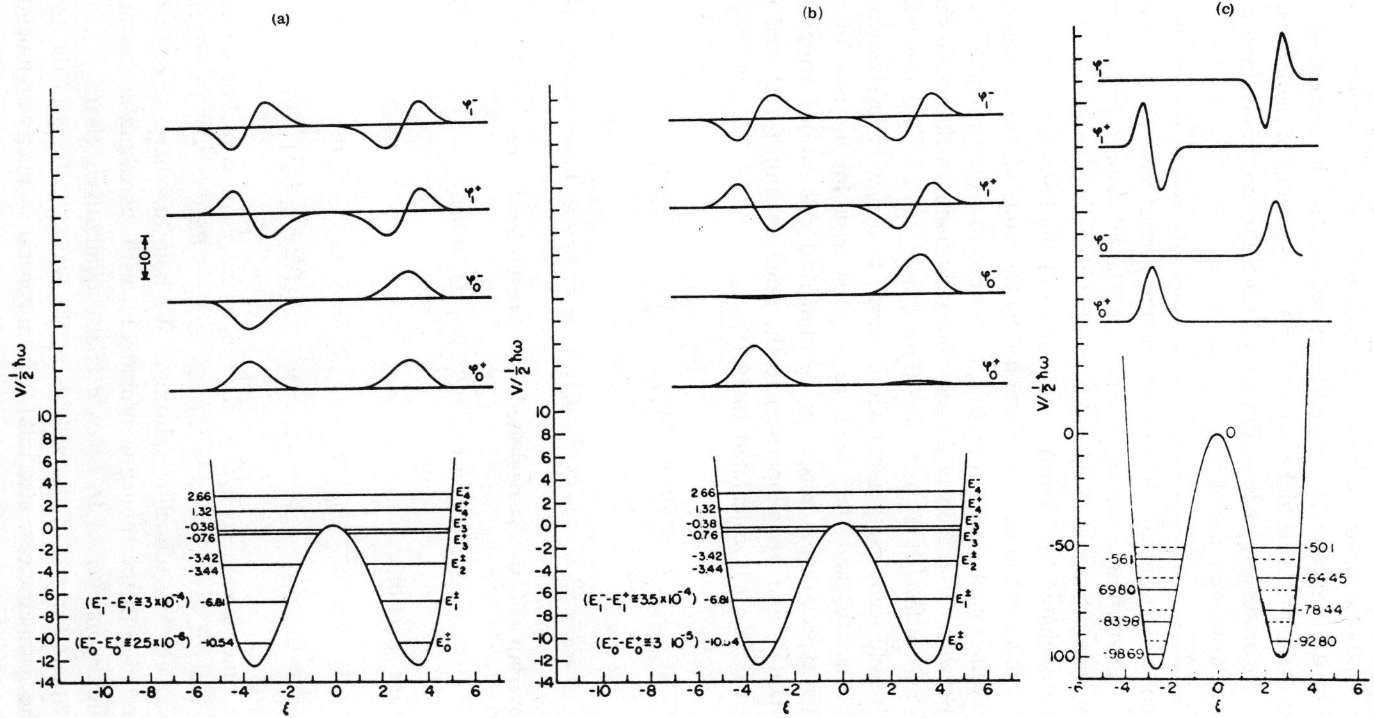

Fig. 2. Double minimum potentials, energy levels, and wave functions. (*a*) Symmetric. (*b*) Slightly asymmetric (only lowest pair localized). (*c*) Very asymmetric. From Imry [8].

However, for the extreme asymmetric case the wave functions are localized; $\beta = 0$, $| 1\rangle$ is concentrated in the left-hand well and $| 2\rangle$ in the right-hand well.

Figure 2 demonstrates our results numerically; potential functions of a polynomial type were used, together with a different variational method ([5]) using up to 80×80 matrices. In Figure 2(*a*) we have a symmetric potential, we note the pairs of levels whose splittings increase with level height (because of the increased tunneling probability). By looking at these essentially exact wave functions one may get some feeling as to the validity of the two-state approximation. In Fig. 2(*b*) we have chosen B such that $\Omega_1 \gg B \gg \Omega_0$; thus only the lowest two levels are localized and all the higher ones are essentially the same as in the symmetric case.

When the number of minima is $N > 2$, the same sort of treatment is possible. Each level of a single well will split into at most N levels. In the case that all the minima are equivalent one may find the relative weights, analogous to α and β, by symmetry considerations ([7]).

A particularly interesting case is that of $N \gg 1$ minima periodically arranged on a straight line. Here one gets groups of N levels which merge for $N \rightarrow \infty$ into a band. Thus one sees that band theory is in some sense a multi-minimum problem. We shall return to this later.

CALCULATION OF THE CROSS SECTIONS FOR EMISSION, ABSORPTION, AND SCATTERING OF RADIATION BY TUNNELING ATOMS IN ONE DIMENSION

The interference effects from two sources of radiation located at two given points are well known. We now turn our attention to the problem of a single source which tunnels quantum mechanically between the two positions. If the two positions are equivalent, we know that in each energy level the probabilities of finding the source particle in either of the two positions are equal. If, however, the energy difference between the two minima is large enough, different levels will be localized in different minima and the probabilities for the two positions become temperature dependent. One should also ask the question whether the radiation from the two levels is coherent.

For mathematical simplicity we assume that the particle moves in the potential of the previous section. For the calculation of transition probabilities we have to evaluate matrix elements of $\exp(i\mathbf{k} \cdot \mathbf{x})$ between the various wave functions. Assuming the particle to move along the x axis and writing $k_x = k$, we have, neglecting overlap integrals:

$$\langle 1 \mid e^{ikx} \mid 1\rangle \approx \alpha^2 e^{-ikx_0}\langle 0l \mid e^{ik(x+x_0)} \mid 0l\rangle + \beta^{ikx_0}\langle 0r \mid e^{ik(x-x_0)} \mid 0r\rangle$$
$$= e^{-k^2\sigma^2}(\alpha^2 e^{-ikx_0} + \beta^2 e^{ikx_0}), \tag{4}$$

$$\langle 2 \mid e^{ikx} \mid 2\rangle \approx e^{-k^2\sigma^2}(\alpha^2 e^{ikx_0} + \beta^2 e^{-ikx_0}), \tag{5}$$

$$\langle 1 \mid e^{ikx} \mid 2\rangle \approx e^{-k^2\sigma^2}\alpha\beta(e^{ikx_0} - e^{ikx_0})$$
$$= 2i\alpha\beta \sin(kx_0)e^{-k^2\sigma^2}. \tag{6}$$

Infrared Absorption

Here we have to calculate the electric dipole matrix element between states $\mid 1\rangle$ and $\mid 2\rangle$; $M = \langle 1 \mid Qx \mid 2\rangle$, where Q is the effective charge of the particle. Since for light $kx_0 \ll 1$, and $\langle 1 \mid \text{const} \mid 2\rangle = 0$, we have, using Eq. (6)

$$M = \frac{Q}{ik}\langle 1 \mid e^{ikx} \mid 2\rangle = \frac{2Q\alpha\beta}{k}\sin kx_0 \approx 2x_0\alpha\beta Q. \tag{7}$$

For a symmetric well $M = 2Qx_0$, which is much larger than $Q\sigma$—the order of magnitude of the transition dipole moment in a usual potential well. When the well becomes asymmetric the intensity will decrease with $\mid \alpha\beta \mid^2$.

The isotope effect is of some interest here. In a symmetric well the substitution of a higher mass will greatly reduce Ω and the transition energy $\Delta E = 2\Omega$ without affecting the intensity. In an asymmetric well if $B \gg \Omega$, the transition energy

$$\Delta E = 2\left[\left(\frac{B}{2}\right)^2 + \Omega^2\right]^{1/2},$$

will change only slightly, while the intensity, determined by

$$\alpha\beta \approx \frac{2\Omega}{B}\left(1 + \frac{4\Omega^2}{B^2}\right)^{-1/2}$$

will drop sharply.

Neutron Scattering

For a scattering process where the neutron changes its momentum from $\mathbf{k}_0$ to $\mathbf{k}'$ and the scattering system changes its state from $\mid i\rangle$ to $\mid f\rangle$ the cross section is proportional to

$$\mid \langle i \mid \sum_n \exp(i\mathbf{K}\cdot\mathbf{x}_n) \mid f\rangle \mid^2,$$

where $\mathbf{x}_n$ is the coordinate of a scattering particle and $\mathbf{K} = \mathbf{k}' - \mathbf{k}_0$ the momentum transfer. This Born approximation result is valid provided that a correct pseudopotential is used for the interaction of the neutron with the particles of the system.

We are interested in cases where the scattering nuclei are protons. Here the cross section is incoherent and thus we have to add the *cross sections* of the particles of our system. We therefore have to calculate the scattering from a single particle. We shall assume each particle to move in the potential of the previous section along the x axis. In an interacting system the potential will only have a meaning as an average over the motions of other particles. Again letting $k_x = k$, we have for the elastic scattering

$$(d\sigma/d\Omega)_{\text{el}} \propto |\langle 1 \mid e^{ikx} \mid 1\rangle|^2 P_1(T) + |\langle 2 \mid e^{ikx} \mid 2\rangle|^2 P_2(T),$$

where $kx = \mathbf{K} \cdot \mathbf{x}$, and $P_1(T)$ and $P_2(T)$ are the populations of the levels 1 and 2. By Eqs. (4) and (5), and since $P_1 + P_2 = 1$

$$(d\sigma/d\Omega)_{\text{el}} \propto e^{-2k^2\sigma^2}(\alpha^4 + \beta^4 + 2\alpha^2\beta^2 \cos 2kx_0). \tag{8}$$

From this interference pattern one can ascertain that tunneling exists and measure x_0. By estimating the ratio of the constant part to the sinusoidal part (as a function of $\mathbf{K}$), the asymmetry may be obtained.

In the inelastic case, where the system goes from level 1 to 2 or vice versa and the neutron respectively loses or gains energy of $E_2 - E_1$, we get from Eq. (6):

$$\frac{d^2\sigma}{d\Omega\, d\omega} \propto |\langle 1 \mid e^{ikx} \mid 2\rangle|^2 = 4\alpha^2\beta^2 \sin^2(kx_0)e^{-2k^2\sigma^2}, \tag{9}$$

i.e., just a sinusoidal variation. Here $2\alpha\beta$ decreases with the increase of the asymmetry, as do the inelastic cross section and the kx_0-dependent part of the elastic cross section.

Recoil-less Emission by Tunneling Impurities (8)

We shall begin by reviewing an argument of Lipkin (9) regarding the Mössbauer effect from an impurity performing a localized oscillation in a crystal. Equivalent results have been obtained for the harmonic case by Lax and Waller (10) and Krivoglaz (11).

Suppose that we have a γ-emitting particle in an isolated potential well with levels $|\, n\rangle$. If we know that the particle is in the nth level, then the recoil-less fraction is given by

$$|f_n|^2 = |\langle n \mid \exp(i\mathbf{K}\cdot\mathbf{x}) \mid n\rangle|^2, \tag{10}$$

where $\mathbf{K}$ is the momentum transfered by the γ ray to the system and $\mathbf{x}$ the coordinate of the source particle. If the "oscillator" is in thermal equilibrium with a heat bath (usually the host crystal) in a temperature T, then one would expect the recoil-less fraction to be given by:

$$f_i = \sum_n P_n |f_n|^2, \tag{11}$$

where P_n is the occupation probability of the nth level. However, Lipkin shows that this is true only when the relaxation time τ_R between the oscillator levels, determined by the interaction with the heat bath, is much longer than the lifetime of the nuclear level, τ_N. If $\tau_N \gg \tau_R$, as is almost always the case ($\tau_N \approx 10^{-7}$–10^{-8} sec; $\tau_R \approx 10^{-10}$–10^{-11} sec), the correct expression for f becomes:

$$f_2 = |\sum_n P_n f_n|^2, \tag{12}$$

i.e., a coherent superposition of *amplitudes*. This can be understood physically in two reciprocal ways. First, if $\tau_N \gg \tau_R$, the nucleus has enough time during a lifetime to sweep through all the oscillator levels. Second, $\tau_N \gg \tau_R$ is equivalent to $\hbar/\tau_R \gg \hbar/\tau_N$, i.e., the relaxational width of the oscillator levels is much larger than the "natural" nuclear linewidth, and thus one does not have to consider transitions between levels that are separated by the relaxation interaction. Lipkin shows that exactly this leads from Eq. (11) to Eq. (12).

For a harmonic oscillator the ratio between f_1 and f_2 turns out to be a "Bessel function enhancement" factor ([9,10]). We shall show that tunneling systems provide an instance where the difference between f_1 and f_2 is extremely prominent. Interesting temperature-dependent interference effects are also obtained.

Taking again the model of the previous section, we have, neglecting higher levels ($\hbar\omega \gg kT$),

$$f_1 = e^{-2k^2\sigma^2}[\alpha^4 + \beta^4 + 2\alpha^2\beta^2 \cos(\mathbf{K}, \mathbf{x}_0)].$$

This is the same as Eq. (8) for the neutron case,

$$f_2 = e^{-2k^2\sigma^2} | e^{-ikx_0}(P_1\alpha^2 + P_2\beta^2) + e^{ikx_0}(P_1\beta^2 + P_2\alpha^2) |^2.$$

For a symmetric well we get:

$$f_{2,\mathrm{sym}} = e^{-2k^2\sigma^2} \cos^2(\mathbf{K}, \mathbf{x}_0),$$

i.e., an interference pattern, temperature independent. When the well becomes asymmetric the interference effects become temperature dependent. For a totally asymmetric case $\beta = 0$, $\alpha = 1$:

$$\begin{aligned} f_{2,\text{asym}} &= e^{-2k^2\sigma^2} \mid P_1 e^{-ikx_0} + P_2 e^{ikx_0} \mid^2 \\ &= e^{-2k^2\sigma^2} \mid (\cos kx_0) + i(P_2 - P_1) \sin kx_0 \mid^2 \\ &= e^{-2k^2\sigma^2}\left[\cos^2 kx_0 + \tanh^2\left(\frac{\Delta E}{2kT}\right) \sin^2 kx_0\right]. \end{aligned} \tag{13}$$

Besides the spatial interference, note that if $kx_0 \approx n\pi/2$ ($n = 0, 1, \ldots$), we get destructive interference between the real parts of the recoil-less amplitudes of levels 1 and 2. This leads to an extremely sharp temperature dependence of f_2. Figure 3 demonstrates the latter effect numerically. The experimental points are for Kr in hydroquinone clathrates ([12]). Here tunneling is probable since the zero-point motion is about 0.2 Å according to IR results ([13]), while the free diameter for the Kr motion is about 1.5 Å. The simple one-dimensional model is probably a gross oversimplification (which still demonstrates the relevant physical points).

Another physical case where a Mössbauer source may perform a tunneling motion is Fe in KCl and KBr.

For the case of N minima at $\mathbf{x}_i$ ($i = 1, \ldots, N$), with the bottom of the ith minimum lying at an energy E_i, it can be shown that if $\mid E_i - E_j \mid \gg \Omega_{ij}$ (tunneling matrix element between minima i and j) and if all the oscillators are otherwise equivalent, then:

$$f_2 = f_{2s} \mid \sum_{i=1}^{N} P_n \exp(i\mathbf{K} \cdot \mathbf{x}_n) \mid^2,$$

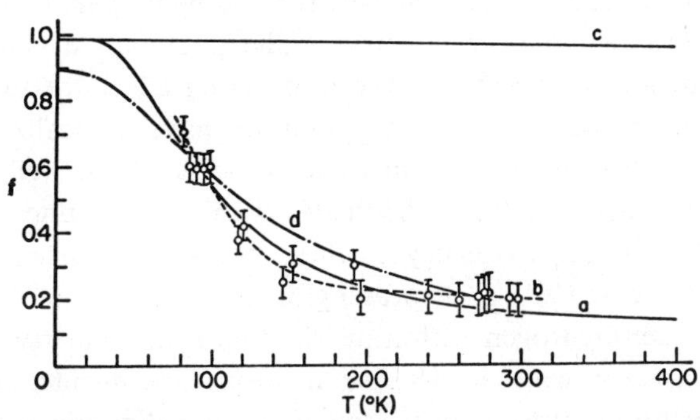

Fig. 3. Recoil-less fraction in an asymmetric potential. From Imry ([8]). (*a*) f_2. (*b*) Experimental. From Burgiel *et al.* ([12]). (*c*) f_1. (*d*) f_2 with some change in the parameters.

where f_{2s} is f_2 of a single oscillator and

$$P_n = \frac{\exp(-E_n/kT)}{\sum_m \exp(-E_m/kT)}.$$

This is a direct generalization of Eq. (13). The interference pattern, reflecting the positions of the minima, is temperature dependent.

REVIEW OF THE EXPERIMENTAL SITUATION. STUDY OF FERROELECTRICITY IN KDP-TYPE CRYSTALS

KH_2PO_4 has a phase transition point at 123°K, below which it is ferroelectric. KD_2PO_4 has the same behavior at 221°K. The large isotope effect hints at a dynamic involvement of the hydrogens in the phase transition mechanism. Neutron diffraction work ([14]) has revealed an ordering of the hydrogens in the ferroelectric phase, whereas they are disordered in the nonferroelectric state. This disorder might be static in the sense that each hydrogen can be in one of two possible positions, without long-range ordering through the crystal. Also, a dynamic situation in which each of the hydrogens actually tunnels between the two positions is possible. The resolution of the diffraction experiment was not enough to decide between these two possibilities. Following the observation of Pelah *et al.* ([15]) that the inelastic neutron scattering spectrum from KDP contained a hydrogen mode at about 400 cm^{-1}, much theoretical and experimental work has been done by the Soreq group on this problem. We shall only describe the conclusions and discuss the results relevant to tunneling.

Each hydrogen has two potential minima along the H bond. The tunneling matrix element is estimated to be $\Omega \approx 200$ cm^{-1} for H and much less for D. The interactions between the hydrogens tend to localize them in an ordered manner by making the potentials asymmetric. This ordering tendency is opposed by the disordering effects of tunneling and thermal agitation. Below the Curie point we have a localized situation, with each particle experiencing an asymmetric potential well. Above the Curie point the asymmetry, B, fluctuates with time around $B = 0$, with nonzero $\langle | B | \rangle$. If the frequency of these fluctuations is low enough, as seems to be the case ([1,16]), it is meaningful to talk about an instantaneous potential for each hydrogen with a distribution of asymmetry values. This picture is consistent with the IR and neutron experiments. Isotopic and chemical substitution were used to isolate the tunneling mode at 400 cm^{-1}. Of particular interest is the change in the IR spectrum through the transition ([1]). Figure 4 shows that the broad tunneling band at 400 cm^{-1} disap-

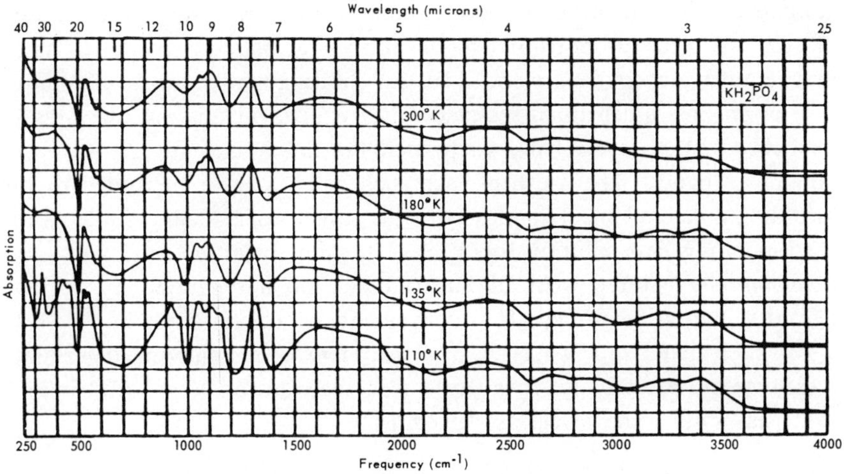

Fig. 4. IR spectra of KH_2PO_4 at the temperatures: 300, 180, 135, and 110°K (Curie point at 123°K). From Imry *et al.* ([1]).

pears below 123°K and only split PO_4 modes remain. This is explained by the fact that the symmetric and nearly symmetric configurations disappear in the ordered phase, and the asymmetric localized configuration has by Eq. (7) a low IR transition probability.

The predictions of Eqs. (8) and (9) for the neutron cross section are interesting. The marked dependence of the inelastic cross section on $\mathbf{K} \cdot \mathbf{x}_0$ was qualitatively detected. Further measurements of the sinusoidal law for a quantitative redetermination of the parameters of the potential are in progress. Since the hydrogen bonds have fixed directions in the crystal, it turns out by simple integration of Eqs. (8) and (9) that the total neutron cross section will depend on the direction of the neutron beam with respect to the crystal axes. This effect was measured ([17]) using single-crystal plates and found to be in agreement with the model. Changes of the total cross section through the transition are found, also in agreement with the IR measurements and the theoretical picture. Some effects, related presumably to collective phenomena in the critical region and not covered by the simple effective potential model, have still to be clarified ([17]).

TUNNELING IN MANY MINIMA AND BAND THEORY. THE BEHAVIOR OF HYDROGEN DISSOLVED IN V

As mentioned during the discussion of multiple-well potentials, the phenomenon of energy bands is related to tunneling. If we take a single oscillator well and let another equivalent one approach it, then each oscillator level will split into two levels of the composite system, the amount of splitting being proportional to the tunneling matrix element Ω. When more equivalent minima are added further splitting may occur. In the simple case of an one-dimensional sequence of N equivalent minima one obtains "bunches" of N levels from each oscillator level, and for $N \to \infty$ one obtains a "band." The width of each such band is also determined by the tunneling matrix elements and it is of the same order of magnitude as the corresponding pair splitting for $N = 2$. A simple example is the Kronig–Penny model: square wells of linear dimension a separated by barriers of width $2b$ and height V_0. It is convenient to take the limit of $V_0 \to \infty$, $b \to 0$ such that $(2m/\hbar^2)(V_0 - E)b^2 = M$. Here M is finite and determines the tunneling probability. It is easily seen that (1) for $N = 2$ and infinite "external walls" the levels are, given by

$$\begin{aligned} &M \sin(\alpha a) + \alpha a \cos(\alpha a) = 0, && \text{even levels,} \\ &\sin \alpha a = 0, && \text{odd levels,} \end{aligned} \tag{14}$$

where $\alpha = (1/\hbar)(2mE)^{1/2}$, and (2) for $N \gg 1$ the allowed energies are given as a function of the wave number k by the solution of:

$$M \sin(\alpha a) + \alpha a \cos(\alpha a) = \alpha a \cos(ka); \tag{15}$$

the bands are given by $| M \sin(\alpha a) + \alpha a \cos(\alpha a) | \leq \alpha a$.

It is easily seen graphically that the width of each band is the same order of magnitude as the splitting for $N = 2$ (the ratio is about 2). For $M \to \infty$ both quantities go to zero ("tight binding"); for $M \to 0$ the particle can move freely in the whole volume of the two or N wells. In the classical case $M = \infty$, there is no tunneling, no splittings, and no band theory.

A system where this band theory might apply for atoms is the α-phase ([2,3]) of VH_x. In this phase of the vanadium–hydrogen system, existing at low hydrogen concentration and high temperature, it is known that hydrogen is quite mobile. If we assume ([2,3]) that the positions available for H in the α-phase are separated by distance of 1 Å and that the frequency of a single oscillator is about 120 meV [the latter parameter is taken from neutron scattering results ([2,3])], then from Eq. (1) one gets $q \approx 2.4$. Thus the band picture is estimated to be as follows. The lowest band is of width ~1 meV, the separation from the next band is ~100 meV, and the width of the latter is ~50 meV. This picture agrees roughly with the neutron scattering results of Rubin *et al.* ([2]) and Kley *et al.* ([3]). The hydrogen modes found by these authors are centered in two bands—the lower is of width ~2–3 meV, the higher of width 70 meV, the two separated by 100 meV. Thus by taking one parameter from the neutron results one can explain the general features and the width of the bands qualitatively. We should stress that these estimates are very rough. Because of the exponential dependence small changes in q will alter the results considerably. The agreement seems, however, to indicate that the band picture is a useful crude approximation providing a physical point of view for VH_x-α.

DISCUSSION OF THE TWO TIME SCALE PROBLEM

We shall make some remarks concerning the tunneling time problem.

There exist physical processes which have more than one characteristic time. A much discussed problem is that of γ emission, encountered for example in Mössbauer effect theory. We know that the nuclear lifetime of typical Mössbauer sources, as given by the uncertainty principle and the measured linewidths, is 10^{-7}–10^{-8} sec. This time also comes into the relaxational effects noted in the discussion of recoil-less emission, and therefore it is of direct physical meaning. However, in γ emission there exists another characteristic time which is believed to be much shorter ($\sim 10^{-18}$ sec). This is the time that it "takes for the γ ray to pass through the nucleus." It is conceivable that although *we* do not know when the emission took place to better than $\sim 10^{-7}$ sec, it is still a fast process. A rough way of stating this would be to say that the duration of the process is short, but its occurrence time is unknown. In fact, the whole Mössbauer effect theory is based on some "impulse approximation," or on the assumption that the momentum transfer process is very fast. Thus this second time is also of relevance and physical meaning in some context. Those points are discussed by Lipkin ([18]).

Let us now consider a particle in a double-minimum well. It takes a time of $\hbar/2\Omega$ for a wave packet to pass from one well to the other. This time might be extremely long. Still, it is conceivable that the tunneling time itself, which is *not* the lifetime of a wave packet in one side, is much shorter. Indeed, phonon induced tunneling theories ([19]) assume that the tunneling particle is able to follow perturbations that are much faster than $2\Omega/\hbar$.

A possible answer to the question, "How much time does it take to cross the barrier?", to the extent that this question is meaningful, might be the following. Tunneling can be looked at as a process where energy is not conserved momentarily. By quantum mechanical energy fluctuation the particle is able to cross the barrier of height V_0. The lifetime of such a fluctuation should be of order $\hbar/V_0$, which is many orders of magnitude shorter, in typical cases, than $\hbar/\Omega$.

REFERENCES

1. Y. Imry, I. Pelah, and E. Wiener, J. Chem. Phys. 43:2332 (1965).
2. R. Rubin, J. Peretti, G. Verdan, and W. Kley, Phys. Letters 14:B100 (1965).
3. W. Kley, J. Peretti, R. Rubin, and G. Verdan, Proc. Brookhaven Symposium on Inelastic Scattering of Neutrons, 1965, BNL 940(C-45), p. 105.
4. R. Blinc and D. Hadzi, Molec. Phys. 1:398 (1958); R. Blinc and D. Hadzi, in: Hydrogen Bonding (D. Hadzi, ed.) Pergamon Press, New York and London, 1959.
5. R. Somorjai and D. F. Hornig, J. Chem. Phys. 36:1980 (1962).
6. E. Merzbacher, Quantum Mechanics, John Wiley, and Sons, New York, 1961, p. 84.
7. M. Gomez, S. P. Bowen, and J. A. Krumhansl, Phys. Rev. 153:100 (1967).
8. Y. Imry, The Dynamics of Kr in Hydroquinone Clathrates, Israel AEC Report IA-1070 (1966).
9. H. J. Lipkin, Ann. Phys. (N.Y.) 26:115 (1964).
10. M. Lax and I. Waller, Phys. Rev. 138A:523 (1965).
11. M. A. Krivoglaz, Eksperim. i. Teor. Fiz. 46:637 (1964) [English translation: Soviet Phys.—JETP 19:432 (1964).
12. Y. Hazony, P. Hillman, M. Pasternak, and S. Ruby, Phys. Letters 2:337 (1962); also, to be published.
13. J. C. Burgiel, H. Meyer, and P. L. Richards, J. Chem. Phys. 43:4291 (1965).
14. G. E. Bacon and R. S. Pease, Proc. Roy. Soc. (London) A230:359 (1955).
15. I. Pelah, I. Lefkovitz, W. Kley, and E. Tunkello, Phys. Rev. Letters 2:94 (1959).
16. V. H. Schmidt and E. A. Uehling, Phys. Rev. 126:447 (1962).
17. Y. Imry, I. Pelah, E. Wiener, and H. Zafrir, Solid State Commun. 5:41 (1967).
18. H. J. Lipkin, Ann. Phys. (N.Y.) 18:182 (1962).
19. J. A. Zussmann, Phys. kondens, Materie 2:146 (1964).

NOTES ADDED IN PROOF

Recently, evidence for off-center positions for Fe in $FeCl_2$ was obtained from Mössbauer effect studies by D. P. Johnson and J. G. Dash, Phys. Rev. 172:983 (1968).

Multiple-minima tunneling was considered in the two-dimensional adsorbed He system by J. G. Dash, in Proceedings of the 11th Low Temperature Conference, St. Andrews, Scotland, August, 1968 (to be published).

Subject Index